Springer Series in Computational Physics

Editors: R. Glowinski   M. Holt   P. Hut
H. B. Keller   J. Killeen   S. A. Orszag   V. V. Rusanov

Springer Series in Computational Physics

Editors: R. Glowinski   M. Holt   P. Hut   H. B. Keller   J. Killeen
S. A. Orszag   V. V. Rusanov

C.A.J. Fletcher

Computational Techniques for Fluid Dynamics 2

Specific Techniques for Different Flow Categories

Second Edition
With 184 Figures

Springer-Verlag

Berlin Heidelberg New York London Paris
Tokyo Hong Kong Barcelona Budapest

Dr. Clive A. J. Fletcher
Department of Mechanical Engineering, The University of Sydney
New South Wales 2006, Australia

Editors

R. Glowinski
Institut de Recherche d'Informatique
et d'Automatique (INRIA)
Domaine de Voluceau
Rocquencourt, B. P. 105
F-78150 Le Chesnay, France

M. Holt
College of Engineering and
Mechanical Engineering
University of California
Berkeley, CA 94720, USA

P. Hut
The Institute for Advanced Study
School of Natural Sciences
Princeton, NJ 08540, USA

H. B. Keller
Applied Mathematics 101-50
Firestone Laboratory
California Institute of Technology
Pasadena, CA 91125, USA

J. Killeen
Lawrence Livermore Laboratory
P. O. Box 808
Livermore, CA 94551, USA

S. A. Orszag
Program in Applied
and Computational Mathematics
Princeton University, 218 Fine Hall
Princeton, NJ 08544-1000, USA

V. V. Rusanov
Keldysh Institute
of Applied Mathematics
4 Miusskaya pl.
SU-125047 Moscow, USSR

ISBN 3-540-53601-9 2. Auflage   Springer-Verlag Berlin Heidelberg New York
ISBN 0-387-53601-9 2nd edition Springer-Verlag New York Berlin Heidelberg

ISBN 3-540-18759-6 1. Auflage Springer-Verlag Berlin Heidelberg New York
ISBN 0-387-18759-6 1st edition Springer-Verlag New York Berlin Heidelberg

Library of Congress Cataloging-in-Publication Data. Fletcher, C. A. J. Computational techniques for fluid dynamics 2 : specific techniques for different flow categories / C. A. J. Fletcher.–2nd ed. p. cm.–(Springer series in computational physics) Includes bibliographical references and index. ISBN 0-387-53601-9 (U.S.) 1. Fluid dynamics–Mathematics. 2. Fluid dynamics–Data processing. 3. Mathematical analysis. I. Title. II. Series. QC151.F58   1991   532'.05'0151–dc20   91-7345

© Springer-Verlag Berlin Heidelberg 1988, 1991
Printed in Germany

Typesetting: Macmillan India Ltd., India
55/3140-543210 – Printed on acid-free paper

Preface to the Second Edition

The purpose and organisation of this book are described in the preface to the first edition (1988). In preparing this edition minor changes have been made, particularly to Chap. 1 (Vol. 1) to keep it reasonably current, and to upgrade the treatment of specific techniques, particularly in Chaps. 12–14 and 16–18. However, the rest of the book (Vols. 1 and 2) has required only minor modification to clarify the presentation and to modify or replace individual problems to make them more effective. The answers to the problems are available in *Solutions Manual for Computational Techniques for Fluid Dynamics* by K. Srinivas and C. A. J. Fletcher, published by Springer-Verlag, Heidelberg, 1991. The computer programs have also been reviewed and tidied up. These are available on an IBM-compatible floppy disc direct from the author.

I would like to take this opportunity to thank the many readers for their usually generous comments about the first edition and particularly those readers who went to the trouble of drawing specific errors to my attention. In this revised edition considerable effort has been made to remove a number of minor errors that had found their way into the original. I express the hope that no errors remain but welcome communication that will help me improve future editions.

In preparing this revised edition I have received considerable help from Dr. K. Srinivas, Nam-Hyo Cho, Zili Zhu and Susan Gonzales at the University of Sydney and from Professor W. Beiglböck and his colleagues at Springer-Verlag. I am very grateful to all of them.

Sydney, March 1991 *C. A. J. Fletcher*

Preface to the First Edition

As indicated in Vol. 1, the purpose of this two-volume textbook is to provide students of engineering, science and applied mathematics with the specific techniques, and the framework to develop skill in using them, that have proven effective in the various branches of computational fluid dynamics.

Volume 1 describes both fundamental and general techniques that are relevant to all branches of fluid flow. This volume contains specific techniques applicable to the different categories of engineering flow behaviour, many of which are also appropriate to convective heat transfer.

The contents of Vol. 2 are suitable for specialised graduate courses in the engineering computational fluid dynamics (CFD) area and are also aimed at the established research worker or practitioner who has already gained some fundamental CFD background. It is assumed that the reader is familiar with the contents of Vol. 1.

The contents of Vol. 2 are arranged in the following way: Chapter 11 develops and discusses the equations governing fluid flow and introduces the simpler flow categories for which specific computational techniques are considered in Chaps. 14 – 18.

Most practical problems involve computational domain boundaries that do not conveniently coincide with coordinate lines. Consequently, in Chap. 12 the governing equations are expressed in generalised curvilinear coordinates for use in arbitrary computational domains. The corresponding problem of generating an interior grid is considered in Chap. 13.

Computational techniques for inviscid flows are presented in Chap. 14 for incompressible, supersonic and transonic conditions. In Chap. 15 methods are described for predicting the flow behaviour in boundary layers.

For many steady flows with a dominant flow direction it is possible to obtain accurate flow predictions, based on reduced forms of the Navier-Stokes equations, in a very efficient manner. Such techniques are developed in Chap. 16. In Chaps. 17 and 18 specific computational methods are discussed for separated flows, governed by the incompressible and compressible Navier-Stokes equations respectively.

In preparing this textbook I have been assisted by many people, some of whom are acknowledged in the Preface of Vol. 1. However, the responsibility for any errors or omissions remaining rests with me. Any comments, criticism and suggestions that will improve this textbook are most welcome and will be gratefully received.

Sydney, October 1987   *C.A.J. Fletcher*

Contents

11. Fluid Dynamics: The Governing Equations

In this chapter, equations will be developed that govern the more common categories of fluid motion. Subsequently, various simplifications of these equations will be presented and the physical significance of these simpler equations discussed. The simplifications often coincide with limiting values of particular nondimensional numbers (Sect. 11.2.5), e.g. incompressible flow is often associated with very small values of the Mach number.

A fluid is categorised as a substance which cannot withstand any attempt to change its shape when at rest. Consequently, a fluid cannot sustain a shear force when at rest, as can a solid. However, a fluid can sustain and transmit a shear force when in motion. The proportionality between the shear force per unit area (or stress) and an appropriate velocity gradient defines the viscosity of the fluid (Sect. 11.1). Fluids include both liquids and gases. The two fluids that occur most often, naturally or in flow machinery, are water (often in the liquid phase) and air.

Fluids, whether liquids or gases, consist of molecules which are individually in a state of random motion. The large-scale motion of a fluid adds a uniform or slowly varying velocity vector to the motion of each molecule. If a large enough sample of molecules is considered, (one cubic millimetre of air at normal temperature (15°C) and pressure (101 kPa) contains approximately 3×10^{16} molecules), the individual molecular motion is not detectable and only the large-scale (macroscopic) motion is perceived. By assuming that the various properties of the fluid in motion, pressure, velocity, etc., vary continuously with position and time (continuum hypothesis) it is possible to derive the equations that govern fluid motion without regard to the behaviour of the individual molecules.

However, for flows at very low density, e.g. re-entry vehicles travelling through the outer parts of the atmosphere, the continuum hypothesis is not appropriate and the molecular nature of the flow must be taken into account. This also dictates the choice of appropriate computational techniques (Bird 1976).

11.1 Physical Properties of Fluids

The thermodynamic state of a small volume of fluid in equilibrium (i.e. uniform in space and time) is defined uniquely by specifying two independent thermodynamic properties e.g. for air, pressure and temperature would be appropriate. Other thermodynamic properties, such as density or internal energy, are then functions of the two primary thermodynamic properties.

For air at normal temperature and pressure the various thermodynamic properties are related by the ideal gas equation

$$p = \varrho R T \ , \tag{11.1}$$

where p is the pressure, measured in kPa, ϱ is the density, measured in kg/m³, T is the (absolute) temperature, measured in K, and R is the gas constant. For air, $R = 0.287$ kJ/kgK. It is not possible to write down a simple algebraic equation of state connecting the thermodynamic properties for water, but the relationship is contained implicitly in steam tables (e.g. van Wylen and Sonntag 1976, pp. 645–669).

The pressure is defined as the force per unit area and has the same dimensions as a stress. The pressure on a surface acts normal to the surface. Pressure is an important property since an integration of the pressure distribution over the surface of an immersed body will determine major forces (e.g. form drag, lift) and moments acting on the body. For fluids at rest the forces acting on a small volume of fluid due to the local pressure gradient are typically balanced by the force due to gravity, which gives rise to the following equation for the hydrostatic pressure:

$$\partial p / \partial z + \varrho g = 0 \ , \tag{11.2}$$

where z is measured in the vertical direction and g is the acceleration due to gravity. Equation (11.2) may also hold for fluids in motion, under certain circumstances. For many geophysical flows the pressure variation in the vertical direction is given approximately by (11.2).

Variation in temperature of a fluid may be due to the processes of heat transfer if the fluid is in contact with a substance at a different temperature or if latent heat release occurs. The temperature variation may also be influenced by the compression of the fluid, which might be due to the motion in high speed flow or due to the weight of the fluid in atmospheric flows.

The density is the mass per unit volume. For gases, changes in density are connected to changes in the pressure and temperature through the ideal gas equation (11.1). However, for liquids very substantial changes in pressure are necessary to alter the density, so that water (in the liquid phase) is often treated as an incompressible (constant density) fluid. The properties of air and water for different values of pressure, temperature and density are shown in Tables 11.1 and 11.2, respectively.

For fluids in motion, the concept of thermodynamic equilibrium must be given a local interpretation so that equations like (11.1) are still valid. But now the properties are functions of position and time, i.e.

$$p = p(x, y, z, t) \ , \quad \varrho = \varrho(x, y, z, t) \quad \text{and} \quad T = T(x, y, z, t) \ .$$

In addition, it is necessary to describe the motion uniquely. Here we use the Eulerian description. That is, the values of the velocities and the thermodynamics

Table 11.1. Properties of air at atmospheric pressure

Temperature, T [K]	Density, ϱ [kg/m³]	Dynamic viscos., $\mu \times 10^5$ [kg/ms]	Thermal conduc., k [W/mK]	Thermal diffus., $\alpha \times 10^5$ [m²/s]	Prandtl number, Pr	Specific heat ratio, γ
100	3.6010	0.6924	0.00925	0.2501	0.770	1.39
300	1.1774	1.983	0.02624	2.216	0.708	1.40
500	0.7048	2.671	0.04038	5.564	0.680	1.39
900	0.3925	3.899	0.06279	14.271	0.696	1.34
1900	0.1858	6.290	0.11700	48.110	0.704	1.28

Table 11.2. Properties of water for saturated conditions

Temperature, T [°C]	Pressure, p [kPa]	Density, ϱ [kg/m³]	Dynamic viscos., $\mu \times 10^5$ [kg/ms]	Thermal conduc., k [W/mK]	Thermal diffus., $\alpha \times 10^5$ [m²/s]	Prandtl number Pr
0.01	0.611	1002.28	179.2	0.552	0.01308	13.6
40	7.384	994.59	65.44	0.628	0.01512	4.34
100	101.35	960.63	28.24	0.680	0.01680	1.74
200	1553.8	866.76	13.87	0.665	0.01706	0.937
300	8581.0	714.26	9.64	0.540	0.01324	1.019

properties are given at fixed locations (x, y, z, t) in the space-time domain. The alternative Lagrangian description follows individual fluid particles treating their position and thermodynamic properties as dependent variables. The connection between the Eulerian and the Lagrangian representation is discussed by von Schwind (1980, p. 22).

For fluids in motion, the ability to transmit a shear force introduces the property of dynamic viscosity. Consider the motion of a plane surface with velocity U parallel to a second stationary plane surface (Fig. 11.1).

Fluid adjacent to the upper surface moves with the velocity U and exerts a resisting force on the plate of τA, where A is the surface area of the upper plate and τ is the shear stress. For a given element in the fluid two shear forces $(\tau \cdot l \cdot 1)$ are felt, to the right at the top and to the left at the bottom. The fluid adjacent to the bottom

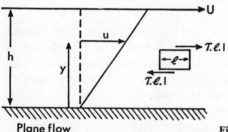

Plane flow

Fig. 11.1. Plane flow parallel to a stationary surface

surface exerts a drag force τA on the lower fixed plate. It is found empirically that the shear stress is directly proportional to the velocity gradient, $\partial u/\partial y$, i.e.

$$\tau = \mu \frac{\partial u}{\partial y} , \tag{11.3}$$

where the constant of proportionality, μ, is the (dynamic) viscosity. Viscosity is measured in kg/ms. For this example, the shear stress, τ, is constant; it follows that the velocity distribution is given by

$$\frac{u}{U} = \frac{y}{h} \tag{11.4}$$

The relationship (11.3) defines Newtonian fluids. Flows involving air or water satisfy (11.3). Non-Newtonian fluids, which do not satisfy (11.3), are described by Crochet et al. (1984).

The viscosity of gases like air is, to a close approximation, a function of temperature alone (for normal temperatures and pressures). For air the viscosity increases approximately like $T^{0.76}$, where T is the absolute temperature. Typical values are given in Table 11.1. The viscosity of liquids like water is a weak function of pressure but a strong function of temperature. In contrast to the behaviour for gases, the viscosity of liquids typically decreases rapidly with increasing temperature. Representative values for water are given in Table 11.2.

For flow problems involving temperature changes, Fourier's law indicates that the local rate of heat transfer is a linear function of the local temperature gradient, i.e.

$$\dot{Q}_i = -k \frac{\partial T}{\partial x_i} , \tag{11.5}$$

where $\dot{Q}_i$ is the rate of the heat transfer per unit area in the x_i direction and k is the thermal conductivity. The similarity in structure between (11.5) and (11.3) is noteworthy. If the temperature of the two plates in Fig. 11.1 were different, (11.5) indicates that there would be a heat transfer through the fluid, given by

$$\dot{Q}_y = -k \frac{\partial T}{\partial y} . \tag{11.6}$$

The thermal conductivity is measured in W/mK. Like viscosity, the thermal conductivity increases with temperature for gases. However, for liquids such as water the thermal conductivity rises slightly with temperature in the range $0°$–$100°$C at a pressure of one atmosphere. Typical values for the thermal conductivities of air and water are shown in Tables 11.1 and 11.2.

Because of the way that viscosity and thermal conductivity appear in the momentum (11.31) and energy (11.38) equations it is convenient to define the kinematic viscosity v and the thermal diffusivity α by

$$v = \frac{\mu}{\varrho} \quad \text{and} \quad \alpha = \frac{k}{\varrho c_p} ,$$

where c_p is the specific heat at constant pressure. Both v and α are diffusivities, controlling the diffusion of momentum (or vorticity) and heat, respectively. Both v and α are measured in m^2/s. For gases like air both v and α increase with temperature (Table 11.1). For liquids like water the kinematic viscosity v falls rapidly with increasing temperature but the thermal diffusivity α increases slightly (Table 11.2).

For a discussion of the fluid properties, particularly in relation to the underlying molecular behaviour, the reader is referred to Lighthill (1963) or Batchelor (1967, pp. 1–60). Eckert and Drake (1972) provide a tabulation of the properties of common fluids.

11.2 Equations of Motion

The general technique for obtaining the equations governing fluid motion is to consider a small control volume through which the fluid moves, and to require that mass and energy are conserved, and that the rate of change of the three components of linear momentum are equal to the corresponding components of the applied force. This produces five equations which, when combined with an equation of state, provide sufficient information for the determination of six variables: p, ϱ, T, u, v, w typically. For flows associated with combustion and some geophysical flows, more than one species will be present. Each new species requires an additional (species conservation) equation. For some flow problems not all six variables will be involved and less than six equations will be required.

11.2.1 Continuity Equation

For an arbitrary control volume V fixed in space and time (Fig. 11.2), conservation of mass requires that the rate of change of mass within the control volume is equal to the mass flux crossing the surface S of V, i.e.

$$\frac{d}{dt} \int_V \varrho \, dV = -\int_S \varrho \mathbf{v} \cdot \mathbf{n} \, dS , \tag{11.7}$$

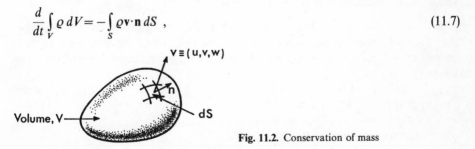

Fig. 11.2. Conservation of mass

where **n** is the unit (outward) normal vector. Using the Gauss (divergence) theorem (Gustafson 1980, p. 35), the surface integral may be replaced by a volume integral. Then (11.7) becomes

$$\int_V \left[\frac{\partial \varrho}{\partial t} + \nabla \cdot (\varrho \mathbf{v}) \right] dV = 0 \tag{11.8}$$

where $\nabla \cdot (\varrho \mathbf{v}) \equiv \operatorname{div} \varrho \mathbf{v}$. Since (11.8) is valid for any size of V, it implies that

$$\frac{\partial \varrho}{\partial t} + \nabla \cdot (\varrho \mathbf{v}) = 0 \ , \tag{11.9}$$

which is the mass-conservation or continuity equation. In Cartesian coordinates (11.9) becomes

$$\frac{\partial \varrho}{\partial t} + \frac{\partial}{\partial x}(\varrho u) + \frac{\partial}{\partial y}(\varrho v) + \frac{\partial}{\partial z}(\varrho w) = 0 \ . \tag{11.10}$$

It is convenient to collect all the density terms together and to write (11.10) as

$$\frac{1}{\varrho} \left(\frac{\partial \varrho}{\partial t} + u \frac{\partial \varrho}{\partial x} + v \frac{\partial \varrho}{\partial y} + w \frac{\partial \varrho}{\partial z} \right) + \left(\frac{\partial u}{\partial x} + \frac{\partial v}{\partial y} + \frac{\partial w}{\partial z} \right) = 0 \ , \tag{11.11}$$

or

$$\frac{1}{\varrho} \left(\frac{D\varrho}{Dt} \right) + \mathscr{D} = 0 \ , \tag{11.12}$$

where D/Dt is called the time derivative following the motion or the material derivative and $\mathscr{D}$ is called the dilatation. For flows of constant density (e.g. incompressible flow) (11.12) reduces to

$$\mathscr{D} \equiv \nabla \cdot \mathbf{v} \equiv \frac{\partial u}{\partial x} + \frac{\partial v}{\partial y} + \frac{\partial w}{\partial z} = 0 \ , \tag{11.13}$$

for both steady and unsteady flow.

11.2.2 Momentum Equations: Inviscid Flow

Newton's second law of motion states that the time rate of change of linear momentum is equal to the sum of the forces acting. For a small element of fluid treated as a closed system (i.e. no flow across its boundaries) Newton's second law is

$$\frac{d}{dt} \int \varrho \mathbf{v} \, dV_{\text{cs}} = \sum \mathbf{F} \ , \tag{11.14}$$

where subscript cs denotes a closed system.

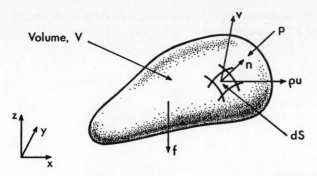

Fig. 11.3. Control volume geometry for Euler's equations

For a control volume V fixed in space and time with flow allowed to occur across the boundaries (Fig. 11.3), the following connection with the closed system is available (Streeter and Wylie 1979, p. 93):

$$\frac{d}{dt}\int \varrho \mathbf{v}\, dV_{\text{cs}} = \int_V \frac{\partial}{\partial t}(\varrho \mathbf{v})\, dV + \int_S \varrho \mathbf{v}(\mathbf{v}\cdot \mathbf{n})\, dS \ . \tag{11.15}$$

In (11.15) $\varrho \mathbf{v}$ is the momentum and $\mathbf{v}\cdot\mathbf{n}$ is the velocity normal to the surface of the control volume. By the Gauss theorem, (11.15) becomes

$$\frac{d}{dt}\int \varrho \mathbf{v}\, dV_{\text{cs}} = \int_V \left(\frac{\partial}{\partial t}(\varrho \mathbf{v}) + \nabla\cdot(\varrho \mathbf{v}\mathbf{v})\right) dV \ . \tag{11.16a}$$

By expanding (11.16a) and making use of (11.9), one obtains

$$\frac{d}{dt}\int \varrho \mathbf{v}\, dV_{\text{cs}} = \int_V \varrho \frac{D\mathbf{v}}{Dt}\, dV \ , \tag{11.16b}$$

where $D\mathbf{v}/Dt = \partial \mathbf{v}/\partial t + \mathbf{v}\cdot\nabla \mathbf{v}$. $D\mathbf{v}/Dt$ is the total time rate of change of $\mathbf{v}$ or the acceleration.

Thus, (11.14) becomes

$$\int_V \varrho \frac{D\mathbf{v}}{Dt}\, dV = \Sigma \mathbf{F} \ , \tag{11.17}$$

i.e. "mass × acceleration = force".

Contributions to the summation $\Sigma \mathbf{F}$ come from forces acting at the surface of the control volume (surface forces) and throughout the volume (volume or body forces). The most common volume force is the force due to gravity and this is the only volume force considered here. The nature of the surface forces depends on whether the fluid viscosity is taken into account or not. Initially, an inviscid fluid will be assumed, in which case the only surface force is due to the pressure, which acts normal to the surface. Thus, the right-hand side of (11.17) can be written

$$\Sigma \mathbf{F} = \int_V \varrho \mathbf{f}\, dV - \int_S p\mathbf{n}\, dS \ , \tag{11.18}$$

where **f** is the volume force per unit mass. Applying the Gauss theorem converts the right-hand side of (11.18) to a volume integral

$$\Sigma \mathbf{F} = \int_V (\varrho \mathbf{f} - \nabla p)\, dV \tag{11.19}$$

and substitution into (11.17) gives

$$\int_V \left(\varrho \frac{D\mathbf{v}}{Dt} - \varrho \mathbf{f} + \nabla p \right) dV = 0 \ . \tag{11.20}$$

Or, for an arbitrarily small volume V,

$$\varrho \frac{D\mathbf{v}}{Dt} = \varrho \mathbf{f} - \nabla p \ . \tag{11.21a}$$

An alternative conservation form, using (11.16) is

$$\frac{\partial}{\partial t}(\varrho \mathbf{v}) + \nabla \cdot (\varrho \mathbf{v} \mathbf{v}) = \varrho \mathbf{f} - \nabla p \ . \tag{11.21b}$$

Equations (11.21) are Euler's equations which are applicable, strictly, to inviscid flow. However, for many flow situations the influence of the viscosity is very small so that (11.21) are then a very accurate approximation.

With gravity as the body force in the negative z direction, Euler's equations (11.21) become in Cartesian coordinates

$$\varrho \left(\frac{\partial u}{\partial t} + u \frac{\partial u}{\partial x} + v \frac{\partial u}{\partial y} + w \frac{\partial u}{\partial z} \right) = -\frac{\partial p}{\partial x} \tag{11.22}$$

$$\varrho \left(\frac{\partial v}{\partial t} + u \frac{\partial v}{\partial x} + v \frac{\partial v}{\partial y} + w \frac{\partial v}{\partial z} \right) = -\frac{\partial p}{\partial y} \tag{11.23}$$

$$\varrho \left(\frac{\partial w}{\partial t} + u \frac{\partial w}{\partial x} + v \frac{\partial w}{\partial y} + w \frac{\partial w}{\partial z} \right) = -\varrho g - \frac{\partial p}{\partial z} \ . \tag{11.24}$$

Equations (11.21–24) are applicable to both compressible and incompressible flow.

11.2.3 Momentum Equations: Viscous Flow

If a balance of forces is considered for a viscous fluid, (11.17) is still valid. In (11.18) the surface stress associated with the pressure above must be replaced by a general stress tensor $\underline{\sigma}$ that can produce a net stress in any direction. The average of the normal stresses is set equal to the negative of the pressure. The remaining contributions to the stress tensor are associated with the viscous nature of the fluid.

---

underlined bold type denotes matrix or tensor

These contributions, considered collectively, constitute the viscous (or deviatoric) stress tensor $\underline{\tau}$ so that $\underline{\sigma} = -p\underline{I} + \underline{\tau}$.

Consequently, for a viscous fluid (11.21a) is replaced by

$$\varrho \frac{D\mathbf{v}}{Dt} = \varrho\mathbf{f} - \nabla p + \nabla \cdot \underline{\tau} \; . \tag{11.25}$$

Or, in Cartesian coordinates,

$$x\text{-mmtm:} \quad \varrho\frac{Du}{Dt} = \varrho f_x - \frac{\partial p}{\partial x} + \frac{\partial \tau_{xx}}{\partial x} + \frac{\partial \tau_{yx}}{\partial y} + \frac{\partial \tau_{zx}}{\partial z} \; ,$$

$$y\text{-mmtm:} \quad \varrho\frac{Dv}{Dt} = \varrho f_y - \frac{\partial p}{\partial y} + \frac{\partial \tau_{xy}}{\partial x} + \frac{\partial \tau_{yy}}{\partial y} + \frac{\partial \tau_{zy}}{\partial z} \; , \tag{11.26}$$

$$z\text{-mmtm:} \quad \varrho\frac{Dw}{Dt} = \varrho f_z - \frac{\partial p}{\partial z} + \frac{\partial \tau_{xz}}{\partial x} + \frac{\partial \tau_{yz}}{\partial y} + \frac{\partial \tau_{zz}}{\partial z} \; .$$

Equations (11.26) replace the Euler equations (11.22–24). However, it is necessary to relate the various viscous stresses to the rates of strain, i.e. the velocity gradients. These relationships are

$$\tau_{xx} = -\frac{2}{3}\mu\mathscr{D} + 2\mu\frac{\partial u}{\partial x} \; ,$$

$$\tau_{yy} = -\frac{2}{3}\mu\mathscr{D} + 2\mu\frac{\partial v}{\partial y} \; ,$$

$$\tau_{zz} = -\frac{2}{3}\mu\mathscr{D} + 2\mu\frac{\partial w}{\partial z} \; , \tag{11.27}$$

$$\tau_{xy} = \tau_{yx} = \mu\left(\frac{\partial u}{\partial y} + \frac{\partial v}{\partial x}\right) \; ,$$

$$\tau_{xz} = \tau_{zx} = \mu\left(\frac{\partial u}{\partial z} + \frac{\partial w}{\partial x}\right)$$

$$\tau_{yz} = \tau_{zy} = \mu\left(\frac{\partial v}{\partial z} + \frac{\partial w}{\partial y}\right) \; ,$$

where the dilatation $\mathscr{D}$ is given by (11.13). Substitution of (11.27) into (11.26) gives the Navier–Stokes equations

$$\varrho\frac{Du}{Dt} = \varrho f_x - \frac{\partial p}{\partial x} - \frac{2}{3}\frac{\partial(\mu\mathscr{D})}{\partial x} + 2\frac{\partial}{\partial x}\left(\mu\frac{\partial u}{\partial x}\right)$$

$$+ \frac{\partial}{\partial y}\left[\mu\left(\frac{\partial u}{\partial y} + \frac{\partial v}{\partial x}\right)\right] + \frac{\partial}{\partial z}\left[\mu\left(\frac{\partial u}{\partial z} + \frac{\partial w}{\partial x}\right)\right] \; , \tag{11.28}$$

$$\varrho\frac{Dv}{Dt}=\varrho f_y-\frac{\partial p}{\partial y}-\frac{2}{3}\frac{\partial(\mu\mathscr{D})}{\partial y}+\frac{\partial}{\partial x}\left[\mu\left(\frac{\partial u}{\partial y}+\frac{\partial v}{\partial x}\right)\right]$$

$$+\frac{2\partial}{\partial y}\left(\mu\frac{\partial v}{\partial y}\right)+\frac{\partial}{\partial z}\left[\mu\left(\frac{\partial v}{\partial z}+\frac{\partial w}{\partial y}\right)\right] , \tag{11.29}$$

$$\varrho\frac{Dw}{Dt}=\varrho f_z-\frac{\partial p}{\partial z}-\frac{2}{3}\frac{\partial(\mu\mathscr{D})}{\partial z}+\frac{\partial}{\partial x}\left[\mu\left(\frac{\partial u}{\partial z}+\frac{\partial w}{\partial x}\right)\right]$$

$$+\frac{\partial}{\partial y}\left[\mu\left(\frac{\partial v}{\partial z}+\frac{\partial w}{\partial y}\right)\right]+2\frac{\partial}{\partial z}\left(\mu\frac{\partial w}{\partial z}\right) . \tag{11.30}$$

Or, in vector notation,

$$\varrho\frac{D\mathbf{v}}{Dt}=\varrho\mathbf{f}-\nabla p-\frac{2}{3}\nabla(\mu\nabla\cdot\mathbf{v})+2\nabla\cdot(\mu\,\text{def}\,\mathbf{v}) , \tag{11.31}$$

where $\text{def}\,\mathbf{v}\equiv\frac{1}{2}\left(\frac{\partial v_i}{\partial x_j}+\frac{\partial v_j}{\partial x_i}\right).$

Equations (11.28–31) are applicable to viscous compressible flow. A more detailed derivation is given by Batchelor (1967). Panton (1984, Chaps. 5 and 6) and Sherman (1990, Chaps. 4 and 5) provide extended discussions of the various terms in the Navier–Stokes equations.

11.2.4 Energy Equation

The first law of thermodynamics states that the time rate of change of internal energy plus kinetic energy is equal to the rate of heat transfer less the rate of work done by the system. For a control volume V this can be written as

$$\int_V \varrho\frac{D}{Dt}(e+0.5q^2)dV=\int_V \varrho\mathbf{f}\cdot\mathbf{v}\,dV+\int_S \mathbf{n}\cdot(\mathbf{v}\underline{\sigma}-\dot{\mathbf{Q}})dS , \tag{11.32}$$

where $\dot{\mathbf{Q}}$ is the surface rate of heat transfer and e is the specific internal energy. Internal sources of heat transfer, e.g. associated with chemical reactions, are not included. The first and second terms on the right-hand side of (11.32) give the work done by the volume and surface forces respectively.

Applying the Gauss theorem and shrinking the volume to zero size gives

$$\varrho\frac{D}{Dt}(e+0.5q^2)-\varrho\mathbf{f}\cdot\mathbf{v}-\nabla\cdot(\mathbf{v}\underline{\sigma})+\nabla\cdot\dot{\mathbf{Q}}=0 . \tag{11.33}$$

However, (11.33) "contains" the mechanical energy equation

$$\varrho\frac{D}{Dt}(0.5q^2)-\varrho\mathbf{f}\cdot\mathbf{v}-\mathbf{v}\cdot\text{div}\,\underline{\sigma}=0 . \tag{11.34}$$

When this is removed the thermal energy equation is left:

$$\varrho\frac{De}{Dt}+p\nabla\cdot\mathbf{v}=\Phi-\nabla\cdot\dot{\mathbf{Q}}\quad\text{or}\tag{11.35}$$

$$\varrho\frac{Dh}{Dt}-\frac{Dp}{Dt}=\Phi-\nabla\cdot\dot{\mathbf{Q}}\ ,\tag{11.36}$$

where the specific enthalpy $h=e+p/\varrho$, and $\Phi(\equiv\underline{\tau}\cdot\nabla\mathbf{v})$ is the dissipation function, which arises from irreversible viscous work. The heat transfer rate $\dot{\mathbf{Q}}$ can be related to the local temperature gradient by (11.5),

$$\dot{\mathbf{Q}}=-k\nabla T\ .\tag{11.37}$$

In Cartesian coordinates (11.36) then becomes

$$\rho\frac{Dh}{Dt}-\frac{Dp}{Dt}=\Phi+\frac{\partial}{\partial x}\left(k\frac{\partial T}{\partial x}\right)+\frac{\partial}{\partial y}\left(k\frac{\partial T}{\partial y}\right)+\frac{\partial}{\partial z}\left(k\frac{\partial T}{\partial z}\right)\ ,\tag{11.38}$$

with Φ given by

$$\Phi=2\mu\left[\left(\frac{\partial u}{\partial x}\right)^2+\left(\frac{\partial v}{\partial y}\right)^2+\left(\frac{\partial w}{\partial z}\right)^2+0.5\left(\frac{\partial u}{\partial y}+\frac{\partial v}{\partial x}\right)^2\right.$$
$$\left.+0.5\left(\frac{\partial v}{\partial z}+\frac{\partial w}{\partial y}\right)^2+0.5\left(\frac{\partial w}{\partial x}+\frac{\partial u}{\partial z}\right)^2\right]$$
$$-\frac{2}{3}\mu\left(\frac{\partial u}{\partial x}+\frac{\partial v}{\partial y}+\frac{\partial w}{\partial z}\right)^2\ .\tag{11.39}$$

It may be noted that the energy equation will be used primarily for flows of air, which can be treated as an ideal gas (11.1). Consequently, the internal energy and enthalpy can be related to the temperature by

$$e=c_{\mathrm{v}}(T-T_{\mathrm{ref}})\quad\text{and}\quad h=c_{\mathrm{p}}(T-T_{\mathrm{ref}})\ ,\tag{11.40}$$

where c_{v} and c_{p} are the specific heats at constant volume and pressure, respectively.

In Chaps. 14 to 18, discretisation will be applied to the various equations governing fluid flow, for example (11.21). However, since the discretisation is carried out on a grid of finite size it is also feasible to carry out the discretisation on the original conservation equation, for example (11.20). An example of this is provided in Sect. 5.2.

The derivation of the equations that govern fluid flow are discussed with special attention given to the equivalent molecular behaviour by Batchelor (1967, Chaps. 2 and 3). The equations governing fluid flow are stated, without derivation, in various coordinate systems by Hughes and Gaylord (1964, Chap. 1).

11.2.5 Dynamic Similarity

To obtain the flow behaviour around bodies of similar shape with minimum computational (or experimental) effort it is desirable to group all the parameters, such as the body length and freestream velocity, into nondimensional numbers. Two flows are dynamically similar if the nondimensional numbers that govern the flows have the same value, even though the parameters contained in the nondimensional numbers have different values. The best way to identify the appropriate nondimensional groups is to nondimensionalise the governing equations and boundary conditions.

For example, in considering the wave motion generated by a ship of length L travelling at a speed U_∞, it would be appropriate to start with the z-momentum equation for viscous incompressible flow

$$\frac{\partial w}{\partial t}+u\frac{\partial w}{\partial x}+v\frac{\partial w}{\partial y}+w\frac{\partial w}{\partial z}+\left(\frac{1}{\varrho}\right)\frac{\partial p}{\partial z}=\left(\frac{\mu}{\varrho}\right)\left(\frac{\partial^2 w}{\partial x^2}+\frac{\partial^2 w}{\partial y^2}+\frac{\partial^2 w}{\partial z^2}\right)-g \ . \tag{11.41}$$

Nondimensional variables are introduced as

$$x^*=x/L \ , \quad y^*=y/L \ , \quad z^*=z/L \quad \text{and} \quad t^*=U_\infty t/L \ ,$$

$$u^*=u/U_\infty \ , \quad v^*=v/U_\infty \ , \quad w^*=w/U_\infty \quad \text{and} \quad p^*=(p-p_\infty)/\varrho U_\infty^2 \ .$$

Then (11.41) becomes

$$\frac{\partial w^*}{\partial t}+u^*\frac{\partial w^*}{\partial x^*}+v^*\frac{\partial w^*}{\partial y^*}+w^*\frac{\partial w^*}{\partial z^*}+\frac{\partial p^*}{\partial z^*}$$

$$=(v/U_\infty L)\left(\frac{\partial^2 w^*}{\partial x^{*2}}+\frac{\partial^2 w^*}{\partial y^{*2}}+\frac{\partial^2 w^*}{\partial z^{*2}}\right)-\frac{gL}{U_\infty^2} \ . \tag{11.42}$$

In (11.42) there are two nondimensional groups,

$$\text{Re}=\frac{U_\infty L}{v} \quad \text{and} \quad \text{Fr}=\frac{U_\infty}{(gL)^{1/2}} \ ; \tag{11.43}$$

Re is the Reynolds number and Fr is the Froude number. Two incompressible viscous flows involving free surfaces are dynamically similar if they have the same values of Re and Fr even though the values of U_∞ or L or v are different for the two flows.

Other nondimensional numbers, such as the Mach number, Prandtl number and specific heat ratio, can be obtained by nondimensionalising the energy equation. When applied to air, which is an ideal gas, the energy equation (11.38) can be written as

$$\varrho c_p \frac{DT}{Dt}-\frac{Dp}{Dt}=\Phi+\frac{\partial}{\partial x}\left(k\frac{\partial T}{\partial x}\right)+\frac{\partial}{\partial y}\left(k\frac{\partial T}{\partial y}\right)+\frac{\partial}{\partial z}\left(k\frac{\partial T}{\partial z}\right) \ . \tag{11.44}$$

The following nondimensional variables are introduced:

$$T^* = T/T_\infty \ , \qquad p^* = p/\varrho_\infty U_\infty^2 \ ,$$
$$\varrho^* = \varrho/\varrho_\infty \ , \qquad \Phi^* = \Phi L^2/\mu_\infty U_\infty^2 \ , \qquad k^* = k/k_\infty \ ,$$

and the other nondimensional variables are as before. Substitution into (11.44) and rearrangement gives

$$\varrho^* \frac{DT^*}{Dt^*} = (\gamma - 1) M_\infty^2 \left(\frac{Dp^*}{Dt^*} + \frac{\Phi^*}{Re} \right) + \left[\frac{\partial}{\partial x^*} \left(k^* \frac{\partial T^*}{\partial x^*} \right) \right.$$
$$\left. + \frac{\partial}{\partial y^*} \left(k^* \frac{\partial T^*}{\partial y^*} \right) + \frac{\partial}{\partial z^*} \left(k^* \frac{\partial T^*}{\partial z^*} \right) \right] \bigg/ (\text{Pr Re}) \ . \qquad (11.45)$$

Equation (11.45) indicates that viscous compressible flow of an ideal gas is governed by at least four nondimensional numbers:

Reynolds number, $Re = U_\infty L/v$

Prandtl number, $\quad Pr = \mu_\infty c_p/k_\infty$

Mach number, $\quad M_\infty = U_\infty/a_\infty = U_\infty/(\gamma R T_\infty)^{1/2}$ $\qquad\qquad (11.46)$

Specific heat ratio, $\gamma = c_p/c_v$.

If viscosity and thermal conductivity are considered to be functions of temperature then a fifth nondimensional number appears. However, the four nondimensional numbers given in (11.46), and the Froude number given in (11.43) will be sufficient to ensure the dynamic similarity of a large class of fluid flows. Of the five nondimensional numbers included in (11.43 and 46), three, namely Re, M and Fr, include the motion of the flow. The other two, Pr and γ, are essentially properties of the fluid in question (see Tables 11.1 and 11.2).

The Reynolds number indicates the relative magnitude of the inertia forces and viscous forces, and depending on the particular flow problem, can take values from close to zero (where inertia forces are negligible) to 10^{10} and beyond for which viscous forces are negligible except adjacent to solid surfaces. Some typical values are provided in Table 11.3.

The Mach number measures the speed of the fluid motion in relation to the speed of sound ($M = u/a$). The Mach number provides a measure of the compressibility or change in density due to the motion. A Mach number less than 0.14 causes less than a 1% change in density due to motion. Mach numbers up to three are common for fighter aircraft and considerably greater for reentry vehicles. Because water is almost incompressible (at temperatures and pressures where motion is likely to be of interest) flow problems characterised by non-small Mach numbers will generally involve gases such as air.

For flows involving free surfaces the Froude number is important. Such flows might involve tidal flows in harbours or estuaries or the motion of ships. The

Table 11.3. Typical Reynolds number

Description	Re
Spermatozoon ($L=0.07$ mm) swimming at max. speed	6×10^{-3}
Water droplet ($D=0.07$ mm) falling through air	6.4×10^{-1}
Wind blowing (10 m/s) over telegraph wires	1×10^3
A cricket or baseball propelled at 35 m/s	2×10^5
A shark ($L=1.5$ m) swimming at maximum speed	8×10^6
Large jet transport aircraft (747) at cruise altitude	7×10^7
Ocean liner (Q.E. II, $L=324$ m) at $U=15$ m/s	4.5×10^9
Planetary boundary layer ($L=1000$ km, $U=20$ m/s)	18×10^{12}

Froude number provides a measure of the relative importance of inertia and gravity forces. For very small Froude numbers, gravity is able to keep the water surface flat and the resistance to motion associated with the generation of surface waves is negligible.

Prandtl number measures the ratio of the diffusivity of momentum and the diffusivity of heat. $Pr = \mu c_p/k = \nu/\alpha$. For air at normal temperature and pressure, $Pr = 0.72$ and falls slightly with increasing temperature (Table 11.1). For water $Pr = 8.1$ at 15°C and falls rapidly to 1.74 at 100°C (Table 11.2). The specific heat ratio for air is about $\gamma = 1.4$ and for water $\gamma = 1.0$.

As well as leading to more efficient computation, the exploitation of dynamic similarity has a long history of permitting the accurate extrapolation of the performance of engineering equipment to other working conditions. A fuller discussion of dynamic similarity is provided by Lighthill (1963) and by Panton (1984, pp. 215–226).

11.2.6 Useful Simplifications

Equations (11.10, 31, 38) with an appropriate equation of state and boundary conditions govern the unsteady three-dimensional motion of a viscous, compressible fluid. However, such an equation system is very complicated and consequently time-consuming to solve even on a state-of-the-art supercomputer.

The historical development of fluid dynamics has introduced many flow categories governed by equation systems considerably simpler but less accurate than that indicated above. The rest of this chapter and Chaps. 14–18 will be structured to suit these simpler flow categories. Broadly, the various categories arise through neglect or limiting values of the flow properties. In turn this often implies that the nondimensional numbers (Sect. 11.2.5) are either limited to certain values or do not appear in the simplified equations.

For engineering flows, Table 11.4 provides an appropriate classification. The classification depends on two properties, density and viscosity. Incompressible flows will normally be associated with flow speeds small compared with the speed of sound (M ≪ 1). Conversely, compressible flows will imply M > 0.1 or temperature differences sufficient to require the solution of the full continuity equation (11.10) instead of (11.13) and the inclusion of the energy equation, e.g. (11.38).

Table 11.4. Flow classification

Density: Viscosity	Incompressible (density constant)	Compressible (density varies)
Inviscid flow ("$\mu = 0$")	Potential flow (if vorticity is zero)	Gas dynamics (with "$k = 0$")
Boundary layer flow (viscosity important close to surface)	Laminar flow (low Re) Turbulent flow (high Re)	Heat transfer (also important)
Separated flow (viscosity important everywhere)	Laminar flow (very low Re) Turbulent flow (mod. to high Re)	Heat transfer (also important)

A consideration of the influence of viscosity suggests three major categories. For the flow around streamlined bodies the gross fluid motion, and particularly the pressure distribution, is accurately predicted by assuming the viscosity of the flow is zero. For compressible, inviscid flows it is worth making a further subdivision (not shown in Table 11.4) based on whether $M \gtrless 1$. For $M > 1$ the governing equations are hyperbolic (Chap. 2) in character and shock waves may appear in the flowfield.

For the flow around streamlined bodies significant viscous effects are confined to thin boundary layers immediately adjacent to the body surface. The frictional force (skin friction drag) on the body is determined by the viscous behaviour in the boundary layer. For a nonzero thermal conductivity, heat transfer will also be determined by the (thermal) boundary layer behaviour. For high Reynolds number boundary-layer flows, viscosity is unable to suppress disturbances that may occur naturally and the flow is turbulent. Consequently, if only the time-average behaviour is of interest it is necessary to include additional empirical parameters or relationships to account for the influence of turbulence on the mean motion (Sects. 11.4.2 and 11.5.2).

For flows around bluff bodies, i.e. a typical automobile, there are regions of sheared flow, on the downstream side of the body, where viscous effects are significant. Unless the Reynolds number is very small the flow in the separated region downstream of the body is turbulent and often unsteady. Typically, for separated flow, the full Navier–Stokes equations, either incompressible or compressible, must be solved.

Figure 11.4 shows an inclined aerofoil or turbine blade. Far from the aerofoil the flow behaves as though it were inviscid. Adjacent to the surface on the windward side a thin boundary layer forms and the boundary layer equations accurately describe the local flow behaviour. The flow separates from the leeward surface of the aerofoil and a large separated flow region exists in which the full Navier–Stokes equations must be solved. For as large a separated region as that

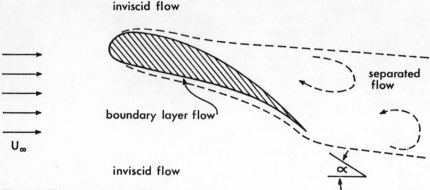

inviscid flow

U_∞

inviscid flow

boundary layer flow

separated flow

$\propto$

Fig. 11.4. Flow around an aerofoil or turbine blade

shown in Fig. 11.4, the local flow would be expected to be unsteady. This example suggests the possibility of solving the different flow regions with appropriate equations and coupling the solutions at the interface. This is the strategy used to obtain the three-dimensional transonic flow about a wing described in Sect. 1.2.2.

The classification shown in Table 11.4 is also useful for analysing internal flows in ducts, turbine casings, diffusers, etc. In Sects. 11.3–11.6 the various flow categories indicated in Table 11.4 will be considered in more detail.

11.3 Incompressible, Inviscid Flow

For this class of flow the density is constant and the viscosity is "zero" (Table 11.4), i.e. viscous effects are neglected. The fluid motion is completely described by the continuity equation, in the form of (11.13) and the Euler equations (11.21).

To permit unique solutions for unsteady flow to be obtained it is necessary to specify initial conditions, $u = u_0(x, y, z)$, $v = v_0(x, y, z)$, $w = w_0(x, y, z)$ and $p = p_0(x, y, z)$. For the flow past an isolated body (Fig. 11.5), a boundary condition of zero normal velocity at the body surface is necessary. At the boundary far from a two-dimensional body, two boundary conditions are required on an inflow boundary, e.g. AD, and one on an outflow boundary, e.g. BC. Typical boundary conditions are shown in Fig. 11.5. This configuration corresponds to an "inviscid" two-dimensional duct.

For any class of flow a line whose tangent is instantaneously parallel to the velocity vector $\mathbf{v}$ is called a streamline. The local slope of the streamline is defined by

$$\frac{dx}{u} = \frac{dy}{v} = \frac{dz}{w} \,. \tag{11.47}$$

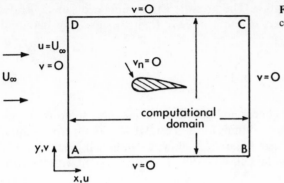

Fig. 11.5. Boundary conditions for incompressible inviscid flow

For steady flow, equations (11.21) can be integrated along a streamline to give the result that

$$\nabla H = \nabla\left(0.5q^2 + \frac{p}{\varrho} + \psi\right) = 0 \ , \tag{11.48}$$

where ψ is a force potential which assumes that the body forces are conservative, i.e. $\mathbf{f} = -\nabla\psi$. When gravity is the body force, acting in the negative z direction, (11.48) becomes

$$H = 0.5q^2 + \frac{p}{\varrho} + gz = \text{const} \ , \tag{11.49}$$

on each streamline. Equation (11.49) is the Bernoulli equation and H is the Bernoulli variable. In (11.49), $0.5q^2$ is the kinetic energy, i.e. $q^2 = \mathbf{v}\cdot\mathbf{v}$. Equation (11.49) is very important because it gives a direct algebraic relationship between the pressure and the velocity.

For flows that are irrotational ($\zeta = \text{curl } \mathbf{v} = 0$), e.g. if the flow far from an immersed body is uniform, H has the same value on all streamlines and consequently (11.49) can be used to compare any two points in the flow domain irrespective of whether they lie on the same streamline or not. For flows that are irrotational (curl $\mathbf{v} = 0$) it is useful to define a velocity potential such that $\mathbf{v} = \nabla\Phi$ or

$$u = \frac{\partial\Phi}{\partial x} \ , \quad v = \frac{\partial\Phi}{\partial y} \quad \text{and} \quad w = \frac{\partial\Phi}{\partial z} \ . \tag{11.50}$$

The continuity equation (11.13) then becomes

$$\nabla^2\Phi = 0 \ , \tag{11.51}$$

i.e. Laplace's equation. Consequently, this class of flow (inviscid, incompressible and irrotational) is called potential flow. Laplace's equation for the velocity potential, along with the specification of zero normal velocity at the surface of an immersed body and specification of the velocity far from the body, completely determines the velocity distribution. The pressure can be obtained, subsequently, from the Euler equations (11.21) or more directly from the unsteady Bernoulli equation for irrotational flow

$$\frac{\partial \Phi}{\partial t} + H = \frac{\partial \Phi}{\partial t} + 0.5q^2 + \frac{p}{\varrho} + gz = \text{const} \ . \tag{11.52}$$

Laplace's equation (11.51) is linear and possesses simple exact solutions which can be superposed to create new solutions for Φ. Through (11.50) superposition of velocity solutions is also permitted. Figure 11.6 shows a two-dimensional source of strength m at location r_s. The potential, satisfying (11.51), associated with the two-dimensional source is

$$\Phi = (m/2\pi)\log(r - r_s) \ . \tag{11.53}$$

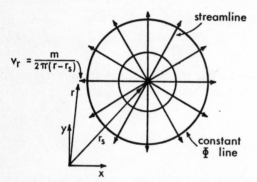

Fig. 11.6. Source flow

The radial and circumferential velocity components are

$$v_r = \left(\frac{m}{2\pi}\right) \Big/ [(x - x_s)^2 + (y - y_s)^2]^{0.5} \ , \qquad v_\theta = 0 \ . \tag{11.54}$$

By combining sources and sinks (negative sources) together it is possible to obtain the flow about closed bodies. Thus, a source and sink in a uniform flow (Fig. 11.7) produce the flow about a Rankine oval. The velocity at any point $P(x, y)$ in the flow can be obtained by combining the formulae for individual sources (11.54) and the freestream to give

$$u = U_\infty + \left(\frac{m}{2\pi}\right)\left(\frac{x + a}{(x + a)^2 + y^2} - \frac{x - a}{(x - a)^2 + y^2}\right) \tag{11.55}$$

$$v = \left(\frac{m}{2\pi}\right)y\left(\frac{1}{(x + a)^2 + y^2} - \frac{1}{(x - a)^2 + y^2}\right) \ . \tag{11.56}$$

In principle, a distribution of sources and sinks along the x-axis with appropriate choices for the strengths will accurately represent the flow behaviour around

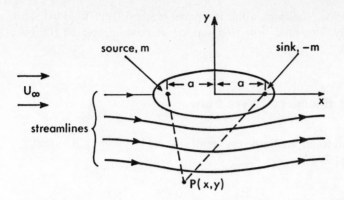

Fig. 11.7. Potential flow about a Rankine oval

streamlined shapes. However, the closely related panel method (Sect. 14.1.1) is usually preferred because it is computationally more efficient.

Other exact solutions of Laplace's equation exist and are useful for modelling special geometries and flows, e.g. lifting aerofoils. The limit in which the source and sink in Fig. 11.7 approach each other and coalesce, such that $\mu = 2am = $ const, defines a doublet of strength μ. A doublet in a freestream provides the solution for the inviscid flow around a two-dimensional circular cylinder (Fig. 11.8)

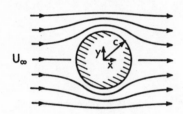

Fig. 11.8. Potential flow about a circular cylinder

$$\Phi = U_\infty x + \frac{\mu}{2\pi}\left(\frac{x}{x^2+y^2}\right) , \tag{11.57}$$

where μ is the doublet strength and $\mu = 2\pi U_\infty c^2$. The velocity components are given by (11.50) as

$$u/U_\infty = 1 - c^2 \frac{x^2-y^2}{(x^2+y^2)^2} \quad \text{and} \tag{11.58}$$

$$v/U_\infty = \frac{-2c^2xy}{(x^2+y^2)^2} . \tag{11.59}$$

Clearly, the velocity components are given as explicit functions of position, and the pressure is also available through (11.49).

In general, potential flow solutions will only give accurate pressure (and velocity) distributions for the flow about streamlined bodies, e.g. aircraft wings and turbine blades at small angles of attack. However, when potential flow is an appropriate approximation, solutions using the panel method (Sect. 14.1) can be obtained very efficiently. Potential flow is described at considerable length by Milne–Thomson (1968).

11.4 Incompressible Boundary Layer Flow

The viscosities of air and water are both very small (Tables 11.1 and 11.2) so that a typical viscous stress (11.27)

$$\tau_{xy} = \mu \left(\frac{\partial u}{\partial y} + \frac{\partial v}{\partial x} \right)$$

will only be large if the velocity gradients are large. For a flow past a streamlined body aligned parallel to the flow direction the viscosity reduces the velocity to zero at the surface, and consequently (Fig. 11.9) the normal velocity gradient is large close to the surface. As a result, viscous forces are only significant in the thin boundary layer that forms close to the surface.

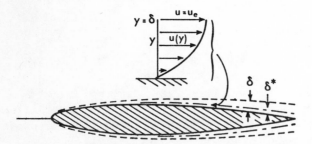

Fig. 11.9. Velocity profile in a boundary layer

For incompressible viscous flow the viscosity is constant and can be brought outside the second derivative terms in (11.28–30). The governing equations for incompressible viscous flow are the continuity equation and the Navier–Stokes equations (11.81).

11.4.1 Laminar Boundary Layer Flow

Consideration of the order of magnitude of the various terms in (11.81) indicates that if $\delta \ll L$, the length of the body, then the governing equations for steady two-dimensional laminar incompressible boundary layer flow over a flat surface are

$$\frac{\partial u}{\partial x} + \frac{\partial v}{\partial y} = 0 \ , \tag{11.60}$$

$$u \frac{\partial u}{\partial x} + v \frac{\partial u}{\partial y} = -\frac{1}{\varrho} \frac{dp_e}{dx} + v \frac{\partial^2 u}{\partial y^2} \tag{11.61}$$

and

$$\frac{\partial p}{\partial y} = 0 \; , \tag{11.62}$$

where p_e is the pressure at the outer edge of the boundary layer.

Equation (11.62) indicates that the pressure remains constant across the boundary layer and becomes a boundary condition either given experimentally or by the inviscid solution (as though $\mu = 0$, everywhere). Through the Bernoulli equation (11.49), assuming no changes in altitude,

$$-\frac{dp_e}{dx} = \varrho u_e \frac{du_e}{dx} \; , \tag{11.63}$$

where u_e is the longitudinal velocity component at the outer edge of the boundary layer. Although the steady form of (11.81) is mixed elliptic/hyperbolic, the equation system (11.60–62) is mixed parabolic/hyperbolic (Sect. 2.3) in character with the x coordinate playing a time-like role. The change in character follows from dropping the term $\partial^2 u/\partial x^2$ in (11.81). Consequently, the equation system (11.60–62) requires initial conditions

$$u(x_0, y) = u_0(y) \; , \tag{11.64}$$

and boundary conditions

$$u(x, 0) = 0 \; , \qquad v(x, 0) = 0 \quad \text{and} \quad u(x, \delta) = u_e(x) \; . \tag{11.65}$$

The parabolic/hyperbolic character of the boundary layer equation system is important because it allows computational solutions to be obtained in a single march in the time-like direction (Chap. 15).

The solution of the boundary layer equations produces the two velocity distributions $u(x, y)$ and $v(x, y)$ from which other important parameters can be obtained.

The nondimensional shear stress at the surface is called the skin friction coefficient c_f and is given by

$$c_f = \frac{\tau_w}{0.5 \varrho u_e^2} = \frac{1}{0.5 \varrho u_e^2} \mu \frac{\partial u}{\partial y}\bigg|_{y=0} \; . \tag{11.66}$$

The frictional drag (in nondimensional form) is obtained by an integration over the skin friction coefficient distribution.

Another important parameter is the displacement thickness, which determines the amount by which the body contour should be displaced to compensate for the loss of mass flow in the boundary layer when the combined inviscid and boundary layer problem is interpreted as an equivalent, purely inviscid problem (Fig. 11.9). Using the displacement thickness to generate a modified body contour produces a more accurate prediction of the pressure distribution (Sect. 14.1.4). The displacement thickness $\delta*$ is defined by

$$\delta* = \int_0^\delta \left(1 - \frac{u}{u_e}\right) dy \ . \tag{11.67}$$

For the equations governing boundary layer flow, it is often possible to transform the equations into a simpler set (Chap. 15). For certain choices of the external velocity distribution $u_e(x)$ the number of independent variables can be reduced by one. This will be illustrated for the boundary layer flow over a flat plate. In this case $u_e(x) = U_\infty$. A new independent variable is introduced as

$$\eta = y \left(\frac{U_\infty}{vx}\right)^{1/2} , \tag{11.68}$$

and u and v are replaced by a single dependent variable f where

$$u = \frac{\partial \psi}{\partial y} \ , \quad v = -\frac{\partial \psi}{\partial x} \quad \text{and} \quad f = \frac{\psi}{(U_\infty vx)^{1/2}} \tag{11.69}$$

Using (11.68, 69), the governing equations (11.60 and 61) become

$$\frac{\partial^3 f}{\partial \eta^3} + 0.5 f \frac{\partial f}{\partial \eta} = 0 \tag{11.70}$$

with boundary conditions

$$\frac{\partial^2 f}{\partial \eta^2} = f = 0 \quad \text{at} \quad \eta = 0 \ ,$$

$$\frac{\partial f}{\partial \eta} = 0 \quad \text{at} \quad \eta = \infty \ . \tag{11.71}$$

Equations (11.70, 71) can be solved, numerically, very accurately. Cebeci and Bradshaw (1977, p. 65) provide a program. From the numerical solution the following expressions can be obtained for the downstream behaviour of the displacement thickness and skin friction coefficient:

$$\delta*/x = 1.72 \, \text{Re}_x^{-0.5} \quad \text{and} \quad c_f = 0.664 \, \text{Re}_x^{-0.5} \ ,$$

where $\text{Re}_x = U_\infty x/v$.

More information on laminar boundary layers and traditional methods of analysis may be obtained from Rosenhead (1963) and Schlichting (1968).

11.4.2 Turbulent Boundary Layer Flow

For $\text{Re}_x \gtrsim 2 \times 10^5$, turbulent fluctuations develop in the boundary layer which the viscosity is unable to suppress. Eventually (further downstream), the mean velocity

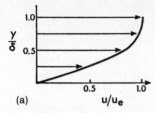

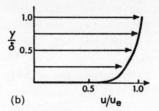

Fig. 11.10a,b. Boundary layer velocity profiles: (**a**) laminar, (**b**) turbulent

profile takes on a much fuller character (Fig. 11.10), due to the more effective mixing associated with turbulence.

Turbulent boundary layers are generally thicker than laminar boundary layers. For a turbulent boundary layer on a flat plate

$$\frac{\delta^*}{x}=0.046\,\mathrm{Re}_x^{-0.2} \quad \text{and} \quad c_f=0.059\,\mathrm{Re}_x^{-0.2} \ . \tag{11.72}$$

Although the instantaneous flow behaviour is unsteady, the gross behaviour of a turbulent boundary layer can usually be obtained from the mean flow. The equations governing the mean flow can be obtained by splitting the instantaneous velocities into the sum of a mean and a fluctuating part, i.e.

$$u=\bar{u}+u' \quad \text{where} \quad \bar{u}=\frac{1}{T}\int\limits_t^{t+T} u\,dt^* \ .$$

Here $\bar{u}$ is the mean value and u' is the turbulent fluctuation. Clearly, $\bar{u}'=0$. It is assumed that the period of the fluctuations is much shorter than T. However, $\bar{u}$, $\bar{v}$, etc., can still be functions of time. The governing equations (11.81) are averaged over the period T and the boundary layer assumption, $\delta \ll L$, is made. Neglecting terms of $O(\delta/L)$ gives the following equations governing steady two-dimensional turbulent boundary layer flow over a flat surface:

$$\frac{\partial\bar{u}}{\partial x}+\frac{\partial\bar{v}}{\partial y}=0 \ , \tag{11.73}$$

$$\bar{u}\frac{\partial\bar{u}}{\partial x}+\bar{v}\frac{\partial\bar{u}}{\partial y}=-\left(\frac{1}{\varrho}\right)\frac{dp_e}{dx}+\frac{1}{\varrho}\left[\frac{\partial}{\partial y}\left(\mu\frac{\partial\bar{u}}{\partial y}-\overline{\varrho u'v'}\right)\right] \ . \tag{11.74}$$

and

$$\frac{\partial\bar{p}}{\partial y}=0 \ . \tag{11.75}$$

This equation system is mixed parabolic/hyperbolic and consequently requires the same initial and boundary conditions (11.64, 65) as for laminar boundary layers. The equation system (11.73–75) is a less accurate approximation to the Navier–

Stokes equations than that for laminar flow (11.60–62). This is because terms of $O(\delta/L)^2$ are neglected in forming the laminar boundary layer equations.

Compared with the laminar boundary layer equations (11.60–62) an additional term, a Reynolds stress, $-\overline{\varrho u'v'}$, appears. This term can be eliminated in favour of mean-flow variables by introducing an eddy viscosity v_T in the construction

$$\varrho v_T \frac{\partial \bar{u}}{\partial y} = -\overline{\varrho u'v'} \; . \tag{11.76}$$

To obtain solutions of (11.73 and 74) it is necessary to relate the eddy viscosity to the mean flow quantities. One way to do this is to introduce a mixing length l so that

$$v_T = l^2 \left| \frac{\partial \bar{u}}{\partial y} \right| \; . \tag{11.77}$$

Different expressions for the mixing length are used in the different parts of the boundary layer. For the inner region, approximately $0 \le \bar{u}/\bar{u}_e \le 0.7$,

$$l = \kappa y [1 - e^{-y^+/A}] \; , \qquad \kappa = 0.41 \; , \tag{11.78}$$

$$y^+ = u_\tau y/v \; , \qquad u_\tau = (0.5 c_f)^{0.5} u_e \; ,$$

and y is measured perpendicular to the wall. In (11.78) the value of A depends on the pressure gradient; for a turbulent boundary layer on a flat plate $(dp_e/dx = 0)$, $A = 26$.

In the outer region, approximately $0.7 \le \bar{u}/\bar{u}_e \le 1.0$, the mixing length is proportional to the boundary layer thickness δ. Typically $l/\delta \approx 0.08$. Alternatively, the Clauser formulation replaces (11.77) in the outer region with

$$v_T = 0.0168 \, \bar{u}_e \delta^* \; . \tag{11.79}$$

The prescription of the eddy viscosity, via algebraic formulae (11.77–79), adds another level of approximation to (11.73–75). However, for boundary layer flow the algebraic eddy viscosity formulae lead to accurate predictions of mean flow quantities, e.g. $\bar{u}(x, y)$.

Discussion of the above algebraic eddy viscosity formulation (11.76–79) and more information on boundary layer flows may be obtained from Cebeci and Bradshaw (1977) and the references cited therein.

11.4.3 Boundary Layer Separation

For the flow past a body, an important consideration is whether the boundary layer separates from the solid surface. Separation will, typically, lead to a large region of slowly fluctuating and/or reversed flow downstream, which will generally invalidate the assumption of a thin boundary layer, on which the derivation of (11.60–62) and (11.73–75) is based. Boundary layer separation is illustrated in

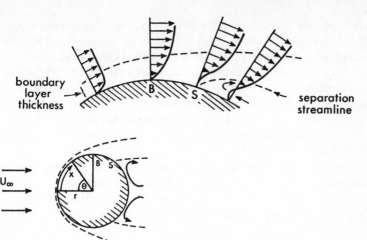

Fig. 11.11. Boundary layer separation from a circular cylinder

Fig. 11.11 for the flow over a two-dimensional circular cylinder. The tangential velocity distribution at the outer edge of the boundary layer is given approximately by the inviscid solution

$$u_e = 2U_\infty \sin\left(\frac{x}{r}\right) , \tag{11.80}$$

at least up to separation. Beyond the shoulder B the velocity u_e decreases with x and consequently the pressure [from the Bernoulli equation (11.49)] increases with x. This adverse pressure gradient retards the fluid in the boundary layer. It is the fluid closest to the wall that is least able (due to the small value of momentum) to withstand the adverse pressure gradient. Consequently, the velocity profile changes until, at separation,

$$\left.\frac{\partial u}{\partial y}\right|_{y=0} = 0 ,$$

and beyond separation $\partial u/\partial y|_{y=0}$ becomes negative. In turn this implies reversed flow adjacent to the wall. Since the shear stress at the wall is given by

$$\tau = \mu \left.\frac{\partial u}{\partial y}\right|_{y=0} ,$$

separation corresponds to a zero shear stress or zero skin friction at the wall.

For the inviscid velocity distribution given by (11.80), solution of the laminar boundary layer equations predicts a separation at about $\theta = 106°$. However, for a viscous (laminar) flow around a circular cylinder, the large separated flow region immediately behind the cylinder modifies the pressure distribution and boundary layer separation occurs at about $\theta = 82°$.

A consideration of the velocity profiles shown in Fig. 11.10 leads to the expectation that a turbulent boundary layer would resist an adverse pressure gradient over a larger distance before separation occurred than would a laminar boundary layer. This is, in fact, the case. For a circular cylinder at a Reynolds number based on diameter, $Re_d > 5 \times 10^6$, the separation occurs at about $\theta = 120°$. Often for the separation of turbulent boundary layers, the flow is unsteady and the separation point undergoes a low frequency oscillation (Simpson 1981). But the above criterion of separation coinciding with a zero wall shear stress is useful for predicting the mean flow.

The derivation of the boundary layer equations from the Navier–Stokes equations on three-dimensional surfaces (Cebeci and Bradshaw, 1977, pp. 315–340), is similar to the derivation of the two-dimensional equations (11.60–62) in that the boundary layer thickness measured normal to the surface is thin compared with a typical dimension in the flow direction. The prediction of three-dimensional separation is much more complicated (Tobak and Peake 1982), since separation of the boundary layer does not necessarily coincide with a zero shear stress at the surface.

Boundary layer flow is an example of a thin shear layer flow. The general class of thin shear layers are found to be governed, at least approximately, by equations equivalent to (11.73 and 74). Wake flows, jets, mixing layers and many developing internal flows can be effectively analysed using thin-shear-layer-type approximations. The importance of the thin shear layer approximation is that it permits an accurate computational solution to be obtained in a single spatial march in the direction of the thin shear layer. Appropriate computational techniques are discussed in Chap. 15.

11.5 Incompressible, Viscous Flow

This class of flow is governed by the continuity and momentum equations in the form

$$\nabla \cdot \mathbf{v} = 0 \ , \qquad \varrho \frac{D\mathbf{v}}{Dt} = \varrho \mathbf{f} - \nabla p + \mu \nabla^2 \mathbf{v} \ . \tag{11.81}$$

For unsteady flow an initial velocity distribution is required that satisfies the continuity equation throughout the computational domain, including the boundaries. It is not necessary (or desirable) to specify the initial pressure distribution, other than at one reference point.

Boundary conditions to suit (11.81) depend on the problem being considered. If a solid surface forms the boundary of the computational domain, it is necessary to set all velocity components equal to the velocity components of the solid surface, i.e., there can be no slip at the fluid/solid interface and no relative motion normal to the surface. For a liquid/liquid interface both velocity and stress must be continuous. Usually the continuity of stress is used as the boundary condition. If a liquid/gas interface forms the boundary of the computational domain lying in the liquid region, then continuity of stress at the liquid/gas interface reduces, for large Reynolds number, to continuity of pressure at the interface (in the absence of surface tension effects).

For the flow about an immersed body, farfield boundary conditions are also required. For inflow and outflow boundaries, it is appropriate to specify all but one of the dependent variables (Table 11.5). However, since viscous terms in the governing equation are usually negligibly small, far from the immersed body, it is also permissible to treat the flow as being locally inviscid so that only one boundary condition is required at an outflow boundary. Specific choices for the boundary conditions are discussed in Sect. 17.1 and from a mathematical perspective by Oliger and Sundstrom (1978) and Gustafsson and Sundstrom (1978).

The finite element formulation (Sects. 5.3–5.5) suggests the following boundary conditions (Gresho 1991),

$$\mathbf{v} = \mathbf{w} \quad \text{on} \quad \Gamma_D \tag{11.81a}$$

$$-p + \mu \frac{\partial u_n}{\partial n} = F_n \quad \text{and} \quad \mu \frac{\partial \mathbf{u}_\tau}{\partial n} = \mathbf{F}_\tau \quad \text{on} \quad \Gamma_N , \tag{11.81b}$$

where $\mathbf{w}$ is a specified velocity field. Thus (11.81a) would be used at solid surfaces and inflow boundaries. Equation (11.81b) is appropriate for outflow boundaries, with F_n and $\mathbf{F}_\tau$ specified and u_n and $\mathbf{u}_\tau$ representing the normal and tangential velocity components. If required F_n and $\mathbf{F}_\tau$ can be related to the normal and surface forces (Gresho 1991). For a high Reynolds number outflow boundary far from an immersed body or for fully-developed duct outflow, it would be appropriate to set $F_n = -p_{\text{spec}}$ and $\mathbf{F}_\tau = 0$. Since the solution, at the outflow, gives $\partial u_n/\partial n \approx 0$, the boundary condition, (11.81b), is often interpreted (in finite difference and finite volume formulations) as specified pressure and $\partial \mathbf{v}/\partial n = 0$.

Table 11.5. Number of farfield boundary conditions for incompressible flow (four variables)

Equation system	Inflow	Outflow
Euler	3	1
Navier–Stokes	3	3

The full equations (11.81) must be solved when regions of separated flow occur. For flows not involving body forces, nondimensionalisation of (11.81) as in (11.42) produces the following equations, in two-dimensional Cartesian coordinates:

$$\frac{\partial u}{\partial x} + \frac{\partial v}{\partial y} = 0 , \tag{11.82}$$

$$\frac{\partial u}{\partial t} + u \frac{\partial u}{\partial x} + v \frac{\partial u}{\partial y} + \frac{\partial p}{\partial x} = \frac{1}{\text{Re}} \left(\frac{\partial^2 u}{\partial x^2} + \frac{\partial^2 u}{\partial y^2} \right) , \tag{11.83}$$

$$\frac{\partial v}{\partial t} + u \frac{\partial v}{\partial x} + v \frac{\partial v}{\partial y} + \frac{\partial p}{\partial y} = \frac{1}{\text{Re}} \left(\frac{\partial^2 v}{\partial x^2} + \frac{\partial^2 v}{\partial y^2} \right) , \tag{11.84}$$

where the density has been absorbed into the Reynolds number. The extension to three dimensions is clear. Solutions to (11.82–84) at small values of Re ($\lesssim 1000$) describe laminar flow. The case of two-dimensional laminar incompressible flow is considered in Sect. 11.5.1.

However, for large values of Re the flow is turbulent. Strictly (11.82–84) in three dimensions would provide the solution, but a very fine grid would be required to accurately represent the finest scales of turbulence, at any realistic Reynolds number. The use of the spectral method (Sect. 5.6) for the direct simulation of turbulence is described by Hussaini and Zang (1987). The wider availability of supercomputers (Chap. 1) has generated considerable interest in large eddy simulation (Rogallo and Moin 1984) in which an unsteady solution of a modified three-dimensional form of (11.82–84) directly captures the large-scale turbulent (eddy) motion and the small-scale (subgrid) turbulent motion is modelled empirically by adding additional terms to the governing equations.

If the large-scale mean motion is steady or of low frequency it is preferable (i.e. computationally more efficient) for engineering purposes to consider time-averaged equations in place of (11.82–84). This approach is pursued further in Sect. 11.5.2.

Computational techniques for flows requiring the full incompressible Navier–Stokes equations are discussed in Sects. 17.1 and 17.2. For flows with a dominant flow direction some reduction of the equations is often possible. This is discussed in Sect. 16.1 and appropriate computational techniques are indicated in Sects. 16.2 and 16.3.3.

11.5.1 Laminar Flow

The equations governing incompressible laminar (viscous) flow are given by (11.82–84). However, for two-dimensional flow it is of interest to consider an alternative formulation in terms of vorticity ζ and stream function ψ.

The following equation is constructed:

$$\frac{\partial}{\partial y}[(11.83)] - \frac{\partial}{\partial x}[(11.84)] \ ,$$

which eliminates the pressure and gives

$$\frac{\partial \zeta}{\partial t} + u\frac{\partial \zeta}{\partial x} + v\frac{\partial \zeta}{\partial y} = \frac{1}{\text{Re}}\left(\frac{\partial^2 \zeta}{\partial x^2} + \frac{\partial^2 \zeta}{\partial y^2}\right) \ , \tag{11.85}$$

where the vorticity

$$\zeta = \frac{\partial u}{\partial y} - \frac{\partial v}{\partial x} \ . \tag{11.86}$$

Strictly, $\zeta = -\zeta_z$, where $\zeta = $ curl **v**. Introducing a stream function ψ such that

$$u = \frac{\partial \psi}{\partial y} \quad \text{and} \quad v = -\frac{\partial \psi}{\partial x} \, , \tag{11.87}$$

and substitution into (11.86) gives

$$\nabla^2 \psi = \zeta \, . \tag{11.88}$$

It may be noted that ψ automatically satisfies (11.82). In the stream function vorticity formulation, the governing equations are (11.85, 87, 88). Elimination of u and v using (11.87) produces a system of equations (11.85, 88) which is parabolic in time and elliptic in space (Sect. 2.1). Initial conditions are provided by setting $\zeta = \zeta_0(x, y, t)$ and solving (11.88) for ψ subject to Dirichlet boundary conditions, $\psi|_c = a$, on the domain boundary c.

Since the system is elliptic in space two boundary conditions are required. If velocity components are given on the boundary, the two boundary conditions specify derivatives of ψ through (11.87). Since ψ only appears in derivative form it is appropriate to fix the value of ψ at one point on the boundary. Consequently the two boundary conditions can be reinterpreted as

$$\psi|_c = a \quad \text{and} \quad \left. \frac{\partial \psi}{\partial n} \right|_c = b \, , \tag{11.89}$$

where n is the direction normal to the boundary c.

It may be noted that no boundary conditions are specified on ζ. This is appropriate for solid surfaces since they are sources of vorticity which is diffused and convected into the flow field (Lighthill 1963). For the flow about an immersed body, farfield boundary conditions such as $\zeta = 0$ may be substituted for $\partial \psi / \partial n|_c = b$ if the flow is locally uniform.

In obtaining computational solutions (Sect. 17.3) it is often desirable to have available equivalent boundary conditions for ζ, particularly at solid surfaces. In the past these have often been constructed from the discrete form of (11.88) to ensure that $\partial \psi / \partial n|_c = b$ is satisfied. However Quartapelle and Valz–Gris (1981) demonstrate that there is no strictly equivalent local boundary condition available for ζ. Instead, the following integral condition on ζ must be satisfied:

$$\iint \zeta \eta \, dx \, dy = \int_c \left(b\eta - a\frac{\partial \eta}{\partial n} \right) ds \, , \tag{11.90}$$

where a and b are as in (11.89) and η is an arbitrary function such that $\nabla^2 \eta = 0$ throughout the computational domain.

Typical boundary conditions can be illustrated by considering the steady flow over a backward-facing step. In Fig. 11.12, AF is an inflow boundary. A bubble of recirculating fluid forms behind the step ED. On boundary AF, u is specified, which fixes ψ through (11.87). On AB, u is set equal to the freestream value and the vorticity is set equal to zero. If $v = 0$ on AB then ψ is fixed and equal to ψ_A.

On BC $\partial^2\zeta/\partial x^2$ is very small and can be deleted from (11.85). This changes the character of the system of governing equations (Sect. 16.1) and leads to the requirement of only one boundary condition on BC. Setting $v=0$ on BC is equivalent to setting $\partial\psi/\partial x=0$. This would not be physically correct close to C. A preferred boundary condition on BC is $\partial v/\partial x=0$. From (11.87 and 88) this is equivalent to $\partial^2\psi/\partial y^2=\zeta$.

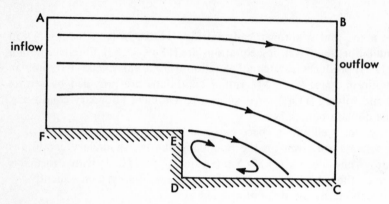

Fig. 11.12. Flow over a backward-facing step

On FEDC, the stream function boundary condition is $\psi=0$. A suitable boundary condition for the vorticity on FEDC is more difficult. The equivalent boundary condition for the primitive variable formulation requires $u=v=0$ on FEDC. For the vorticity the usual form of the boundary condition at a solid surface is obtained from (11.86) or (11.88). Thus, since $\partial v/\partial x=0$ on FE, the vorticity boundary condition becomes $\zeta_{FE}=\partial u/\partial y_{FE}$. On DE the boundary condition is $\zeta_{DE}=-\partial v/\partial x_{DE}$. As noted above, these boundary conditions are not strictly equivalent to the velocity boundary conditions.

In principle, the vorticity formulation can be used in three dimensions. But it then has three components and the stream function must be replaced with a three-component vector potential (Richardson and Cornish 1977). A more recent application to internal flows is described by Wong and Reizes (1984). Vorticity formulations suitable for three-dimensional flows are considered briefly in Sect. 17.4.

If the pressure is required the following Poisson equation is formed from $\partial(11.83)/\partial x+\partial(11.84)/\partial y$ as

$$\frac{\partial^2 P}{\partial x^2}+\frac{\partial^2 P}{\partial y^2}=2\left[\frac{\partial^2\psi}{\partial x^2}\frac{\partial^2\psi}{\partial y^2}-\left(\frac{\partial^2\psi}{\partial x\partial y}\right)^2\right], \tag{11.91}$$

with Neumann boundary conditions on P determined from (11.83 and 84).

For steady flow (11.91) is solved once and for all after the ψ solution has been obtained. For unsteady flow (11.91) must be solved at every time step if the pressure is required, e.g. free-surface flows.

The computational implementation of the stream function vorticity formulation is discussed in Sects. 17.3 and 17.4.

11.5.2 Turbulent Flow

For incompressible flows that are turbulent the use of the three dimensional equivalent of (11.82–84) would be too expensive for engineering design calculations. For most practical calculations, the mean motion is of primary interest. This can be obtained by first averaging the equations over a small time T (as in Sect. 11.4.2). This process produces the time-averaged governing equations

$$\frac{\partial \bar{u}}{\partial x} + \frac{\partial \bar{v}}{\partial y} = 0 , \tag{11.92}$$

$$\varrho\left(\frac{\partial \bar{u}}{\partial t} + \bar{u}\frac{\partial \bar{u}}{\partial x} + \bar{v}\frac{\partial \bar{u}}{\partial y}\right) + \frac{\partial \bar{p}}{\partial x} = \frac{\partial}{\partial x}\left(\mu\frac{\partial \bar{u}}{\partial x} - \overline{\varrho u'u'}\right) + \frac{\partial}{\partial y}\left(\mu\frac{\partial \bar{u}}{\partial y} - \overline{\varrho u'v'}\right) , \tag{11.93}$$

$$\varrho\left(\frac{\partial \bar{v}}{\partial t} + \bar{u}\frac{\partial \bar{v}}{\partial x} + \bar{v}\frac{\partial \bar{v}}{\partial y}\right) + \frac{\partial \bar{p}}{\partial y} = \frac{\partial}{\partial x}\left(\mu\frac{\partial \bar{v}}{\partial x} - \overline{\varrho u'v'}\right) + \frac{\partial}{\partial y}\left(\mu\frac{\partial \bar{v}}{\partial y} - \overline{\varrho v'v'}\right) , \tag{11.94}$$

where $\bar{u}$, $\bar{v}$ are mean values and u', v' are turbulent fluctuations. For three-dimensional flows additional Reynolds stress $-\overline{\varrho u'w'}$ $-\overline{\varrho v'w'}$ and $-\overline{\varrho w'w'}$ appear in the equivalent of (11.93 and 94).

The time-averaged equations can be solved if the Reynolds stresses can be related to mean flow quantities. In Sect. 11.4.2, this was done by introducing an eddy viscosity v_T, letting $-\overline{\varrho u'v'} = \varrho v_T \partial u/\partial y$, and introducing algebraic formulae for the eddy viscosity v_T, etc. However, although this is effective for boundary layer flow, where the local production of turbulent energy is approximately equal to the rate of dissipation, it may not be effective for more complicated turbulent flows, where the transport of turbulence quantities is also important.

An alternative approach is to construct (differential) transport equations for some of the turbulence quantities and to model higher-order terms, which turn out to be triple correlations. Here we indicate the form of a typical two-equation turbulence model, the k–ε model (Launder and Spalding 1974).

In the k–ε model differential equations are introduced for the turbulent kinetic energy k and the rate of dissipation of turbulent energy ε, where

$$k = 0.5(\overline{u'u'} + \overline{v'v'} + \overline{w'w'}) = 0.5(\overline{u_i'u_i'}) \quad \text{and}$$

$$\varepsilon = v_T\left(\frac{\partial u_i'}{\partial x_j}\right)\left(\frac{\partial u_i'}{\partial x_j}\right) .$$

Because of the complexity of the equations, a Cartesian tensor notation (Aris 1962) has been adopted so that the structure of the equations is still easily recognisable.

The governing equations for k and ε are

$$\varrho\frac{Dk}{Dt}=\frac{\partial}{\partial x_j}\left(\frac{\mu_T}{\sigma_k}\frac{\partial k}{\partial x_j}\right)+\mu_T\left(\frac{\partial u_i}{\partial x_j}+\frac{\partial u_j}{\partial x_i}\right)\frac{\partial u_i}{\partial x_j}-\varrho\varepsilon \tag{11.95}$$

and

$$\varrho\frac{D\varepsilon}{Dt}=\frac{\partial}{\partial x_j}\left(\frac{\mu_T}{\sigma_\varepsilon}\frac{\partial\varepsilon}{\partial x_j}\right)+\frac{C_{\varepsilon1}\mu_T\varepsilon}{k}\left(\frac{\partial u_i}{\partial x_j}+\frac{\partial u_j}{\partial x_i}\right)\frac{\partial u_i}{\partial x_j}-\frac{\varrho C_{\varepsilon2}\varepsilon^2}{k}\ , \tag{11.96}$$

where the left-hand sides of (11.95 and 96) represent transport (11.12) of k and ε, respectively. The three terms on the right-hand sides of (11.95 and 96) represent diffusion, production and dissipation, respectively. These two equations are derived from the unsteady Navier–Stokes equations with the introduction of the diffusive terms, the neglect of terms representing viscous dissipation and the modification of some other terms.

From the local values of k and ε, a local (turbulent) eddy viscosity μ_T can be evaluated as

$$\mu_T=\frac{C_\mu\varrho k^2}{\varepsilon}\ , \tag{11.97}$$

and the eddy viscosity is used to relate the Reynolds stresses, e.g. in (11.93 and 94), to the mean quantities by

$$-\overline{\varrho u_i' u_j'}=\mu_T\left(\frac{\partial u_i}{\partial x_j}+\frac{\partial u_j}{\partial x_i}\right)-\frac{2}{3}\varrho k\delta_{ij}\ . \tag{11.98}$$

In equations (11.95–97) the following values are used for the empirical constants:

$$C_\mu=0.09\ ,\quad C_{\varepsilon1}=1.45\ ,\quad C_{\varepsilon2}=1.90\ ,\quad \sigma_k=1.0\ ,\quad \sigma_\varepsilon=1.3\ . \tag{11.99}$$

The use of (11.95 and 96) implies that $\mu_T\gg\mu$. This is clearly invalid close to a solid wall where the turbulent fluctuations are suppressed by the presence of the wall. Therefore, adjacent to walls special wall functions are introduced (Launder and Spalding, 1974; Patel et al. 1985) that typically assume a logarithmic dependence of the tangential velocity component on the normal coordinate and that the production of turbulent kinetic energy is equal to the dissipation in the log-law region. This is equivalent, in the simplest form, to introducing a mixing-length eddy viscosity formulation (Sect. 11.4.2) adjacent to a wall. The use of the special wall functions provide boundary conditions on k and ε away from the wall. Alternatively additional terms (Patel et al. 1985) are added to (11.95 and 96) and boundary conditions $k=0$, $\partial\varepsilon/\partial n=0$ are applied at the wall.

The k–ε turbulence model is suitable for computing free shear layers, boundary layers, duct flows and separated flows; although predictions of far-wake unconfined separated flows overestimate the turbulence production (Rodi 1982). The major weakness of the k–ε model is the assumption of an isotropic eddy viscosity (11.98). This can be avoided by introducing a separate partial differential equation for each Reynolds stress. However, this increases the computational cost substantially.

An effective intermediate model assumes that the transport of the individual Reynolds stresses is proportional to the transport of k (11.95). This reduces the differential equations for the Reynolds stresses to algebraic equations. The details of the algebraic stress model and other turbulence models are provided by Rodi (1980).

11.6 Compressible Flow

This category introduces the additional complication of variable density and temperature. The classification into inviscid, boundary layer and separated flows (Table 11.4) is also useful for compressible flows. The discussion of compressible flows is here related to typical engineering problems, e.g. the flow around turbine blades. In compressible flow density changes are usually associated with high speed (large Mach number) or with large temperature differences. From a computational perspective large temperature differences imply that the energy equation will also be coupled into the solution process.

11.6.1 Inviscid Compressible Flow

Inviscid compressible flow (gas dynamics in Table 11.4) is governed by the continuity equation (11.9), the Euler equations (11.21), and the energy equation (11.35) with the right-hand side set to zero ("$\mu=0$, $k=0$"). At a solid surface the normal velocity component is set to zero to provide a boundary condition. For the flow about an immersed body the number of farfield boundary conditions is indicated in Table 11.6. In supersonic flow boundary conditions are required for each characteristic pointing into the computational domain. The compatibility conditions (Sect. 2.5.1) provide the form of the boundary conditions. Specific choices will be indicated in Sect. 14.2 where appropriate computational techniques are discussed. Appropriate boundary conditions are considered from a mathematical perspective by Oliger and Sundstrom (1978) and from a wave propagation or characteristic perspective by Thompson (1990).

For the steady flow of an inviscid ($\mu=0$), nonconducting ($k=0$) fluid the energy equation may be integrated along a streamline to give the result

$$H=\left(0.5q^2+e+\frac{p}{\varrho}+\psi\right)=\text{const} ,\tag{11.100}$$

where q is the total velocity, e is the specific internal energy and ψ is a force potential introduced to represent body forces, i.e. $\mathbf{f}=-\nabla\psi$. It can be seen that (11.100) is equivalent to (11.48) with the addition of the internal energy e.

From Crocco's theorem (Liepmann and Roshko 1957, p. 193), if a steady flow is irrotational (zero vorticity) and isentropic throughout, H takes the same values on all streamlines. For flows where an isentropic relationship between p and ϱ is readily available, an alternative form of (11.100), in which the internal energy does not appear, is

$$H = 0.5q^2 + \int \frac{1}{\varrho} dp + \psi = \text{const} . \tag{11.101}$$

For isentropic, irrotational flow it is useful to define a velocity potential Φ, i.e. (11.50), such that

$$u = \frac{\partial \Phi}{\partial x} , \quad v = \frac{\partial \Phi}{\partial y} \quad \text{and} \quad w = \frac{\partial \Phi}{\partial z} . \tag{11.102}$$

As a result the Euler equations (11.21) are satisfied identically and the continuity (11.10) and energy (11.101) equations, for steady flow with no body forces, can be combined to give

$$\left(\frac{u^2}{a^2} - 1 \right) \frac{\partial^2 \Phi}{\partial x^2} + \left(\frac{v^2}{a^2} - 1 \right) \frac{\partial^2 \Phi}{\partial y^2} + \left(\frac{w^2}{a^2} - 1 \right) \frac{\partial^2 \Phi}{\partial z^2}$$

$$+ \frac{2}{a^2} \left(uv \frac{\partial^2 \Phi}{\partial x \partial y} + vw \frac{\partial^2 \Phi}{\partial y \partial z} + uw \frac{\partial^2 \Phi}{\partial x \partial z} \right) = 0 , \tag{11.103}$$

where $a = (\partial p / \partial \varrho)^{0.5}$ is the speed of sound. This is the speed at which acoustic waves (pressure waves of small amplitude) propagate in a compressible medium. For an ideal gas, like air, $a = (\gamma p / \varrho)^{0.5}$. For incompressible flow, $a = \infty$, and (11.103) reverts to Laplace's equation (11.51).

Equation (11.101) provides the relationship between a and q. For an ideal gas in which there is a negligible change in the body force, (11.101) gives

$$a^2 + 0.5(\gamma - 1)q^2 = a_\infty^2 + 0.5(\gamma - 1)q_\infty^2 , \tag{11.104}$$

where γ is the specific heat ratio and the subscript ∞ indicates known reference conditions. Equations (11.102–104) are the governing equations for this class of flows. This system can be reduced to a single differential equation after substitution of (11.102 and 104) into (11.103). However, the resulting equation is highly nonlinear.

At a solid surface the boundary condition of no normal flow is given by $\partial \phi / \partial n = 0$. For transonic flow (11.103) is elliptic far away from an immersed body. Consequently a Dirichlet boundary condition for ϕ is imposed on all farfield boundaries.

If shock waves are not present, or are weak, e.g. transonic flow, physically accurate computational solutions of (11.102–104) can be obtained much more economically than the solution of the continuity, Euler and inviscid energy equations in terms of the primitive variables (u, v, w, ϱ, p, T). Appropriate computational techniques to solve (11.102–104) are discussed in Sect. 14.3. The main application of (11.102–104) is to predict the flow behaviour around streamlined bodies, such as aircraft wings and turbine blades at small angles of attack for which the flow is unseparated.

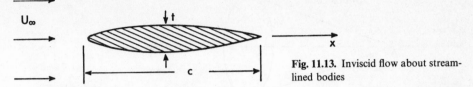

Fig. 11.13. Inviscid flow about stream-lined bodies

For thin bodies in a uniform stream of velocity U_∞ directed along the x-axis (Fig. 11.13), it is conceptually useful to think of the body as introducing small velocity perturbations u', v', and w' to the freestream velocity U_∞

Thus, defining

$$u = U_\infty + u' \ , \qquad v = v' \quad \text{and} \quad w = w' \quad \text{with} \quad u', v', w' \ll U_\infty, \tag{11.105}$$

equation (11.104) can be reduced to

$$\left(\frac{a}{a_\infty}\right)^2 = 1 - (\gamma - 1)M_\infty^2 \frac{u'}{U_\infty} \ , \tag{11.106}$$

where $M_\infty = U_\infty / a_\infty$. As long as $M_\infty < 3$, (11.106) indicates that $a \approx a_\infty$. Making use of (11.105 and 106) gives the following approximate equation, in place of (11.103):

$$(1 - M_\infty^2) \frac{\partial^2 \phi}{\partial x^2} + \frac{\partial^2 \phi}{\partial y^2} + \frac{\partial^2 \phi}{\partial z^2} = \left(\frac{(\gamma + 1)M_\infty^2}{U_\infty} \frac{\partial \phi}{\partial x}\right) \frac{\partial^2 \phi}{\partial x^2} \ , \tag{11.107}$$

where ϕ is the perturbation (or disturbance) potential associated with the velocity perturbations, i.e.

$$\Phi = U_\infty x + \phi \quad \text{and} \quad u' = \frac{\partial \phi}{\partial x} \ , \quad \text{etc.} \tag{11.108}$$

Equation (11.107) is clearly simpler than (11.103) at the price of accepting restrictions on the geometry ($t \ll c$) of the body. However, many practical aerofoils and turbine blades meet these restrictions anyway. Computational techniques to solve (11.107) for transonic flow are indicated in Sect. 14.3.2.

For subsonic ($M_\infty < 1$) or supersonic ($M_\infty > 1$) flow (11.107) can be simplified further to

$$(1 - M_\infty^2) \frac{\partial^2 \phi}{\partial x^2} + \frac{\partial^2 \phi}{\partial y^2} + \frac{\partial^2 \phi}{\partial z^2} = 0 \ . \tag{11.109}$$

Equation (11.109) is linear and is very similar to Laplace's equation (11.51), which governs incompressible potential flow. For $M_\infty < 1$ (11.109) is elliptic. For $M_\infty > 1$ (11.109) is hyperbolic and this implies that discontinuities in normal derivatives of the velocity components can occur on characteristics.

For locally supersonic flow, i.e. $M > 1$, (11.103) is also hyperbolic. The characteristics in supersonic inviscid flow are called Mach lines. The angle μ between the Mach cone (envelope of Mach lines) and the local flow direction (Fig. 11.14) is

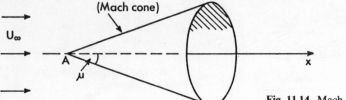

Fig. 11.14. Mach lines in supersonic flow

related to the local Mach number by

$$\mu = \sin^{-1}\left(\frac{1}{M}\right) .$$

That is, as M increases, the Mach cone lies closer to the local flow direction. Any disturbances at A can only influence the part of the flow inside the Mach cone directed downstream from A.

If M is locally increasing in the flow direction, successive Mach lines spread out in moving further from A. However, if M is locally decreasing in the flow direction, successive Mach lines would appear to cross each other. This cannot ocur in reality and a shock wave develops, which can be interpreted, heuristically, as a coalescence of Mach lines. A typical Mach number distribution around an aircraft wing (two dimensional section) is shown in Fig. 11.15.

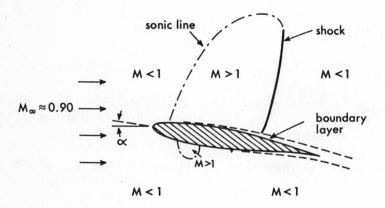

Fig. 11.15. Mach number distribution about a wing

The jump conditions across the shock in the normal direction require conservation of mass, momentum and energy and are called the Rankine–Hugoniot conditions (Liepmann and Roshko 1957, p. 64). However, entropy increases across the shock. A typical form of the Rankine–Hugoniot relation is

$$\frac{u_1}{u_2} = \frac{\varrho_2}{\varrho_1} = \left(\frac{\gamma+1}{\gamma-1} + \frac{p_1}{p_2}\right) \bigg/ \left(1 + \frac{\gamma+1}{\gamma-1}\frac{p_1}{p_2}\right) , \qquad (11.110)$$

$$\frac{p_2}{p_1} = 1 + \left[\frac{2\gamma}{\gamma+1}\right](M_1^2 - 1) \ , \quad \text{and}$$

$$M_2^2 = \frac{1 + 0.5(\gamma - 1)M_1^2}{\gamma M_1^2 - 0.5(\gamma - 1)} \ ,$$

where stations 1 and 2 are upstream and downstream of the shock, respectively, and u_1 and u_2 are the velocity components normal to the shock.

It may be noted that although the Rankine–Hugoniot relations connect two states of an inviscid, nonconducting fluid the local behaviour within the shock is modified by viscosity and thermal conductivity (Fig. 11.16). Here x is a local coordinate normal to the shock. However, when determining the behaviour of an otherwise inviscid, nonconducting fluid, it is usually possible to treat the very severe gradients across the shock as discontinuities without affecting the behaviour away from the shock.

If the shock is not weak, $M_1 \gtrsim 1.1$, it is necessary to solve continuity, the Euler equations and the inviscid energy equation to accurately predict the flow behaviour (Sect. 14.2).

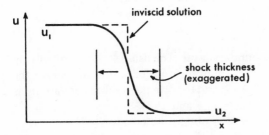

Fig. 11.16. Velocity distribution within a shock

11.6.2 Compressible Boundary Layer Flow

The computation of the compressible boundary layer equations follows the same path as for the incompressible boundary layer equations, namely that the thickness of the boundary layer is assumed to be small compared with a characteristic dimension in the flow direction. However, the energy equation must also be considered and it is necessary to define a thermal boundary layer, across which the temperature undergoes a rapid variation, in an analogous manner to the velocity boundary layer (Schlichting 1968).

For compressible boundary layer flow the only simplification possible for the continuity equation (11.10) is that terms may be deleted for either steady flow or two-dimensional flow. Thus, for steady, two-dimensional compressible laminar boundary layer flow the continuity equation (11.10) takes the form

$$\frac{\partial}{\partial x}(\varrho u) + \frac{\partial}{\partial y}(\varrho v) = 0 \ . \tag{11.111}$$

The steady x-momentum equation (11.28) reduces to

$$\varrho\left(u\frac{\partial u}{\partial x}+v\frac{\partial u}{\partial y}\right)=-\frac{dp_e}{dx}+\frac{\partial}{\partial y}\left(\mu\frac{\partial u}{\partial y}\right) . \tag{11.112}$$

This equation may be compared with (11.61). Here, the density is a function of position and the viscosity is typically a function of temperature. The energy equation (11.38) for the flow in a thin steady thermal boundary layer takes the form

$$\varrho c_p\left(u\frac{\partial T}{\partial x}+v\frac{\partial T}{\partial y}\right)=u\frac{dp_e}{dx}+\frac{\partial}{\partial y}\left(k\frac{\partial T}{\partial y}\right)+\mu\left(\frac{\partial u}{\partial y}\right)^2 , \tag{11.113}$$

where c_p is the specific heat at constant pressure and k is the thermal conductivity. The equation system (11.111–113) is mixed parabolic/hyperbolic and, consequently, requires initial conditions

$$u(x_0, y)=u_0(y) , \qquad T(x_0, y)=T_0(y) \tag{11.114}$$

and boundary conditions

$$u(x, 0)=0 , \quad v(x, 0)=0 , \quad T(x, 0)=T_w(x) \ \text{ or } \ k\frac{\partial T}{\partial y}(x, 0)=-\dot{Q}_w(x)$$

$$u(x, \delta)=u_e(x), \quad T(x, \delta)=T_e(x) , \tag{11.115}$$

where $y=\delta$ denotes the edge of the boundary layer. The usual strategy for solving the equation system (11.111–115) is to introduce a transformation to remove the explicit appearance of the density (Sect. 15.2) and to solve the resulting equations as an equivalent incompressible system (Sects. 11.4 and 15.1).

The extension of the above equations to obtain the compressible mean-flow turbulent boundary layer equations follows the path indicated in Sect. 11.4.2. However, additional products of terms arise associated with density and temperature fluctuations (Schlichting 1968, Chap. 13; Cebeci and Bradshaw 1984, Chap. 3). For compressible boundary layer flow straightforward extensions of the algebraic eddy viscosity turbulence model, described in Sect. 11.4.2, are available and effective (Cebeci and Bradshaw 1984, Chap. 6).

11.6.3 Compressible Viscous Flow

For compressible viscous fluids with extensive regions of separated flow, (e.g. Fig. 11.17), it is necessary to solve the full equations, i.e. (11.10, 26 and 33). Usually, only limited simplification is possible, e.g. neglect of the body force terms. For the analysis of external flows where compressibility effects are mainly due to motion the relevant equations can be expressed conveniently in conservation form. Equation (11.10) is already in conservation form. Equation (11.26) is put into conservation form by adding $\mathbf{v}\times$(11.10) to the left-hand sides. Equation (11.33) is put into conservation form by adding $(e+\frac{1}{2}\mathbf{v}\cdot\mathbf{v})\times$(11.10) to the left-hand side.

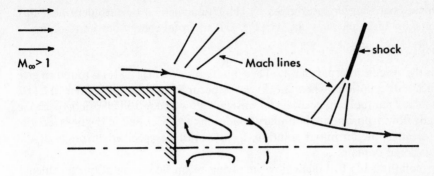

Fig. 11.17. Base flow

In three dimensions the governing equations can be written compactly as a single vector equation as

$$\frac{\partial \mathbf{q}}{\partial t} + \frac{\partial \mathbf{F}}{\partial x} + \frac{\partial \mathbf{G}}{\partial y} + \frac{\partial \mathbf{H}}{\partial z} = 0 \ , \tag{11.116}$$

where

$$\mathbf{q} \equiv \begin{bmatrix} \varrho \\ \varrho u \\ \varrho v \\ \varrho w \\ E \end{bmatrix} , \quad \mathbf{F} = \begin{bmatrix} \varrho u \\ \varrho u^2 + p - \tau_{xx} \\ \varrho uv - \tau_{xy} \\ \varrho uw - \tau_{xz} \\ (E + p - \tau_{xx})u - \tau_{yx}v - \tau_{zx}w + \dot{Q}_x \end{bmatrix} ,$$

$$\mathbf{G} \equiv \begin{bmatrix} \varrho v \\ \varrho uv - \tau_{yx} \\ \varrho v^2 + p - \tau_{yy} \\ \varrho vw - \tau_{yz} \\ (E + p - \tau_{yy})v - \tau_{xy}u - \tau_{zy}w + \dot{Q}_y \end{bmatrix} ,$$

$$\mathbf{H} \equiv \begin{bmatrix} \varrho w \\ \varrho uw - \tau_{zx} \\ \varrho vw - \tau_{zy} \\ \varrho w^2 + p - \tau_{zz} \\ (E + p - \tau_{zz})w - \tau_{xz}u - \tau_{yz}v + \dot{Q}_z \end{bmatrix} , \tag{11.117}$$

The form of (11.116, 117) includes both laminar and turbulent compressible flow. For laminar flow τ_{xx}, etc., are given by (11.27). For turbulent flow τ_{xx}, etc., will also include terms representing the Reynolds stresses (Sect. 11.5.2). $\dot{Q}_x$, $\dot{Q}_y$ and $\dot{Q}_z$ are directional heat transfer rates given by (11.37) augmented by turbulent heat flux gradients, for turbulent flow. In (11.117), E is the total energy per unit volume

$$E = \varrho[e + 0.5(u^2 + v^2 + w^2)] \ , \tag{11.118}$$

and e is the specific internal energy. The conservation form (11.116) is found to give more accurate solutions when shock waves occur. The equation system (11.116, 117) is mixed parabolic/hyperbolic for unsteady flow and mixed elliptic/hyperbolic for steady flow. Appropriate boundary conditions are discussed in the next section. For unsteady flow an initial solution, $\mathbf{q} = \mathbf{q}_0$, must be specified throughout the computational domain.

For compressible turbulent flow involving mean flow separation, turbulence models like the k–ε model (Sect. 11.5.2) are required. Compressible forms of the k–ε model can be developed but the additional terms associated with compressibility have little influence (Marvin 1983) for Mach numbers up to 5. However, if density changes are due to large temperature differences in the computational domain then additional terms are required (Rodi 1980).

Computational techniques for solving (11.116 and 117) are discussed in Chap. 18. For flows with a dominant flow direction some reduction is possible in the complexity of (11.116 and 117). This is discussed in Sect. 16.1 and appropriate computational techniques are indicated in Sects. 16.3.1, 16.3.2 and 16.3.7.

11.6.4 Boundary Conditions for Compressible Viscous Flow

Here, boundary conditions are considered for a body in an unbounded fluid, with the flow parallel to the x-axis far from the body (Fig. 11.18).

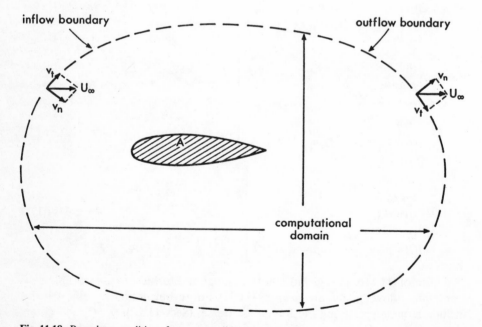

Fig. 11.18. Boundary conditions for compressible viscous flow

Two types of boundary conditions can be distinguished; those at the solid/fluid interface A, and those far from the body. For the solid surface A it is necessary that

$$\mathbf{v} = 0 \quad \text{and} \quad T = T_s \quad \text{or} \quad k\frac{\partial T}{\partial n} = -\dot{Q}_s \, , \tag{11.119}$$

i.e., zero relative velocity and either specified temperature or heat transfer rate.

The farfield boundary conditions are more difficult to specify in a way that facilitates computation. It is necessary to differentiate between inflow and outflow boundary conditions, which can be determined by considering the sign of the normal velocity.

The unsteady *Euler* equations are hyperbolic and it is straightforward to construct characteristics and then to require that as many boundary conditions must be specified as there are characteristics entering the domain (Chu 1978). For three-dimensional flow with two thermodynamic variables (the equation of state provides the other) this implies five boundary conditions must be specified at inflow if the flow is supersonic (Table 11.6).

Oliger and Sundstrom (1978) obtain comparable results for the unsteady Euler equations, but are able to extend their analysis to the equations governing viscous compressible flow. The required number of boundary conditions are shown in Table 11.6. Gustafsson and Sundstrom (1978) have examined the two-dimensional compressible Navier–Stokes equations, based on (11.36) with viscous dissipation Φ ignored, instead of (11.33). Gustafsson and Sundstrom refer to the unsteady compressible Navier–Stokes equations as being an incompletely parabolic system. Gustafsson and Sundstrom demonstrate that the number of boundary conditions shown in Table 11.6 are necessary and sufficient to produce a well-posed problem. They also provide specific combinations of boundary conditions that produce a well-posed problem. Broadly the compatibility conditions associated with the characteristics of the equivalent Euler equations are combined so that the problem remains well-posed according to the energy method. This provides suitable Dirichlet boundary conditions. The additional boundary conditions required by the compressible Navier–Stokes equations are then imposed as Neumann boundary conditions. Nordstrom (1989), using a similar method to Gustafsson and Sundstrom, obtains a subsonic outflow boundary condition that is similar to (11.81b). Nordstrom demonstrates that the use of boundary conditions based on the energy method leads to faster convergence to the steady state for a two-dimensional subsonic test problem.

Table 11.6. Number of farfield boundary conditions to be specified

| Equation System | Compressible (5 variables) | | | |
| | Inflow | | Outflow | |
	Supersonic	Subsonic	Supersonic	Subsonic
Euler	5	4	0	1
Navier–Stokes	5	5	4	4

For many flows at high Reynolds number the solution, far away from the body, behaves as though it were governed by the Euler equations rather than the Navier–Stokes equations. This suggests choosing boundary conditions accordingly and this often works in practice. Strictly, where this leads to an under-prescription one may fail to obtain a unique solution. An over-prescription of boundary conditions usually produces a solution with a severe unphysical boundary layer adjacent to the boundary in question. In a sense the farfield boundary conditions should be chosen so as to make the boundary appear transparent to the solution, i.e., the same solution would be obtained if the boundary location were moved farther from the body. A more detailed discussion of the boundary conditions for viscous compressible flow is provided by Peyret and Taylor (1983, pp. 312–316).

11.7 Closure

The properties of air and water have been discussed briefly. In engineering fluid dynamics, air and water are the two fluids that occur most often. Both fluids are characterised by small values of viscosity. However, although air is easily compressed, water, in the liquid phase, is effectively incompressible.

Central to this chapter has been the presentation of the equations and boundary conditions governing fluid dynamics. The governing equations are derived by requiring that, for a small control volume, mass and energy are conserved and that the time rate of change of linear momentum is equal to the net force. Nondimensionalisation of the governing equations leads to the appearance of nondimensional numbers and the concept of dynamic similarity. Two of the more important nondimensional numbers are the Reynolds number and Mach number.

To provide more manageable sets of governing equations a classification based on the viscosity and density (Table 11.4) has been introduced. The classification, which is particularly useful for engineering fluid dynamics, draws attention to the categories of inviscid, boundary layer and separated flows for both incompressible and compressible fluids.

To suit complicated computational domains, it is desirable to express the governing equations in generalised curvilinear coordinates; this is taken up in Chap. 12. In Chaps. 14–18, the computational techniques, introduced in Chaps. 3–10, will be extended to handle the more complicated governing equations, and related boundary conditions, developed in this chapter.

11.8 Problems

Physical Properties of Fluids (Sect. 11.1)

11.1 Plot the variation with temperature of v, α and Pr for air at $p = 100$ kPa and $p = 500$ kPa. Comment on any differences.

11.2 Plot the variation with temperature of v, α and Pr of water for saturated conditions, Table 11.2. Compare the behaviour with that of air.

11.3 Plot the variation of μ with T shown in Table 11.1 and compare with

(a) $\mu = \mu_{300}(T/300)^{0.76}$

(b) Sutherland's law,

$$\mu = \frac{1.458 \times 10^{-6} T^{1.5}}{110.4 + T} ,$$

where $T[\text{K}]$ is the absolute temperature.

Equations of Motion (Sect. 11.2)

11.4 Show that the continuity equation in cylindrical polar coordinates is

$$\frac{\partial \varrho}{\partial t} + \frac{1}{r}\frac{\partial}{\partial r}(\varrho r v_r) + \frac{1}{r}\frac{\partial}{\partial \theta}(\varrho v_\theta) + \frac{\partial}{\partial z}(\varrho v_z) = 0$$

where v_r, v_θ and v_z are the velocity components in the r, θ and z directions.

11.5 For incompressible laminar flow show that the steady axisymmetric Navier–Stokes equations (momentum) can be written

$$v_r \frac{\partial v_r}{\partial r} + v_z \frac{\partial v_r}{\partial z} + \frac{1}{\varrho}\frac{\partial p}{\partial r} = v\left(\nabla^2 v_r - \frac{v_r}{r^2}\right)$$

$$v_r \frac{\partial v_z}{\partial r} + v_z \frac{\partial v_z}{\partial z} + \frac{1}{\varrho}\frac{\partial p}{\partial z} = v\nabla^2 v_z ,$$

assuming no body force and $v_\theta = 0$. Velocity components v_r and v_z are in the radial (r) and axial (z) directions, respectively.

11.6 Show that the steady energy equation for an incompressible fluid with constant thermal conductivity can be written in spherical coordinates as

$$\varrho \frac{De}{Dt} = \Phi + k\nabla^2 T ,$$

and that

$$\frac{De}{Dt} \equiv v_r \frac{\partial e}{\partial r} + \frac{v_\theta}{r}\frac{\partial e}{\partial \theta} + \frac{v_\phi}{r\sin\theta}\frac{\partial e}{\partial \phi}$$

and

$$\nabla^2 T \equiv \frac{1}{r^2}\frac{\partial}{\partial r}\left(r^2 \frac{\partial T}{\partial r}\right) + \frac{1}{r^2 \sin\theta}\frac{\partial}{\partial \theta}\left(\sin\theta \frac{\partial T}{\partial \theta}\right) + \frac{1}{r^2 \sin^2\theta}\frac{\partial^2 T}{\partial \phi^2} .$$

11.7 Starting with the two-dimensional form of the equation governing viscous compressible flow (11.116, 117) introduce the nondimensionalisation after

(11.41) with

$$p^* = \frac{p}{\varrho_\infty U_\infty^2}, \quad \varrho^* = \frac{\varrho}{\varrho_\infty}, \quad T^* = \frac{T}{T_\infty}, \quad \mu^* = \frac{\mu}{\mu_\infty}, \quad e^* = \frac{e}{U_\infty^2} \quad \text{and}$$

$$k^* = \frac{k}{k_\infty}.$$

Show that the resulting nondimensional equations have the same appearance as the dimensional equations except for the shear stresses and heat transfer rates which become, e.g.

$$\tau_{xx}^* = \mu^* \frac{\left(\dfrac{2\partial u^*}{\partial x^*} - \dfrac{2}{3} \mathscr{D}^* \right)}{\text{Re}} \quad \text{and}$$

$$Q_x^* = -k^* \frac{\dfrac{\partial T^*}{\partial x^*}}{(\gamma - 1)M_\infty^2 \, \text{Pr} \, \text{Re}}$$

where $\text{Re} = \varrho_\infty U_\infty L / \mu_\infty$, $M_\infty^2 = U_\infty^2 / \gamma R T_\infty$ and $\text{Pr} = \mu_\infty c_p / k_\infty$.
Comment on whether the present nondimensional form of the equations is suitable for
a) hypersonic flow, $M_\infty > 5$,
b) low speed flow with large temperature differences appearing through the boundary conditions.

Incompressible, Inviscid Flow (Sect. 11.3)

11.8 For two-dimensional irrotational, inviscid and incompressible flow the governing equations are

$$\frac{\partial u}{\partial x} + \frac{\partial v}{\partial y} = 0, \quad \frac{\partial u}{\partial y} - \frac{\partial v}{\partial x} = 0.$$

By introducing the stream function ψ with

$$u = \frac{\partial \psi}{\partial y}, \quad v = -\frac{\partial \psi}{\partial x}$$

the governing equations reduce to Laplace's equation $\nabla^2 \psi = 0$.
Comment on the suitability of the stream function approach compared with the velocity potential approach (11.51) with regard to
a) boundary conditions,
b) extension to three dimensions.

11.9 For potential flow around a two-dimensional circular cylinder show that the surface pressure coefficient, $C_p = (p - p_\infty)/\frac{1}{2}\varrho U_\infty^2$, is given by

$$C_p = 1 - 4\sin^2\theta \; ,$$

where θ is measured from the front stagnation point. Comment on the approximate range of validity of θ for the above expression in a) laminar flow, b) turbulent flow.

Incompressible Boundary Layer Flow (Sect. 11.4)

11.10 For laminar boundary layer flow past a two-dimensional wedge (Sect. 15.1.2) the velocity at the outer edge of the boundary layer is given by

$$u_e = Cx^{\beta/(2-\beta)}$$

where β is the wedge angle. By introducing the similarity variable $\eta = y\{u_e/[(2-\beta)xv]\}^{1/2}$ and the dependent variable $f = \psi/[(2-\beta)u_e vx]^{1/2}$, where ψ is the stream function, show that the governing equations (11.60 and 61) reduce to

$$\frac{\partial^3 f}{\partial \eta^3} + f\frac{\partial^2 f}{\partial \eta^2} + \beta\left[1 - \left(\frac{\partial f}{\partial \eta}\right)^2\right] = 0 \; .$$

11.11 Show that by integrating (11.61) across the boundary layer it is possible to obtain the momentum integral equation

$$\frac{d\theta}{dx} = 0.5c_f - (H+2)\theta\left(\frac{du_e}{dx}\right)\bigg/u_e \; , \quad \text{where}$$

$$\theta = \int_0^\delta u/u_e(1 - u/u_e)dy \quad \text{and} \quad H = \delta^*/\theta \; .$$

How would the above momentum integral equation change for the equations governing a two-dimensional turbulent boundary layer (11.73, 74 and 76)?

Incompressible, Viscous Flow (Sect. 11.5)

11.12 By introducing the Bernoulli variable $H = p + \varrho(u^2 + v^2)/2$, show that the steady counterparts of (11.83 and 84) reduce to

$$v\zeta + \frac{\partial H}{\partial x} = \frac{1}{Re}\frac{\partial \zeta}{\partial y} \; ,$$

$$-u\zeta + \frac{\partial H}{\partial y} = -\frac{1}{Re}\frac{\partial \zeta}{\partial x} \; ,$$

where the vorticity $\zeta = \partial u / \partial y - \partial v / \partial x$. For the external flow around an isolated body, what do the above equations reduce to in the "inviscid" region away from the body? Can this be exploited to generate an efficient computational method?

11.13 Derive (11.91). What would be the corresponding form in three dimensions, with the right-hand side expressed in terms of the velocity components?

Compressible Flow (Sect. 11.6)

11.14 For inviscid compressible flow derive the two-dimensional equivalent of (11.103 and 104) from (11.10, 101 and 102).

11.15 For steady two-dimensional flow apply the boundary layer assumptions, $\delta \ll L$, to the energy equation (11.38) to obtain the boundary layer energy equation (11.113).

11.16 For two-dimensional viscous compressible flow of air, use (11.116) and (11.117) to express the conservation form of the momentum equations in terms of u, v, ϱ, p and μ.

11.17 For two-dimensional viscous flow of air use (11.116) and (11.117) to obtain the equivalent of (11.34) with the material derivative of the kinetic energy expressed in conservation form. Using the result obtain the equivalent of (11.35) and (11.36). There is no need to expand τ_{xx} etc.

12. Generalised Curvilinear Coordinates

The computation of flowfields in and around complex shapes such as ducts, engine intakes, complete aircraft or automobiles, etc., involves computational boundaries that do not coincide with coordinate lines in physical space. For finite difference methods, the imposition of boundary conditions for such problems has required a complicated interpolation of the data on local grid lines and, typically, a local loss of accuracy in the computational solution.

Such difficulties motivate the introduction of a mapping or transformation from physical (x, y, z) space to a generalised curvilinear coordinate (ξ, η, ζ) space. The generalised coordinate domain is constructed so that a computational boundary in physical space coincides with a coordinate line in generalised coordinate space.

The use of generalised coordinates implies that a distorted region in physical space is mapped into a rectangular region in the generalised coordinate space (Fig. 12.1). The governing equations are expressed in terms of the generalised coordinates as independent variables and the discretisation is undertaken in the generalised coordinate space. Thus the computation is performed in the generalised coordinate space, effectively.

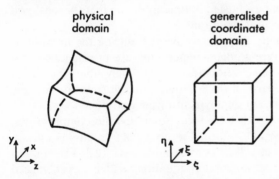

physical domain

generalised coordinate domain

Fig. 12.1. Correspondence of the physical and generalised-coordinate domains

For example, to compute the flow in a two-dimensional curved duct it would be appropriate to make the walls of the duct coincide with lines of constant η (Fig. 12.2).

Location along the duct wall, say from A to B or D to C, then corresponds to specific values of ξ in the computational domain. Corresponding points on AB and CD, connected by a particular η line, will have the same value ξ_j but different η values (η_1 on $A'B'$ and η_{KMAX} on $C'D'$). At a particular point (j, k) along this η line,

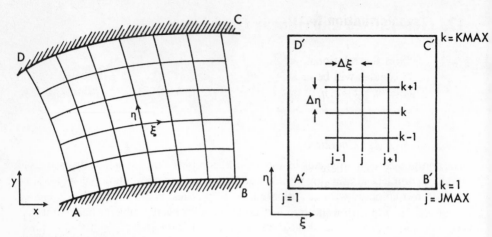

Fig. 12.2. Two-dimensional curved duct

$\xi = \xi_j$ and $\eta = \eta_k$. A corresponding point, $x = x(\xi_j, \eta_k)$ and $y = y(\xi_j, \eta_k)$, exists in the physical domain.

The concept of generalised coordinates suggests additional possibilities. First, the computational grid in generalised-coordinate space can correspond to a moving grid in physical space as would be appropriate for an unsteady flow with boundary movement.

The mapping between physical and generalised-coordinate space permits grid lines to be concentrated in parts of the physical domain where severe gradients are expected. If the severe-gradient regions change with time, e.g. shock-wave propagation, the physical grid can be adjusted in time to ensure that the local grid is sufficiently refined to obtain accurate solutions.

Even for steady flows the ability to relocate the grid during the iteration is useful. For initially unknown severe-gradient regions the use of generalised coordinates allows the iterative solution development to adaptively adjust the grid to resolve the flow behaviour in all parts of the computational domain to the same accuracy. The result is a more efficient deployment of the grid points.

The use of generalised coordinates introduces some specific complications. Firstly, it is necessary to consider what form the governing equations take in generalised coordinates. This aspect will be considered in Sect. 12.3. The computational solution of a typical partial differential equation written in generalised coordinates will be demonstrated in Sect. 12.4.

The governing equations, expressed in generalised coordinates, will contain additional terms that define the mapping between the physical space and the generalised-coordinate space. These additional terms (transform parameters) in the form of derivatives, e.g. $\partial x / \partial \xi$, usually need to be discretised (Sect. 12.2). This introduces an additional source of error in the solution (Sect. 12.2.4). The origin of the transformation parameters is indicated in Sect. 12.1.

12.1 Transformation Relationships

In this section the relationships between the physical (x, y, z) and computational (ξ, η, ζ) coordinates will be established. The corresponding time-dependent transformations $(x, y, z, t \rightarrow \xi, \eta, \zeta, \tau)$ are given by Steger (1978) and Thompson et al. (1985, Chap. 3).

12.1.1 Generalised Coordinates

It is assumed that there is a unique, single-valued relationship between the generalised coordinates and the physical coordinates, which can be written as

$$\xi = \xi(x, y, z) , \qquad \eta = \eta(x, y, z) \quad \text{and} \quad \zeta = \zeta(x, y, z) \tag{12.1}$$

and by implication, $x = x(\xi, \eta, \zeta)$, etc. The specific relationship is established once the physical grid is created (Chap. 13).

Given the functional relationships, $\xi = \xi(x, y, z)$, the governing equations can be transformed into corresponding equations containing partial derivatives with respect to ξ, η and ζ.

As an example, first derivatives of the velocity components, u, v and w, with respect to x, y and z become

$$
\begin{bmatrix}
\dfrac{\partial u}{\partial x} & \dfrac{\partial u}{\partial y} & \dfrac{\partial u}{\partial z} \\[2ex]
\dfrac{\partial v}{\partial x} & \dfrac{\partial v}{\partial y} & \dfrac{\partial v}{\partial z} \\[2ex]
\dfrac{\partial w}{\partial x} & \dfrac{\partial w}{\partial y} & \dfrac{\partial w}{\partial z}
\end{bmatrix}
=
\begin{bmatrix}
\dfrac{\partial u}{\partial \xi} & \dfrac{\partial u}{\partial \eta} & \dfrac{\partial u}{\partial \zeta} \\[2ex]
\dfrac{\partial v}{\partial \xi} & \dfrac{\partial v}{\partial \eta} & \dfrac{\partial v}{\partial \zeta} \\[2ex]
\dfrac{\partial w}{\partial \xi} & \dfrac{\partial w}{\partial \eta} & \dfrac{\partial w}{\partial \zeta}
\end{bmatrix}
\begin{bmatrix}
\dfrac{\partial \xi}{\partial x} & \dfrac{\partial \xi}{\partial y} & \dfrac{\partial \xi}{\partial z} \\[2ex]
\dfrac{\partial \eta}{\partial x} & \dfrac{\partial \eta}{\partial y} & \dfrac{\partial \eta}{\partial z} \\[2ex]
\dfrac{\partial \zeta}{\partial x} & \dfrac{\partial \zeta}{\partial y} & \dfrac{\partial \zeta}{\partial z}
\end{bmatrix},
\tag{12.2}
$$

where the Jacobian matrix, $\underline{J}$, of the transformation is

$$
\underline{J} \equiv
\begin{bmatrix}
\dfrac{\partial \xi}{\partial x} & \dfrac{\partial \xi}{\partial y} & \dfrac{\partial \xi}{\partial z} \\[2ex]
\dfrac{\partial \eta}{\partial x} & \dfrac{\partial \eta}{\partial y} & \dfrac{\partial \eta}{\partial z} \\[2ex]
\dfrac{\partial \zeta}{\partial x} & \dfrac{\partial \zeta}{\partial y} & \dfrac{\partial \zeta}{\partial z}
\end{bmatrix}.
\tag{12.3}
$$

In principle, if an analytic relationship $\xi = \xi(x, y, z)$ is available, the elements of $\underline{J}$ can be evaluated directly. In practice, an explicit analytic relationship is not usually available and it is more convenient to work with the inverse Jacobian matrix, $\underline{J}^{-1}$, given by

$$
\underline{J}^{-1} \equiv
\begin{bmatrix}
\dfrac{\partial x}{\partial \xi} & \dfrac{\partial x}{\partial \eta} & \dfrac{\partial x}{\partial \zeta} \\[2mm]
\dfrac{\partial y}{\partial \xi} & \dfrac{\partial y}{\partial \eta} & \dfrac{\partial y}{\partial \zeta} \\[2mm]
\dfrac{\partial z}{\partial \xi} & \dfrac{\partial z}{\partial \eta} & \dfrac{\partial z}{\partial \zeta}
\end{bmatrix} .
\tag{12.4}
$$

The elements of $\underline{J}^{-1}$ can be related to the elements of $\underline{J}$ by noting that

$$
\underline{J} = \frac{\text{Transpose of Cofactor } (\underline{J}^{-1})}{|\underline{J}^{-1}|} .
\tag{12.5}
$$

The determinant of the inverse Jacobian $|\underline{J}^{-1}|$ is given by

$$
|\underline{J}^{-1}| = x_\xi(y_\eta z_\zeta - y_\zeta z_\eta) - x_\eta(y_\xi z_\zeta - y_\zeta z_\xi) + x_\zeta(y_\xi z_\eta - y_\eta z_\xi) ,
\tag{12.6}
$$

where $x_\xi \equiv \partial x/\partial \xi$, etc. In two dimensions (ξ, η), (12.6) simplifies $(z_\zeta = 1, \, y_\zeta = x_\zeta = 0)$ to

$$
|\underline{J}^{-1}| = x_\xi y_\eta - x_\eta y_\xi .
$$

Using (12.5 and 6) the elements of $\underline{J}$ in (12.3) can be expressed as

$$
\xi_x = \frac{y_\eta z_\zeta - y_\zeta z_\eta}{|\underline{J}^{-1}|} , \qquad
\xi_y = \frac{x_\zeta z_\eta - x_\eta z_\zeta}{|\underline{J}^{-1}|} , \qquad
\xi_z = \frac{x_\eta y_\zeta - x_\zeta y_\eta}{|\underline{J}^{-1}|} ,
$$

$$
\eta_x = \frac{y_\zeta z_\xi - y_\xi z_\zeta}{|\underline{J}^{-1}|} , \qquad
\eta_y = \frac{x_\xi z_\zeta - x_\zeta z_\xi}{|\underline{J}^{-1}|} , \qquad
\eta_z = \frac{x_\zeta y_\xi - x_\xi y_\zeta}{|\underline{J}^{-1}|} ,
\tag{12.7}
$$

$$
\zeta_x = \frac{y_\xi z_\eta - y_\eta z_\xi}{|\underline{J}^{-1}|} , \qquad
\zeta_y = \frac{x_\eta z_\xi - x_\xi z_\eta}{|\underline{J}^{-1}|} , \qquad
\zeta_z = \frac{x_\xi y_\eta - x_\eta y_\xi}{|\underline{J}^{-1}|} ,
$$

where $|\underline{J}^{-1}|$ is given by (12.6).

Once a grid in the physical domain has been constructed (Chap. 13), the discretised form of the elements, e.g. x_ξ, of the inverse Jacobian are evaluated, as in Sect. 12.2. Equations (12.7) are then used to evaluate the elements, e.g. ξ_x, of the Jacobian matrix (12.3). This facilitates the discretisation of the governing equations in generalised coordinates since they have a more compact structure (Sect. 12.3) when expressed in terms of ξ_x, etc., rather than x_ξ, etc.

12.1.2 Metric Tensor and the Physical Features of the Transformation

In order to link generalised coordinates, orthogonal and conformal coordinates it is appropriate to introduce the metric tensor g_{ij}, which is related to the Jacobian matrix $\boldsymbol{J}$ in (12.3). Initially tensor notation (Aris 1962) will be used.

It will be assumed that the physical domain (Fig. 12.1) is represented by Cartesian coordinates x^i ($\equiv x, y, z$), $i = 1, 3$, and the computational domain by generalised coordinates ξ^i ($\equiv \xi, \eta, \zeta$), $i = 1, 3$.

The small distance Δs between two points in physical space can be written in terms of the coordinate displacements as

$$\Delta s^2 = \sum_{k=1}^{3} \Delta x^k \, \Delta x^k \ . \tag{12.8}$$

The increments in the physical coordinates Δx^k can be related to changes in the generalised coordinates $\Delta \xi^i$ by

$$\Delta x^k = \frac{\partial x^k}{\partial \xi^i} \, \Delta \xi^i \quad \text{(summation over } i \text{ implied)} \ . \tag{12.9}$$

Consequently the small distance Δs, related to generalised coordinates, becomes

$$\Delta s^2 = \sum_{k=1}^{3} \left(\frac{\partial x^k}{\partial \xi^i} \, \Delta \xi^i \right) \left(\frac{dx^k}{d\xi^j} \, \Delta \xi^j \right)$$

$$= g_{ij} \, \Delta \xi^i \, \Delta \xi^j \quad \text{(summation over } i \text{ and } j \text{ implied)} \ , \tag{12.10}$$

where

$$g_{ij} = \sum_{k=1}^{3} \frac{\partial x^k}{\partial \xi^i} \frac{\partial x^k}{\partial \xi^j} \ . \tag{12.11}$$

The metric tensor g_{ij} relates the contributions to the distance Δs to small changes in the generalised coordinates $\Delta \xi^i$. The metric tensor is discussed at greater length by Aris (1962, p. 142). In two dimensions the distances measured along ξ and η grid lines are given by $\Delta s_\xi = g_{11}^{1/2} \, \Delta \xi$ and $\Delta s_\eta = g_{22}^{1/2} \, \Delta \eta$, respectively (Fig. 12.3).

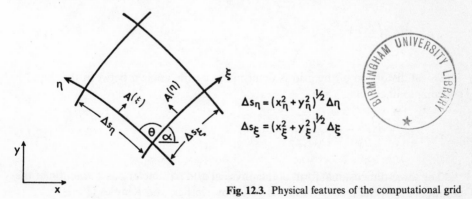

$$\Delta s_\eta = (x_\eta^2 + y_\eta^2)^{1/2} \, \Delta \eta$$
$$\Delta s_\xi = (x_\xi^2 + y_\xi^2)^{1/2} \, \Delta \xi$$

Fig. 12.3. Physical features of the computational grid

In two dimensions it is convenient to write (12.11) as a matrix

$$
\underline{g} = \begin{bmatrix} (x_\xi^2 + y_\xi^2) & (x_\xi x_\eta + y_\xi y_\eta) \\ (x_\xi x_\eta + y_\xi y_\eta) & (x_\eta^2 + y_\eta^2) \end{bmatrix} = \frac{1}{|\underline{J}|^2} \begin{bmatrix} (\eta_x^2 + \eta_y^2) & -(\xi_x \eta_x + \xi_y \eta_y) \\ -(\xi_x \eta_x + \xi_y \eta_y) & (\xi_x^2 + \xi_y^2) \end{bmatrix},
$$
(12.12)

where $x_\xi \equiv \partial x / \partial \xi$ as above, and $|\underline{J}|$ is the determinant of the Jacobian matrix (12.3).

The metric tensor written in matrix form is related to the inverse Jacobian (12.6) by

$$
\underline{g} = (\underline{J}^{-1})^T \underline{J}^{-1} .
$$

Taking determinants gives

$$
|\underline{g}|^{1/2} = |\underline{J}^{-1}| .
$$
(12.13)

This can be easily demonstrated by direct substitution, particularly in two dimensions.

The metric tensor g_{ij} and the various transformation parameters, x_ξ, etc., on which it depends, can be interpreted in relation to physical features of the computational grid (Fig. 12.3). The various formulae will be developed here, in two dimensions. The grid cell area (Fig. 12.3) is given by

$$
\text{Area} = |\underline{g}|^{1/2} \Delta \xi \Delta \eta ,
$$
(12.14)

which, from (12.13), gives a physical interpretation of the inverse Jacobian determinant. The physical orientation of the computational grid (tangent to a ξ coordinate line) relative to the x-axis is given by the direction cosine

$$
\cos \alpha = \frac{x_\xi}{(g_{11})^{1/2}} .
$$
(12.15)

The grid aspect ratio AR is given by the ratio of the magnitude of the tangent vectors (with $\Delta \xi = \Delta \eta$)

$$
\text{AR} = \frac{\Delta s_\eta}{\Delta s_\xi} = \left(\frac{g_{22}}{g_{11}} \right)^{1/2} .
$$
(12.16)

The local distortion of the grid is determined by the angle θ between the ξ and η coordinate lines. Thus

$$
\cos \theta = \frac{g_{12}}{(g_{11} g_{22})^{1/2}} .
$$
(12.17)

The three-dimensional form of the physical grid parameters, as a function of the components of the metric tensor, is given by Kerlick and Klopfer (1982).

It is useful to associate an "area" vector with each side of the cell shown in Fig. 12.3. The magnitude of the area vector is equal to the length of the side and the direction is normal to the side in the direction of increasing ξ and η respectively. These vectors are truly area vectors in three dimensions.

The Cartesian components of the area vectors are given by the direction cosines multiplied by the magnitude of the area vectors. Thus (from Fig. 12.3),

$$\mathbf{A}^{(\xi)} = \{\Delta s_\eta \sin\beta, \quad \Delta s_\eta \cos\beta\}$$
$$\mathbf{A}^{(\eta)} = \{-\Delta s_\xi \sin\alpha, \quad \Delta s_\xi \cos\alpha\} \ . \tag{12.18}$$

The area vectors can be linked to the coordinate transformation as follows. Forming an area vector matrix and evaluating $\sin\beta$ etc. gives

$$\underline{A} = \begin{bmatrix} y_\eta \Delta\eta & -x_\eta \Delta\eta \\ -y_\xi \Delta\xi & x_\xi \Delta\xi \end{bmatrix}, \tag{12.19}$$

where $\mathbf{A}^{(\xi)}$ provides the first row and $\mathbf{A}^{(\eta)}$ the second. If $\Delta\xi = \Delta\eta = 1$ (the usual assumption when generating the grid, Sect. 12.4.1), (12.19) can be re-expressed using (12.17) as

$$\underline{A} = |\underline{J}^{-1}|\underline{J} \ . \tag{12.20}$$

Thus the area vector matrix is an appropriately normalised Jacobian of the transformation. This connection is exploited to evaluate the transformation parameters, ξ_x etc., in Sect. 12.2.3. The above discussion of area vectors extends directly to three dimensions where the area vector of each face of the cell is conveniently described by three Cartesian components.

12.1.3 Restriction to Orthogonal and Conformal Coordinates

The use of generalised coordinates permits quite arbitrary geometries to be considered. However, it is well-known that the accuracy of the solution is degraded by grid distortion. For high accuracy the grid should be orthogonal or near-orthogonal. For orthogonal coordinate systems some of the transformation terms disappear and the equations simplify. If the coordinate system is also conformal the governing equations simplify further. The use of completely orthogonal or con-formal coordinate systems implies relatively simple boundary shapes for the computational domains and some restriction on the disposition of the grid points.

A two-dimensional orthogonal grid must have $\theta = 90°$ (Fig. 12.3), or, from (12.17),

$$g_{12} = x_\xi x_\eta + y_\xi y_\eta = 0 \ . \tag{12.21}$$

In three dimensions the orthogonality condition becomes

$$g_{ij} = 0 \ , \quad i \neq j \ . \tag{12.22}$$

If the coordinate system is orthogonal, i.e. the metric tensor contains only diagonal terms g_{ii}, it is the convention to define

$$h_i = (g_{ii})^{1/2} , \quad i = 1, 3 \quad \text{(no summation)} .$$ (12.23)

The terms h_i can be interpreted as scale factors since a small change in the ξ^i coordinate, on an orthogonal grid, produces a scaled overall movement given by

$$\Delta s = h_i \Delta \xi^i \quad \text{(no summation)} .$$ (12.24)

In two dimensions the condition of orthogonality implies, from (12.12), that

$$x_\eta = -y_\xi \text{AR} \quad \text{and} \quad y_\eta = x_\xi \text{AR} ,$$ (12.25)

where AR is the grid aspect ratio (12.16).

If AR $= 1$ then (12.25) reduces to the Cauchy–Riemann conditions and the grid is conformal. If AR is constant, but not equal to unity a simple scaling of ξ or η will produce a related conformal coordinate system.

The level of complexity in the governing equations in the various coordinate systems can be appreciated by considering Laplace's equation. In two-dimensional Cartesian coordinates, this is

$$\frac{\partial^2 T}{\partial x^2} + \frac{\partial^2 T}{\partial y^2} = 0 .$$ (12.26)

In generalised coordinates (ξ, η), (12.26) can be written

$$\frac{\partial}{\partial \xi} \left(\frac{g_{22}}{g^{1/2}} \frac{\partial T}{\partial \xi} - \frac{g_{12}}{g^{1/2}} \frac{\partial T}{\partial \eta} \right) + \frac{\partial}{\partial \eta} \left(-\frac{g_{21}}{g^{1/2}} \frac{\partial T}{\partial \xi} + \frac{g_{11}}{g^{1/2}} \frac{\partial T}{\partial \eta} \right) = 0 ,$$ (12.27)

where g_{22}, etc., are given by (12.12) and $g^{1/2}$ is evaluated from

$$|\underline{g}|^{1/2} \equiv g^{1/2} = x_\xi y_\eta - x_\eta y_\xi .$$ (12.28)

The term g is the determinant of the matrix $\underline{g}$. To keep the structure of the equations more compact this notation is used instead of $|\underline{g}|$, as in (12.14).

Substitution for the various terms in (12.27) gives, after some manipulation, the conservation form (Sect. 12.3.2)

$$-\left(\frac{\nabla^2 \xi T}{J} \right)_\xi - \left(\frac{\nabla^2 \eta T}{J} \right)_\eta + \left[\left\{ \frac{(\xi_x^2 + \xi_y^2) T}{J} \right\}_{\xi\xi} + \left\{ \frac{2(\xi_x \eta_x + \xi_y \eta_y) T}{J} \right\}_{\xi\eta} \right.$$
$$\left. + \left\{ \frac{(\eta_x^2 + \eta_y^2) T}{J} \right\}_{\eta\eta} \right] = 0 .$$ (12.29)

In (12.29) J is the determinant of $\underline{J}$ (previously denoted by $|\underline{J}|$) and $\nabla^2 \xi \equiv \xi_{xx} + \xi_{yy}$, etc. It will be seen that (12.29) involves a mixed second derivative and first derivatives, not present in Cartesian coordinates. In addition the terms $\nabla^2 \xi$, $\nabla^2 \eta$

involve second derivative transformation parameters, which are often difficult to evaluate accurately (Sect. 12.2.4).

If the coordinate system is orthogonal, $g_{12}=g_{21}=0$ and (12.27) simplifies to

$$\frac{\partial}{\partial \xi}\left(\frac{h_2}{h_1}\frac{\partial T}{\partial \xi}\right)+\frac{\partial}{\partial \eta}\left(\frac{h_1}{h_2}\frac{\partial T}{\partial \eta}\right)=0 \ , \tag{12.30}$$

where the orthogonal scale factors are given by

$$h_1=(g_{11})^{1/2}=(x_\xi^2+y_\xi^2)^{1/2} \quad \text{and} \quad h_2=(g_{22})^{1/2}=(x_\eta^2+y_\eta^2)^{1/2} \ .$$

The equivalent conservation form of (12.30) is the same as (12.29) except that the mixed second derivative disappears. For a conformal grid $h_1=h_2$ and the grid parameters satisfy the Cauchy–Riemann conditions

$$x_\eta=-y_\xi \quad \text{and} \quad y_\eta=x_\xi \ . \tag{12.31}$$

Since $h_1=h_2$, (12.30) reverts to Laplace's equation

$$T_{\xi\xi}+T_{\eta\eta}=0 \ , \tag{12.32}$$

i.e. the equation is structurally no more complex than in Cartesian coordinates.

The choice between conformal, orthogonal or generalised coordinates is usually dictated by the nature of the computational boundaries. If the geometry is simple enough and a grid can be constructed that is able to place grid points in regions of severe gradients, conformal coordinates should be used since they imply fewer terms in the governing equations and, consequently, a more economical algorithm. In three dimensions completely conformal or orthogonal grids are not usually possible, so that the equations must be expressed in generalised coordinates, although with some simplification in particular coordinate directions or local computational regions.

12.2 Evaluation of the Transformation Parameters

If an analytic mapping from physical space (x, y) to the computational space (ξ, η) is available, as often occurs with simple conformal transformations, the transformation parameters, x_ξ, etc., can be evaluated exactly. More typically the mapping is only defined at grid points and the transformation parameters must be evaluated numerically. For convenience this will be illustrated here for two-dimensional grids. The extension to three-dimensional grids is straightforward.

The equations in generalised coordinates are discretised in the (ξ, η) domain and the mapping is usually arranged so that a uniform rectangular grid is specified in the (ξ, η) domain (Fig. 12.2).

The numerical evaluation of the transformation parameters can be performed by any convenient means of discretisation (Chaps. 3 and 5). Central difference formulae are used in Sect. 12.2.1, a finite element evaluation is described in Sect. 12.2.2 and a finite volume evaluation via area vectors is discussed in Sect. 12.2.3. But it is recommended that the same means of discretisation be used for the evaluation of the transformation parameters and the derivatives in the governing equations.

12.2.1 Centred-Difference Formulae

The evaluation of the transformation parameters is carried out most conveniently in terms of the variables x_ξ, y_η, etc. Thus for the point P (Fig. 12.4) a centred difference evaluation of x_ξ, etc., gives

$$x_\xi \approx \frac{x_{j+1,k} - x_{j-1,k}}{\xi_{j+1,k} - \xi_{j-1,k}} \ ,$$

$$x_\eta \approx \frac{x_{j,k+1} - x_{j,k-1}}{\eta_{j,k+1} - \eta_{j,k-1}} \ , \tag{12.33}$$

$$y_\xi \approx \frac{y_{j+1,k} - y_{j-1,k}}{\xi_{j+1,k} - \xi_{j-1,k}} \ ,$$

$$y_\eta \approx \frac{y_{j,k+1} - y_{j,k-1}}{\eta_{j,k+1} - \eta_{j,k-1}} \ .$$

For the general case of a second-order equation transformed into generalised coordinates (Sect. 12.3.2) certain second-order transformation parameters are required. For example

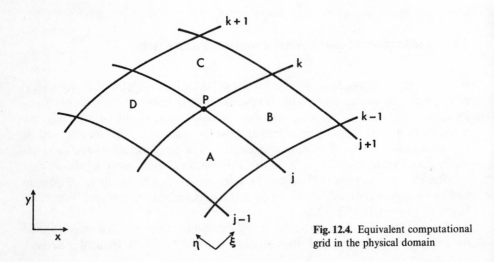

Fig. 12.4. Equivalent computational grid in the physical domain

$$x_{\xi\xi} \approx \frac{x_{j-1,k} - 2x_{j,k} + x_{j+1,k}}{\Delta\xi^2} \; ,$$

$$x_{\xi\eta} \approx \frac{x_{j+1,k+1} - x_{j-1,k+1} + x_{j-1,k-1} - x_{j+1,k-1}}{4\Delta\xi\Delta\eta} \; , \tag{12.34}$$

$$x_{\eta\eta} \approx \frac{x_{j,k-1} - 2x_{j,k} + x_{j,k+1}}{\Delta\eta^2} \; ,$$

where a uniform (ξ, η) grid is assumed, i.e. $\Delta\xi = \xi_{j+1} - \xi_j = \xi_j - \xi_{j-1}$ etc. In a similar way parameters $y_{\xi\xi}$, etc., can be evaluated.

Once the basic transformation parameters have been evaluated, using (12.33 and 34), the inverse parameters ξ_x, etc., can be obtained from (12.7). Typical governing equations for fluid flow (Sect. 12.3.3) generally have a more compact structure when written with terms like ξ_x appearing explicitly. Second derivative inverse parameters, e.g. ξ_{xx}, can be related to the expressions in (12.34). The construction of such relations is indicated in (12.81). Given an evaluation of x_ξ, etc., the grid related parameters, g_{ij}, α, AR and θ, follow from (12.12, 15, 16 and 17), respectively.

The transformation parameters could be evaluated with higher-order formulae than those given by (12.33 and 34). Normally the discretisation of the transform parameters will use the same formulae as used to discretise the derivative terms in the governing equations. This aspect will be pursued in Sect. 12.2.4.

If the governing equations are solved on an orthogonal grid, terms proportional to the off-diagonal elements, g_{ij}, of the metric tensor may be deleted. For a two-dimensional orthogonal grid, $g_{12} = x_\xi x_\eta + y_\xi y_\eta = 0$. If the transformation parameters, x_ξ, etc., are evaluated numerically it is important to establish that the discrete evaluation of g_{ij} is zero. Thus for a two-dimensional orthogonal grid, g_{12} could be evaluated, using centred differences, as

$$g_{12} = (x_{j+1,k} - x_{j-1,k})(x_{j,k+1} - x_{j,k-1})$$
$$+ (y_{j+1,k} - y_{j-1,k})(y_{j,k+1} - y_{j,k-1}) = 0 \; . \tag{12.35}$$

The geometric interpretation of (12.35) is shown in Fig. 12.5. Discrete orthogonality requires that lines AB and CD are perpendicular.

Consequently, if the transformation parameters are to be evaluated numerically on an orthogonal grid it is important that the grid is constructed (Sect. 13.2.4) so that *discrete* orthogonality is achieved.

12.2.2 Finite Element Evaluation

Equations (12.33, 34) also arise if the finite element method is used to evaluate the transformation parameters. Using the isoparametric construction (Sect. 5.5.3) on four contiguous elements (A, B, C and D in Fig. 12.4) with bilinear interpolation and averaging the value x_ξ, etc., at P it is possible to obtain (12.33). If the

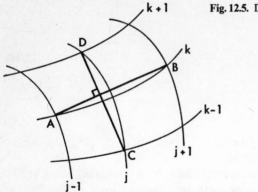

Fig. 12.5. Discrete orthogonality

isoparametric construction is used on one quadratic Lagrange element, with P as the internal node, both (12.33 and 34) can be obtained.

To obtain better control of the Jacobian distribution it is useful to consider a rectangular but non-uniform grid in (ξ, η) space (Fig. 12.6) in which r_ξ and r_η determine the grid growth. In this more general situation, first derivative transformation parameters like x_ξ are still given by (12.33), but (12.34) are replaced by

$$x_{\xi\xi} = \frac{2}{1+r_\xi}\left\{ \frac{x_{j-1,k} - \left(1+\dfrac{1}{r_\xi}\right)x_{j,k} + \dfrac{x_{j+1,k}}{r_\xi}}{\Delta\xi^2} \right\},$$

$$x_{\xi\eta} = \frac{x_{j+1,k+1} - x_{j-1,k+1} + x_{j-1,k-1} - x_{j+1,k-1}}{(1+r_\xi)(1+r_\eta)\Delta\xi\Delta\eta}, \qquad (12.36)$$

$$x_{\eta\eta} = \frac{2}{1+r_\eta}\left\{ \frac{x_{j,k-1} - \left(1+\dfrac{1}{r_\eta}\right)x_{j,k} + \dfrac{x_{j,k+1}}{r_\eta}}{\Delta\eta^2} \right\}.$$

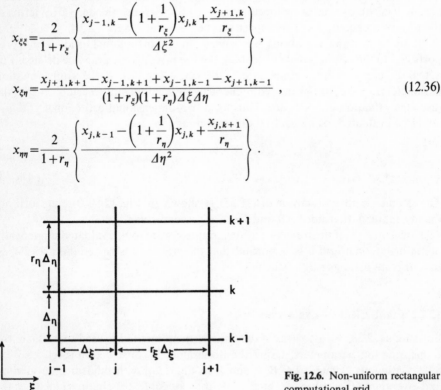

Fig. 12.6. Non-uniform rectangular computational grid

12.2.3 Finite Volume Evaluation

The transformation parameters, ξ_x etc., can be evaluated directly, without first determining x_ξ etc. and using (12.7). This is achieved with the aid of a finite volume evaluation of the area vector matrix, (12.20).

For example ξ_x is obtained as

$$\xi_x = \frac{A_{11}}{|\underline{J}^{-1}|} \, , \tag{12.37}$$

with A_{11} approximated by $0.5\{A^{(\xi)}_{1,AD} + A^{(\xi)}_{1,BC}\}$. From Fig. 12.7 and Eq. (12.18),

$$A_{11} = 0.5(y_D - y_A + y_C - y_B) \, , \tag{12.38}$$

i.e. (12.38) provides a discrete evaluation of y_η with $\Delta\eta = 1$. The term $|\underline{J}^{-1}|$ in (12.37) is the "volume" of the finite volume which can be calculated from

$$|\underline{J}^{-1}| = \tfrac{1}{3}\underline{A} \cdot \mathbf{x}^{cf} \, , \tag{12.39}$$

where $\mathbf{x}^{cf}$ is the vector of the centroids of the faces of the finite volume. Once the components of the area vectors have been determined from the grid, the various transformation parameters which are components of $\underline{J}$ follow directly from (12.20).

However when a finite volume discretisation (Sect. 5.2) is used it is not necessary (or natural) to evaluate the transformation parameters explicitly at the grid points, (j, k). It is more appropriate to carry out the discretisation by an integration over the control volume.

For example, in Sect. 5.2, the following equation,

$$\int_{ABCD} \left(\frac{\partial \bar{q}}{\partial t} + \frac{\partial \bar{F}}{\partial x} + \frac{\partial \bar{G}}{\partial y} \right) dx\, dy = 0 \, , \tag{12.40}$$

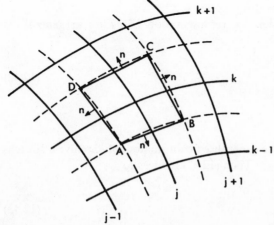

Fig 12.7. Two-dimensional finite volume

is evaluated in discrete form by (5.28). In the present notation (5.28) can be written

$$|\underline{J}^{-1}|\frac{dq_{j,k}}{dt}+\sum_{AB}^{DA}\mathbf{H}\cdot\mathbf{A}=0 \; , \tag{12.41}$$

where $\mathbf{H}=(\bar{F}, \bar{G})$. In (12.41) $\mathbf{H}\cdot\mathbf{A}$ is the scalar product of $\mathbf{H}$ and the area vectors. This is the flux of H through the appropriate finite volume face in the direction normal to the face.

Consequently when used with a finite volume discretisation it is appropriate to evaluate the area vectors, once and for all, instead of the transformation coefficients. The discrete equations are then expressed in terms of the components of the area vectors. Thus the area vectors can be interpreted as performing the role of the transformation parameters in carrying information of the local grid orientation and distortion. In the present context area vectors provide the essential link between finite volume discretisation and finite difference discretisation of the generalised coordinate governing equations.

Burns and Wilkes (1987) exploit the above properties of area vectors to construct an efficient finite volume formulation for internal turbulent flows. They suggest that, if the grid is distorted, the use of numerically evaluated area vectors will produce more accurate solutions than the use of numerically evaluated transformation parameters. The role of area vectors and the connection between finite difference and finite volume discretisation of the compressible flow equation is discussed by Vinokur (1989).

12.2.4 Additional Errors Associated with the Use of Generalised Coordinates

An effective way of determining, a priori, the influence of generalised coordinates on the solution error is to develop an expression for the truncation error and to expect, on a sufficiently refined grid, that factors causing an increase in the truncation error will also increase the solution error. However, as is clear from Sect. 10.1.5, the correspondence is not precise.

As an illustration we consider the representation of $\partial T/\partial x$ in generalised coordinates as

$$\frac{\partial T}{\partial x}=\xi_x T_\xi+\eta_x T_\eta \; , \tag{12.42}$$

where $T_\xi\equiv\dfrac{\partial T}{\partial\xi}$, etc.

We will further assume that the generalised coordinates only involve a stretching in the x direction, so that $\eta=y$. Then (12.42) reduces to

$$\frac{\partial T}{\partial x}=\frac{T_\xi}{x_\xi} \; . \tag{12.43}$$

physical domain

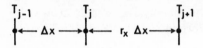

computational domain

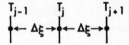

Fig. 12.8. One-dimensional mapping

Using centred difference formulae, (12.43) can be discretised as (Fig. 12.8)

$$\frac{\partial T}{\partial x} = \frac{T_\xi}{x_\xi} \approx \frac{T_{j+1} - T_{j-1}}{x_{j+1} - x_{j-1}} \ . \tag{12.44}$$

A Taylor series expansion of the various terms in (12.44) about node j in the computational domain gives

$$\frac{T_{j+1} - T_{j-1}}{x_{j+1} - x_{j-1}} = \frac{T_\xi \left(1 + \dfrac{\Delta \xi^2}{6} \dfrac{T_{\xi\xi\xi}}{T_\xi} + \cdots \right)}{x_\xi \left(1 + \dfrac{\Delta \xi^2}{6} \dfrac{x_{\xi\xi\xi}}{x_\xi} + \cdots \right)} \ .$$

Assuming that $\Delta \xi$ is small, this can be rewritten as

$$\frac{T_{j+1} - T_{j-1}}{x_{j+1} - x_{j-1}} = \frac{T_\xi}{x_\xi} \left[1 + \frac{\Delta \xi^2}{6} \left(\frac{T_{\xi\xi\xi}}{T_\xi} - \frac{x_{\xi\xi\xi}}{x_\xi} \right) + \cdots \right] \ . \tag{12.45}$$

Clearly (12.45) is consistent and at first sight appears to be second-order accurate in ξ. However, further algebraic expansion of (12.45) gives

$$\frac{T_{j+1} - T_{j-1}}{x_{j+1} - x_{j-1}} = T_x + \frac{\Delta \xi^2}{6} \left[T_{xxx}(x_\xi)^2 + 3 T_{xx} x_{\xi\xi} \right] + \cdots \tag{12.46}$$

The transformation parameters x_ξ and $x_{\xi\xi}$ are evaluated as

$$x_\xi = \frac{x_{j+1} - x_{j-1}}{2\Delta \xi} + O(\Delta \xi^2) = 0.5(1 + r_x)\frac{\Delta x}{\Delta \xi} + O(\Delta \xi^2) \ ,$$

$$x_{\xi\xi} = \frac{x_{j-1} - 2x_j + x_{j+1}}{\Delta \xi^2} + O(\Delta \xi^2) = (r_x - 1)\frac{\Delta x}{\Delta \xi^2} + O(\Delta \xi^2) \ .$$

Substitution in (12.46) gives

$$\frac{T_{j+1} - T_{j-1}}{x_{j+1} - x_{j-1}} = T_x + \left[\frac{1}{6}\left(\frac{1 + r_x}{2} \right)^2 \Delta x^2 T_{xxx} + \frac{(r_x - 1)}{2} \Delta x T_{xx} \right] + \cdots \tag{12.47}$$

The term involving $x_{\xi\xi}$ in (12.46) has introduced a first-order error into an expression that would be second-order with a regular grid ($r_x = 1$) in the physical domain. To ensure that (12.44) is accurate to $O(\Delta x^2)$ requires that $r_x = 1 + O(\Delta x)$. That is, for second-order accuracy in the evaluation of T_x using three-point centred difference formulae, it is necessary that the grid in the physical domain grow only slowly. It is also clear that the use of a non-uniform grid has introduced diffusive and dispersive (Sect. 9.2) terms.

It is of interest to compare the above result (12.46, 47) with the truncation error in representing $\partial T/\partial x$ with x_ξ evaluated exactly. Thus (12.44) is replaced by

$$\frac{\partial T}{\partial x} \approx \frac{T_{j+1} - T_{j-1}}{2\Delta\xi\, x_\xi}$$

and the Taylor series expansion of the right-hand side leads to

$$\frac{T_{j+1} - T_{j-1}}{2\Delta\xi\, x_\xi} = T_x + \left(\frac{\Delta\xi^2}{6}\right)\left[\frac{x_{\xi\xi\xi}}{x_\xi} T_x + 3x_{\xi\xi} T_{xx} + (x_\xi)^2 T_{xxx}\right] + \cdots , \qquad (12.48)$$

in place of (12.46). Thus the use of exact evaluation of x_ξ introduces an *additional* term, and a dominant term, in the truncation error. Clearly the numerical evaluation of x_ξ, with the same discretisation as T_ξ, is to be preferred since it leads to a cancellation of the $(x_{\xi\xi\xi}/x_\xi) T_x$ term in the truncation error and this implies a smaller solution error.

In addition the use of higher-order formulae or exact evaluation of x_ξ and $x_{\xi\xi}$ may lead to a failure to satisfy certain metric identities (Thompson et al. 1985, p. 185), thereby introducing a spurious source term, if the equations are discretised in conservation form (see below). However x_ξ and $x_{\xi\xi}$ are evaluated, the use of a rapidly growing grid produces large magnitudes for x_ξ and $x_{\xi\xi}$ which increases the truncation error of evaluating $\partial T/\partial x$.

At a more general level it is clear that the truncation error now depends on the transformation parameters as well as the grid size and higher-order derivatives, as is the case for a uniform grid.

A discretisation of $\partial^2 T/\partial x^2$ in the one-dimensional computational domain, Fig. 12.8, can be obtained as

$$\frac{\partial^2 T}{\partial x^2} = \left(\frac{2\Delta\xi}{x_{j+1} - x_{j-1}}\right)^2 \left[\frac{(T_{j-1} - 2T_j + T_{j+1})}{2\Delta\xi^2}\right.$$
$$\left. - \frac{(x_{j-1} - 2x_j + x_{j+1})}{\Delta\xi^2}\left(\frac{T_{j+1} - T_{j-1}}{x_{j+1} - x_{j-1}}\right)\right]. \qquad (12.49)$$

A Taylor series expansion about node j indicates the finite difference expression on the right-hand side is of the form

$$\text{RHS} = T_{xx}[1 - 2(r_x - 1)^2] + \cdots . \qquad (12.50)$$

That is, it is consistent with error of $O(\Delta x^2)$ as long as $r_x = 1 + O(\Delta x)$. However, if this doesn't hold, the discretisation of T_{xx} in the computational domain is actually inconsistent. This indicates that the accuracy of representing second derivatives using generalised coordinates falls off more rapidly with grid growth than for discretisations of first derivatives.

The use of generalised coordinates in multidimensions introduces additional terms in the truncation error if the grid is not orthogonal. Typically these errors are proportional to $\cos \theta$ (Thompson et al. 1985), which is given by (12.17). However, it is generally accepted that departures from orthogonality of up to $45°$ can be tolerated.

For computing flow involving shocks or containing severe gradients it is often desirable to express the equations in conservation form (12.62) and to discretise in a way that maintains the conservation form. When used with generalised coordinates additional difficulties arise. This can be illustrated by the discretisation of the first derivative T_x. Thus

$$J^{-1} T_x = (T y_\eta)_\xi - (T y_\xi)_\eta = T_\xi y_\eta - T_\eta y_\xi \ . \tag{12.51}$$

The first equality expresses T_x in conservation form in the computational coordinates. The second equality is a non-conservative form. Using centred differences the first equality becomes (assuming $\Delta\xi = \Delta\eta = 1$)

$$(J^{-1} T_x)_{j,k} \approx 0.5\,[(T y_\eta)_{j+1,k} - (T y_\eta)_{j-1,k} - (T y_\xi)_{j,k+1} + (T y_\xi)_{j,k-1}] \ . \tag{12.52}$$

If the transformation parameters are also evaluated by centred differences, (12.52) becomes

$$(J^{-1} T_x)_{j,k} \approx 0.25\,[T_{j+1,k}(y_{j+1,k+1} - y_{j+1,k-1}) - T_{j-1,k}(y_{j-1,k+1} - y_{j-1,k-1})$$
$$- T_{j,k+1}(y_{j+1,k+1} - y_{j-1,k+1}) + T_{j,k-1}(y_{j+1,k-1} - y_{j-1,k-1})] \ . \tag{12.53}$$

If T is uniform it can be seen that the right-hand side of (12.53) will be zero as required.

However, if y_η and y_ξ in (12.52) are evaluated analytically there is no guarantee that the right-hand side of (12.51) will be zero when T is uniform. Consideration of the non-conservative form shown in (12.51) indicates that $T_x = 0$ if T is uniform, however y_η and y_ξ are evaluated.

The importance of the conservation form of the governing equations prompts the following advice (Thompson 1984). The transformation parameters, y_η, etc., should be evaluated numerically rather than analytically and with the same discretisation formulae as used to represent derivatives of the dependent variables.

The other important result to appear from the error analysis is the need to keep the grid growth parameters, e.g. r_x in Fig. 12.8, close to unity, particularly if the governing equation involves second derivatives in the physical variables. Strict grid orthogonality reduces the number of terms in the transformed governing equations, and is therefore more economical. But it does not significantly increase the accuracy compared with the use of a near-orthogonal grid.

12.3 Generalised Coordinate Structure of Typical Equations

In this section the relationships developed in Sect. 12.1.1 will be used to express typical partial differential equations in generalised coordinates. Two-dimensional first and second-order partial differential equations of general form are considered in Sects. 12.3.1 and 12.3.2. The formulae developed are applied to some of the specific equations governing fluid flow in Sect. 12.3.3.

12.3.1 General First-Order Partial Differential Equation

Here a general first-order equation will be transformed into an equivalent generalised coordinate form. The equations governing inviscid fluid flow, when written in conservation form, have a structure no more complicated than the present example.

The following two-dimensional equation will be considered:

$$q_t + F_x + G_y = 0 \ , \tag{12.54}$$

where $F_x \equiv \partial F / \partial x$, etc.

It will be assumed that the grid in the physical domain (Fig. 12.1) does not change with time. That is, the generalised coordinates (ξ, η) are a function of (x, y) only. Thus

$$\xi = \xi(x, y) \ , \quad \eta = \eta(x, y) \ . \tag{12.55}$$

The more general situation where the grid is time dependent, i.e. $\xi = \xi(x, y, t)$, $\eta = \eta(x, y, t)$, is considered by Steger (1978) and by Thompson et al. (1985).

Introducing

$$\frac{\partial}{\partial x} = \xi_x \frac{\partial}{\partial \xi} + \eta_x \frac{\partial}{\partial \eta} \quad \text{and}$$

$$\frac{\partial}{\partial y} = \xi_y \frac{\partial}{\partial \xi} + \eta_y \frac{\partial}{\partial \eta} \ ,$$

allows (12.54) to be written as

$$\left(\frac{q}{J} \right)_t + \frac{\xi_x F_\xi}{J} + \frac{\xi_y G_\xi}{J} + \frac{\eta_x F_\eta}{J} + \frac{\eta_y G_\eta}{J} = 0 \ , \tag{12.56}$$

where

$$J = \xi_x \eta_y - \xi_y \eta_x = \frac{1}{x_\xi y_\eta - x_\eta y_\xi} \ . \tag{12.57}$$

The parameter J, defined by (12.57), is the determinant of the transformation Jacobian (12.3).

Terms like $\xi_x F_\xi / J$ can be written as, using (12.7),

$$\frac{\xi_x F_\xi}{J} = \left(\frac{\xi_x F}{J}\right)_\xi - F\left(\frac{\xi_x}{J}\right)_\xi = \left(\frac{\xi_x F}{J}\right)_\xi - F(y_\eta)_\xi \ . \tag{12.58}$$

Substitution into (12.56) causes terms like $(y_\eta)_\xi$ to cancel out and the result is

$$q_t^* + F_\xi^* + G_\eta^* = 0 \ , \quad \text{where} \tag{12.59}$$

$$q^* = \frac{q}{J} \ , \quad F^* = \frac{\xi_x F + \xi_y G}{J} \quad \text{and} \quad G^* = \frac{\eta_x F + \eta_y G}{J} \ . \tag{12.60}$$

It is apparent that the structure of (12.59) is the same as in (12.54) once new dependent variables, q^*, F^* and G^*, have been introduced.

Comparing (12.54) and (12.59), a direct transformation of the spatial terms can be written down,

$$F_x + G_y = J(F_\xi^* + G_\eta^*) \ , \tag{12.61}$$

where F^* and G^* are given in (12.60).

12.3.2 General Second-Order Partial Differential Equation

In this subsection, (12.54) will be extended to second order in two ways. First,

$$q_t + F_x + G_y = R_{xx} + S_{xy} + T_{yy} \ . \tag{12.62}$$

To suit the structure that appears in fluid flow equations with position-dependent fluid properties, the following equation is also considered:

$$q_t + F_x + G_y = (\alpha R_x)_x + (\beta S_y)_x + (\delta S_x)_y + (\gamma T_y)_y \ , \tag{12.63}$$

where α, β, δ and γ may be functions of (x, y).

The transformation (12.61) is used in two stages to convert the second-order terms in (12.62) into equivalent generalised coordinate terms. First (12.62) is written as

$$q_t + F_x + G_y = \text{RHS} \quad \text{and} \tag{12.64}$$

$$\text{RHS} = [R_x + \varepsilon S_y]_x + [(1 - \varepsilon)S_x + T_y]_y \ , \tag{12.65}$$

where ε is a parameter which determines how the cross-derivative term S_{xy} is distributed between the two bracketed terms. Using (12.61), (12.65) can be written as

$$\frac{\text{RHS}}{J} = P_\xi + Q_\eta \ , \quad \text{where} \tag{12.66}$$

$$P = \frac{\xi_x}{J}(R_x + \varepsilon S_y) + \frac{\xi_y}{J}[(1-\varepsilon)S_x + T_y] \quad \text{and} \tag{12.67}$$

$$Q = \frac{\eta_x}{J}(R_x + \varepsilon S_y) + \frac{\eta_y}{J}[(1-\varepsilon)S_x + T_y] \quad \text{or}$$

$$P = \frac{[\xi_x R + (1-\varepsilon)\xi_y S]_x + [\xi_x \varepsilon S + \xi_y T]_y}{J} - \frac{\xi_{xx}R + \xi_{xy}S + \xi_{yy}T}{J} \;.$$

Using (12.53) again, P can be written as

$$P = \left[\frac{\xi_x^2 R + \xi_x \xi_y S + \xi_y^2 T}{J}\right]_\xi + \left[\frac{\xi_x \eta_x R + \eta_x \xi_y S + \varepsilon JS + \xi_y \eta_y T}{J}\right]_\eta$$
$$- \frac{\xi_{xx}R + \xi_{xy}S + \xi_{yy}T}{J} \;. \tag{12.68}$$

Similar manipulation allows Q to be put in the form $Q = A_\xi + B_\eta + C$. Evaluation of (12.66) and combination with (12.61) permits the second-order partial differential equation (12.62) to be written in strong conservation form as

$$q_t^* + F_\xi^{**} + G_\eta^{**} = R_{\xi\xi}^* + S_{\xi\eta}^* + T_{\eta\eta}^* \;, \quad \text{where} \tag{12.69}$$

$$q^* = q/J \;,$$

$$F^{**} = F^* + \frac{\xi_{xx}R + \xi_{xy}S + \xi_{yy}T}{J} \;,$$

$$G^{**} = G^* + \frac{\eta_{xx}R + \eta_{xy}S + \eta_{yy}T}{J} \;,$$

$$R^* = \frac{\xi_x^2 R + \xi_x \xi_y S + \xi_y^2 T}{J} \;, \tag{12.70}$$

$$S^* = \frac{2\xi_x \eta_x R + (\xi_x \eta_y + \xi_y \eta_x)S + 2\xi_y \eta_y T}{J} \;, \quad \text{and}$$

$$T^* = \frac{\eta_x^2 R + \eta_x \eta_y S + \eta_y^2 T}{J} \;.$$

Although the parameter ε was introduced to derive (12.69 and 70) it cancels out of the final equations. The same strong conservation structure has been retained in (12.69) as in (12.62), at the expense of more complicated dependent variables. The treatment of the second derivative terms, R_{xx}, etc., now gives a contribution to first derivative terms in ξ and η.

It was noted in Sect. 12.2.4 that unless terms like $x_{\xi\xi}$ associated with grid stretching are small the accuracy of the discretisation, and by implication the solution, may be seriously degraded. This effect shows up in the terms like ξ_{xx} in F^{**} and G^{**}. Terms like ξ_{xx} can be related directly to terms like $x_{\xi\xi}$; an example is provided by (12.89, 90).

For the more general second-order partial differential equation, (12.63), the two-stage implementation of (12.61) again produces (12.69) but with the following replacing (12.70):

$$F^{**} = F^* + [(\alpha\xi_x)_x R + (\beta\xi_x)_y S + (\delta\xi_y)_x S + (\gamma\xi_y)_y T]\left(\frac{1}{J}\right) ,$$

$$G^{**} = G^* + [(\alpha\eta_x)_x R + (\beta\eta_x)_y S + (\delta\eta_y)_x S + (\gamma\eta_y)_y T]\left(\frac{1}{J}\right) ,$$

$$R^* = [\alpha\xi_x^2 R + (\beta+\delta)\xi_x\xi_y S + \gamma\xi_y^2 T]\left(\frac{1}{J}\right) , \qquad (12.71)$$

$$S^* = [2\alpha\xi_x\eta_x R + (\beta+\delta)(\xi_x\eta_y + \xi_y\eta_x)S + 2\gamma\xi_y\eta_y T]\left(\frac{1}{J}\right) ,$$

$$T^* = [\alpha\eta_x^2 R + (\beta+\delta)\eta_x\eta_y S + \gamma\eta_y^2 T]\left(\frac{1}{J}\right) .$$

12.3.3 Equations Governing Fluid Flow

The continuity and Euler equations are first-order partial differential equations, whereas the momentum and energy equations for viscous flow are second-order partial differential equations.

The continuity equation (11.10) can be written directly in the form of (12.59) by defining

$$q = \varrho, \quad F = \varrho u \quad \text{and} \quad G = \varrho v .$$

Then F^* and G^* become

$$F^* = \frac{\varrho(\xi_x u + \xi_y v)}{J} \quad \text{and} \quad G^* = \frac{\varrho(\eta_x u + \eta_y v)}{J} . \qquad (12.72)$$

It is useful to introduce the contravariant velocities U^c and V^c, the velocity components along ξ and η lines respectively, by

$$U^c = \xi_x u + \xi_y v \quad \text{and} \quad V^c = \eta_x u + \eta_y v . \qquad (12.73)$$

Then the continuity equation in generalised coordinates becomes

$$\left(\frac{\varrho}{J}\right)_t + \left(\frac{\varrho U^c}{J}\right)_\xi + \left(\frac{\varrho V^c}{J}\right)_\eta = 0 , \qquad (12.74)$$

which has a structure very similar to that in Cartesian coordinates.

For incompressible, viscous flow the x-momentum equation (11.83) in non-dimensional conservation form can be written

$$u_t + (u^2 + p)_x + (uv)_y = \frac{1}{\text{Re}}(u_{xx} + u_{yy}) . \qquad (12.75)$$

This corresponds to (12.62) if

$$q=u , \quad F=u^2+p , \quad G=uv , \quad R=T=u/\mathrm{Re} \quad \text{and} \quad S=0 .$$

As a result (12.69 and 70) become

$$q_t^* + F_\xi^{**} + G_\eta^{**} = \frac{1}{\mathrm{Re}}(R_{\xi\xi}^* + S_{\xi\eta}^* + T_{\eta\eta}^*) , \quad \text{where} \tag{12.76}$$

$$q^* = u/J ,$$

$$F^{**} = \left[uU^c + \left(\frac{\xi_{xx}+\xi_{yy}}{\mathrm{Re}} \right)u + \xi_x p \right]\left(\frac{1}{J} \right) ,$$

$$G^{**} = \left[uV^c + \left(\frac{\eta_{xx}+\eta_{yy}}{\mathrm{Re}} \right)u + \eta_x p \right]\left(\frac{1}{J} \right) , \tag{12.77}$$

$$R^* = \frac{(\xi_x^2+\xi_y^2)u}{J} ,$$

$$S^* = \frac{2(\xi_x\eta_x+\xi_y\eta_y)u}{J} , \quad \text{and}$$

$$T^* = \frac{(\eta_x^2+\eta_y^2)u}{J} .$$

On an orthogonal grid, the above expression for $S^*=0$, which simplifies the implementation of approximate factorisation schemes (Sect. 8.2).

It was noted in Sect. 12.3.2 that the appearance of terms like ξ_{xx} in F^{**} and G^{**} in (12.77) may introduce an error if significant grid stretching is permitted. However, it is clear that for large values of the Reynolds number, Re, in (12.77) this problem will not be as serious since, for many situations,

$$U^c \gg \frac{\xi_{xx}+\xi_{yy}}{\mathrm{Re}} , \quad \text{etc.} \tag{12.78}$$

For two-dimensional laminar compressible flow the x-momentum equation (11.26) can be written in conservation form as

$$(\varrho u)_t + (\varrho u^2 + p)_x + (\varrho uv)_y = \frac{\partial \tau_{xx}}{\partial x} + \frac{\partial \tau_{xy}}{\partial y} , \tag{12.79}$$

where the viscous stresses τ_{xx} and τ_{xy} are given by

$$\tau_{xx} = \frac{4}{3}\mu u_x - \frac{2}{3}\mu v_y \quad \text{and} \quad \tau_{xy} = \mu u_y + \mu v_x . \tag{12.80}$$

Substitution into (12.79) gives

$$(\varrho u)_t + (\varrho u^2 + p)_x + (\varrho uv)_y = \left(\frac{4}{3}\mu u_x \right)_x - \left(\frac{2}{3}\mu v_y \right)_x + (\mu v_x)_y + (\mu u_y)_y . \tag{12.81}$$

This equation has the same structure as (12.63) with

$$q=\varrho u \ , \quad F=\varrho u^2+p \ , \quad G=\varrho uv \ , \quad R=T=u \ , \quad S=v \ , \quad \text{and} \quad (12.82)$$

$$\alpha=\frac{4\mu}{3} \ , \quad \beta=-\frac{2\mu}{3} \ , \quad \delta=\gamma=\mu \ .$$

Consequently (12.81) can be transformed into generalised coordinates in the form of (12.69) with

$$q^*=\frac{\varrho u}{J} \ ,$$

$$F^{**}=\left\{\varrho uU^c+\left[\frac{4}{3}(\mu\xi_x)_x+(\mu\xi_y)_y\right]u+\left[-\frac{2}{3}(\mu\xi_x)_y+(\mu\xi_y)_x\right]v+\xi_x p\right\}\Big/J \ ,$$

$$G^{**}=\left\{\varrho uV^c+\left[\frac{4}{3}(\mu\eta_x)_x+(\mu\eta_y)_y\right]u+\left[-\frac{2}{3}(\mu\eta_x)_y+(\mu\eta_y)_x\right]v+\eta_x p\right\}\Big/J \ ,$$

$$(12.83)$$

$$R^*=\left[\left(\frac{4}{3}\xi_x^2+\xi_y^2\right)\mu u+\left(\frac{\mu}{3}\right)\xi_x\xi_y v\right]\left(\frac{1}{J}\right)$$

$$S^*=\left[2\left(\frac{4}{3}\xi_x\eta_x+\xi_y\eta_y\right)\mu u+\left(\frac{\mu}{3}\right)(\xi_x\eta_y+\xi_y\eta_x)v\right]\left(\frac{1}{J}\right)$$

$$T^*=\left[\left(\frac{4}{3}\eta_x^2+\eta_y^2\right)\mu u+\left(\frac{\mu}{3}\right)\eta_x\eta_y v\right]\left(\frac{1}{J}\right) \ .$$

The governing equations for fluid dynamics in conservation form and generalised coordinates are discussed by Eiseman and Stone (1980). The use of generalised coordinates in solving flow problems is illustrated in Sects. 15.4.2 and 18.4.

12.4 Numerical Implementation of Generalised Coordinates

To solve practical problems using generalised coordinates requires the casting of the governing equation in generalised coordinates (Sect. 12.3), the (typically) numerical evaluation of grid-related parameters (Sect. 12.2) and the construction and solution of the discretised equations. These various processes are illustrated for Laplace's equation which will be solved in a generalised-coordinate finite difference form for the domain previously considered to illustrate the finite volume method (Sect. 5.2.3).

12.4.1 LAGEN: Generalised Coordinate Laplace Equation

Solutions to Laplace's equation

$$\frac{\partial^2\phi}{\partial x^2}+\frac{\partial^2\phi}{\partial y^2}=0 \ , \tag{12.84}$$

will be sought in the domain shown in Fig. 12.9, with the following Dirichlet boundary conditions:

on WX, $\quad \phi = 0$,

on XY, $\quad \phi = \dfrac{\sin\theta}{r_{XY}}$, $\hspace{4cm}$ (12.85)

on YZ, $\quad \phi = \dfrac{1}{r_{YZ}}$,

on ZW, $\quad \phi = \dfrac{\sin\theta}{r_{WZ}}$.

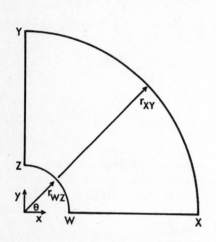 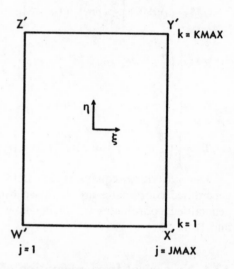

Fig. 12.9. Solution domain for (12.84, 85)

Equations (12.84, 85) have an exact solution,

$$\phi = \frac{\sin\theta}{r} , \hspace{4cm} (12.86)$$

which will be used to assess the accuracy of the computational solutions.

Laplace's equation, in generalised coordinates, is available from (12.29). Thus (12.84) becomes

$$-\left\{\left(\frac{\nabla^2\xi}{J}\right)\phi\right\}_\xi - \left\{\left(\frac{\nabla^2\eta}{J}\right)\phi\right\}_\eta + \left\{\left(\frac{\alpha}{J}\right)\phi\right\}_{\xi\xi} + \left\{\left(\frac{\beta}{J}\right)\phi\right\}_{\xi\eta} + \left\{\left(\frac{\gamma}{J}\right)\phi\right\}_{\eta\eta} = 0 ,$$
$$(12.87)$$

where

$$\alpha = \xi_x^2 + \xi_y^2 , \quad \beta = 2(\xi_x \eta_x + \xi_y \eta_y) , \quad \gamma = \eta_x^2 + \eta_y^2 ,$$
$$J = \xi_x \eta_y - \xi_y \eta_x \quad \text{and} \quad \nabla^2 \xi = \xi_{xx} + \xi_{yy} , \quad \text{etc.}$$

If the various terms, e.g. (α/J), in (12.87) can be evaluated at each grid point, the discretisation of (12.87) using centred differences is quite straightforward.

As indicated in Sect. 12.2, it is easier to start from the numerical evaluation of the transformation parameters x_ξ, etc., given by (12.33 and 34). The various terms in (12.87) can be expressed in terms of x_ξ, etc. The conversion of terms $\alpha/J, \beta/J, \gamma/J$ is available from (12.12) as

$$\text{GTT} \equiv \frac{\alpha}{J} = \frac{x_\eta^2 + y_\eta^2}{J^{-1}} ,$$

$$\text{GWT} \equiv \frac{\beta}{J} = -\frac{2(x_\xi x_\eta + y_\xi y_\eta)}{J^{-1}} , \tag{12.88}$$

$$\text{GWW} \equiv \frac{\gamma}{J} = \frac{x_\xi^2 + y_\xi^2}{J^{-1}} ,$$

where the determinant of the inverse Jacobian $J^{-1} = x_\xi y_\eta - x_\eta y_\xi$. The notation GTT, etc., is introduced to suit the program LAGEN (Fig. 12.10).

Everything on the right-hand side of (12.88) can be evaluated using (12.33). Starting from the two-dimensional form of (12.7) differentiation of ξ_x gives

$$\xi_{xx} = \frac{\xi_x y_{\xi\eta} + \eta_x y_{\eta\eta}}{J^{-1}} - \frac{\xi_x^2 J_\xi^{-1} + \xi_x \eta_x J_\eta^{-1}}{J^{-1}} . \tag{12.89}$$

Obtaining similar expressions for ξ_{yy} and further manipulation allows the terms $(\nabla^2 \xi/J)$ and $(\nabla^2 \eta/J)$ to be written

$$\text{DELZI} \equiv \frac{\nabla^2 \xi}{J} = \frac{\text{GTT}(x_\eta y_{\xi\xi} - y_\eta x_{\xi\xi})}{J^{-1}} + \frac{\text{GWT}(x_\eta y_{\xi\eta} - y_\eta x_{\xi\eta})}{J^{-1}}$$
$$+ \frac{\text{GWW}(x_\eta y_{\eta\eta} - y_\eta x_{\eta\eta})}{J^{-1}} \tag{12.90}$$

$$\text{DELET} \equiv \frac{\nabla^2 \eta}{J} = \frac{\text{GTT}(y_\xi x_{\xi\xi} - x_\xi y_{\xi\xi})}{J^{-1}} + \frac{\text{GWT}(y_\xi x_{\xi\eta} - x_\xi y_{\xi\eta})}{J^{-1}}$$
$$+ \frac{\text{GWW}(y_\xi x_{\eta\eta} - x_\xi y_{\eta\eta})}{J^{-1}} . \tag{12.91}$$

Everything on the right-hand sides of (12.90, 91) can be evaluated using (12.33, 34, 88). With GTT, etc., evaluated from (12.88–91) at each nodal point, (12.87) is discretised using three-point centred difference formulae to give

$$-0.5[(DELZI.\phi)_{j+1,k} - (DELZI.\phi)_{j-1,k}]$$
$$-0.5[(DELET.\phi)_{j,k+1} - (DELET.\phi)_{j,k-1}]$$
$$+[(GTT.\phi)_{j-1,k} - 2(GTT.\phi)_{j,k} + (GTT.\phi)_{j+1,k}]$$
$$+0.25[(GWT.\phi)_{j+1,k+1} - (GWT.\phi)_{j-1,k+1} + (GWT.\phi)_{j-1,k-1}$$
$$-(GWT.\phi)_{j+1,k-1}]$$
$$+[(GWW.\phi)_{j,k-1} - 2(GWW.\phi)_{j,k} + (GWW.\phi)_{j,k+1}] = 0 \ . \tag{12.92}$$

In forming (12.92) the grid size in the computational domain is $\Delta\xi = \Delta\eta = 1$. Equation (12.92) is solved iteratively using successive over-relaxation (Sect. 6.3). Equation (12.92) provides an estimate of $\phi_{j,k}^{n+1}$ as

$$\phi_{j,k}^* = \{-0.5[(DELZI.\phi^n)_{j+1,k} - (DELZI.\phi^n)_{j-1,k}]$$
$$-0.5[(DELET.\phi^n)_{j,k+1} - (DELET.\phi^n)_{j,k-1}]$$
$$+[(GTT.\phi^n)_{j-1,k} + (GTT.\phi^n)_{j+1,k} + (GWW.\phi^n)_{j,k-1} + (GWW.\phi^n)_{j,k+1}]$$
$$+0.25[(GWT.\phi^n)_{j+1,k+1} - (GWT.\phi^n)_{j-1,k+1} + (GWT.\phi^n)_{j-1,k-1}$$
$$-(GWT.\phi^n)_{j+1,k-1}]\} \cdot \{2(GTT_{j,k} + GWW_{j,k})\}^{-1} \ , \tag{12.93}$$

and the improved solution is

$$\phi_{j,k}^{n+1} = \phi_{j,k}^n + \lambda(\phi_{j,k}^* - \phi_{j,k}^n) \ , \tag{12.94}$$

where λ is the relaxation parameter. The above formulation has been incorporated into a computer program, LAGEN. A listing of LAGEN and subroutines GRID, TRAPA and ITER are provided in Figs. 12.10–13. The main parameters used by program LAGEN are given in Table 12.1.

After reading in the various parameters, LAGEN calls GRID to determine the grid points which are stored in augmented arrays, XG, YG. The grid points in the computational domain (Fig. 12.9) are augmented by additional rows of points ($j=0$, JMAX+1, $k=0$, KMAX+1) lying just outside the computational domain. This permits the transformation parameters x_ξ, etc., to be evaluated on the boundary using the same formulae (12.33, 34) as in the interior.

The various grid point parameters GTT, etc., in (12.93) are evaluated in subroutine TRAPA and the iterative solution of (12.93 and 94) is carried out in subroutine ITER. LAGEN subsequently generates the solution output.

Typical output for a 6×6 grid is shown in Fig. 12.14. For this particular case the grid is orthogonal and the terms β/J and $\nabla^2\eta/J$ in (12.87) are zero. The convergence rate on this grid (Table 12.2) is seen to be approximately second-order and the solution accuracies are similar to those achieved by the finite volume method (Table 5.24). The effect of grid distortion on the generalised-coordinate

```
 1   C     LAGEN APPLIES THE FINITE DIFFERENCE METHOD TO LAPLACES
 2   C     EQUATION IN GENERALISED COORDINATES ON A MODIFIED POLAR GRID.
 3   C     THE DISCRETISED EQUATION IS SOLVED BY SOR
 4   C
 5         DIMENSION XG(23,23),YG(23,23),GWW(21,21),GWT(21,21),GTT(21,21),
 6        1DELZI(21,21),DELET(21,21),PHI(21,21),PHIX(21,21)
 7         COMMON /GRIDP/XG,YG,PHIX,PHI
 8         COMMON /TRAPP/GWW,GWT,GTT,DELZI,DELET
 9   C
10         OPEN(1,FILE='LAGEN.DAT')
11         OPEN(6,FILE='LAGEN.OUT')
12         READ(1,1)JMAX,KMAX,NMAX,IEX
13         READ(1,2)RW,RX,RY,RZ,THEB,THEN,EPS,OM
14       1 FORMAT(8I5)
15       2 FORMAT(8E10.3)
16   C
17         WRITE(6,3)
18         WRITE(6,4)JMAX,KMAX,NMAX,IEX,EPS,OM
19         WRITE(6,5)RW,RX,RY,RZ,THEB,THEN
20       3 FORMAT(' LAPLACE EQUATION BY GEN. COORD. FDM',//)
21       4 FORMAT(' JMAX=',I2,' KMAX=',I2,' NMAX=',I5,' IEX=',I2,
22        1'  EPS=',E10.3,'   OM=',F5.3)
23       5 FORMAT('   RW=',F5.3,'   RX=',F5.3,'   RY=',F5.3,'   RZ=',F5.3,
24        15X,'  THEB=',F5.1,'   THEN=',F5.1,//)
25         AKM = KMAX - 2
26         AJM = JMAX - 2
27   C
28         CALL GRID(JMAX,KMAX,THEB,THEN,RW,RX,RY,RZ)
29   C
30         DO 8 J = 1,JMAX
31         PHI(J,1) = 0.
32         PHI(J,KMAX) = PHIX(J,KMAX)
33       8 CONTINUE
34         DO 9 K = 1,KMAX
35         PHI(1,K) = PHIX(1,K)
36         PHI(JMAX,K) = PHIX(JMAX,K)
37       9 CONTINUE
38   C
39         CALL TRAPA(JMAX,KMAX)
40   C
41         CALL ITER(JMAX,KMAX,NMAX,N,OM,EPS)
42   C
43   C     COMPARE SOLUTION WITH EXACT
44   C
45      10 SUM = 0.
46         DO 15 K = 1,KMAX
47         WRITE(6,11)K
48      11 FORMAT(/,' K=',I2)
49         DO 12 J = 1,JMAX
50         DIF = PHI(J,K) - PHIX(J,K)
51      12 SUM = SUM + DIF*DIF
52         WRITE(6,13)(PHI(J,K),J=1,JMAX)
53         WRITE(6,14)(PHIX(J,K),J=1,JMAX)
54      13 FORMAT(' PHI=',10F7.4)
55      14 FORMAT(' PHX=',10F7.4)
56      15 CONTINUE
57         RMS = SQRT(SUM/AJM/AKM)
58         WRITE(6,16)N,RMS
59      16 FORMAT(/,' CONVERGED AFTER ',I3,' STEPS,   RMS=',E12.5)
60         STOP
61         END
```

Fig. 12.10. Listing of program LAGEN

Fig. 12.11. Listing of subroutine GRID

```
1
2         SUBROUTINE GRID(JMAX,KMAX,THEB,THEN,RW,RX,RY,RZ)
3 C
4 C       SET THE AUGMENTED GRID, INITIAL AND EXACT PHI
5 C
6         DIMENSION XG(23,23),YG(23,23),PHIX(21,21),PHI(21,21)
7         COMMON /GRIDP/XG,YG,PHIX,PHI
8         JMAP = JMAX - 1
9         KMAP = KMAX - 1
10        AJM = JMAP
11        AKM = KMAP
12        DRWX = (RX - RW)/AJM
13        DRZY = (RY - RZ)/AKM
14        DTH = (THEN-THEB)/AKM
15        PI = 3.1415927
16        KPP = KMAX + 2
17        JPP = JMAX + 2
18 C
19 C      SET XG, YG, EXACT AND INITIAL PHI
20 C
21        DO 7 K = 1,KPP
22        AK = K - 2
23        THK = (THEB + AK*DTH)*PI/180.
24        CK = COS(THK)
25        SK = SIN(THK)
26        DR = DRWX + (DRZY - DRWX)*AK/AKM
27        RWZ = RW + (RZ - RW)*AK/AKM
28        DO 6 J = 1,JPP
29        AJ = J - 2
30        R = RWZ + AJ*DR
31        XG(J,K) = R*CK
32        YG(J,K) = R*SK
33        IF(K .EQ. 1 .OR. K .EQ. KPP)GOTO 6
34        IF(J .EQ. 1 .OR. J .EQ. JPP)GOTO 6
35        JM = J-1
36        KM = K - 1
37        PHIX(JM,KM) = SK/R
38        PHI(JM,KM) = PHIX(JM,KM)
39  6 CONTINUE
40  7 CONTINUE
41        RETURN
42        END
```

Table 12.1. Parameters used in program LAGEN

Parameter	Description
JMAX	Number of points in the radial direction
KMAX	Number of points in the circumferential direction
NMAX	Maximum number of iterations
RW	r_W (Fig. 12.9)
RX	r_X (Fig. 12.9)
RY	r_Y (Fig. 12.9)
RZ	r_Z (Fig. 12.9)
THEB	θ_{WX} (Fig. 12.9)
THEN	θ_{ZY} (Fig. 12.9)
EPS	Tolerance for convergence of SOR
OM	Relaxation parameter λ in (12.94)
PHI	ϕ
PHIX	Exact solution for ϕ, (12.86)
PHD	ϕ^* in (12.93) (subroutine ITER)
XG, YG	Augmented grid coordinates (subroutine GRID)
RMS	$\phi^* - \phi_{rms}^n$, $\ \|\phi^{n+1} - \phi_{ex}\|_{rms}$

```
1
2          SUBROUTINE TRAPA(JMAX,KMAX)
3  C
4  C       FROM GRID COORDINATES CALCULATES TRANSFORM PARAMETERS
5  C
6          DIMENSION XG(23,23),YG(23,23),PHIX(21,21),PHI(21,21)
7          DIMENSION GWW(21,21),GWT(21,21),GTT(21,21),DELZI(21,21),
8         1DELET(21,21)
9          COMMON /GRIDP/XG,YG,PHIX,PHI
10         COMMON /TRAPP/GWW,GWT,GTT,DELZI,DELET
11         DO 2 K = 1,KMAX
12         KP = K+1
13         KPP = K + 2
14         DO 1 J = 1,JMAX
15         JP = J + 1
16         JPP = J + 2
17 C
18 C       BASIC TRANSFORM PARAMETERS
19 C
20         XZI = 0.5*(XG(JPP,KP) - XG(J,KP))
21         YZI = 0.5*(YG(JPP,KP) - YG(J,KP))
22         XET = 0.5*(XG(JP,KPP) - XG(JP,K))
23         YET = 0.5*(YG(JP,KPP) - YG(JP,K))
24         XZZ = XG(J,KP) - 2.*XG(JP,KP) + XG(JPP,KP)
25         YZZ = YG(J,KP) - 2.*YG(JP,KP) + YG(JPP,KP)
26         XEE = XG(JP,K) - 2.*XG(JP,KP) + XG(JP,KPP)
27         YEE = YG(JP,K) - 2.*YG(JP,KP) + YG(JP,KPP)
28         XZE = 0.25*(XG(JPP,KPP)-XG(J,KPP)+XG(J,K)-XG(JPP,K))
29         YZE = 0.25*(YG(JPP,KPP)-YG(J,KPP)+YG(J,K)-YG(JPP,K))
30         AJ = XZI*YET - XET*YZI
31 C
32 C       MODIFIED METRIC TENSOR COEFFICIENTS, G11,G12,G22
33 C
34         GWW(J,K) = (XZI*XZI + YZI*YZI)/AJ
35         GWT(J,K) = -2.*(XZI*XET + YZI*YET)/AJ
36         GTT(J,K) = (XET*XET + YET*YET)/AJ
37 C
38 C       MODIFIED DEL**2ZI AND DEL**2ETA
39 C
40         DUM = GTT(J,K)*(XET*YZZ-YET*XZZ)/AJ
41         DUM = DUM + GWT(J,K)*(XET*YZE-YET*XZE)/AJ
42         DELZI(J,K) = DUM + GWW(J,K)*(XET*YEE-YET*XEE)/AJ
43         DUM = GTT(J,K)*(YZI*XZZ - XZI*YZZ)/AJ
44         DUM = DUM + GWT(J,K)*(YZI*XZE-XZI*YZE)/AJ
45         DELET(J,K) = DUM + GWW(J,K)*(YZI*XEE-XZI*YEE)/AJ
46       1 CONTINUE
47       2 CONTINUE
48         RETURN
49         END
```

Fig. 12.12. Listing of subroutine TRAPA

```
1
2         SUBROUTINE ITER(JMAX,KMAX,NMAX,N,OM,EPS)
3  C
4  C      ITERATE USING SOR APPLIED TO DISCRETISED EQUATIONS
5  C
6         DIMENSION GWW(21,21),GWT(21,21),GTT(21,21),DELZI(21,21),
7        1DELET(21,21),PHI(21,21),PHIX(21,21),XG(23,23),YG(23,23)
8         COMMON /GRIDP/XG,YG,PHIX,PHI
9         COMMON /TRAPP/GWW,GWT,GTT,DELZI,DELET
10        KMAP = KMAX-1
11        JMAP = JMAX-1
12        AKM = KMAP-1
13        AJM = JMAP-1
14 C
15        DO 3 N = 1,NMAX
16        SUM = 0.
17        DO 2 K = 2,KMAP
18        KM = K - 1
19        KP = K + 1
20        DO 1 J = 2,JMAP
21        JM = J - 1
22        JP = J + 1
23        PHD = -0.5*(DELZI(JP,K)*PHI(JP,K)-DELZI(JM,K)*PHI(JM,K))
24        PHD = PHD-0.5*(DELET(J,KP)*PHI(J,KP)-DELET(J,KM)*PHI(J,KM))
25        PHD = PHD + GTT(JM,K)*PHI(JM,K) + GTT(JP,K)*PHI(JP,K)
26        PHD = PHD + GWW(J,KM)*PHI(J,KM) + GWW(J,KP)*PHI(J,KP)
27        PHD = PHD + 0.25*(GWT(JP,KP)*PHI(JP,KP)-GWT(JM,KP)*PHI(JM,KP)
28       1 + GWT(JM,KM)*PHI(JM,KM) - GWT(JP,KM)*PHI(JP,KM))
29        PHD = 0.5*PHD/(GTT(J,K)+GWW(J,K))
30        DIF = PHD - PHI(J,K)
31        SUM = SUM + DIF*DIF
32        PHI(J,K) = PHI(J,K) + OM*DIF
33      1 CONTINUE
34      2 CONTINUE
35        RMS = SQRT(SUM/AJM/AKM)
36        IF(RMS .LT. EPS)RETURN
37      3 CONTINUE
38        WRITE(6,4)NMAX,RMS
39      4 FORMAT(' CONVERGENCE NOT ACHIEVED IN',I5,' STEPS,  RMS=',
40       1E12.5)
41        RETURN
42        END
```

Fig. 12.13. Listing of subroutine ITER

Table 12.2. Generalised-coordinate solution errors with grid refinement $(r_W = r_Z = 0.1, r_Y = 1.0, \theta_{WX} = 0, \theta_{ZY} = 90, \lambda = 1.5)$

Case	GRID	$\|\phi - \phi_{ex}\|_{rms}$	No. of iterations to convergence
A,	6×6	0.1338	15
$r_X = 1.00$	11×11	0.0473	19
	21×21	0.0138	53
B,	6×6	0.2541	15
$r_X = 2.00$	11×11	0.1176	21
	21×21	0.0430	66

LAPLACE EQUATION BY GEN. COORD. FDM   **Fig. 12.14.** Typical output from LAGEN

```
JMAX= 6 KMAX= 6 NMAX=  100 IEX= 0    EPS=   .100E-04   OM=1.500
RW= .100    RX=1.000    RY=1.000   RZ= .100 THEB= .0   THEN= 90.0


K= 1
PHI=  .0000   .0000   .0000   .0000   .0000   .0000
PHX=  .0000   .0000   .0000   .0000   .0000   .0000

K= 2
PHI= 3.0902 1.2454   .7453   .5209   .3931   .3090
PHX= 3.0902 1.1036   .6718   .4828   .3768   .3090

K= 3
PHI= 5.8779 2.3498 1.4048   .9834   .7444   .5878
PHX= 5.8779 2.0992 1.2778   .9184   .7168   .5878

K= 4
PHI= 8.0902 3.1813 1.9002 1.3351 1.0166   .8090
PHX= 8.0902 2.8893 1.7587 1.2641   .9866   .8090

K= 5
PHI= 9.5106 3.6228 2.1688 1.5352 1.1804   .9511
PHX= 9.5106 3.3966 2.0675 1.4860 1.1598   .9511

K= 6
PHI=10.0000 3.5714 2.1739 1.5625 1.2195 1.0000
PHX=10.0000 3.5714 2.1739 1.5625 1.2195 1.0000

CONVERGED AFTER  15 STEPS,    RMS=   .13384E+00
```

solution is also indicated in Table 12.2 (case B, $r_x = 2.00$). As might be expected the accuracy and the convergence rate with grid refinement are both lower than for case A.

12.5 Closure

The use of generalised coordinates permits finite difference methods to be effective in computational domains with complicated boundary shapes, primarily by making the boundaries coincide with particular generalised coordinate lines and, thereby, avoiding a local interpolation to implement the boundary conditions.

In generalised coordinates the governing equations (Sect. 12.3) include additional terms that contain the mapping information between the irregular grid in the physical domain and the regular grid in the computational domain. The number of additional terms in the governing equations are reduced if orthogonal or conformal grids can be constructed for the particular computational domain.

The discretisation of the governing equations in generalised coordinates introduces the additional problem (usually) of discretising the transformation parameters. It is generally recommended that the same formulae be used as for discretising derivatives of the dependent variables.

That the discretisation is undertaken on a typically uniform computational grid may imply that higher accuracy can be achieved. Although this is true in the computational domain it is not generally true in the physical domain. If the grid growth or stretching parameter (r_x in Fig. 12.8) is not small then a reduction in accuracy is to be expected.

In general this problem is more severe if the governing equation contains second or higher derivatives. However, for flow problems the terms responsible for the additional second-derivative reduction in accuracy are typically multiplied by 1/Re. Consequently for high Reynolds number flows this additional error does not usually have a large effect.

The implementation of generalised-coordinate finite difference schemes (Sect. 12.4) is no more complicated than the implementation of the finite volume method and produces solutions of comparable accuracy.

12.6 Problems

Transformation Relationships (Sect. 12.1)

12.1 For a two-dimensional transformation derive the equivalent of (12.7) by direct multiplication of $\underline{J}\,\underline{J}^{-1} = \underline{I}$.

12.2 Show by direct substitution that (12.13) is true.

12.3 Use (12.15–17) to show that the transformation parameters can be expressed in terms of α, AR, θ and J as

$$x_\xi = \frac{\cos\alpha}{(\text{AR } J \sin\theta)^{1/2}}, \quad y_\xi = -\frac{\sin\alpha}{(\text{AR } J \sin\theta)^{1/2}},$$

$$x_\eta = \cos(\theta-\alpha)\left(\frac{\text{AR}}{J\sin\theta}\right)^{1/2}, \quad y_\eta = \sin(\theta-\alpha)\left(\frac{\text{AR}}{J\sin\theta}\right)^{1/2}.$$

Evaluation of the Transformation Parameters (Sect. 12.2)

12.4 Starting from the two-dimensional equivalent of (12.7) derive the equations

$$\xi_{xx} = \frac{\xi_x y_{\xi\eta} + \eta_x y_{\eta\eta}}{J^{-1}} - \frac{\xi_x^2 J_\xi^{-1} + \xi_x \eta_x J_\eta^{-1}}{J^{-1}}$$

and

$$\xi_{xy} = \frac{\eta_y y_{\eta\eta} + \xi_y y_{\xi\eta}}{J^{-1}} - \frac{\xi_x \xi_y J_\xi^{-1} + \xi_x \eta_y J_\eta^{-1}}{J^{-1}}.$$

12.5 Derive (12.46).

12.6 For a one-dimensional grid, equivalent to that in Fig. 12.8, with physical and computational grid growth ratios r_x and r_ξ, respectively, show that the equivalent of (12.47) is

$$\frac{T_{j+1}-T_{j-1}}{x_{j+1}-x_{j-1}}=T_x+\left(\frac{(r_\xi-1)}{2}\frac{(1+r_x)}{(1+r_\xi)}+\frac{(r_\xi^3+1)}{(r_\xi+1)^2}\frac{(r_x-r_\xi)}{r_\xi}\right)\varDelta x\,T_{xx}$$

$$+\left(\frac{(r_\xi^3+1)}{(r_\xi+1)^3}\frac{(1+r_x)^2}{6}\right)\varDelta x^2\,T_{xxx}\ .$$

Hence derive a relationship between r_x and r_ξ for second-order accuracy. Would this be a practical way of choosing r_ξ for arbitrary $0.8\leq r_x\leq 1.20$, to give second-order accuracy?

Generalised Coordinate Structure of Typical Equations (Sect. 12.3)

12.7 Transform the equations

$$u_x+v_y=0 \quad \text{and} \quad u_y-v_x=0$$

into generalised coordinates on a conformal grid and show that

$$U_\xi^*+V_\eta^*=0 , \qquad U_\eta^*-V_\xi^*=0 ,$$

where $U^*=J^{-1}(\xi_y v+\eta_y u)$ and $V^*=J^{-1}(\eta_y v-\xi_y u)$.

12.8 Transform the vorticity transport equation

$$(u\zeta)_x+(v\zeta)_y-\frac{1}{\text{Re}}(\zeta_{xx}+\zeta_{yy})=0$$

into generalised coordinates. How would the resulting equation simplify for i) an orthogonal grid, ii) a conformal grid?

12.9 The following equations govern two-dimensional steady incompressible turbulent boundary layer flow (Sect. 11.4.2)

$$u_x+v_y=0 ,$$

$$uu_x+vu_y+p_{\text{ex}}=\frac{1}{\text{Re}}\frac{\partial}{\partial y}\left[(1+v_T)\frac{\partial u}{\partial y}\right] ,$$

where v_T is an eddy viscosity (11.76). Convert these equations to generalised coordinates and decide whether additional terms can be dropped consistent with the boundary layer assumption (Sect. 11.4.1). Assume the boundary layer develops on a surface of constant η.

Numerical Implementation (Sect. 12.4)

12.10 Obtain solutions using LAGEN for the following cases: i) 6×6 grid, ii) 11×11 grid, iii) 21×21 grid for parameter values $r_w=r_z=0.1$, $r_x=1.0$, $r_y=3.00$, $\theta_{wx}=0$, $\theta_{zy}=90$, $\lambda=1.5$. Compare the accuracy and convergence rate with the results shown in Table 12.2 and with FIVOL applied to the same computational domain.

12.11 Modify subroutine GRID so that grid lines can vary like $a_0 + a_1 r + a_2 r^2$ in the radial direction and like $b_0 + b_1 \theta + b_2 \theta^2$, in the circumferential direction. Adjust the parameters a_0, a_1, a_2, b_0, b_1 and b_2 to place more grid lines close to WZ and ZY. Determine whether this produces more accurate solutions for the same number of grid points than the solutions presented in Table 12.2.

12.12 For the solution domain shown in Fig. 12.9 the generalised coordinates (ξ, η) and the physical coordinates (x, y) can be related analytically by noting that

$$\theta = \theta_{WX} + \eta(\theta_{ZY} - \theta_{WX}) ,$$

$$r = r_W + \xi(r_X - r_W) + \eta(r_Z - r_W) + \xi\eta[(r_Y - r_Z) - (r_X - r_W)]$$

and $x = r \cos\theta$, $y = r \sin\theta$.

These relationships permit the transformation parameters, ξ_x, etc., to be determined analytically. Replace the numerical evaluation of ξ_x, etc., in subroutine TRAPA by analytic evaluations and determine what the effect is on solution accuracy and convergence rate.

12.13 Application of the group finite element method (Sect. 10.3) with bilinear interpolating functions on rectangular elements (Sect. 5.3.3) produces the following equations in place of (12.92):

$$-M_\eta \otimes L_\xi(\text{DELZI}.\phi)_{j,k} - M_\xi \otimes L_\eta(\text{DELET}.\phi)_{j,k}$$

$$+ M_\eta \otimes L_{\xi\xi}(\text{GTT}.\phi)_{j,k} + L_\xi \otimes L_\eta(\text{GWT}.\phi)_{j,k} + M_\xi \otimes L_{\eta\eta}(\text{GWW}.\phi)_{j,k} = 0 ,$$

where $M_\xi = M_\eta = \{1/6, 2/3, 1/6\}$, $L_\xi = L_\eta^T = 0.5\{-1, 0, 1\}$

and $L_{\xi\xi} = L_{\eta\eta}^T = \{1, -2, 1\}$

Thus the term $M_\eta \otimes L_{\xi\xi}(\text{GTT}.\phi)_{j,k} = \frac{1}{6}[(\text{GTT}.\phi)_{j-1,k+1}$

$$-2(\text{GTT}.\phi)_{j,k+1} + (\text{GTT}.\phi)_{j+1,k+1}]$$

$$+\frac{2}{3}[(\text{GTT}.\phi)_{j-1,k} - 2(\text{GTT}.\phi)_{j,k} + (\text{GTT}.\phi)_{j+1,k}]$$

$$+\frac{1}{6}[(\text{GTT}.\phi)_{j-1,k-1} - 2(\text{GTT}.\phi)_{j,k-1} + (\text{GTT}.\phi)_{j+1,k-1}] .$$

Construct an SOR algorithm based on the above equation in place of (12.93) and obtain solutions for i) 6×6 grid, ii) 11×11 grid, iii) 21×21 grid for the parameter values given in Table 12.2. Compare the accuracy and convergence rate.

13. Grid Generation

In this chapter grid generation will be discussed in relation to the establishment of the correspondence between points (x, y) in the irregular physical domain and points (ξ, η) in the regular computational domain. A conceptual approach to grid generation is to fix the values of ξ and η on the physical boundaries first. Subsequently interior points are located by determining the intersection of co-ordinate lines of opposite families drawn between corresponding boundary points.

In a sense the problem of grid generation can be posed as a boundary value problem: given $\xi = \xi_b(x, y)$ and $\eta = \eta_b(x, y)$ on the boundary ∂R, generate $\xi = \xi(x, y)$ and $\eta = \eta(x, y)$ in the region R bounded by ∂R (Fig. 13.1). The physical coordinates (x, y), typically Cartesian, are the independent variables, and the generalised coordinates (ξ, η) are the dependent variables.

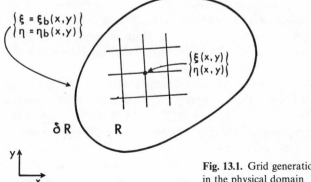

Fig. 13.1. Grid generation as a boundary value problem in the physical domain

In practice the grid is generated with less computational effort by working in the computational domain. Thus fixing the location of the points on the boundary gives $x = x_b(\xi, \eta)$ and $y = y_b(\xi, \eta)$. The generation of the grid in the interior is expressed as the following boundary value problem: given $x = x_b(\xi, \eta)$ and $y = y_b(\xi, \eta)$ on ∂R (Fig. 13.2) generate $x = x(\xi, \eta)$ and $y = y(\xi, \eta)$ in the region R bounded by ∂R.

Since the interior points in the computational domain form a regular grid and the boundaries coincide with coordinate lines, the determination of $x(\xi, \eta)$, $y(\xi, y)$ is easier than working in the irregular physical domain, particularly if a partial differential equation is to be solved to generate the solution, $x(\xi, \eta)$, $y(\xi, \eta)$, as in Sect. 13.2.6.

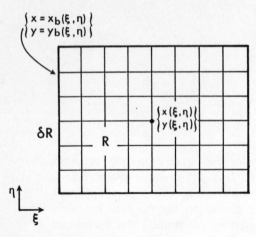

$$\begin{cases} x = x_b(\xi, \eta) \\ y = y_b(\xi, \eta) \end{cases}$$

Fig. 13.2. Grid generation as a boundary value problem in the computational domain

$$\begin{cases} x(\xi, \eta) \\ y(\xi, \eta) \end{cases}$$

δR

R

η

ξ

The specification of the above boundary value problem has been based on Dirichlet boundary conditions. However, it is often desirable to introduce Neuman boundary conditions. For example these would be appropriate if it were required that the coordinate lines intersect the physical boundary normally. This implies that on certain segments of the boundary, the angle of intersection (90° typically) of the coordinate lines with the boundary is specified, and the location of the intersection points on the boundary are determined as part of the overall solution to the boundary value problem. An obvious extension is to a mixed boundary condition specification where it is desirable to maintain control of both the location of the grid points on the boundary and the orthogonality of the grid at the boundary.

The two broad approaches to solving the boundary value problem for the interior grid point locations are either to solve a partial differential equation (Sect. 13.2) or to interpolate the interior region (Sect. 13.3). Before describing specific techniques, typical topological correspondences between the physical and computational domains will be considered (Sect. 13.1).

In defining the relationship between points in the physical and computational domains, i.e. $x = x(\xi, \eta)$ and $y = y(\xi, \eta)$, it is necessary that there be a one-to-one correspondence. It would be unacceptable for a single point in the physical domain to map into two points in the computational domain, and vice-versa. This is equivalent to the requirement that coordinate lines of the same family must not cross and that coordinate lines of different families may only cross once.

Once the mapping, $x = x(\xi, \eta)$ and $y = y(\xi, \eta)$, has been established, the requirement of a one-to-one mapping can be determined by evaluating the determinant of the transformation Jacobian, $|J|$, from (12.3). For the mapping to be one-to-one, $|J|$ must be finite and non-zero. Depending on how the grid has been generated, $|J|$ can be evaluated at each grid point, analytically or numerically (Sect. 12.2), to check for a one-to-one mapping. Such a check can be readily coded into a computer program. During development, computer plotting of the grid will quickly locate any points where the mapping is double-valued.

13.1 Physical Aspects

In this section typical boundary mappings between the physical and computational domains will be indicated for simply connected and multiply connected regions. A particular choice for the gross boundary correspondence can have a marked influence on the distortion of the interior grid.

13.1.1 Simply-Connected Regions

A simply connected region implies that any closed contour in the region can be shrunk to zero length without containing a boundary of the region.

The simplest type of mapping is to require that a region defined by four curves is transformed to a rectangle in the computational domain. The mapping shown in Fig. 12.2 is of this type.

If the physical domain has an overall shape, it may be appropriate to preserve, approximately, the same overall orientation in the computational domain. Thus Fig. 13.3 indicates that a distorted L-shaped region in the physical domain can be mapped into a regular L-shaped region in the computational domain. In Fig. 13.3 only sufficient of the interior grid is shown to indicate the gross features. By preserving the overall shape approximately, it is easier to avoid excessive grid distortion [i.e. lack of orthogonality (12.21)].

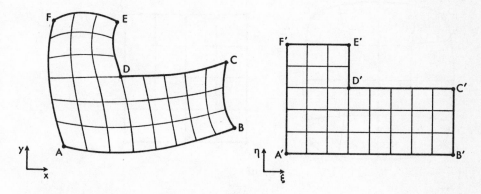

Fig. 13.3. L-shaped region with gross shape preserved

This same L-shaped region in the physical domain could be mapped into a rectangular region in the computational domain (Fig. 13.4). However, now slope discontinuities at A and D in the physical domain occur on lines of constant η (at A' and D') in the computational domain.

It is easier to set up the solution algorithm for the domain shown in Fig. 13.4 than for that shown in Fig. 13.3, but the greater distortion in the physical domain close to A and D implies that the local solution accuracy may be more seriously affected than for the grid in Fig. 13.3.

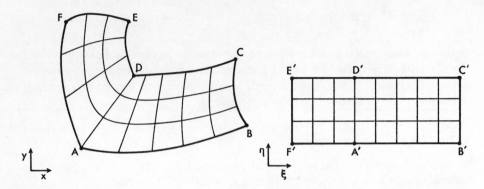

Fig. 13.4. L-shaped region transformed to a rectangle

A converse situation occurs when a corner point in the computational domain corresponds to a continuous part of the physical boundary (Fig. 13.5). Clearly, in this case, the grid adjacent to points A, B, etc., in the physical domain is rather distorted, with an expected deterioration of the local solution accuracy. Also, traversing the boundary through A', in the computational domain, changes from an η coordinate line to a ξ coordinate line. This may require special procedures to set up the discretised equations.

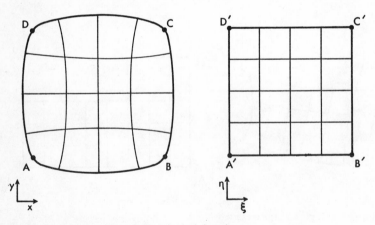

Fig. 13.5. Fictitious corners in the physical domain

Generally the more complicated the boundary in the physical domain, the more choices in laying out the computational grid. For example, the same physical domain shown in Figs. 13.6–8 can be mapped in a number of different ways.

For the bump shown in Fig. 13.6 the introduction of a concentrated grid for small values of η will resolve the solution close to B and D, but implies a fine grid adjacent to A and E. However, if BCD represented an obstruction in a viscous flow with AG as an inlet, points B and D would be stagnation points. If the interest were

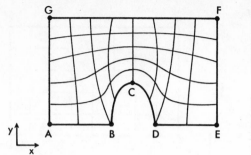
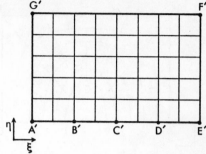

Fig. 13.6. Single bump flattened

on the maximum velocity or shear adjacent to C it might be acceptable to have a relatively coarse and distorted grid adjacent to B and D, as in Fig. 13.6.

In Fig. 13.7 points B and D are collapsed onto the same point in the computational domain. This grid allows more grid points to be clustered close to BCD, with fewer total grid points, than does the grid in Fig. 13.6. However, a more distorted grid is produced close to C in Fig. 13.7. If a flow with GF as the inlet and outlet through AB and DE is being modelled the grid configuration in Fig. 13.7 is possibly acceptable.

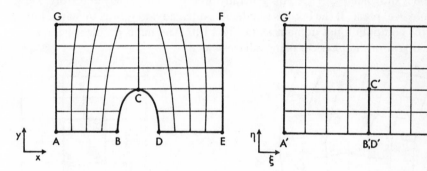

Fig. 13.7. Single bump transformed to a plate

If points G and F are in a freestream well away from an obstruction BCD then a preferred boundary mapping might be as shown in Fig. 13.8. This mapping has some similarity with that shown in Fig. 13.6, except that here the most severely distorted part of the grid occurs at G and F. Since the flow is uniform at G and F the errors due to the locally distorted grid can be kept smaller than for the grid configuration shown in Fig. 13.6.

The way the boundary is mapped from the physical to the computational domains should take into account the resulting grid distortion is relation to the expected flow solution. That is, local grid distortions have less impact on the global accuracy, if they occur in regions where the flow is uniform and if they can be removed from the region of interest.

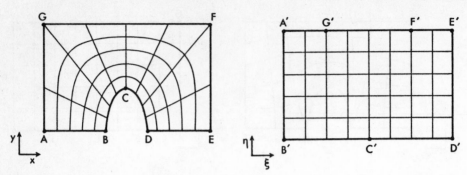

Fig. 13.8. Single bump flattened with good local resolution

13.1.2 Multiply-Connected Regions

An example of a multiply connected region is the external flow around one or more obstacles, e.g. a turbine blade or aerofoil and flap. An obstruction in a duct, e.g. flow through a heat exchanger, is another example. The preferred boundary mapping typically depends on the shape of the obstruction.

For a bluff body it may be appropriate to map the body to a square or rectangle in the computational plane and, thereby, retain a multiply connected region. Such a situation is shown in Fig. 13.9. This generally produces a relatively undistorted grid in the physical plane. If the body is slender it is often advantageous to map it into a slit in the computational domain as in Fig. 13.10. The main problem here is the distorted grid in the physical plane adjacent to A.

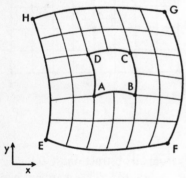
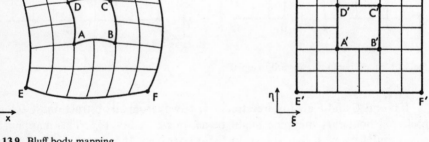

Fig. 13.9. Bluff body mapping

An effective way of overcoming the problem is to introduce a branch cut in the physical plane to generate a simply connected computational domain, as shown in Fig. 13.11. This type of grid is called an O-grid due to the gridline pattern in the physical domain. The introduction of the branch cut in the physical domain implies that grid points on $A'I'$ and $C'D'$ in the computational domain coincide. Contiguous points across the branch cut may be required when evaluating derivatives on or adjacent to $A'I'$ and $C'D'$.

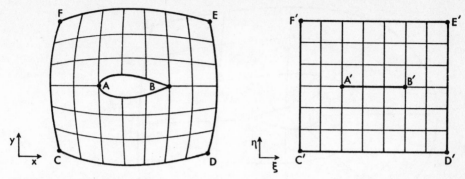

Fig. 13.10. Slender body mapping

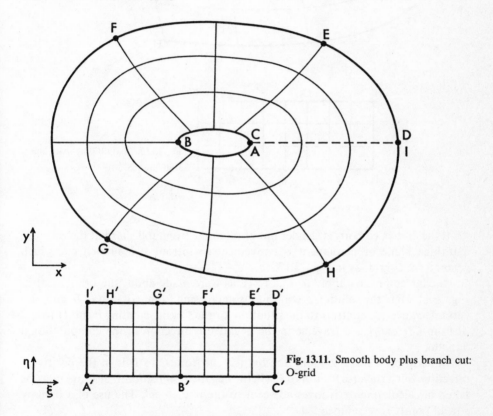

Fig. 13.11. Smooth body plus branch cut: O-grid

However, care must be taken, in crossing through boundary $A'I'$ and reentering $C'D'$, that the ξ, η coordinate directions appropriate to the $A'I'$ boundary are retained. It is conceptually useful to extend $A'I'$ and $C'D'$ as overlapping regions. At the coding level this can be done by defining dummy rows of points outside $A'I'$ and $C'D'$ that are used to compute derivatives and are upgraded when the corresponding points at the other end of the computational domain are upgraded.

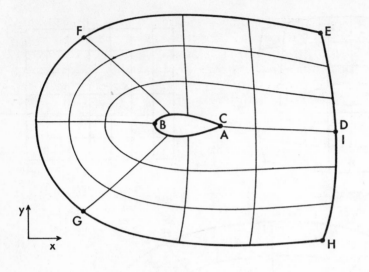

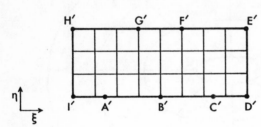

Fig. 13.12. Smooth body with sharp trailing edge: C-grid

If the immersed body is slender but blunt at the front and sharp at the rear, e.g. a turbine blade or aerofoil, it is convenient to introduce a branch cut which generates a C-grid, as in Fig. 13.12.

Similar comments apply to Fig. 13.12 as were made about the branch cut in Fig. 13.11. Here the solution is continuous in crossing from $A'I'$ into $C'D'$ and vice versa. However, in contrast to the situation for the O-grid, crossing from AI to CD or from CD to AI is a traverse in the negative η direction in the computational domain.

In addition a traverse along the branch cut from C to D is in the positive ξ direction, but a traverse from A to I is in the negative ξ direction. Thus care must be taken in calculating derivatives adjacent to the branch cut. The use of a dummy row of points is recommended.

The occurrence of multiple isolated bodies typically requires either multiple embedded rectangles as in Fig. 13.9 or multiple branch cuts as in Fig. 13.12. Some of the choices are discussed by Thompson (1982, pp. 1–31).

The extension of the above concepts to three dimensions is straightforward, although the book-keeping associated with contiguous surfaces, etc., can become tedious. Typical examples are provided by Rubbert and Lee (1982) and Thomas (1982).

It has been implicit in the above description that the grid being generated is not time-dependent. However for time-dependent problems or for resolving initially unknown severe gradients, e.g. associated with internal shocks, it is desirable to allow the grid to change with time or, equivalently, to adapt to the developing solution. The added level of complexity that this involves is reviewed by Thompson (1984) and described in more detail by Thompson et al. (1985, Chap. 11).

13.2 Grid Generation by Partial Differential Equation Solution

In transforming the governing equations for fluid dynamics into generalised coordinates it was pointed out (Sect. 12.1.3) that the governing equations have a simpler structure if the grid is conformal or orthogonal.

For both classes of grid the transformation between the physical and computational domain can be obtained by solving a partial differential equation; (13.2) for conformal grids and (13.24) for orthogonal grids. With no restriction on the grid the transformation can be obtained by solving a Poisson equation (13.35).

13.2.1 Conformal Mapping: General Considerations

For conformal transformations it is possible to write the relationship between the physical (x, y) and computational domains as, in two dimensions,

$$\begin{bmatrix} dx \\ dy \end{bmatrix} = \begin{bmatrix} h\cos\alpha & -h\sin\alpha \\ h\sin\alpha & h\cos\alpha \end{bmatrix} \begin{bmatrix} d\xi \\ d\eta \end{bmatrix} . \tag{13.1}$$

The scale factor h is related to the components of the metric tensor by (12.23), i.e. $h = g_{11}^{1/2} = g_{22}^{1/2}$. The angle α is the angle between the tangent to the ξ coordinate line and the x-axis (12.15). Clearly once h and α are known for a particular conformal transformation, $x_\xi (= h\cos\alpha)$, etc., follow directly from (13.1).

When a conformal transformation is used the computational grid (ξ, η) is linked to the physical grid (x, y) by Laplace's equations

$$\xi_{xx} + \xi_{yy} = 0 , \qquad \eta_{xx} + \eta_{yy} = 0 , \tag{13.2}$$

and the Cauchy–Riemann conditions $\xi_x = \eta_y$ and $\xi_y = -\eta_x$. Since simple exact solutions to (13.2) exist it is possible to construct the solutions $\xi(x, y)$ and $\eta(x, y)$ by superposition and by complex transformations (Milne-Thomson 1968).

Using complex variables, $z = x + iy$ and $\zeta = \xi + i\eta$, a conformal transformation can be expressed symbolically as $Z = F(\zeta)$ or, in a more useful form,

$$dZ = H\, d\zeta \quad \text{or} \quad Z = \int H\, d\zeta , \quad \text{where} \tag{13.3}$$

$$H = he^{i\alpha} = h(\cos\alpha + i\sin\alpha) . \tag{13.4}$$

Thus, from (13.1), H contains the transformation parameters x_ξ, etc.

Traditionally (Milne-Thomson 1968), conformal mapping has been used to obtain potential flow solutions (Sect. 11.3) about relatively complicated shapes given the flow behaviour about a simple shape, such as a circle with a unit radius (unit circle).

Here conformal mapping is used as a grid generation technique, with no restriction on the type of flow. In practice the generated gridlines may be chosen to coincide with the streamlines of an equivalent potential flow problem. This feature often aids the stability of the computational method used to solve the more general problem.

As a complete technique for grid generation, conformal mapping may be interpreted as consisting of two stages:

i) the construction of a single mapping or sequence of mappings to obtain the correspondence between boundary points in the physical and computational domains,
ii) the generation of the interior points in the physical domain given the boundary correspondence from stage i).

It is understood that, as with the general philosophy of grid generation, the computational domain is typically a simple rectangle in which interior points are laid out on a regular grid.

Two approaches will be considered. The first approach (Sect. 13.2.2) is appropriate to streamlined shapes like aerofoils or turbine blades which can be transformed via a sequence of mappings into a unit rectangle. The second approach (Sect. 13.2.3) uses a one-step mapping based on a Schwarz–Christoffel transformation of a polygon with N straight sides into a straight line. Various modifications of the Schwarz–Christoffel transformation are now available which make it applicable to quite general shapes.

13.2.2 Sequential Conformal Mapping

The sequence is often built around the von Karman–Trefftz transformation (Milne-Thomson 1968, p. 199),

$$\frac{Z'-a}{Z'-b}=\left(\frac{Z-A}{Z-B}\right)^{1/k} ,$$
(13.5)

where a, b in the Z'-plane and A and B in the Z-plane are chosen to suit the geometry considered. The von Karman–Trefftz transformation maps an aerofoil in the Z-plane into a near-circle in the Z' plane. This technique will be illustrated for the aerofoil shown in Fig. 13.13.

The parameters in (13.5) are chosen as

$$A=Z_t , \quad B=Z_n , \quad a=-b=\frac{Z_n-Z_t}{2k}$$
(13.6)

and $k=2-\tau/\pi$.

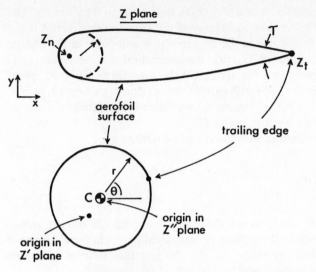

Fig. 13.13. Sequential mapping for an aerofoil

The location Z_t is the trailing edge and Z_n is a point midway between the nose of the aerofoil and its centre of curvature. The transformation is singular at these two points. The parameter k is related to the trailing edge included angle τ by (13.6). The choice of parameters given by (13.6) maps the aerofoil in the Z plane to a near-circle in the Z' plane (Fig. 13.13), approximately centred at C. It can be seen that the angle τ has been expanded to $180°$ in the Z' plane.

It is convenient to execute (13.5) in the following sequence:

$$\omega = \frac{Z-A}{Z-B} , \tag{13.7a}$$

$$v = \omega^{1/k} , \tag{13.7b}$$

$$Z' = \frac{a-bv}{1-v} , \tag{13.7c}$$

Equations (13.7a and c) involve only linear transformations. However, (13.7b) introduces the complication that multiple values of v exist for each value of ω. If the aerofoil has a cusped trailing-edge, $\tau = 0$, and two values of v arise for each value of ω. For the more general case of a non-zero value of τ an infinite number of values of v are available for each value of ω.

Computational strategies for selecting the correct value of v are discussed by Ives (1982, p. 114). Broadly the transformation is tracked to the point of interest from a 'safe' point such as upstream infinity in the physical (Z) plane.

The near-circle in the Z' plane is conveniently transformed to a near-circle in the Z'' plane centred at the origin by

$$Z'' = Z' - C . \tag{13.8}$$

The point C is the approximate centroid of the near-circle in the Z' plane. This mapping is included to improve the convergence of the Theodorsen–Garrick (1933) transformation from the near-circle in the Z'' plane to a unit circle in the ζ plane.

To implement the Theodorsen–Garrick transformation it is necessary to be able to define arbitrary points on the surface of the near-circle in the Z'' plane. Introducing polar coordinates, $Z'' = r(\theta)\exp(i\theta)$, it is customary (Ives 1976) to fit $\ln r$ as a function of θ with a periodic cubic spline. Cubic spline fitting is discussed by Ahlberg et al. (1967) and Forsythe et al. (1977).

The Theodorsen–Garrick transformation can be written as

$$\frac{dZ''}{d\zeta} = \exp\left[\sum_{j=0}^{N} (A_j + iB_j)\zeta^j \right] . \tag{13.9}$$

The coefficients A_j and B_j in (13.9) are chosen by mapping $2N$ equally spaced points around the unit circle in the ζ plane to equivalent points in the Z'' plane. A particularly efficient technique, based on the discrete fast Fourier transform (Cooley and Tukey 1965), is described by Ives (1976).

The region outside of the unit circle in the ζ plane (the intermediate computational domain) is mapped to the inside of a rectangle ($1 \le R \le R_{max}, 0 \le \beta \le 2\pi$) by letting

$$\zeta = re^{i\phi} ,$$

and setting

$$R = \exp(\ln r) \quad \text{and} \quad \beta = \phi . \tag{13.10}$$

Consequently the sequence of transformations (13.7–10) maps the region exterior to an isolated streamlined body in the (x, y) plane to the interior of a rectangle in the (R, β) plane.

In principle a uniform grid can be laid out in the (R, β) plane and the inverse mapping used to provide the corresponding grid in the physical domain. However, Ives (1982, p. 128) notes that although the establishment of the boundary correspondence between the physical and computational domains, by the above sequence of transformations, is relatively efficient, the inverse transformation often is not. Therefore, for the general case of sequential mapping, Ives recommends that a fast elliptic solver (Temperton 1979) be used to generate the interior grid by solving

$$x_{\xi\xi} + x_{\eta\eta} = 0 \quad \text{and} \quad y_{\xi\xi} + y_{\eta\eta} = 0 , \tag{13.11}$$

with the boundary values already determined by stage i).

The sequential mapping procedure described above can be extended to multiple isolated bodies, e.g. an aerofoil and flap (Ives 1976).

13.2.3 One-Step Conformal Mapping

In this section a computationally efficient implementation of the Schwarz–Christoffel transformation due to Davis (1979) is described. This particular implementation can be extended to bodies with curved sides. The traditional Schwarz–Christoffel transformation allows a region bounded by a simple closed polygon in the physical plane Z to be mapped into the upper half of the transform plane ω. The polygon coincides with the real axis in the ω plane.

By introducing a branch cut, Fig. 13.14, it is possible to consider the region between the polygon and infinity in the Z plane as being the bounded region. Consequently the Schwarz–Christoffel transformation can be used to obtain exterior grids. However, in this section the Schwarz–Christoffel transformation will be described for an internal flow geometry, a two-dimensional duct (Fig. 13.15).

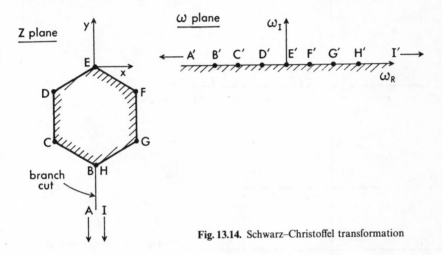

Fig. 13.14. Schwarz–Christoffel transformation

In the conventional form (Milne-Thomson 1968, p. 277) the Schwarz–Christoffel transformation is written

$$\frac{dZ}{d\omega} = M \prod_{j=1}^{N} (\omega - b_j)^{-\alpha_j/\pi} \; , \tag{13.12}$$

where the α_js are the angles turned through at each corner (anticlockwise positive). The b_js are unknown locations on the real axis in the transform plane; three of the b_js may be chosen arbitrarily. M is a complex constant, typically related to the geometry of the physical domain.

The body in the physical domain does not need to be closed. Thus the region inside an arbitrary duct, Fig. 13.15, can be mapped to the upper half plane by (13.12) so that Z_N, at the top of the inlet to the duct, is mapped to a downstream point in the ω plane.

In principle a uniform grid can be laid out in the ω plane and the inverse mapping used to generate the corresponding grid in the physical plane. However,

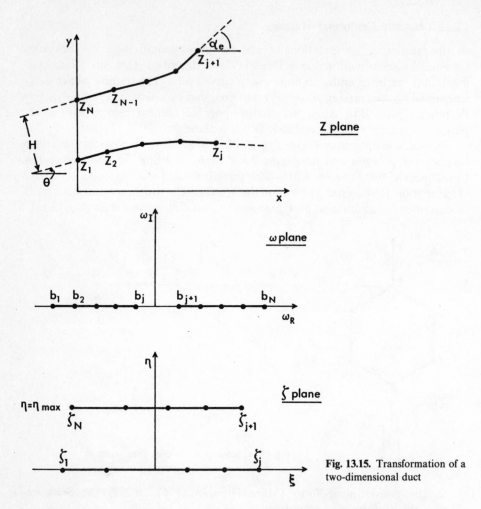

Fig. 13.15. Transformation of a two-dimensional duct

for a duct flow it is convenient to make a second transformation from the upper half ω plane to a straight duct parallel to the real axis in the ζ plane (Fig. 13.15).

The Schwarz–Christoffel mapping for the duct from the Z plane to the ω plane is given by

$$\frac{dZ}{d\omega} = \frac{M}{\omega}\left(\prod_{j=1}^{N}(\omega - b_j)^{-\alpha_j/\pi}\right)\omega^{-\alpha_e/\pi} . \tag{13.13}$$

Equation (13.13) has been written so that it involves only angles and poles actually in the duct. In (13.13), α_j are the corner angles of the jth segment in the Z plane. M is related to the duct height and orientation to the x-axis, and can be obtained explicitly (13.17). The b_js are poles in ω plane, corresponding to the corners in the physical plane. These are unknown and must be determined iteratively by repeatedly integrating (13.13).

The mapping from the ω plane to the ζ plane is given by

$$\zeta = -\frac{1}{\pi}\ln\omega + i \ . \tag{13.14}$$

If it is assumed that the duct extends far upstream as a straight duct (equivalent to $\omega \to \infty$), (13.13) becomes

$$\frac{dZ}{d\omega} = \frac{M}{\omega} \ . \tag{13.15}$$

Integrating (13.15) and combining with (13.14) gives

$$Z = \pi M(i - \zeta) + Z_0 \ . \tag{13.16}$$

Applied to the upper and lower duct surface, (13.16) produces the following expression for M:

$$Z_u - Z_1 = iHe^{i\theta} = -\pi Mi \ , \tag{13.17}$$

where H and θ are defined in Fig. 13.15.

In principle, (13.13) could be integrated numerically, to generate the grid, if the pole locations b_j were known. However, difficulties would be experienced in the neighbourhood of b_j due to the singular nature of (13.13) at $\omega = b_j$. The integration path can be located at $\omega + i\varepsilon$ in the ω plane which is equivalent to marching just inside the duct axis. However, a better way (Davis 1979) is to use a composite, second-order marching scheme which incorporates the analytic integration at each pole. This scheme is given by

$$\frac{Z_{k+1} - Z_k}{\zeta_{k+1} - \zeta_k} = \frac{M}{w_{k+1/2}^{(1+\alpha_e/\pi)}} \prod_{j=1}^{N} \left(\frac{(\omega_{k+1} - b_j)^{1-\alpha_j/\pi} - (\omega_k - b_j)^{1-\alpha_j/\pi}}{(\omega_{k+1} - \omega_k)^{1-\alpha_j/\pi}} \right) \ . \tag{13.18}$$

A corresponding finite difference representation of (13.14) can be written

$$\omega_{k+1} - \omega_k = -\pi\omega_{k+1/2}(\zeta_{k+1} - \zeta_k) \ . \tag{13.19}$$

Combining (13.18) and (13.19) provides a direct link between the physical plane Z and the computational plane ζ which is valid for any integration path in the computational domain.

Since the pole locations b_j are unknown, (13.18) must be integrated to obtain $Z_j^v - Z_{j-1}^v$, where v is the iteration index. But the converged values, Z_j^c, etc., are known, in the physical plane. Therefore the unknown ζ_j^{v+1} are upgraded from

$$\zeta_j^{v+1} = \zeta_{j-1}^{v+1} + \frac{|Z_j^c - Z_{j-1}^c|}{|Z_j^v - Z_{j-1}^v|}(\zeta_j^v - \zeta_{j-1}^v) \ , \tag{13.20}$$

and the upgraded b_j^{v+1} are obtained from (13.14) as

$$b_j^{v+1} = \exp[\pi(\mathrm{i} - \zeta_j^{v+1})] \ . \tag{13.21}$$

The whole process (13.18–21) is repeated until Z_j^v is sufficiently close to Z_j^c for all j. Typically this takes 10–15 iterations for agreement to five decimal places, and appears to be independent of the number of segments, N. Further details are given by Anderson et al. (1982, p. 507).

The repeated integrations to establish the correct b_j locations are along the computational boundaries ($\eta = 0$ and η_{max} in Fig. 13.15) and generate the correspondence between boundary points in the physical and computational domains (stage i), Sect. 13.2.1). Subsequently, with known b_j values, (13.18) is numerically integrated along lines of constant ξ and η in the computational domain to generate the interior grid in the physical domain.

An interesting feature of the present approach is that in the computational domain the potential flow in the duct is given by

$$\phi + \mathrm{i}\psi = \zeta \ , \tag{13.22}$$

where ϕ is the velocity potential, ψ is the streamfunction and there is a unit velocity. Thus constant values of η and ξ will give streamlines and isopotential lines. Consequently the corresponding grid lines in the physical plane will correspond to the streamlines and isopotential lines for potential flow through the actual duct. For a viscous flow through the duct the gridlines will still be a reasonable approximation to the actual streamlines.

The general construction of conformal mappings via the Schwarz–Christoffel transformation and approximate methods of numerical integration are surveyed by Trefethen (1980).

Traditionally the Schwarz–Christoffel transformation has been applied to straight-sided domains, with discontinuous changes in body slope associated with the turning angles α_i in Fig. 13.14. However, Woods (1961) has extended the Schwartz–Christoffel transformation to bodies with curved surfaces, replacing (13.12) with

$$\frac{dZ}{d\omega} = M \exp\left[\frac{1}{\pi}\int \ln(\omega - b)d\beta\right] , \tag{13.23}$$

where b is a segment of the real axis in the ω plane and $d\beta$ is the corresponding incremental angle turned through in the Z plane. Davis (1979) provides a numerical implementation of (13.23) and obtains appropriate mappings for the region outside an aerofoil in the physical plane to a rectangle in the ω (computational) plane.

Although conformal mapping is not available in three dimensions, flow geometries in three dimensions have been successfully analysed by using a sequence of two-dimensional transformations at successive values of the third coordinate (Moretti 1980).

13.2.4 Orthogonal Grid Generation

Orthogonal coordinate systems require the metric parameters $g_{ij} = 0$, $i \neq j$. This implies that the physical (x, y) and computational (ξ, η) domains must be related, in two dimensions, by

$$\left(\xi_x \frac{h_1}{h_2}\right)_x + \left(\xi_y \frac{h_1}{h_2}\right)_y = 0 \quad \text{and} \quad \left(\eta_x \frac{h_2}{h_1}\right)_x + \left(\eta_y \frac{h_2}{h_1}\right)_y = 0 \qquad (13.24)$$

in the interior and

$$h_1 \xi_x = h_2 \eta_y \quad \text{and} \quad h_1 \xi_y = -h_2 \eta_x \qquad (13.25)$$

on the boundary. The ratio h_2/h_1 is the grid aspect ratio (12.16) where

$$h_1 = (x_\xi^2 + y_\xi^2)^{1/2} \quad \text{and} \quad h_2 = (x_\eta^2 + y_\eta^2)^{1/2} . \qquad (13.26)$$

Orthogonal grids may be constructed from an existing or preliminary non-orthogonal grid by generating orthogonal trajectories. This permits points to be specified on three surfaces (say EA, ABC and CD in Fig. 13.16), with points on the fourth surface (ED in Fig. 13.16) determined by the orthogonal trajectory.

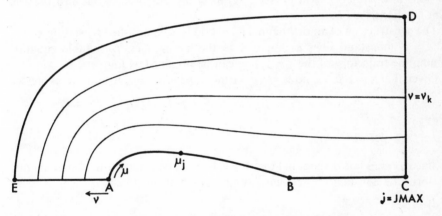

Fig. 13.16. Preliminary configuration for orthogonal trajectory construction

If an orthogonal grid is required with points specified on all four surfaces it is possible (Thompson 1984) to construct a conformal grid and to apply separate one-dimensional stretchings in the ξ and η directions. Equation (13.44) provides an effective stretching function. Alternatively (13.24) can be solved with an iteratively prescribed form, $h_2/h_1 = f(\xi, \eta)$. This technique will be described in Sect. 13.2.6.

If the orthogonal grid is only required close to a specific boundary (say ABC in Fig. 13.16) it is possible to construct a locally orthogonal grid by specifying a metric-related parameter, say J, in addition to the orthogonality constraint, $g_{12} = 0$. This produces a hyperbolic system of equations that can be marched away from the specified boundary to construct the grid. Steger and Sorenson (1980) describe such a technique.

In this section the orthogonal trajectory method will be described in more detail. The starting point is to lay out a family of curves, as in Fig. 13.16. The location along a particular member ($v = v_k$) of the family may be determined by a simple shearing transformation, e.g.

$$x = (1 - \mu')EA^x(v_k) + \mu'CD^x(v_k) \ ,$$

$$y = (1 - \mu')EA^y(v_k) + \mu'CD^y(v_k) \ , \quad \text{and} \tag{13.27}$$

$$\mu' = (\mu - \mu_1)/(\mu_{JMAX} - \mu_1) \ .$$

The prescribed functions $EA^x(v)$, etc., determine the grid point distribution on EA and CD (Fig. 13.16) and the clustering of grid lines adjacent to ABC. If the $v = v_k$ lines must also be orthogonal to EA and CD a cubic representation in μ', equivalent to (13.49), can be introduced (Eiseman 1982a, p. 209) to replace (13.27). The distribution of points on ABC ($\mu = \mu_j$) is prescribed.

The construction of an orthogonal grid (ξ, η) requires defining trajectories starting at $\mu = \mu_j$, finishing at the target boundary ED and intersecting each intervening coordinate line ($v = v_k$) at right angles. Clearly not only are the locations of the interior grid points determined by this process but also the grid point locations on ED.

The construction of an orthogonal (ξ, η) grid is accomplished by setting $\eta_k = v_k$ and $\xi = \xi_j$ (constant) lines orthogonal to the $\eta = \eta_k$ lines. A suitable equation defining the trajectory of the $\xi = \xi_j$ line can be obtained as follows.

Given that $x = x(\mu, v)$ and $y = y(\mu, v)$, the slope of a $v = v_k$ line can be expressed as

$$\left.\frac{dy}{dx}\right|_{v = v_k} = \frac{\partial y/\partial \mu}{\partial x/\partial \mu} \ . \tag{13.28}$$

Since $\eta = v$, a line orthogonal to a $v = v_k$ line is a constant ξ line. Consequently the slope of a constant ($\xi = \xi_j$) line can be written, from (13.28),

$$\left.\frac{dy}{dx}\right|_{\xi = \xi_j} = -\frac{\partial x/\partial \mu}{\partial y/\partial \mu} \ . \tag{13.29}$$

But since $\xi = \xi(\mu, v)$ and $\eta = v$,

$$\left.\frac{dy}{dx}\right|_{\xi = \xi_j} = \frac{[(\partial y/\partial \mu)(d\mu/dv)_{\xi = \xi_j} + \partial y/\partial v]}{[(\partial x/\partial \mu)(d\mu/dv)_{\xi = \xi_j} + \partial x/\partial v]} \ . \tag{13.30}$$

Equating (13.29) and (13.30) produces an ordinary differential equation defining the trajectory of the $\xi = \xi_j$ line in the (μ, v) plane. That is,

$$\left.\frac{d\mu}{dv}\right|_{\xi = \xi_j} = -\left(\frac{\partial x}{\partial v}\frac{\partial x}{\partial \mu} + \frac{\partial y}{\partial \mu}\frac{\partial y}{\partial v}\right) \Big/ \left[\left(\frac{\partial x}{\partial \mu}\right)^2 + \left(\frac{\partial y}{\partial \mu}\right)^2\right] \ . \tag{13.31}$$

A comparison with (12.12) indicates that (13.31) can be written in the general form

$$\left.\frac{d\mu}{dv}\right|_{\xi=\xi_j} = -\frac{g_{12}}{g_{11}} \ , \tag{13.32}$$

where g_{12} and g_{11} are components of the metric associated with mapping from the physical (x, y) plane to the non-orthogonal (μ, v) plane. Clearly the right-hand side of (13.31) can be evaluated since the (μ, v) grid is defined (Fig. 13.16).

Initial values for (13.31) are given by $\mu = \mu_j$ on ABC. Typically (13.31) is integrated numerically. At the intersection with each $v = v_k$ line, the physical coordinates (x, y) follow from the equation defining the (μ, v) grid, e.g. (13.27).

Equation (13.32) is a characteristic equation (Sect. 2.2) for the hyperbolic equation

$$\frac{\partial \xi}{\partial v} - \frac{g_{12}}{g_{11}} \frac{\partial \xi}{\partial \mu} = 0 \ , \tag{13.33}$$

so that the orthogonal trajectory method is essentially a method of characteristics. The method is discussed at greater length by Eiseman (1982).

Generally strictly orthogonal grids are restricted to two dimensions, since it is not possible to construct completely orthogonal grids in three dimensions and still retain sufficient control over boundary grid point locations (Eiseman 1982). However, it is possible to construct grids that are orthogonal to specific surfaces and to construct three-dimensional grids from a stack of two-dimensional orthogonal grids. In this case the grid distribution in the third direction may not be very smooth (Thompson 1984).

13.2.5 Near-Orthogonal Grids

Although the use of a strictly orthogonal grid permits certain terms in the governing equations (Sect. 12.1.3) to be dropped, a near-orthogonal grid is very easy to construct and will avoid errors due to grid distortion.

The following is an effective technique for generating a near-orthogonal grid. It is assumed that the family of grid lines shown in Fig. 13.16 has been constructed. The procedure starts with specified points, $\mu = \mu_j$ on ABC, as in Sect. 13.2.4. The present procedure is a predictor-corrector scheme designed to obtain a point on the $v = v_2$ line that is approximately orthogonal (Fig. 13.17) to the point (μ_j, v_1). The normal to the $v = v_1$ line is drawn to the intersection of the $v = v_2$ line. At the intersection point the normal, $-dx/dy|_{v_2, n}$, is calculated and the intersection point moved until this normal passes through the original point (μ_j, v_1). The final point (μ_j, v_2) on the v_2 line, in "orthogonal correspondence" with the point (μ_j, v_1), is taken as the average of the two intersection points. This is equivalent to using a characteristic direction defined by

$$\left.\frac{dy}{dx}\right|_{av} = 0.5\left(-\left.\frac{dx}{dy}\right|_{v_1, n} - \left.\frac{dx}{dy}\right|_{v_2, n}\right) \ . \tag{13.34}$$

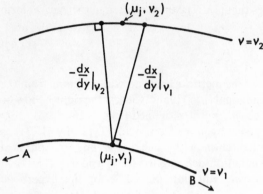

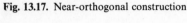

Fig. 13.17. Near-orthogonal construction

Once all the grid points have been set on the $v = v_2$ line the process is repeated to establish the near-orthogonal grid points on the $v = v_3$ gridline. The process is continued until the outer boundary, e.g. *ED* in Fig. 13.16, is reached.

Typical code to construct the predicted intersection with the $v = v_2$ grid line is provided in subroutine SURCH (Fig. 13.29). The construction described here was originated by McNally (1972).

13.2.6 Solution of Elliptic Partial Differential Equations

In this section more general techniques for coordinate generation will be considered, which do not necessarily produce grids that are conformal or orthogonal. However, the present techniques do permit more control over the clustering of interior grids points.

As indicated at the beginning of Chap. 13 the problem of interior grid point generation can be posed as a boundary value problem, and preferably in the computational (ξ, η) domain. Since an elliptic partial differential equation is to be solved it is necessary to specify the grid point locations or the local grid line slopes at the boundaries, as boundary conditions.

The most common partial differential equation used for grid generation is a Poisson equation in the form

$$\frac{\partial^2 \xi}{\partial x^2} + \frac{\partial^2 \xi}{\partial y^2} = P(\xi, \eta) , \qquad \frac{\partial^2 \eta}{\partial x^2} + \frac{\partial^2 \eta}{\partial y^2} = Q(\xi, \eta) , \qquad (13.35)$$

where P and Q are known functions used to control interior grid clustering.

The use of an elliptic partial differential equation to generate the interior grid points brings with it certain advantages. First, the grid will be smoothly varying even if the boundary of the domain has a slope discontinuity. By contrast if a hyperbolic partial differential equation were used to generate the interior grid any slope discontinuities at the boundary would also appear in the interior grid.

Elliptic equations like (13.35) satisfy the maximum principle, for reasonable values of P and Q. That is, the maximum and minimum values of ξ and η must occur on the boundary. Thompson (1982) indicates that this normally guarantees a one-to-one mapping. However, extreme choices for P and Q may cause a local grid overlap.

The actual solution of (13.35) is carried out in the computational (ξ, η) domain. In this domain (13.35) transform to

$$\alpha \frac{\partial^2 x}{\partial \xi^2} - 2\beta \frac{\partial^2 x}{\partial \xi \partial \eta} + \gamma \frac{\partial^2 x}{\partial \eta^2} + \delta \left(P \frac{\partial x}{\partial \xi} + Q \frac{\partial x}{\partial \eta} \right) = 0 \ , \tag{13.36a}$$

$$\alpha \frac{\partial^2 y}{\partial \xi^2} - 2\beta \frac{\partial^2 y}{\partial \xi \partial \eta} + \gamma \frac{\partial^2 y}{\partial \eta^2} + \delta \left(P \frac{\partial y}{\partial \xi} + Q \frac{\partial y}{\partial \eta} \right) = 0 \ , \tag{13.36b}$$

where $\alpha = g_{22}, \beta = g_{12}, \gamma = g_{11}$ and $\delta = g$, the determinant of the metric tensor (12.12).

To specify boundary conditions to suit (13.36) it is useful to consider the specific example shown in Fig. 13.18. On contour $ABC(A'B'C')$, $\eta = \eta_1$ and $x = x_{ABC}(\xi)$, $y = y_{ABC}(\xi)$, for $\xi_1 \leq \xi \leq \xi_2$, where the functional relations, $x_{ABC}(\xi)$ and $y_{ABC}(\xi)$, are

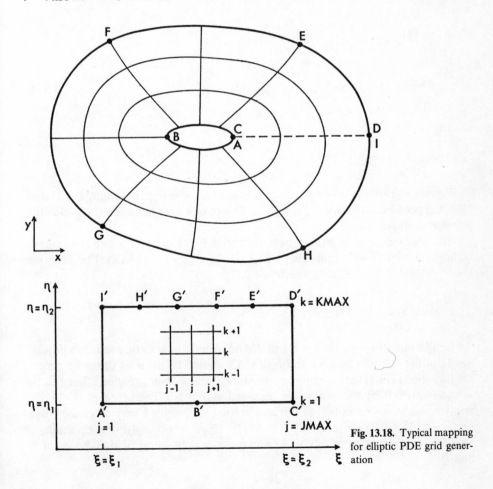

Fig. 13.18. Typical mapping for elliptic PDE grid generation

known and specify the distribution of grid points on ABC. In a similar way, on contour $DFI(D'F'I')$, $\eta = \eta_2$ and $x = x_{DFI}(\xi)$, $y = y_{DFI}(\xi)$, for $\xi_1 \leqq \xi \leqq \xi_2$, where $x_{DFI}(\xi)$ and $y_{DFI}(\xi)$ specify the distribution of grid points on DFI. The boundary grid point specification is facilitated by the one-dimensional stretching function described in Sect. 13.3.1.

It is emphasised that boundary conditions must not be specified on $A'I'$ or $C'D'$ since the corresponding lines in the physical plane are interior lines (and coincide).

Equation (13.36a) can be discretised using centred difference formulae to give

$$
\begin{aligned}
&\alpha'(x_{j-1,k} - 2x_{j,k} + x_{j+1,k}) - 0.5\beta'(x_{j+1,k+1} - x_{j-1,k+1} - x_{j+1,k-1} + x_{j-1,k-1}) \\
&+ \gamma'(x_{j,k-1} - 2x_{j,k} + x_{j,k+1}) + 0.5\delta' P(x_{j+1,k} - x_{j-1,k}) \\
&+ 0.5\delta' Q(x_{j,k+1} - x_{j,k-1}) = 0 \;,
\end{aligned} \tag{13.37}
$$

where

$$
\begin{aligned}
\alpha' &= 0.25[(x_{j,k+1} - x_{j,k-1})^2 + (y_{j,k+1} - y_{j,k-1})^2] \;, \\
\beta' &= 0.25[(x_{j+1,k} - x_{j-1,k})(x_{j,k+1} - x_{j,k-1}) \\
&\quad + (y_{j+1,k} - y_{j-1,k})(y_{j,k+1} - y_{j,k-1})] \;, \\
\gamma' &= 0.25[(x_{j+1,k} - x_{j-1,k})^2 + (y_{j+1,k} - y_{j-1,k})^2] \;, \\
\delta' &= [(x_{j+1,k} - x_{j-1,k})(y_{j,k+1} - y_{j,k-1}) \\
&\quad - (x_{j,k+1} - x_{j,k-1})(y_{j+1,k} - y_{j-1,k})]^2/16 \;.
\end{aligned} \tag{13.38}
$$

Equation (13.36b) is discretised in an equivalent manner. In forming (13.37 and 38) it has been assumed that $\Delta\xi = \Delta\eta = 1$. This choice does not affect the grid in the physical domain.

The appearance of the branch cut (AI/CD) in Fig. 13.18 causes some increase in coding complexity for points on $A'I'(j=1)$ or $C'D'(j=J\,\mathrm{MAX})$. For example $\partial^2 x/\partial\xi^2$ located on $A'I'$ is evaluated as

$$
\frac{\partial^2 x}{\partial\xi^2} = x_{J\mathrm{MAX}-1,k} - 2x_{1,k} + x_{2,k} \;.
$$

In addition, the solution $j=1$ and $J\mathrm{MAX}$ is identical, so that iterative adjustments to the solution on $j=1$ and $j=J\mathrm{MAX}$ should be made at the same time.

Equations (13.37) are nonlinear simultaneous algebraic equations that can be solved iteratively by the techniques discussed in Sect. 6.3. Thompson et al. (1977b) apply point SOR and note that the acceleration parameter λ may be greater than unity if $(\alpha')^2 > (0.5\delta' P)^2$ and $(\gamma')^2 > (0.5\delta' Q)^2$. Not surprisingly, the optimum λ and the number of iterations to convergence is a function of the choice of P and Q.

The following choice for P and Q is recommended by Thompson et al. (1977a):

$$P(\xi, \eta) = -\sum_{l=1}^{L} a_l \operatorname{sgn}(\xi - \xi_l) \exp(-c_l |\xi - \xi_l|)$$
$$-\sum_{m=1}^{M} b_m \operatorname{sgn}(\xi - \xi_m) \exp[-d_m[(\xi - \xi_m)^2 + (\eta - \eta_m)^2]^{1/2}] \qquad (13.39)$$

$$Q(\xi, \eta) = -\sum_{l=1}^{L} a_l \operatorname{sgn}(\eta - \eta_l) \exp(-c_l |\eta - \eta_l|)$$
$$-\sum_{m=1}^{M} b_m \operatorname{sgn}(\eta - \eta_m) \exp[-d_m \{(\xi - \xi_m)^2 + (\eta - \eta_m)^2\}^{1/2}] , \qquad (13.40)$$

where coefficients a_l, b_m, c_l and d_m are chosen to generate appropriate grid clustering. The sgn function has the property

$$\operatorname{sgn}(x) = 1 \qquad \text{if } x \text{ is positive,}$$
$$= 0 \qquad \text{if } x = 0,$$
$$= -1 \qquad \text{if } x \text{ is negative.}$$

The first term in (13.39) has the effect of moving $\xi = \text{const}$ lines towards the $\xi = \xi_l$ line, and the first term in (13.40) has the effect of moving $\eta = \text{const}$ lines towards the $\eta = \eta_l$ line. Thus the choice $\eta_l = \eta_1$ (Fig. 13.18) and a large value of a_l would tend to cluster lines close to the surface ABC. The second terms in (13.39, 40) attract $\xi = \text{const}$ lines and $\eta = \text{const}$ lines to the point (ξ_m, η_m). For á slender body the second terms could be used to concentrate points close to the leading and trailing edges, B and A/C in Fig. 13.18.

The use of large values of P and Q to generate clustering slows down the convergence and restricts the choice of initial (x, y) values from which convergence can be achieved. Consequently it is recommended that a solution is obtained initially with little or no clustering, as this situation has a wider radius of convergence. The converged solution is used as the starting solution for P and Q corresponding to increased clustering. This procedure can be continued until the required clustering is achieved.

Thompson et al. (1977a, b) provide many examples of grids generated using the above formulation. A computer listing, TOMCAT, is provided by Thompson et al. (1977b). The extension to three dimensions and the generalisation of (13.35) to provide greater grid control is discussed by Thompson (1982). More recent developments are indicated by Thompson (1984) and Thompson et al. (1985).

It is possible to modify the Thompson technique to generate orthogonal grids with specified boundary grid points and some control over the interior grid point distribution. This is achieved by defining

$$\frac{h_2}{h_1} = \frac{g_{22}^{1/2}}{g_{11}^{1/2}} = f(\xi, \eta) ,$$

where $f(\xi, \eta)$ is to be specified, in principle. The particular choice

$$P = \frac{1}{h_1 h_2} \frac{\partial f}{\partial \xi} \quad \text{and} \quad Q = \frac{1}{h_1 h_2} \frac{\partial}{\partial \eta} \left(\frac{1}{f} \right) \tag{13.41}$$

allows (13.36) to be written

$$\frac{\partial}{\partial \xi} \left(f \frac{\partial x}{\partial \xi} \right) + \frac{\partial}{\partial \eta} \left(\frac{1}{f} \frac{\partial x}{\partial \eta} \right) = 0 \; , \quad \frac{\partial}{\partial \xi} \left(f \frac{\partial y}{\partial \xi} \right) + \frac{\partial}{\partial \eta} \left(\frac{1}{f} \frac{\partial y}{\partial \eta} \right) = 0 \; . \tag{13.42}$$

Equations (13.42) are the governing equations for generating orthogonal grids. The function $f(\xi, \eta)$ may be chosen on the boundary. The interior values of $f(\xi, \eta)$ are typically (Ryskin and Leal 1983) generated by transfinite interpolation (Sect. 13.3.4) of the boundary values.

In practice a non-orthogonal grid is generated initially with specified $x(\xi, \eta)$, $y(\xi, \eta)$ on the boundaries. The definition $f = (g_{22}/g_{11})^{1/2}$ and (12.12) are used to evaluate $f(\xi, \eta)$ on the boundary. Transfinite interpolation generates corresponding interior values of $f(\xi, \eta)$. The discretised version of (13.42), equivalent to (13.37), is iterated for a few steps. The modified $x(\xi, \eta)$, $y(\xi, \eta)$ solution allows a new $f(\xi, \eta)$ to be calculated on the boundary, and the whole process is repeated until the grid no longer changes. Ryskin and Leal present typical grids requiring 50–100 iterations for convergence. Most of the interior point control comes from the boundary point distribution; the transfinite interpolation step provides some additional adjustment.

13.3 Grid Generation by Algebraic Mapping

Algebraic mapping techniques interpolate the boundary data to generate the interior grid. The explicit interpolation may be in one dimension (Sects. 13.3.2 and 13.3.3) or multidimensions (Sect. 13.3.4). A key requirement is that the generated grid should be well-conditioned i.e. smoothly varying, close to orthogonal and with local grid aspect ratios close to unity. For fluid flow problems the solution is often changing rapidly close to a particular surface. It is important to construct a grid that is orthogonal or near-orthogonal adjacent to such a surface. All three methods to be described in this section have this capability.

The distribution of grid points in the interior is mainly governed by the stretching functions on the boundaries. Consequently the two-boundary (Sect. 13.3.2) and multisurface (Sect. 13.3.3) techniques are still able to generate smoothly varying near-orthogonal grids (Fig. 13.30) with only one-dimensional explicit interpolation.

The distribution of points along the boundary of the domain is handled effectively by defining normalized one-dimensional stretching functions along boundary segments, typically corresponding to each side of the computational rectangle in the (ξ, η) plane. A suitable one-dimensional stretching function is described in Sect. 13.3.1. Boundary stretching functions are applicable whether the interior grid is generated by solving a partial differential equation (Sect. 13.2) or by an algebraic mapping (present section).

13.3.1 One-Dimensional Stretching Functions

One-dimensional stretching functions are widely used for distributing points along a particular boundary so that specific regions of the domain can be resolved accurately. For the viscous flow around an isolated symmetric body (Fig. 13.16) it would be appropriate to introduce one-dimensional stretching functions on AE and CD so that grid points are clustered close to ABC to resolve the high gradients expected in that area.

For relatively simple geometries it may be possible to combine one-dimensional boundary stretching functions with a simple shearing transformation, e.g. (13.27), to establish the interior grid.

It is desirable to express the dependent and independent variables in the stretching function in normalised form. For a one-dimensional stretching function applied to EA in Fig. 13.16 an appropriate normalised independent variable would be

$$\eta^* = \frac{\eta - \eta_A}{\eta_E - \eta_A} \tag{13.43}$$

so that $0 \leq \eta^* \leq 1$ as $\eta_A \leq \eta \leq \eta_E$.

An effective stretching function due to Roberts (1971), and modified by Eiseman (1979), is

$$s = P\eta^* + (1 - P)\left(1 - \frac{\tanh[Q(1 - \eta^*)]}{\tanh Q}\right) , \tag{13.44}$$

where P and Q are parameters to provide grid point control. P effectively provides the slope of the distribution, $s \approx P\eta^*$, close to $\eta^* = 0$. Q is called a damping factor by Eiseman and controls the departure from the linear s versus η^* behaviour. Small values of Q cause small departures from linearity. However, if P is close to unity the departure from linearity will be small and will occur only for η^* close to unity.

Once s is obtained it is used to specify the distribution of x and y. For example, defining

$$\frac{x - x_A}{x_E - x_A} = f(s), \quad \frac{y - y_A}{y_E - y_A} = g(s) , \tag{13.45}$$

generates $x(s)$ and $y(s)$ directly. A simple choice would be $f(s) = g(s) = s$, so that (13.45) gives

$$x = x_A + s(x_E - x_A), \quad y = y_A + s(y_E - y_A) . \tag{13.46}$$

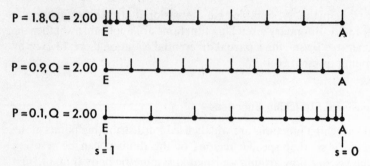

Fig. 13.19. Grid distributions using (13.44)

Typical distributions of points on EA (Fig. 13.16) using (13.46), for various values of P and Q, are shown in Fig. 13.19. For values of $P > 1.0$ it is possible to cluster points close to E. However, this requirement could be handled better by defining $\eta^* = (\eta - \eta_E)/(\eta_A - \eta_E)$ in (13.43) and $f(s) = g(s) = 1 - s$ in (13.45).

An alternative two-parameter stretching function is given by Vinokur (1983). The two parameters are the slopes $ds/d\eta^*$ at each end of the interval $\eta^* = 0$ and $\eta^* = 1.0$. These are appropriate since multiple contiguous stretching functions can be used to cover a particular boundary with continuity of s and $\partial s/\partial \eta^*$ at the interfaces. However Vinokur's stretching function cannot be reduced to a single equation like (13.44), so coding is slightly more complicated.

13.3.2 Two Boundary Technique

This method will be illustrated for a curved two-dimensional channel, Fig. 13.20. It will be assumed that stretching functions, $s_{AD}(\eta^*)$ and $s_{BC}(\eta^*)$, have been defined, to control the distribution of points on the inlet and outlet boundaries. The normalised parameter $\eta^* = (\eta - \eta_1)/(\eta_2 - \eta_1)$. Equations equivalent to (13.44) could be used to generate $s_{AD}(\eta^*)$ and $s_{BC}(\eta^*)$.

To obtain the value s between surfaces AD and BC a simple linear interpolation is recommended. Thus

$$s = s_{AD} + \xi^*(s_{BC} - s_{AD}) , \tag{13.47}$$

where $\xi^* = (\xi - \xi_1)/(\xi_2 - \xi_1)$.

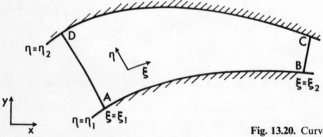

Fig. 13.20. Curved two-dimensional channel

In a similar way the distribution of grid points along AB and CD are controlled by one-dimensional stretching functions $r_{AB}(\xi^*)$ and $r_{DC}(\xi^*)$. If r_{AB} and r_{DC} are interpreted as normalised coordinates measured along the surface then $x_{AB}(r_{AB})$, $y_{AB}(r_{AB})$ follow directly; and similarly for x_{DC}, y_{DC}.

The two-boundary technique provides a means of interpolating the interior between the two boundaries, AB and DC. In so doing the interior grid is completely specified. A simple interpolation is provided by

$$x(\xi, \eta) = (1-s)x_{AB}(r_{AB}) + sx_{DC}(r_{DC}) \quad \text{and}$$

$$y(\xi, \eta) = (1-s)y_{AB}(r_{AB}) + sy_{DC}(r_{DC}) \ , \tag{13.48}$$

where s is given by (13.47). Considerable control over the clustering of the grid points in the interior can be obtained through the boundary stretching functions, s_{AD}, s_{BC}, r_{AB} and r_{DC}.

A difficulty with the interpolation given by (13.48) is that grids adjacent to the surface may become distorted if corresponding boundary points (x_{AB}, y_{AB}) and (x_{DC}, y_{DC}) are out of alignment. By replacing (13.48) with

$$x(\xi, \eta) = \mu_1(s)x_{AB}(r_{AB}) + \mu_2(s)x_{DC}(r_{DC}) + T_1\mu_3(s)\left(\frac{dy_{AB}}{dr_{AB}}(r_{AB})\right)$$

$$+ T_2\mu_4(s)\left(\frac{dy_{DC}}{dr_{DC}}(r_{DC})\right) \ , \tag{13.49}$$

$$y(\xi, \eta) = \mu_1(s)y_{AB}(r_{AB}) + \mu_2(s)y_{DC}(r_{DC}) - T_1\mu_3(s)\left(\frac{dx_{AB}}{dr_{AB}}(r_{AB})\right)$$

$$- T_2\mu_4(s)\left(\frac{dx_{DC}}{dr_{DC}}(r_{DC})\right) \ ,$$

where

$$\mu_1(s) = 2s^3 - 3s^2 + 1 \ , \qquad \mu_2(s) = -2s^3 + 3s^2 \ ,$$

$$\mu_3(s) = s^3 - 2s^2 + s \ , \qquad \mu_4(s) = s^3 - s^2 \ , \tag{13.50}$$

a grid is generated which is locally orthogonal to the boundaries AB and DC (Fig. 13.20).

The parameters T_1 and T_2 are used to control how far into the grid interior orthogonality is enforced. Choosing too large a value for T_1 and T_2 may cause a doule-valued mapping in the interior (Smith 1982).

Typical grids generated by the two-boundary technique are indicated in Figs. 13.21 and 13.22. Figure 13.21 indicates how $r_{AB}(\xi)$ and $r_{DC}(\xi)$ control the spacing in the ξ direction. Grid point concentration corresponds to a small slope; a coarser grid corresponds to a large slope of $r_{AB}(\xi)$ and $r_{DC}(\xi)$. Figure 13.22 has a more uniform distribution in the ξ direction, but has a concentration of grid lines close to the $\eta^* = 0$, due to the $s(\eta^*)$ distribution.

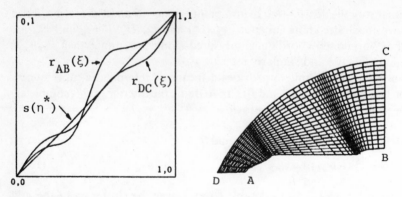

Fig. 13.21. Effect of boundary control functions r_{AB} and r_{DC}

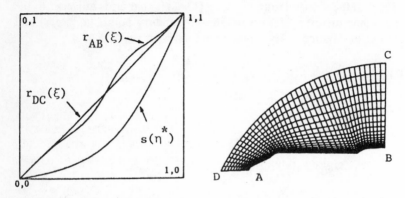

Fig. 13.22. Effect of control function $s(\eta^*)$

The extension of the two-boundary method to include interpolation between two surfaces for three-dimensional grid generation is described by Smith (1982).

13.3.3 Multisurface Method

Additional control over the interior grid distribution can be obtained if intermediate surfaces are introduced, between boundary surfaces AB and CD in Fig. 13.20, on which the $x_i(r_j)$ and $y_i(r_j)$ behaviour is specified.

By connecting corresponding points (same r_j value) on adjacent surfaces a sequence of directions are specified. In the multisurface method of Eiseman (1979) it is the sequence of directions which are interpolated. This provides two direct advantages.

First by adjusting the grid point correspondence between a bounding surface, say AB, and its neighbouring intermediate surface it is possible to make the grid locally orthogonal at the boundary.

Second, the grid distribution in s is obtained by integrating the interpolation of the sequence of directions. This provides a very smooth s distribution and, incidentally, does not require that the interior grid interpolates the intermediate surfaces. In principle there is no limit to the number of intermediate surfaces. In practice good control over the interior grid can be obtained with two intermediate surfaces.

The distribution of points on the ith surface will be combined into a single vector function $\mathbf{Z}_i(r)$ with components $x_i(r)$ and $y_i(r)$, in two dimensions. A sequence of surfaces is shown in Fig. 13.23. In general there will be $N-2$ intermediate surfaces.

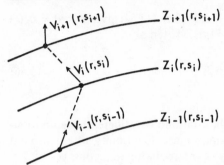

Fig. 13.23. Intermediate surfaces $\mathbf{Z}_i$ and tangent vectors, $\mathbf{V}_i$

The parameter r defines the location in all the surfaces simultaneously. However, different choices for the functions $\mathbf{Z}_i(r)$ will allow adjustment of the relative orientation of (x_i, y_i) on each surface for a single choice of r. It is assumed that corresponding points (x_i, y_i) on each surface, associated with a particular value of $r = r_j$, are joined by straight lines between surfaces (Fig. 13.23).

The tangents to these straight lines between the surfaces define a family of vector functions, $\mathbf{V}_i(r)$ for $i = 1, \ldots, N-1$. The tangent vector functions $\mathbf{V}_i$ can be connected to the surface vector functions $\mathbf{Z}_i$ by defining

$$\mathbf{V}_i(r) = A_i[\mathbf{Z}_{i+1}(r) - \mathbf{Z}_i(r)] , \quad i = 1, \ldots, N-1 . \tag{13.51}$$

The parameters A_i will be determined later so that the final grid interpolation fits properly into the interval $0 \leq s \leq 1$. An interpolation through the family of semi-discrete tangent vector functions $\mathbf{V}_i(r)$ provides a tangent vector function $\mathbf{V}(r, s)$ that is continuous in both r and s. Thus

$$\mathbf{V}(r,s) = \sum_{i=1}^{N-1} \psi_i(s) \mathbf{V}_i(r) , \tag{13.52}$$

where $\psi_i(s)$ are interpolating functions, to be determined, such that $\psi_i(s_k) = 1$ if $i = k$ and is zero if $i \neq k$. However, from the way that the $\mathbf{V}_i(r)$ have been constructed, it is clear that

$$\frac{\partial \mathbf{Z}}{\partial s}(r, s) = \mathbf{V}(r, s) = \sum_{i=1}^{N-1} \psi_i(s) \mathbf{V}_i(r) , \tag{13.53}$$

where $\mathbf{Z}(r, s)$ is the continuous function that will generate the grid in the physical plane for given values of r and s (and hence ξ and η). $\mathbf{Z}(r, s)$ is obtained by integrating (13.53) over the interval $0 \leq s \leq 1$. This interval corresponds to $\eta_1 \leq \eta \leq \eta_2$ in Fig. 13.20. Thus, with the aid of (13.51),

$$\mathbf{Z}(r, s) = \mathbf{Z}_1(r) + \sum_{i=1}^{N-1} A_i G_i(s)[\mathbf{Z}_{i+1}(r) - \mathbf{Z}_i(r)] \ , \tag{13.54}$$

where

$$G_i(s) = \int_0^s \psi_i(s')ds' \ . \tag{13.55}$$

The parameters A_i are chosen so that $A_i G_i(1) = 1$. Then (13.54) gives $\mathbf{Z}(r, s) = \mathbf{Z}_N(r)$ when $s = 1$, as required. Consequently (13.54) is written as

$$\mathbf{Z}(r, s) = \mathbf{Z}_1(r) + \sum_{i=1}^{N-1} \frac{G_i(s)}{G_i(1)}[\mathbf{Z}_{i+1}(r) - \mathbf{Z}_i(r)] \ , \tag{13.56}$$

which is the general multisurface transformation.

The interpolating functions $\psi_i(s)$ must be continuously differentiable up to an order that is one less than the level of smoothness (continuity of derivatives) required in the grid. An appropriate family of interpolating functions, ψ_i, is

$$\psi_i(s) = \prod_{\substack{l=1 \\ l \neq i}}^{N-1} (s - s_l) \ . \tag{13.57}$$

The simplest case within the present structure is $N = 2$. For this case (13.56) becomes

$$\mathbf{Z}(r, s) = \mathbf{Z}_1(r) + s[\mathbf{Z}_2(r) - \mathbf{Z}_1(r)] \ . \tag{13.58}$$

There are no intermediate surfaces and the mapping (13.58) is equivalent to the linear two-boundary formula (13.48). A typical grid, for $N = 2$, is shown in Fig. 13.24a.

For the case $N = 3$ one intermediate surface is introduced and (13.56), based on (13.57), takes the form

$$\mathbf{Z}(r, s) = (1 - s)^2 \mathbf{Z}_1(r) + 2s(1 - s)\mathbf{Z}_2(r) + s^2 \mathbf{Z}_3(r) \ . \tag{13.59}$$

A typical grid generated by this mapping is shown in Fig. 13.24b. The ellipse in Fig. 13.24 has axis lengths 1.00 and 0.25. The outer rectangle with smooth corners is of length 8 in the x direction and 4.8 in the y direction. For the $N = 2$ case it is apparent that interior grid points are distributed along straight lines in the $s(\eta)$ direction. For the $N = 3$ case an intermediate surface (surface 2) is located at a distance $\Delta s = 0.1$ from the surface of the ellipse (surface 1). Points on surface 2 are adjusted so that a line joining points on surfaces 1 and 2 with the same r value is perpendicular to the surface of the ellipse. This has the effect of producing a grid which is locally orthogonal to the surface of the ellipse.

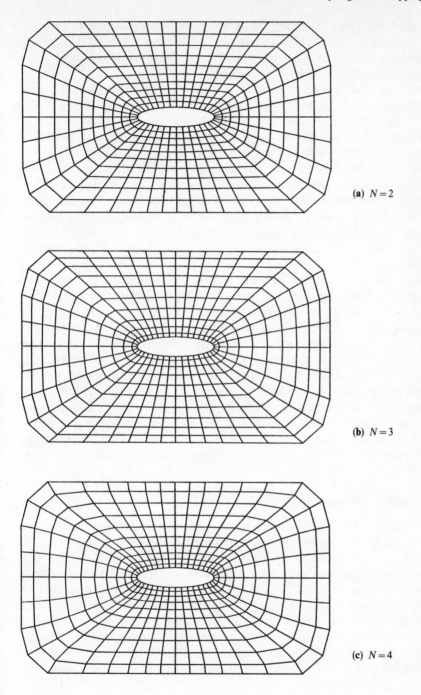

(a) $N = 2$

(b) $N = 3$

(c) $N = 4$

Fig. 13.24. Typical grid generated by the multi-surface method; (a) $N = 2$, (b) $N = 3$, (c) $N = 4$

The use of N surfaces (counting the boundary surfaces) provides N degrees of freedom. Two of the degrees of freedom are utilised in requiring the η grid to match the bounding surface specifications $\mathbf{Z}_1(r)$ and $\mathbf{Z}_N(r)$. Other degrees of freedom can be used to control the character of the interior grid.

For example, with $N=4$ (two interior surfaces) it is possible to generate a grid that is orthogonal to both bounding surfaces at $\eta=\eta_1$ and $\eta=\eta_2$. In Fig. 13.24c surface 3 is located at a distance $\Delta s=0.1$ inside the outer rectangle (surface 4). Points on surface 3 are adjusted to generate a grid locally orthogonal to the surface of the rectangle. The $N=4$ case is discussed at greater length in Sect. 13.4.1.

It should be emphasized that the interior surfaces are introduced to control the interior grid distribution and shape and do not coincide with the final grid location. The introduction of interior surfaces gives a level of control that is additional to that obtained by distributing points in the ξ direction on the bounding surfaces (Sect. 13.3.1), through the choice $r_{AB}(\xi)$ and $r_{DC}(\xi)$, as in Figs. 13.21 and 13.22. The ability to distribute points in the s direction, as in (13.47) is also available with the multisurface method.

Eiseman (1982a, b) has extended the multisurface method to make it more effective in three dimensions. In three dimensions better control over the interior grid distribution can be obtained if the interpolating functions, ψ_i in (13.57), are given a local rather than a global interpretation. Eiseman (1982a) provides examples where the interior mapping can be made locally cartesian (in physical space) away from $\eta=\eta_1$ in Fig. 13.20, whilst still retaining the boundary-fitting capability of having a grid which is locally orthogonal at the surfaces AB and DC.

In three dimensions the bounding surfaces, e.g. $\mathbf{Z}_1$ and $\mathbf{Z}_n$, will depend on two parameters r and t and do not need to be planar, e.g. $\mathbf{Z}_1$ might coincide with the surface of an automobile. This requires that the coordinate system is at least C^2. That is, continuous derivatives up to order 2 exist. This implies that the interpolating functions, ψ_i in (13.57), must be C^1 functions. Eiseman (1982b) discusses the use of such local functions in the multisurface method.

13.3.4 Transfinite Interpolation

Both the two-boundary and the multi-surface techniques interpolate in only one direction (s or η) and assume that a continuous mapping is available on the bounding surfaces $\eta=\eta_1$ and $\eta=\eta_2$ in the other (r or ξ) direction.

However, with transfinite interpolation (Gordon and Hall 1973) it is possible to specify continuous mappings $\mathbf{Z}_{AB}(\xi,\eta_1)$ on AB, $\mathbf{Z}_{DC}(\xi,\eta_2)$ on DC and in addition, $\mathbf{Z}_{AD}(\xi_1,\eta)$ on AD and $\mathbf{Z}_{BC}(\xi_2,\eta)$ on BC (Fig. 13.20). In the interior an interpolation in both ξ and η, or equivalently r and s, is introduced.

As in the two-boundary and multisurface methods the parametric coordinates (r, s) are introduced as an intermediate step in obtaining the transformation $\mathbf{Z}(\xi,\eta)$ to construct the grid in the physical plane. It is assumed that r and s are normalised coordinates, i.e.

$$0 \le r \le 1 \quad \text{as} \quad \xi_1 \le \xi \le \xi_2 \, ,$$
$$0 \le s \le 1 \quad \text{as} \quad \eta_1 \le \eta \le \eta_2 \, . \tag{13.60}$$

The following blending (interpolating) functions are defined:

$$\phi_j(r)=\delta_{jr} , \quad j=0,1 \quad \text{and} \quad \psi_k(s)=\delta_{ks} , \quad k=0,1 , \tag{13.61}$$

where

$$\delta_{jr}=1 \quad \text{if } j=r \quad \text{and} \quad \delta_{ks}=1 \text{ if } k=s$$
$$=0 \quad \text{if } j\neq r \qquad\qquad =0 \text{ if } k\neq s .$$

Thus $\phi_0=1$, $\phi_1=0$ on AD; $\phi_0=0$, $\phi_1=1$ on BC, $\psi_0=1$, $\psi_1=0$ on AB and $\psi_0=0$, $\psi_1=1$ on CD.

An interpolation in the r direction would be

$$\mathbf{Z}_r(r,s)=\phi_0(r)\mathbf{Z}_{AD}(0,s)+\phi_1(r)\mathbf{Z}_{BC}(1,s) , \tag{13.62}$$

where $\mathbf{Z}_{AD}$, $\mathbf{Z}_{BC}$ are the continuous mappings between the (ξ,η) and (x,y) planes on the two boundaries $\xi=\xi_1$ and $\xi=\xi_2$. $\mathbf{Z}_r(r,s)$ is the continuous mapping produced by interpolating between $\mathbf{Z}_{AD}$ and $\mathbf{Z}_{BC}$ for intermediate values of r.

In a similar way

$$\mathbf{Z}_s(r,s)=\psi_0(s)\mathbf{Z}_{AB}(r,0)+\psi_1(s)\mathbf{Z}_{CD}(r,1) . \tag{13.63}$$

$\mathbf{Z}_r$ and $\mathbf{Z}_s$ are equivalent mappings to those used in the two-boundary and simplest ($N=2$) multisurface methods.

To obtain two-dimensional interpolation a product interpolation can be defined as

$$\mathbf{Z}_{rs}(r,s)=\mathbf{Z}_r\cdot\mathbf{Z}_s . \tag{13.64}$$

The product interpolation agrees with the boundary functions, $\mathbf{Z}_{AB}$, etc., only at the four corners $(0,0)$, $(0,1)$, $(1,0)$ and $(1,1)$. This is the type of interpolation used in two-dimensional finite element methods (Sect. 5.3.3).

To achieve exact matching with the mapping functions everywhere on the boundaries of a two-dimensional domain it is necessary to define a Boolean sum interpolation,

$$\mathbf{Z}(r,s)=\mathbf{Z}_r(r,s)+\mathbf{Z}_s(r,s)-\mathbf{Z}_{rs}(r,s) . \tag{13.65}$$

This construction is central to transfinite interpolation (Gordon and Hall 1973).

In practice (13.65) is implemented in two stages. In the first stage

$$\mathbf{Z}_r(r,s) = \sum_{j=0}^{1} \phi_j(r)\mathbf{Z}_b(j,s) , \tag{13.66}$$

where b indicates the appropriate boundary, AD or BC. In the second stage

$$\mathbf{Z}(r,s) = \mathbf{Z}_r(r,s)+ \sum_{k=0}^{1} \psi_k(s)[\mathbf{Z}_b(r,k)-\mathbf{Z}_r(r,k)] . \tag{13.67}$$

The blending functions ϕ_j and ψ_k can be chosen in much the same way as in the two-boundary or multisurface methods. The choice

$$\phi_0(r)=1-r \ , \quad \phi_1(r)=r \ , \quad \psi_0(s)=1-s \ , \quad \psi_1(s)=s \tag{13.68}$$

produces a transfinite bilinear interpolation (13.67), which suffers from the same problem as (13.48). Although they can cluster points through the form of the boundary functions Z_{AB}, etc., (13.67 and 68) cannot make the grid orthogonal adjacent to boundary surfaces.

The extension of the transfinite interpolation method can follow paths parallel to either the two-surface or multisurface methods. Gordon and Thiel (1982) discuss the introduction of interior surfaces to obtain better control of the interior grid. Eriksson (1982) specifies parametric derivatives, e.g. $\partial^n Z/\partial s^n$, to provide near-orthogonal smoothly varying grids. Eriksson (1982) prefers to specify derivatives up to $n=3$ rather than formally enforcing orthogonality as in (13.49). Eriksson claims that this provides more precise control over the grid distribution, particularly in three dimensions.

The transfinite interpolation method extends naturally to three dimensions. The implementation algorithm (13.66, 67) then has a third stage

$$Z(r, s, t) = Z_2(r, s, t) + \sum_{i=0}^{1} \omega_i(t)[Z_b(r, s, i) - Z_2(r, s, i)] \ , \tag{13.69}$$

where $Z_2(r, s, t)$ is equivalent to $Z(r, s)$ in (13.67) and $\omega_i(t)$ are blending functions with equivalent properties to $\phi_j(r)$ and $\psi_k(s)$. Rizzi and Eriksson (1981) describe the application of three-dimensional transfinite interpolation to the problem of wing/body combinations. Rizzi and Eriksson define data only on boundary surfaces. To obtain more control of the interior grid they specify both the mapping Z and out-of-surface derivatives $\partial^n Z/\partial \xi^n$, etc., on the wing surface, with $n=1, 3$. In the spanwise direction a simple linear interpolation, equivalent to (13.68), is used. In the coordinate direction wrapping around the wing, equivalent to ξ in Fig. 13.12, no interpolation is introduced. The grid distribution in this direction is controlled by the boundary specifications. The transfinite interpolation concept is sufficiently flexible that different orders of interpolation can be introduced in the different parametric coordinates r, s and t.

13.4 Numerical Implementation of Algebraic Mapping

In this section some of the techniques discussed previously will be combined to produce a computer program capable of generating a grid between two bounding curves.

For the domain shown in Fig. 13.25 the bounding surfaces consist of a symmetric slender body extended downstream, ABC, and a farfield boundary, FED. Between these two boundaries half of a C-grid (Fig. 13.12) is to be generated.

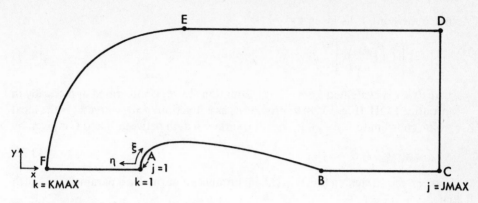

Fig. 13.25. Computational domain for algebraic mapping

Because of the inherent symmetry the complete C-grid can be obtained by reflection about the x-axis.

As indicated at the beginning of Chap. 13, grid generation is split into two parts. First grid point locations on all boundaries are determined. The one-dimensional stretching function (13.44) is used to control the distribution on the boundaries.

Subsequently the interior grid is generated by the multisurface technique. Two intermediate surfaces, Z_2, and Z_3, are introduced, one each adjacent to the bounding surfaces ABC and FED. The parametric (r) correspondence of surfaces Z_2 and Z_3 to their neighbouring bounding surface is adjusted so that grid lines intersect the bounding surfaces orthogonally. The mechanism of choosing $x(r)$, $y(r)$ on surfaces Z_2 and Z_3 requires an orthogonal projection, conceptually similar to the near-orthogonal grid construction discussed in Sect. 13.2.5.

13.4.1 ALGEM: Grid Generation for a Streamlined Body

The surface AB in Fig. 13.25 represents an aerofoil profile of the NACA–$00't'$ family. The t indicates the percentage thickness as a two digit number. Thus a NACA-0012 aerofoil would imply a symmetric aerofoil with a 12% thickness based on a unit chord. The profile of the NACA–$00't'$ family is given by

$$y = t(a_1 x^{1/2} + a_2 x + a_3 x^2 + a_4 x^3 + a_5 x^4) , \qquad (13.70)$$

where

$$a_1 = 1.4779155 , \quad a_2 = -0.624424 , \quad a_3 = -1.727016 , \quad a_4 = 1.384087 ,$$

$a_5 = -0.510563$ and t is the aerofoil thickness.

It is assumed that $0 \le \xi \le 1$ over the physical surface ABC and that incrementing j increments ξ by $\Delta\xi$.

To relate the physical coordinates (x, y) of surface ABC to the computational coordinate ξ it is necessary to introduce a surface measure, r. On the surface AB the surface measure, r_A is given by

$$r_A = \int_0^{x_A} \left[1 + \left(\frac{dy}{dx} \right)^2 \right]^{1/2} dx \ , \tag{13.71}$$

where dy/dx is evaluated from (13.70). Equation (13.71) is integrated numerically in subroutine FOIL (Fig. 13.28) to provide r_A as a function of x_A, where x_A is the local aerofoil coordinate, $0 \leq x_A \leq 1$. The total surface length between A and C is given by

$$r_{AC,\text{max}} = r_{AB} + x_C - x_B \ . \tag{13.72}$$

However, the stretching function (13.44) produces a normalised parameter r_{AC}^n such that

$$0 \leq r_{AC}^n \leq 1 \quad \text{as} \quad 0 \leq \xi \leq 1 \ . \tag{13.73}$$

Consequently the physical surface coordinate $r_{AC,D}$ is obtained as

$$r_{AC,D} = r_{AC}^n \, r_{AC,\text{max}} \ , \tag{13.74}$$

where r_{AC}^n is evaluated in subroutine STRECH (Fig. 13.27) and $r_{AC,\text{max}}$ is given by (13.72). For $r_{AC,D} < r_{AB}$ the physical coordinates $(x(j), y(j))$ of the bounding surface segment AB are obtained by interpolating r_A and x_A and using (13.70) to give the corresponding y. This is carried out in subroutine FOIL when INT $= 1$. For $r_{AC,D} > r_{AB}$ the surface coordinates follow from a linear interpolation of $x_B \leq x \leq x_C$.

The physical coordinates of ABC are denoted by $XS(1, J)$, $YS(1, J)$ in the program ALGEM (Fig. 13.26), since they are the x and y components of surface $\mathbf{Z}_1$ in Sect. 13.3.3.

In a similar manner the bounding surface $XS(4, J)$, $YS(4, J)$, corresponding to boundary FED, is generated by interpolating a normalised surface coordinate r_{FD}^n to give the surface coordinate $r_{FD,D}$. The surface coordinate $r_{FD,D}$ is measured along the circular arc FE and the straight line ED.

The grid point locations on AF and CD (Fig. 13.25) depend on the stretching parameters s_{AF} and s_{CD}. However, the physical coordinates of the grid points on AF and CD are obtained explicitly as part of the multisurface algorithm (13.76).

The multisurface method (Sect. 13.3.3) is implemented here with four surfaces. The bounding surfaces ABC and FED constitute surfaces $\mathbf{Z}_1$ and $\mathbf{Z}_4$. Initially surfaces $\mathbf{Z}_2$ and $\mathbf{Z}_3$ are constructed by linearly interpolating surfaces $\mathbf{Z}_1$ and $\mathbf{Z}_4$ by

$$\mathbf{Z}_2^i = \mathbf{Z}_1 + s_2(\mathbf{Z}_4 - \mathbf{Z}_1) \quad \text{and} \quad \mathbf{Z}_3^i = \mathbf{Z}_1 + s_3(\mathbf{Z}_4 - \mathbf{Z}_1) \ . \tag{13.75}$$

At this stage surfaces $\mathbf{Z}_2$ and $\mathbf{Z}_3$ are in the correct physical location but do not have the correct r correspondence. If the multisurface algorithm (13.76) were implemented with $\mathbf{Z}_2^i$ and $\mathbf{Z}_3^i$, given by (13.75), the resulting η grid lines would be straight lines joining locations with the same j index on ABC and FED (as in Fig. 13.24, $N = 2$).

Therefore the dependence of $\mathbf{Z}_2$ and $\mathbf{Z}_3$ on r must be altered so that lines through points on $\mathbf{Z}_2$ and corresponding points (same r) on $\mathbf{Z}_1$, are orthogonal to $\mathbf{Z}_1$, and lines connecting $\mathbf{Z}_3$ and $\mathbf{Z}_4$ are orthogonal to $\mathbf{Z}_4$. The nature of the

```
 1 C
 2 C       ALGEM APPLIES A MODIFIED MULTI(4)-SURFACE TECHNIQUE TO
 3 C       THE GENERATION OF A GRID ABOUT A NACA-00'T' AEROFOIL
 4 C       AT ZERO INCIDENCE. THE UPPER HALF GRID IS GENERATED
 5 C
 6         DIMENSION X(51,51),Y(51,51),XB(6),YB(6),XS(4,51),YS(4,51)
 7         DIMENSION RAC(51),RFD(51),SAF(51),SCD(51),SH(4),XA(51),RA(51)
 8         COMMON RA,XA
 9 C
10         OPEN(1,FILE='ALGEM.DAT')
11         OPEN(6,FILE='ALGEM.OUT')
12         READ(1,1)JMAX,KMAX,IPR,IRFL,T,S2,S3,AW
13       1 FORMAT(4I5,4E10.3)
14         READ(1,2)PAC,QAC,PFD,QFD,PAF,QAF,PCD,QCD
15       2 FORMAT(8E10.3)
16         WRITE(6,3)
17         WRITE(6,4)JMAX,KMAX,IPR,T,S2,S3,AW
18       3 FORMAT(' MULTISURFACE GRID GENERATION')
19       4 FORMAT(' JMAX KMAX=',2I3,'  IPR=',I2,
20      1'       T=',E10.3,5X,' S2,S3=',2E10.3,'  AW=',F6.3,//)
21         WRITE(6,5)PAC,QAC,PFD,QFD
22         WRITE(6,6)PAF,QAF,PCD,QCD
23       5 FORMAT(' PAC=',E10.3,' QAC=',E10.3,' PFD=',E10.3,' QFD=',E10.3)
24       6 FORMAT(' PAF=',E10.3,' QAF=',E10.3,' PCD=',E10.3,' QCD=',E10.3)
25 C
26 C       DEFINE CORNER POINTS OF THE BOUNDARY
27 C
28         DATA XB/2.00,3.00,5.00,5.00,2.25,0.00/
29         DATA YB/0.00,0.00,0.00,2.25,2.25,0.00/
30         PI = 3.14159265
31 C
32 C       GENERATE STRETCHING FUNCTIONS
33 C
34         CALL STRECH(JMAX,PAC,QAC,RAC)
35 C
36         CALL STRECH(JMAX,PFD,QFD,RFD)
37 C
38         CALL STRECH(KMAX,PAF,QAF,SAF)
39 C
40         CALL STRECH(KMAX,PCD,QCD,SCD)
41 C
42         WRITE(6,7)(RAC(J),J=1,JMAX)
43         WRITE(6,8)(RFD(J),J=1,JMAX)
44         WRITE(6,9)(SAF(J),J=1,JMAX)
45         WRITE(6,10)(SCD(J),J=1,JMAX)
46       7 FORMAT(' RAC=',18F7.4)
47       8 FORMAT(' RFD=',18F7.4)
48       9 FORMAT(' SAF=',18F7.4)
49      10 FORMAT(' SCD=',18F7.4,/)
50 C
51 C       OBTAIN SURFACE COORDINATES OF BODY, AB
52 C
53         CALL FOIL(0,T,RAB,XD,YD)
54 C
```

Fig. 13.26. Listing of program ALGEM

multisurface algorithm (Sect. 13.3.3) then ensures the grid adjacent to surfaces Z_1 and Z_4 is orthogonal.

Since $Z_1(r)$ is known it is sufficient to take each grid point, $XS(1, J)$, $YS(1, J)$, and to project a straight line normal to Z_1 through $XS(1, J)$, $YS(1, J)$ until it intersects Z_2. The intersection point becomes $XS(2, J)$, $YS(2, J)$. This construction is equivalent to the predictor stage shown in Fig. 13.17. Since a given node J fixes r on Z_1, point $XS(2, J)$, $YS(2, J)$ also has the same r value and is orthogonal to Z_1 at $XS(1, J)$, $YS(1, J)$. The orthogonal adjustment of the grid points on Z_2 is carried out in subroutine SURCH (Fig. 13.29). The grid points on surface Z_3, $XS(3, J)$ and

```
55          RACMX = RAB + XB(3) - XB(2)
56          RFE = 0.5*PI*(YB(5) - YB(1))
57          RFDMX = RFE + XB(4) - XB(5)
58          IF(IPR .EQ. 0)GOTO 13
59          WRITE(6,11)(XA(L),L=1,51)
60          WRITE(6,12)(RA(L),L=1,51)
61       11 FORMAT(' XA=',18F7.4)
62       12 FORMAT(' RA=',18F7.4)
63    C
64    C     GENERATE BOUNDING SURFACES 1 AND 4
65    C
66       13 DO 17 J = 1,JMAX
67          RACD = RAC(J)*RACMX
68          IF(RACD .LT. RAB)GOTO 14
69          XS(1,J) = XB(2) + (RACD - RAB)*(XB(3) - XB(2))/(RACMX-RAB)
70          YS(1,J) = 0.
71          GOTO 15
72    C
73       14 CALL FOIL(1,T,RACD,XD,YD)
74    C
75          XS(1,J) = XD*(XB(2)-XB(1)) + XB(1)
76          YS(1,J) = YD*(XB(2)-XB(1)) + YB(1)
77       15 RFDD = RFD(J)*RFDMX
78          IF(RFDD .LT. RFE)GOTO 16
79          XS(4,J) = XB(5) + (RFDD-RFE)*(XB(4)-XB(5))/(RFDMX-RFE)
80          YS(4,J) = YB(5)
81          GOTO 17
82       16 RR = YB(5) - YB(1)
83          THE = RFDD/RR
84          XS(4,J) = XB(5) - RR*COS(THE)
85          YS(4,J) = RR*SIN(THE)
86       17 CONTINUE
87    C
88    C     SURCH GENERATES SURFACES 2 AND 3 SO THAT THE GRID
89    C     ADJACENT TO SURFACES 1 AND 4 IS ORTHOGONAL
90    C
91          CALL SURCH(JMAX,S2,S3,XS,YS)
92    C
93          DO 21 L = 1,4
94          WRITE(6,18)L
95       18 FORMAT(10H SURFACE   ,I1)
96          DO 21 J = 1,JMAX,18
97          JA = J
98          JB = JA + 17
99          WRITE(6,19)(XS(L,JC),JC=JA,JB)
100         WRITE(6,20)(YS(L,JC),JC=JA,JB)
101      19 FORMAT(' XS=',18F7.4)
102      20 FORMAT(' YS=',18F7.4,/)
103      21 CONTINUE
104   C
105   C     GENERATE INTERIOR GRID
106   C
```

Fig. 13.26. (cont.) Listing of program ALGEM

$YS(3, J)$, are adjusted to be orthogonal to surface $\mathbf{Z}_4$ at points $XS(4, J)$ $YS(4, J)$. This is also carried out in subroutine SURCH.

Given the coordinates of the surfaces, $\mathbf{Z}_1$ to $\mathbf{Z}_4$, the multisurface algorithm for four surfaces (Eiseman 1979) is implemented as

$$x(J, K) = \sum_{L=1}^{4} SH(L) XS(L, J) , \qquad y(J, K) = \sum_{L=1}^{4} SH(L) YS(L, J) , \qquad (13.76)$$

where

$$SH(1) = (1-s)^2(1-a_1 s) ,$$
$$SH(2) = s(1-s)^2(a_1+2) , \qquad (13.77)$$

```
107        A1 = 2./(3.*AW-1.)
108        A2 = 2./(3.*(1.-AW) - 1.)
109        AJM = JMAX - 1
110        DZI = 1./AJM
111        DO 24 K = 1,KMAX
112        DO 23 J = 1,JMAX
113        AJ = J - 1
114        ZI = AJ*DZI
115        S = SAF(K) + ZI*(SCD(K)-SAF(K))
116        SH(1) = (1.-S)**2*(1.-A1*S)
117        SH(2) = (1.-S)**2*S*(A1+2.)
118        SH(3) = (1.-S)*S*S*(A2+2.)
119        SH(4) = S*S*(1.-A2*(1.-S))
120        X(J,K) = 0.
121        Y(J,K) = 0.
122        DO 22 L = 1,4
123        X(J,K) = X(J,K) + SH(L)*XS(L,J)
124        Y(J,K) = Y(J,K) + SH(L)*YS(L,J)
125     22 CONTINUE
126     23 CONTINUE
127     24 CONTINUE
128 C
129 C     REFLECT GRID ABOUT X-AXIS
130 C
131        IF(IRFL .EQ. 0)GOTO 28
132        JMAP = JMAX - 1
133        DO 27 K = 1,KMAX
134        DO 25 J = 1,JMAX
135        JA = 2*JMAX - J
136        JB = JA - JMAP
137        X(JA,K) = X(JB,K)
138        Y(JA,K) = Y(JB,K)
139     25 CONTINUE
140        DO 26 J = 1,JMAP
141        JA = 2*JMAX - J
142        X(J,K) = X(JA,K)
143        Y(J,K) = -Y(JA,K)
144     26 CONTINUE
145     27 CONTINUE
146        JMAX = JMAX + JMAP
147 C
148     28 DO 33 K = 1,KMAX
149        WRITE(6,29)K,SAF(K),SCD(K)
150     29 FORMAT(' K=',I3,5X,' SAF=',E10.3,' SCD=',E10.3)
151        DO 32 J = 1,JMAX,18
152        JA = J
153        JB = JA + 17
154        WRITE(6,30)(X(JC,K),JC=JA,JB)
155        WRITE(6,31)(Y(JC,K),JC=JA,JB)
156     30 FORMAT(' X=',18F7.4)
157     31 FORMAT(' Y=',18F7.4,/)
158     32 CONTINUE
159     33 CONTINUE
160        STOP
161        END
```

Fig. 13.26. (cont.) Listing of program ALGEM

$$SH(3) = s^2(1-s)(a_2+2) \ ,$$

$$SH(4) = s^2(1-a_2(1-s)) \ , \quad \text{and}$$

$$a_1 = 2/(3a_w - 1) \quad \text{and} \quad a_2 = 2/(2 - 3a_w) \ . \tag{13.78}$$

In (13.77 and 78) s is a normalised parameter in the η direction. To provide more control s is linearly interpolated in the ξ direction by

$$s = s_{AF}(K) + \xi(J) \left[s_{CD}(K) - s_{AF}(K) \right] \ . \tag{13.79}$$

```
1         SUBROUTINE STRECH(N,P,Q,S)
2  C
3  C      COMPUTES ONE-DIMENSIONAL STRETCHING FUNCTION,
4  C      S = P*ETA + (1.-P)*(1.-TANH(Q*(1.-ETA))/TANH(Q)),
5  C      FOR GIVEN CONTROL PARAMETERS, P AND Q.
6  C
7         DIMENSION S(51)
8         AN = N-1
9         DETA = 1./AN
10        TQI = 1./TANH(Q)
11 C
12        DO 1 L = 1,N
13        AL = L - 1
14        ETA = AL*DETA
15        DUM = Q*(1. - ETA)
16        DUM = 1. - TANH(DUM)*TQI
17        S(L) = P*ETA + (1.-P)*DUM
18      1 CONTINUE
19        RETURN
20        END
```

Fig. 13.27. Listing of subroutine STRECH

Table 13.1. Parameters used in program ALGEM

Parameter	Description
$JMAX$, $KMAX$	number of points in the ξ and η directions
IRFL	.GT.0, reflect grid about the x-axis
T	aerofoil thickness
S2, S3	preliminary interpolation parameters for surfaces Z_2, Z_3, (13.75)
AW	interior grid uniformity parameter a_w, (13.78)
PAC, QAC	stretching control parameters for AC (similarly for FD, AF and CD)
RAC	r_{AC}^n
RACMX	$r_{AC, MAX}$ in (13.72)
RACD	$r_{AC, D}$
XA, RA	aerofoil axial and surface coordinates x_A and r_A, (13.71)
XD, YD	interpolated aerofoil coordinates returned by FOIL
XB, YB	boundary corner points A, B, C, D, E and F, Fig. 13.25
XS, YS	multisurface coordinates
X, Y	grid points to be generated
S	interpolating parameter s, (13.77)
SH	weighting functions, (13.76)
EM1→EM4	tangents to surfaces 1→4 (SURCH)
XS2, YS2	surface 2 coordinates after orthogonalisation (SURCH)
XS3, YS3	surface 3 coordinates after orthogonalisation (SURCH)

The present grid generation scheme is coded as the program ALGEM (Fig. 13.26) and subroutines FOIL (Fig. 13.27), STRECH (Fig. 13.28) and SURCH (Fig. 13.29). The various parameters used in the program are described in Table 13.1. A typical grid for a NACA-0018 aerofoil is shown in Fig. 13.30. The ability of the method to cluster points and to generate an orthogonal grid at the boundaries is clearly evident. This example has been constructed to illustrate the method and would not necessarily be suitable for a computational solution.

```
1            SUBROUTINE FOIL(INT,T,RAB,X,Y)
2   C
3   C    SURFACE PROFILE IS NACA-00'T' AEROFOIL
4   C    IF INT = 0,   NUMERICALLY INTEGRATE TO OBTAIN SURFACE
5   C    COORDINATE
6   C    IF INT = 1,   INTERPOLATE SURFACE COORDINATES TO OBTAIN
7   C    CORRESPONDING (X,Y)
8   C
9            DIMENSION A(5),XA(51),RA(51)
10           COMMON RA,XA
11           DATA A/1.4779155,-0.624424,-1.727016,1.384087,-0.510563/
12           PI = 3.14159265
13           IF(INT .EQ. 1)GOTO 2
14  C
15  C    NUMERICALLY INTEGRATE TO OBTAIN RA(L) AS A FUNCTION OF XA(L)
16  C
17           RA(1) = 0.
18           XA(1) = 0.
19           XA(2) = (A(1)/(1./T - A(2)))**2
20           RA(2) = 0.5*PI*XA(2)
21           DUM = 3.*A(4) + 4.*A(5)*XA(2)
22           DUM = 2.*A(3) + DUM*XA(2)
23           DUM = T*(0.5/SQRT(XA(2)) + A(2) + DUM*XA(2))
24           FLP = SQRT(1. + DUM*DUM)
25           DO 1 L = 2,50
26           AL = L
27           LP = L + 1
28           XA(LP) = 0.02*AL
29           DX = XA(LP) - XA(L)
30           FL = FLP
31           DUM = 3.*A(4) + 4.*A(5)*XA(LP)
32           DUM = 2.*A(3) + DUM*XA(LP)
33           DUM = T*(0.5/SQRT(XA(LP)) + A(2) + DUM*XA(LP))
34           FLP = SQRT(1. + DUM*DUM)
35           RA(LP) = RA(L) + 0.5*(FL + FLP)*DX
36        1 CONTINUE
37           RAB = RA(51)
38           RETURN
39  C
40  C    INTERPOLATE RA(L) TO OBTAIN X CORRESPONDING TO RAB
41  C    SUBSEQUENTLY OBTAIN Y FROM ANALYTIC NACA-00'T' PROFILE
42  C
43        2 DO 3 L = 2,51
44           IF(RAB .GT. RA(L))GOTO 3
45           LM = L - 1
46           X = XA(LM) + (XA(L)-XA(LM))*(RAB-RA(LM))/(RA(L)-RA(LM))
47           IF(X .LT. 1.0E-06)X=1.0E-06
48           DUM = A(4) + A(5)*X
49           DUM = A(3) + DUM*X
50           DUM = A(2) + DUM*X
51           Y = T*(A(1)*SQRT(X) + DUM*X)
52           RETURN
53        3 CONTINUE
54           WRITE(6,4)RAB,RA(1),RA(51)
55        4 FORMAT(' RAB OUTSIDE RANGE',5X,' RAB=',E10.3,
56          1' RA(1)=',E10.3,' RA(51)=',E10.3)
57           RETURN
58           END
```

Fig. 13.28. Listing of subroutine FOIL

```
1          SUBROUTINE SURCH(JMAX,S2,S3,XS,YS)
2 C
3 C        GENERATES SURFACES 2 AND 3 TO CREATE ORTHOGONAL
4 C        BOUNDARY GRIDS
5 C
6          DIMENSION XS2(51),YS2(51),XS3(51),YS3(51),XS(4,51),YS(4,51)
7          JMAP = JMAX - 1
8 C
9 C        PRELIMINARY GENERATION OF SURFACES 2 AND 3
10 C
11         DO 1 J = 1,JMAX
12         DXS = XS(4,J) - XS(1,J)
13         DYS = YS(4,J) - YS(1,J)
14         XS(2,J) = XS(1,J) + S2*DXS
15         YS(2,J) = YS(1,J) + S2*DYS
16         XS(3,J) = XS(1,J) + S3*DXS
17         YS(3,J) = YS(1,J) + S3*DYS
18       1 CONTINUE
19 C
20 C        PROJECT ORTHOGONALLY FROM SURFACE 1 ONTO SURFACE 2
21 C
22         DO 9 J = 2,JMAP
23         IF(ABS(XS(1,J+1)-XS(1,J-1)) .GT. 1.0E-06)GOTO 2
24         EM1 = 1.0E06*(YS(1,J+1)-YS(1,J-1))
25         GOTO 3
26       2 EM1 = (YS(1,J+1)-YS(1,J-1))/(XS(1,J+1)-XS(1,J-1))
27       3 IF(ABS(XS(2,J)-XS(2,J-1)) .GT. 1.0E-06)GOTO 4
28         EM2 = 1.0E+06*(YS(2,J)-YS(2,J-1))
29         GOTO 5
30       4 EM2 = (YS(2,J)-YS(2,J-1))/(XS(2,J)-XS(2,J-1))
31       5 X2 = (EM1*(YS(1,J)-YS(2,J)+EM2*XS(2,J))+XS(1,J))/(1.+EM1*EM2)
32         Y2 = YS(2,J) + EM2*(X2 - XS(2,J))
33         STJM = SQRT((X2-XS(2,J-1))**2 + (Y2-YS(2,J-1))**2)
34         SJJM = SQRT((XS(2,J)-XS(2,J-1))**2 + (YS(2,J)-YS(2,J-1))**2)
35         IF(STJM .LT. SJJM)GOTO 8
36         IF(ABS(XS(2,J+1)-XS(2,J)) .GT. 1.0E-06)GOTO 6
37         EM2 = 1.0E+06*(YS(2,J+1)-YS(2,J))
38         GOTO 7
39       6 EM2 = (YS(2,J+1)-YS(2,J))/(XS(2,J+1)-XS(2,J))
40       7 X2 = (EM1*(YS(1,J)-YS(2,J)+EM2*XS(2,J))+XS(1,J))/(1.+EM1*EM2)
41         Y2 = YS(2,J) + EM2*(X2-XS(2,J))
42       8 XS2(J) = X2
43         YS2(J) = Y2
44       9 CONTINUE
45 C
46 C        PROJECT ORTHOGONALLY FROM SURFACE 4 ONTO SURFACE 3
47 C
48         DO 17 J = 2,JMAP
49         IF(ABS(XS(4,J+1)-XS(4,J-1)) .GT. 1.0E-06)GOTO 10
50         EM4 = 1.0E+06*(YS(4,J+1)-YS(4,J-1))
51         GOTO 11
52      10 EM4 = (YS(4,J+1) - YS(4,J-1))/(XS(4,J+1) - XS(4,J-1))
53      11 IF(ABS(XS(3,J) - XS(3,J-1)) .GT. 1.0E-06)GOTO 12
54         EM3 = 1.0E+06*(YS(3,J)-YS(3,J-1))
55         GOTO 13
```

Fig. 13.29. Listing of subroutine SURCH

```
56     12 EM3 = (YS(3,J)-YS(3,J-1))/(XS(3,J)-XS(3,J-1))
57     13 X3 = (EM4*(YS(4,J)-YS(3,J)+EM3*XS(3,J))+XS(4,J))/(1.+EM3*EM4)
58        Y3 = YS(3,J) + EM3*(X3 - XS(3,J))
59        STJM = SQRT((X3-XS(3,J-1))**2 + (Y3-YS(3,J-1))**2)
60        SJJM = SQRT((XS(3,J)-XS(3,J-1))**2 + (YS(3,J)-YS(3,J-1))**2)
61        IF(STJM .LT. SJJM)GOTO 16
62        IF(ABS(XS(3,J+1)-XS(3,J)) .GT. 1.0E-06)GOTO 14
63        EM3 = 1.0E+06*(YS(3,J+1)-YS(3,J))
64        GOTO 15
65     14 EM3 = (YS(3,J+1)-YS(3,J))/(XS(3,J+1)-XS(3,J))
66     15 X3 = (EM4*(YS(4,J)-YS(3,J)+EM3*XS(3,J))+XS(4,J))/(1.+EM3*EM4)
67        Y3 = YS(3,J) + EM3*(X3 - XS(3,J))
68     16 XS3(J) = X3
69        YS3(J) = Y3
70     17 CONTINUE
71 C
72 C     STORE SURFACE 2 AND 3 LOCATIONS
73 C
74        DO 18 J = 2,JMAP
75        XS(2,J) = XS2(J)
76        YS(2,J) = YS2(J)
77        XS(3,J) = XS3(J)
78        YS(3,J) = YS3(J)
79     18 CONTINUE
80        RETURN
81        END
```

Fig. 13.29. (cont.) Listing of subroutine SURCH

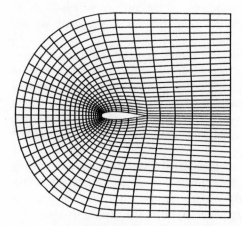

Fig. 13.30. C-grid generated by program ALGEM

It is recommended that $s_2 = 0.100$ and $s_3 = 0.900$ be used to give control over the orthogonal grid at the boundaries. The parameter a_w affects the uniformity of the interior grid. This parameter is set, typically in the range 0.5 to 0.6, most easily with $P_{AC} = P_{FD} = P_{AF} = P_{CD} = 1.0$. For this situation the boundary stretching functions are linear and a_w can be adjusted to give the required grid point distribution.

13.5 Closure

In this chapter various techniques for grid generation have been considered. If the geometry of the physical domain permits a conformal grid to be constructed, this should be exploited, as the structure of the governing equations is then simpler. However, conformal grids sometimes generate extreme clustering and extreme sparsity of the grid. In this case one-dimensional stretching functions (Sect. 13.3.1) can produce a more even grid but at the expense of generating an orthogonal rather than a conformal grid.

In the more general case it is desirable to arrange the gross boundary correspondence (Sect. 13.1) so that severely distorted grids or sparse grids occur well away from the region of interest and preferably in a uniform flow region.

Where possible grid points on all boundaries should be specified since, through the use of one-dimensional stretching functions, this provides a direct means of controlling the distribution of interior points.

Strict orthogonality with adequate control of the grid point distribution is difficult to achieve, particularly when the transformation parameters x_ξ, etc., are evaluated numerically. It is recommended that grids with near-orthogonal distributions, particularly adjacent to boundaries, be generated so as to minimise truncation errors.

The main advantage of generating grids by solving elliptic partial differential equations, such as (13.36), is that discontinuities on the boundary data are not transmitted into the interior and the smoothness of the interior grid is likely to allow numerical evaluation of the transformation parameter x_ξ, etc., with smaller truncation errors.

The main advantage of algebraic grid generation schemes is good control over the interior grid, particularly in relation to generating locally orthogonal grids at boundaries, and the computational efficiency of the grid generation process. This advantage is likely to be particularly significant where the grid must be regenerated during the solution development to obtain a more accurate solution. Techniques for constructing adaptive grids are discussed by Thompson (1985), Eiseman (1987) and Kim and Thompson (1990).

Where grids are constructed as part of an external flow it is usually the region close to the body surface that must be predicted most accurately. Once the distribution of grid points on the body surface has been specified, hyperbolic partial differential equations may be marched away from the surface to obtain the grid in a very efficient manner (Thompson et al. 1985, Chap. 7). However there is often restricted control of the grid point distribution in the farfield.

In addition any discontinuity in the body surface will be propagated into the computational domain. The problem of grid discontinuities can be substantially alleviated by performing a few relaxation sweeps of an elliptic grid generation scheme (Sect. 13.2.6) as a post-processor. This concept is incorporated into the composite grid generation code, EAGLE (Thompson 1988). Post-processing to improve local smoothness and orthogonality can also be achieved efficiently via optimisation procedures (Kennon and Dulikravich 1986).

The grid generation techniques discussed in this chapter have been relevant to relatively simple physical domain that can be readily transformed into rectangular computational domain. Such procedures can be extended to very complex domains, such as the external flow about a complete aircraft, by decomposing the complex domain into linked multiblocks (e.g. Thompson 1988; Yu et al. 1990). The grid generation techniques described here are applicable to individual blocks. The impact of such domain decomposition strategies on the overall algorithm behaviour are beyond the scope of this book.

The multiblock philosophy of grid generation is an effective way of retaining the (rectangular) structure inherent in many grid generation procedures. The exploitation of such structure, at the algorithm level, is often a significant contributor to overall computational efficiency.

However if this structure is not deliberately exploited alternative advantages are available. First there is more flexibility in constructing the grid in complex physical domains and in locally adapting the grid to severe solution gradients. Often triangular (two-dimensional) and tetrahedral (three-dimensional) grids form the basis of unstructured grid generation techniques. Thus the finite element method is well matched to the use of unstructured grids (e.g. Mavriplis 1990; Nakahashi and Egami 1991).

13.6 Problems

Grid Generation by PDE Solution (Sect. 13.2)

13.1 Apply the Joukowski mapping,

$$\frac{Z'+c}{Z'-c}=\left(\frac{Z-2c}{Z+2c}\right)^{1/2} , \tag{13.80}$$

to a NACA-0012 aerofoil [coordinates given by (13.70) and the subroutine FOIL]. The parameter c in (13.80) is the approximate radius of the near-circle in the Z' plane corresponding to the aerofoil in the Z plane with a unit chord. Using (13.6) c is related to the aerofoil nose radius r_N by

$$c=0.25-\frac{r_N}{8} , \tag{13.81}$$

with r_N obtained from (13.70) as

$$r_N=\left(\frac{a_1}{1/t-a_2}\right)^2 . \tag{13.82}$$

The origin for the aerofoil coordinates is $\{1-2c, 0.\}$ so that the trailing edge coordinates are $\{2c, 0\}$; the leading edge coordinates are $\{-(2c+0.5r_N), 0\}$.
  i) For equal increments along the aerofoil chord obtain coordinates on the surface of the near-circle corresponding to points on the surface of the aerofoil. Equations (13.7) are useful for this.
 ii) Interpolate the near-circle surface data at equal angular increments and use the inverse transformation (13.80) to obtain corresponding points on the aerofoil surface. Comment on the distribution of points on the aerofoil surface.

13.2   The Schwarz–Christoffel transformation,

$$\frac{dZ}{d\zeta} = \frac{h}{\pi} \frac{(\zeta+1)^{1/2}}{(\zeta-1)^{1/2}} \; , \tag{13.83}$$

transforms a step of height h in the Z plane into a flat surface (the real axis) in the ζ plane (Milne-Thomson 1968, p. 285). Equation (13.83) can be integrated analytically to give the inverse mapping

$$t = \log(\zeta + \sqrt{\zeta^2 - 1}) \; , \tag{13.84}$$
$$Z = \frac{h}{\pi}(t + \sinh t) \; .$$

Potential flow over the step is given by

$$\phi + i\psi = \frac{hU_\infty}{\pi} \zeta \; , \tag{13.85}$$

where U_∞ is the velocity far upsteam of the step. Take the lines of constant potential (ϕ) and stream function (ψ) to define the grid in the ζ plane. Use the inverse mapping (13.84) to obtain corresponding grid points in the physical (Z) plane.

13.3   Discuss the modification of program ALGEM to construct an orthogonal grid using (13.31). To achieve this an interpolation is made of $(XS(1, J)$, $YS(1, J))$ and $(XS(4, J)$, $YS(4, J))$ for equal increments of s to generate nine intermediate surfaces in the same manner that $\mathbf{Z}_2$ and $\mathbf{Z}_3$ are created initially at the beginning of the subroutine SURCH. The nine intermediate surfaces and the two bounding surfaces define the non-orthogonal (μ, v) grid of Sect. 13.2.4 if $\mu = (J-1)/(J\text{MAX}-1)$ and $v = (K-1)/(K\text{MAX}-1)$.

Starting from surface ABC in Fig. 13.25, (13.31) is integrated numerically to give $\mu_j + \Delta\mu$ on the v_k grid line that corresponds to an orthogonal grid. The surface $XS(v_k)$, $YS(v_k)$ is interpolated to obtain the orthogonal grid point at $(\mu_j + \Delta\mu, v_k)$. This process is repeated until the outer surface FED is reached.

i) Discuss the control over the orthogonal grid point distribution by the deployment of points on ABC and by the number and disposition of the intermediate (v_k) surfaces.

13.4   Repeat Problem 13.3 for the near-orthogonal grid construction described in Sect. 13.2.5.

13.5   Discuss the modification of program ALGEM to generate the interior grid using the Poisson equations of Sect. 13.2.6. Surfaces $\{XS(1, J), YS(1, J)\}$ and $\{XS(4, J), YS(4, J)\}$ are used to generate boundary points on ABC and FED (Fig. 13.25). An equivalent construction is used to obtain boundary points on AF and CD.

i) Equation (13.37) is solved using an SOR algorithm with $P = Q = 0$ and a uniform distribution of boundary points.

ii) Discuss the control over the interior grid using the boundary point stretching functions.

Grid Generation by Algebraic Mapping (Sect. 13.3)

13.6 Modify the program ALGEM to obtain solutions with two ($N = 2$) and three ($N = 3$) multisurfaces. For $N = 3$ make the grid orthogonal to surface ABC only. Compare the interior grid quality with the $N = 4$ case.

13.7 Modify the program ALGEM to obtain the grid between an ellipse and a rectangle and recreate Fig. 13.24.

13.8 Modify the program ALGEM to implement (13.49 and 50) for $N = 2$.

13.9 Obtain Vinokur (1983) and modify subroutine STRECH to incorporate the Vinokur stretching functions as an option. Use this option to cluster grid points close to A and B. That is, split ABC into two segments AB and BC with fine grids close to A and B and continuity of grid-point spacing at B. The grid on FDE may also need some adjustment. Observe the effect on the interior grid for $N = 2$, 3 and 4.

13.10 Discuss the modification of the program ALGEM to introduce transfinite interpolation:
   i) For boundary function evaluation, equivalent to (13.48) on all surfaces.
   ii) Boundary function evaluation on AF and CD. Equation (13.49) on ABC and FED, Fig. 13.25.

14. Inviscid Flow

In this chapter the basic computational techniques developed in Chaps. 3–10 will be extended to construct effective computational methods for inviscid flow. Sects. 11.3 and 11.6.1 provide an appropriate framework for this process. Computational techniques will be selected on the basis of those that are considered to be the most effective without regard for the need to achieve a comprehensive review. This often means that newer methods are described at the expense of older but less efficient methods.

Inviscid flows of engineering interest are governed by the continuity equation (11.10), the Euler equations (11.22–24) and the inviscid energy equation, i.e. (11.38) with the right-hand side set to zero.

The different subcategories of inviscid flow permit particular equation systems to be exploited computationally. Some of these equation systems are shown in Table 14.1. Generally an equation system appearing higher in the list can be computed more efficiently than one appearing lower in the list.

The linearised potential equation is solved efficiently by the panel method (Sect. 14.1) and is accurate for subsonic flow but is generally less accurate for supersonic flow. The full potential equation forms the basis of most transonic computations (Sect. 14.3) as long as only weak shock waves occur.

Table 14.1. Equation systems for inviscid flow

Equation system	Subsonic $(M_\infty \leq 0.7)$	Transonic $(0.7 \leq M_\infty \leq 1.2)$	Supersonic $(M_\infty \geq 1.2)$
Linearised potential equation (11.109)	Requires slender body ('exact' if $M_\infty = 0$)	Not applicable	Requires slender body and weak shocks
Full potential equation (11.103, 104)	Generally applicable	Applicable if weak shocks occur	Not applicable if strong shocks occur
Steady Euler equations (11.22–24) plus $\partial/\partial t = 0$	Generally applicable	Generally applicable	Allows efficient marching schemes (Sect. 14.2.4)
Unsteady Euler equations (11.22–24)	Generally applicable	Generally applicable	Generally applicable

For flows that are everywhere supersonic it is possible to march the steady Euler equations in the approximate flow direction (Sect. 14.2.4). For steady flows featuring mixed subsonic/supersonic regions and strong shocks it is necessary to integrate the unsteady Euler equations until a steady-state solution is obtained (Sects. 14.2.8 and 14.2.9). Both Euler categories also require computation of the continuity and energy (if applicable) equations. If strong shock-waves are present it is often necessary to introduce special procedures (Sects. 14.2.6 and 14.2.7).

14.1 Panel Method

Many practical flows are closely approximated by the assumption of irrotational, as well as inviscid and incompressible, flow. Consequently, as indicated in Sect. 11.3, the governing equations can be reduced to the Laplace equation for the velocity potential,

$$\nabla^2 \Phi = 0 \ , \tag{14.1}$$

with boundary conditions specifying Φ or $\partial\Phi/\partial n$ on all boundaries. The external flow around a streamlined, isolated body (Fig. 14.1) is accurately represented by the solution of (14.1). Of practical interest is the pressure distribution at the surface of the body. This leads directly to the lift force on the body and provides pressure (or equivalently velocity) boundary conditions for the equations governing boundary layer flow (Sect. 11.4 and Chap. 15).

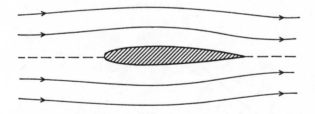

Fig. 14.1. Flow around a streamlined body

Although (14.1) can be solved by the finite difference, finite element or spectral methods, more effective methods are available which exploit the possibility of superposing simple exact solutions of (14.1) in such a way that the boundary conditions are satisfied. An additional feature of such an approach is that the effective computational domain is the surface of the body (Fig. 14.1) rather than the whole region external to the surface (as in the finite difference method). This produces an economical algorithm and permits complicated body shapes to be analysed with relative ease.

In the aircraft industry these techniques are called panel methods (Rubbert and Saaris 1972) although, as indicated in Sect. 14.1.3, panel methods can be interpreted as boundary element methods. Panel methods are widely used in the aircraft industry (Kraus 1978) and automobile industry (Paul and LaFond 1983).

Here we will describe the panel method for the flow about a streamlined non-lifting, two-dimensional body (e.g. as in Fig. 14.1), initially.

14.1.1 Panel Method for Inviscid Incompressible Flow

The panel method takes its name from the subdivision of the surface of the body into a number of contiguous panels (Fig. 14.2) associated with which are source densities of strength σ_j to be determined as an intermediate part of the solution process. An individual source panel (Fig. 14.3) is closely related to an isolated source (Sect. 11.3). A source panel of density σ produces a velocity normal to itself of 0.5σ on each side. The relationship between an individual source panel and an isolated source is discussed by Kuethe and Chow (1976, p. 107).

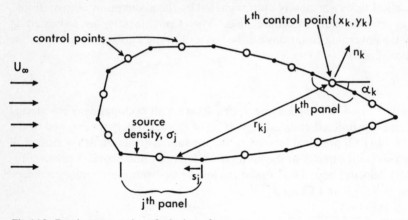

Fig. 14.2. Panel representation of a body surface

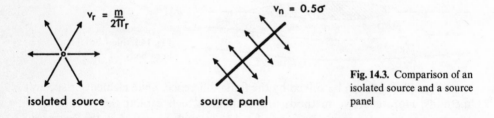

Fig. 14.3. Comparison of an isolated source and a source panel

The distribution of source panels in a uniform stream of velocity U_∞ parallel to the x-axis (Fig. 14.2) produces a potential $\Phi(x_k, y_k)$ given by

$$\Phi(x_k, y_k) = U_\infty x_k + \frac{1}{2\pi} \sum_{j=1}^{N} \sigma_j \int \ln r_{kj}\, ds_j \,, \tag{14.2}$$

where

$$r_{kj} = [(x_k - x_j)^2 + (y_k - y_j)^2]^{1/2} \tag{14.3}$$

and $\sigma_j \int ds_j$ is the source strength of the jth panel. Equation (14.2), with (14.3), satisfies (14.1). The source densities σ_j are to be chosen to satisfy the boundary condition of no flow through the body surface.

The velocity components are given by $\nabla \Phi$, (11.50). In particular, the boundary condition of zero normal velocity at the body surface becomes

$$v_n = \frac{\partial \Phi}{\partial n_k}(x_k, y_k)$$

$$= -U_\infty \sin \alpha_k + \frac{1}{2\pi} \sum_{j=1}^{N} \sigma_j \int \frac{\partial}{\partial n_k}(\ln r_{kj}) \, ds_j = 0 \ , \tag{14.4}$$

where α_k is the slope of the body surfaces at the kth control point (typically the midpoint of the kth panel, Fig. 14.2). Thus (14.4) represents a linear relationship between the source densities σ_j after the integrals have been evaluated. For the particular case $k = j$, the integral can be evaluated analytically, i.e.

$$\int_k \frac{\partial}{\partial n_k}(\ln r_{kk}) \, ds_k = \pi \ . \tag{14.5}$$

For $j \neq k$ the integrals can be evaluated as functions of the nodal points (x_k, y_k; x_j, y_j). Specific formulae will be indicated below, (14.13).

Equation (14.4) is repeated for each control point producing a linear system of equations

$$\underline{A} \sigma = R \ , \tag{14.6}$$

where a component of $\underline{A}$ is

$$A_{kj} = 0.5 \delta_{kj} + \frac{1}{2\pi} \int \frac{\partial}{\partial n_k}(\ln r_{kj}) \, ds_j \ , \tag{14.7}$$

a component of $\mathbf{R}$ is

$$R_k = U_\infty \sin \alpha_k \tag{14.8}$$

and σ is the vector of unknown source densities. The system of linear equations (14.6) can be solved directly or iteratively (Chap. 6).

Once the distribution of source densities is determined the velocity components due to the presence of the body can be obtained from

$$u(x, y) = \frac{1}{2\pi} \sum_{j=1}^{N} \sigma_j \int \frac{x - x_j}{(x - x_j)^2 + (y - y_j)^2} \, ds_j \tag{14.9}$$

and

$$v(x, y) = \frac{1}{2\pi} \sum_{j=1}^{N} \sigma_j \int \frac{y - y_j}{(x - x_j)^2 + (y - y_j)^2} \, ds_j \ , \tag{14.10}$$

and the complete velocity field is just $\mathbf{q} = (U_\infty + u, v)$.

If the velocity components are evaluated at the control points on the body surface the surface pressure distribution follows directly from the Bernoulli equation (11.49)

$$C_\mathrm{p} = \frac{p - p_\infty}{0.5\varrho U_\infty^2} = 1 - \left(\frac{q}{U_\infty}\right)^2 .$$

A typical surface pressure distribution is shown in Fig. 14.4. Agreement is seen to be very good with 23 elements spanning half of a symmetric body. Agreement could be made even better by redistributing the panels to place more in the nose region of the aerofoil and less over the midsection. To suit the available experimental data the present computation is modified by the Prandtl–Glauert transformation appropriate to $M_\infty = 0.4$. The incorporation of the Prandtl–Glauert transformation in the panel method is described in Sect. 14.1.6.

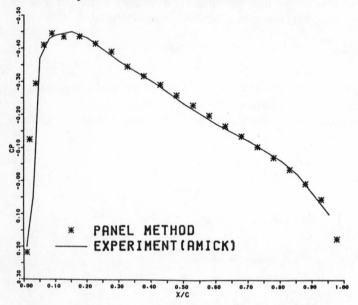

Fig. 14.4. Pressure distribution for NACA-0012 aerofoil at $\alpha = 0°$ and $M_\infty = 0.40$

In many problems the solution (velocity or pressure) is only required at the control points. Then the evaluation of (14.4) duplicates much of the evaluation of (14.9 and 10). Consequently Hess (1975, p. 156) recommends the following more efficient procedure.

The contributions to (14.9, 10) are evaluated to give $v_{kj} = (u_{kj}, v_{kj})$, the velocity components induced at control point (x_k, y_k) due to a unit source density over the jth panel. Equations (14.9 and 10) are then written as

$$u(x_k, y_k) = \sum_{j=1}^{N} u_{kj} \sigma_j , \qquad v(x_k, y_k) = \sum_{j=1}^{N} v_{kj} \sigma_j . \tag{14.11}$$

The components u_{kj} and v_{kj} can be evaluated analytically as

$$u_{kj} = q^t_{kj} \cos \alpha_j - q^n_{kj} \sin \alpha_j \ , \qquad v_{kj} = q^n_{kj} \cos \alpha_j + q^t_{kj} \sin \alpha_j \ , \tag{14.12}$$

where

$$q^t_{kj} = \ln \left(\frac{(\xi_k + 0.5 \Delta s_j)^2 + \eta_k^2}{(\xi_k - 0.5 \Delta s_j)^2 + \eta_k^2} \right) ,$$

$$q^n_{kj} = 2 \tan^{-1} \left(\frac{\eta_k \Delta s_j}{\xi_k^2 + \eta_k^2 - \left(\frac{\Delta s_j}{2}\right)^2} \right) . \tag{14.13}$$

In (14.13) Δs_j is the length of the jth panel and (ξ, η) is a local coordinate system based on the jth panel (Fig. 14.5). Approximate formulae that are much more economical to evaluate are available (Hess 1975, p. 156) if (x_k, y_k) is far from (x_j, y_j). The introduction of such approximate formulae has a negligible effect on the overall solution accuracy.

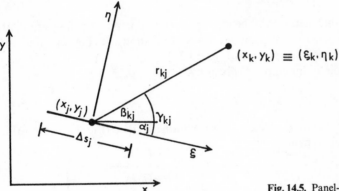

Fig. 14.5. Panel-based coordinate system

In (14.13) q^n_{kj} and q^t_{kj} are the velocity components induced at the control point (x_k, y_k) due to a unit source density over the jth panel, in directions normal and tangential to the jth panel, respectively. If $\mathbf{n}_k$ is the unit normal to the kth panel, A_{kj} in (14.7) is evaluated as

$$A_{kj} = \mathbf{n}_k \cdot \mathbf{v}_{kj} \ . \tag{14.14}$$

After (14.6) has been solved, the evaluation of (14.9 and 10) at the control points is replaced by (14.11). Thus the relatively expensive evaluation of $\mathbf{v}_{kj}$ is done only once. The other computationally expensive part of the present procedure is the solution of (14.6).

It may be noted that the solution for Φ, given by (14.2), satisfies the governing equation (14.1) exactly as $N \to \infty$. The source densities σ_j are chosen to satisfy the boundary condition of zero normal velocity at the body surface (14.4). For locations remote from the body the logarithmic function in (14.2) ensures that the solution reverts to a flow of velocity U_∞ parallel to the x-axis, i.e.

$$\Phi = U_\infty x_k \ .$$

The system of equations (14.6) has large diagonal elements but is not strictly diagonally dominant. However, iterative techniques, such as those discussed in Sect. 6.3, are effective and recommended (Hess 1975, p. 159) if the number of elements exceeds about 1000. For smaller number of elements, direct methods (Sect. 6.2) are more efficient. Since matrix $\underline{A}$ is full the computation time to solve (14.6), by a direct method, will increase like $O(N^3)$. Thus this part of the calculation dominates the execution time for large N.

However, for a given execution time the panel method produces solutions of significantly higher accuracy than does a finite difference, or finite element, method applied on a conventional grid surrounding the body. This also implies that the panel method would have far fewer unknowns (source densities) than finite difference nodal unknowns for the same execution time.

14.1.2 PANEL: Numerical Implementation

In this section the panel method will be implemented to compute the flow about an ellipse. The overall structure of program PANEL is shown in Fig. 14.6 and the listings of the subroutines are given in Figs. 14.7–11.

The surface profile of an ellipse is given by

$$x^2 + (y/b)^2 = 1 \quad \text{or} \quad x = \cos\theta \ , \quad y = b\sin\theta \ , \tag{14.15}$$

where b is the minor semi-axis length. In subroutine BODY (Fig. 14.8), (14.15) is evaluated for equal increments of θ.

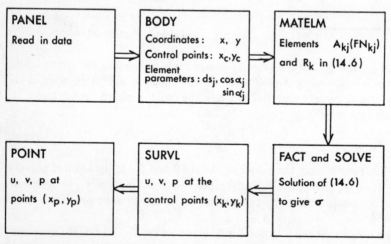

Fig. 14.6. Structure of program PANEL

```
1   C
2   C       PANEL CALCULATES VELOCITY AND PRESSURES ABOUT
3   C       AN ARBITRARY CLOSED BODY USING THE PANEL METHOD.
4   C
5           DIMENSION X(50),Y(50),XC(50),YC(50),DS(50),FN(50,50)
6         1 ,FT(50,50),RHS(50),SDE(50),CI(50),SI(50),AA(50,50),IKS1(50)
7           COMMON X,Y,XC,YC,DS,FN,FT,RHS,PI,CPI,CI,SI
8         1 ,UINF,VINF,SDE
9   C
10          OPEN(1,FILE='PANEL.DAT')
11          OPEN(6,FILE='PANEL.OUT')
12          READ(1,1)N,IPR,UINF,VINF,FMN,B
13        1 FORMAT(2I5,4E10.3)
14  C
15          WRITE(6,2)N,B
16          WRITE(6,3)UINF,VINF,FMN
17        2 FORMAT(1X,'PANEL METHOD WITH ',I2,' ELEMENTS,'5X,
18        1 'ELLIPSE MINOR SEMI-AXIS =',F6.3,/)
19        3 FORMAT(1X,'ONSET VELOCITY COMPONENTS = ',2F6.3,
20        1 2X,'FREESTREAM MACH NUMBER = ',F6.3,//)
21  C
22          M=N+1
23          PI = 3.14159265
24          CPI=2.0/PI
25  C
26  C       CALCULATE COORDINATES OF BODY AND CONTROL POINTS.
27  C
28          CALL BODY(N,M,IPR,FMN,B)
29  C
30  C       CONSTRUCT THE MATRIX EQUATION.
31  C
32          CALL MATELM(N,IPR)
33  C
34  C       TRANSFER FN INTO AA
35  C
36          DO 5 K = 1,N
37          DO 4 J = 1,N
38        4 AA(K,J) = FN(K,J)
39        5 SDE(K) = RHS(K)
40  C
41  C       FACTORISE AA INTO L.U
42  C
43          CALL FACT(N,AA,IKS1)
44  C
45  C       SOLVE FOR THE SOURCE DENSITIES, SDE(K)
46  C
47          CALL SOLVE(N,AA,IKS1,SDE)
48  C
49  C       CALCULATE VELOCITY AND PRESSURE AT THE BODY SURFACE
50  C
51          CALL SURVL(N,B,FMN)
52  C
53  C       CALCULATE FLOW AT GIVEN POINTS.
54  C
55          CALL POINT(N,FMN)
56  C
57          STOP
58          END
```

Fig. 14.7. Listing of program PANEL

```
1              SUBROUTINE BODY(N,M,IPR,FMN,B)
2    C
3    C        CALCULATES BODY AND CONTROL POINT COORDINATES
4    C        FOR AN ELLIPSE WITH MINOR SEMI-AXIS, B
5    C
6              DIMENSION X(50),Y(50),XC(50),YC(50),DS(50),FN(50,50)
7             1 ,FT(50,50),RHS(50),SDE(50),CI(50),SI(50)
8              COMMON X,Y,XC,YC,DS,FN,FT,RHS,PI,CPI,CI,SI
9             1 ,UINF,VINF,SDE
10   C
11   C        BODY POINTS
12   C
13             FAC = SQRT(1.0 - FMN*FMN)
14             NHLFF = N/2 + 1
15             NHH=NHLFF+1
16             AN = NHLFF - 1
17             DTH = PI/AN
18   C
19             DO 2 I=1,NHLFF
20             AI = I - 1
21             TH = AI*DTH
22             TH = PI - TH
23             X(I) = COS(TH)
24             Y(I) = B*SIN(TH)
25   C
26   C        PRANDTL-GLAUERT TRANSFORMATION
27   C
28           2 Y(I) = Y(I)*FAC
29   C
30   C        REFLECT FOR COORDINATES OF LOWER HALF.
31   C
32             DO 3 I=NHH,N
33             X(I)=X(N+2-I)
34           3 Y(I) = - Y(N+2-I)
35             X(M)=X(1)
36             Y(M)=Y(1)
37   C
38   C        PLACE CONTROL POINTS AT THE CENTER OF PANELS.
39   C
40             DO 4 I=1,N
41             XC(I)=(X(I)+X(I+1))*0.5
42           4 YC(I)=(Y(I)+Y(I+1))*0.5
43   C
44   C        CALCULATE PANEL SPANS,COS AND SINE OF ANGLES.
45   C
46             DO 5 I=1,N
47             SX=X(I+1)-X(I)
48             SY=Y(I+1)-Y(I)
49             DS(I)=SQRT(SX*SX+SY*SY)
50             CI(I)=(X(I+1)-X(I))/DS(I)
51           5 SI(I)=(Y(I+1)-Y(I))/DS(I)
52             IF(IPR .EQ. 0)RETURN
53   C
54             WRITE(6,6)
55           6 FORMAT(2X,'ELEMENT PARAMETERS')
56             DO 7 I = 1,N
57           7 WRITE(6,8)I,X(I),Y(I),XC(I),YC(I),DS(I),CI(I),SI(I)
58           8 FORMAT(2X,'I=',I2' X,Y=',2F8.4,' XC,YC= ',2F8.4,' SPAN= ',
59             1F8.4,' CI,SI=',2F8.4)
60             RETURN
61             END
```

Fig. 14.8. Listing of subroutine BODY

```
1              SUBROUTINE MATELM(N,IPR)
2    C
3    C        CALCULATES MATRIX ELEMENTS AND RHS.
4    C
5              DIMENSION X(50),Y(50),XC(50),YC(50),DS(50),FN(50,50)
6          1  ,FT(50,50),RHS(50),SDE(50),CI(50),SI(50)
7              COMMON X,Y,XC,YC,DS,FN,FT,RHS,PI,CPI,CI,SI
8          1  ,UINF,VINF,SDE
9    C
10             DO 2 K=1,N
11             DO 1 J=1,N
12             IF(K.EQ.J) FN(K,J)=2.*PI
13             IF(K.EQ.J) FT(K,J)=0.0
14             IF(K.EQ.J) GO TO 1
15             DYJ=SI(J)*DS(J)
16             DXJ=CI(J)*DS(J)
17             SPH=DS(J)*0.5
18             XD=XC(K)-XC(J)
19             YD=YC(K)-YC(J)
20             RKJ=SQRT(XD*XD+YD*YD)
21             BKJ=ATAN2(YD,XD)
22             ALJ=ATAN2(DYJ,DXJ)
23             GKJ=ALJ-BKJ
24             ZIK=RKJ*COS(GKJ)
25             ETK=-RKJ*SIN(GKJ)
26             R1S=((ZIK+SPH)**2)+ETK*ETK
27             R2S=((ZIK-SPH)**2)+ETK*ETK
28             QT=ALOG(R1S/R2S)
29             DEN=ZIK*ZIK+ETK*ETK-SPH*SPH
30             GNM=ETK*DS(J)
31             QN=2.0*ATAN2(GNM,DEN)
32             UKJ=QT*CI(J)-QN*SI(J)
33             VKJ=QT*SI(J)+QN*CI(J)
34             FN(K,J)=-UKJ*SI(K)+VKJ*CI(K)
35             FT(K,J)=UKJ*CI(K)+VKJ*SI(K)
36         1  CONTINUE
37             RHS(K)=UINF*SI(K)-VINF*CI(K)
38         2  CONTINUE
39   C
40             IF(IPR .LE. 1)RETURN
41         3  WRITE(6,4)
42         4  FORMAT(2X,'MATRIX ELEMENTS = NORMAL VELOCITY COMPONENTS')
43             DO 5 K=1,N
44         5  WRITE(6,8) K,(FN(K,J),J=1,N)
45             WRITE(6,6)
46         6  FORMAT(2X,'TANGENTIAL VELOCITY COMPONENTS')
47             DO 7 K=1,N
48         7  WRITE(6,8)K,(FT(K,J),J=1,N)
49         8  FORMAT(2X,I5,(10F10.5))
50             WRITE(6,9)
51         9  FORMAT(2X,'RIGHT HAND SIDE')
52             WRITE(6,10)(RHS(K),K=1,N)
53        10  FORMAT(2X,10F10.5)
54             RETURN
55             END
```

Fig. 14.9. Listing of subroutine MATELM

```
1          SUBROUTINE SURVL(N,B,FMN)
2  C
3  C      CALCULATES VELOCITIES AND PRESSURE AT THE CONTROL POINTS
4  C      QEX IS THE EXACT VELOCITY AT THE SURFACE OF THE ELLIPSE
5  C
6          DIMENSION X(50),Y(50),XC(50),YC(50),DS(50),FN(50,50)
7         1,FT(50,50),RHS(50),SDE(50),CI(50),SI(50)
8          COMMON X,Y,XC,YC,DS,FN,FT,RHS,PI,CPI,CI,SI
9         1,UINF,VINF,SDE
10 C
11         FAC = SQRT(1. - FMN*FMN)
12         GAM = 1.4
13         C1 = 0.5*(GAM-1.)*FMN*FMN
14         C2 = 0.5*GAM*FMN*FMN
15         GMP = GAM/(GAM-1.)
16         WRITE(6,1)
17       1 FORMAT(2X,'VELOCITY AND PRESSURE AT THE CONTROL POINTS')
18         DO 4 K=1,N
19         QTS=0.0
20         QNS=0.0
21         DO 2 J=1,N
22         QTS=QTS+FT(K,J)*SDE(J)
23       2 QNS=QNS+FN(K,J)*SDE(J)
24 C
25         QNK = QNS + VINF*CI(K) - UINF*SI(K)
26         QTK = QTS + VINF*SI(K) + UINF*CI(K)
27         UU=UINF-QNS*SI(K)+QTS*CI(K)
28         VV=VINF+QNS*CI(K)+QTS*SI(K)
29         UU = UU/FAC/FAC
30         VV = VV/FAC
31         PP=1.-UU*UU-VV*VV
32         IF(FMN .GT. 0.05)PP = ((1.+C1*PP)**GMP-1.)/C2
33 C
34         DUM = B*B*XC(K)
35         DUM = YC(K)*YC(K) + DUM*DUM
36         QEX = (1. + B)*YC(K)/SQRT(DUM)
37 C
38         WRITE(6,3)XC(K),YC(K),QNK,QTK,UU,VV,PP,QEX
39       3 FORMAT(1X,'XC,YC=',2F6.3,'  QN,QT=',2F6.3,
40         1 '  U,V=',2F6.3,'  P=',F6.3,'  QEX=',F6.3)
41       4 CONTINUE
42         RETURN
43         END
```

Fig. 14.10. Listing of subroutine SURVL

To allow program PANEL to be extended to subsonic inviscid flow a Prandtl–Glauert transformation (Sect. 14.1.6) is made to the y coordinate as $y_{inc} = y(1 - M_\infty^2)^{1/2}$. Thus (x, y_{inc}) gives the surface profile of an equivalent body in incompressible flow for the actual flow with freestream Mach number M_∞.

The contributions to the matrix elements A_{kj} in (14.6) are evaluated in subroutine MATELM (Fig. 14.9) via (14.12–14). The elements A_{kj} are stored in FN_{kj}. The tangential velocity increments corresponding to unit source densities are stored in FT_{kj} for subsequent use in the subroutine SURVL (Fig. 14.10). The contributions to R_k in (14.8) are calculated in MATELM under the slightly more general assumption of a two-component freestream velocity (U_∞, V_∞).

Subroutines FACT (Fig. 6.15) and SOLVE (Fig. 6.16) are used to solve (14.6) for the source densities σ. The velocities and pressure at the control points are evaluated in the subroutine SURVL (Fig. 14.10) using the equivalent of (14.11), but based on FT and FN.

```
1         SUBROUTINE POINT(N,FMN)
2 C
3 C       CALCULATES THE FLOW AT GIVEN POINTS, (XP,YP)
4 C
5         DIMENSION X(50),Y(50),XC(50),YC(50),DS(50),FN(50,50)
6        1,FT(50,50),RHS(50),SDE(50),CI(50),SI(50)
7         COMMON X,Y,XC,YC,DS,FN,FT,RHS,PI,CPI,CI,SI
8        1,UINF,VINF,SDE
9 C
10        FAC = SQRT(1. - FMN*FMN)
11        GAM = 1.4
12        C1 = 0.5*(GAM-1.)*FMN*FMN
13        C2 = 0.5*GAM*FMN*FMN
14        GMP = GAM/(GAM-1.)
15    1   CONTINUE
16        READ(1,2) XP,YP
17    2   FORMAT(2F8.5)
18        YP = YP*FAC
19        RPS = XP*XP + YP*YP
20        IF(RPS .LT. 1.0E-04)RETURN
21 C
22        UU=UINF
23        VV=VINF
24        DO 3 J=1,N
25        DYJ=SI(J)*DS(J)
26        DXJ=CI(J)*DS(J)
27        SPH=DS(J)*0.5
28        XD=XP-XC(J)
29        YD=YP-YC(J)
30        R=SQRT(XD*XD+YD*YD)
31        BET=ATAN2(YD,XD)
32        ALJ=ATAN2(DYJ,DXJ)
33        GAM=ALJ-BET
34        ZI=R*COS(GAM)
35        ET=-R*SIN(GAM)
36        R1S=((ZI+SPH)**2)+ET*ET
37        R2S=((ZI-SPH)**2)+ET*ET
38        QT=ALOG(R1S/R2S)
39        DEN=ZI*ZI+ET*ET-SPH*SPH
40        GN = ET*DS(J)
41        QN = 2.0*ATAN2(GN,DEN)
42        UJ=QT*CI(J)-QN*SI(J)
43        VJ=QT*SI(J)+QN*CI(J)
44        UU=UU+UJ*SDE(J)
45        VV=VV+VJ*SDE(J)
46    3   CONTINUE
47 C
48        YP = YP/FAC
49        UU = UU/FAC/FAC
50        VV = VV/FAC
51        PP=1.-UU*UU-VV*VV
52        IF(FMN .GT. 0.05)PP = ((1.+C1*PP)**GMP-1.)/C2
53        WRITE(6,4)XP,YP,UU,VV,PP
54    4   FORMAT(/,2X,'FLOW AT X,Y=',2F6.3,'   U,V=',2F6.3,'   P=',
55    1   F6.3)
56        GO TO 1
57        END
```

Fig. 14.11. Listing of subroutine POINT

The velocity components and pressure at specified points (x_p, y_p) external to the body are calculated in the subroutine POINT (Fig. 14.11) using the equivalent of (14.11–13). The coordinates (x_p, y_p) are read from the input data file on logical unit 1.

The parameters used by PANEL are described in Table 14.2 and typical output for the flow about an ellipse with a minor semi-axis length $b = 0.5$ units is shown in Fig. 14.12. At each control point the velocity components normal and tangential to the local body slope are given by QN and QT in Fig. 14.12. The normal component QN is zero. It may be recalled that this boundary condition (14.4) is used to determine the source densities. The tangential component QT is compared with the exact tangential component QEX in Fig. 14.12 at the surface of the ellipse

Table 14.2. Parameters used in program PANEL

Parameter	Description
AA	matrix $\underline{A}$ in (14.6)
B	minor semi-axis length of the ellipse
FMN	freestream Mach number M_∞
FN(K, J)	induced velocity normal to panel k at (x_k, y_k) due to unit σ_j
FT(K, J)	induced velocity tangential to panel k at (x_k, y_k) due to unit σ_j
IPR	>0 print X, Y, XC, YC, DS, CI, SI in BODY
	$=1$ print FN, FT, RHS in MATELM
M	number of element end points
N	number of elements
RHS	vector $\mathbf{R}$ in (14.6)
SDE	source density vector $\boldsymbol{\sigma}$
UINF, VINF	freestream velocity components U_∞ and V_∞
X, Y	coordinates of panel end points
XC, YC	coordinates of panel control points
CI(J), SI(J)	$\cos \alpha_j$, $\sin \alpha_j$, Fig. 14.5
DS(J)	panel length (span), ds_j
ALJ	α_j, Fig. 14.5, MATELM and POINT
BKJ	β_{kj}, Fig. 14.5, MATELM; BET in POINT
GKJ	γ_{kj}, Fig. 14.5, MATELM; GAM in POINT
RKJ	r_{kj}, Fig. 14.5, MATELM; R in POINT
QN, QT	q_{kj}^n, q_{kj}^t in (14.13), MATELM and POINT
QNK, QTK	velocity components normal and tangential to the kth panel at (x_k, y_k)
UKJ, VKJ	u_{kj}, v_{kj} in (14.12), MATELM; UJ, VJ in POINT
UU, VV, PP	u, v, p at point (x_k, y_k) in SURVL and point (x_p, y_p) in POINT
QE	exact tangential velocity at the surface of the ellipse, $q_{t, ex}$

$$q_{t, ex} = \frac{(1+b)y}{(y^2 + b^4 x^2)^{1/2}} \,. \tag{14.16}$$

Agreement could be made better by introducing more panels, particularly in the nose ($YC \approx 0$) and shoulder ($YC \approx 0.5$) regions. Also shown in Fig. 14.12 are the Cartesian velocity components and pressure at the control points and at a typical off-body point (0, 1.0).

```
PANEL METHOD WITH 20 ELEMENTS,    ELLIPSE MINOR SEMI-AXIS =  .500

ONSET VELOCITY COMPONENTS = 1.000  .000  FREESTREAM MACH NUMBER =   .000

VELOCITY AND PRESSURE AT THE CONTROL POINTS
XC,YC= -.976  .077  QN,QT=  .000  .448  U,V=  .135  .427  P=  .800  QEX=  .453
XC,YC= -.880  .224  QN,QT=  .000 1.067  U,V=  .762  .748  P= -.139  QEX= 1.071
XC,YC= -.698  .349  QN,QT=  .000 1.342  U,V= 1.200  .600  P= -.801  QEX= 1.342
XC,YC= -.448  .440  QN,QT=  .000 1.455  U,V= 1.410  .359  P=-1.117  QEX= 1.454
XC,YC= -.155  .488  QN,QT=  .000 1.497  U,V= 1.492  .118  P=-1.240  QEX= 1.495
XC,YC=  .155  .488  QN,QT=  .000 1.497  U,V= 1.492 -.118  P=-1.240  QEX= 1.495
XC,YC=  .448  .440  QN,QT=  .000 1.455  U,V= 1.410 -.359  P=-1.117  QEX= 1.454
XC,YC=  .698  .349  QN,QT=  .000 1.342  U,V= 1.200 -.600  P= -.801  QEX= 1.342
XC,YC=  .880  .224  QN,QT=  .000 1.067  U,V=  .762 -.748  P= -.139  QEX= 1.071
XC,YC=  .976  .077  QN,QT=  .000  .448  U,V=  .135 -.427  P=  .800  QEX=  .453
XC,YC=  .976 -.077  QN,QT=  .000 -.448  U,V=  .135  .427  P=  .800  QEX= -.453
XC,YC=  .880 -.224  QN,QT=  .000-1.067  U,V=  .762  .748  P= -.139  QEX=-1.071
XC,YC=  .698 -.349  QN,QT=  .000-1.342  U,V= 1.200  .600  P= -.801  QEX=-1.342
XC,YC=  .448 -.440  QN,QT=  .000-1.455  U,V= 1.410  .359  P=-1.117  QEX=-1.454
XC,YC=  .155 -.488  QN,QT=  .000-1.497  U,V= 1.492  .118  P=-1.240  QEX=-1.495
XC,YC= -.155 -.488  QN,QT=  .000-1.497  U,V= 1.492 -.118  P=-1.240  QEX=-1.495
XC,YC= -.448 -.440  QN,QT=  .000-1.455  U,V= 1.410 -.359  P=-1.117  QEX=-1.454
XC,YC= -.698 -.349  QN,QT=  .000-1.342  U,V= 1.200 -.600  P= -.801  QEX=-1.342
XC,YC= -.880 -.224  QN,QT=  .000-1.067  U,V=  .762 -.748  P= -.139  QEX=-1.071
XC,YC= -.976 -.077  QN,QT=  .000 -.448  U,V=  .135 -.427  P=  .800  QEX= -.453

 FLOW AT X,Y=    .000 1.000  U,V= 1.257     .000 P= -.579
```

Fig. 14.12. Typical output from program PANEL

14.1.3 Connection with the Boundary Element Method

The panel method described in Sect. 14.1.1 is particularly effective for the flow about isolated bodies in a uniform stream. However, for internal flows, e.g. the flow around an obstacle in a channel, an alternative formulation based on Green's theorem is often more convenient. This alternative technique seeks the solution Φ directly, without introducing the intermediate source panel distribution.

For two-dimensional problems the potential at any point (x_k, y_k), in the general domain, can be related to the values of Φ and $\partial\Phi/\partial n$ on the computational boundary S, i.e.

$$\Phi(x_k, y_k) = \frac{1}{2\pi}\left[\int_S (\ln r_{kj})\frac{\partial\Phi}{\partial n}(s)\,ds - \int_S \frac{\partial}{\partial n}(\ln r_{kj})\Phi(s)\,ds \right], \tag{14.17}$$

where (x_j, y_j) lies on S and r_{kj} is given by (14.3).

When (x_k, y_k) is restricted to the computational boundary S, (14.17) provides a compatibility condition between $\Phi(s)$ and $\partial\Phi(s)/\partial n$. For the previous example of flow around an isolated body, $\partial\Phi(s)/\partial n$ is known and (14.17) provides a Fredholm integral equation of the second kind for $\Phi(s)$. Typically the local behaviour of $\Phi(s)$ is assumed to be represented by the summation of one-dimensional interpolating functions (Sect. 5.3), $N_j(\xi)$, i.e.

$$\Phi(s) = \sum_j N_j(\xi)\Phi_j , \tag{14.18}$$

where ξ is an element coordinate and Φ_j are the nodal values of $\Phi(s)$.

This is the boundary element method (Brebbia 1978). Substitution of (14.18) into (14.17), and the requirement that the resulting equation is exactly satisfied at the nodes, generates a linear system of equations, equivalent to (14.6) but directly for Φ_j.

Often, for internal flows, Φ is given on part of the domain and $\partial\Phi/\partial n$ is given on another part of the domain. By introducing (14.18) for the unknown $\Phi(s)$ and a comparable trial solution for the unknown $\partial\Phi(s)/\partial n$ it is possible to solve the problem directly, since (14.17) ensures compatibility of $\Phi(s)$ and $\partial\Phi(s)/\partial n$. The flow around a circular cylinder in a channel is of this type, if Φ is interpreted as the stream function. The required steps to obtain the solution are provided by Fletcher (1984, pp. 180–183).

The above Green's formula method can be recast as a generalised 'source' panel method by continuing the solution beyond the computational boundary. The details are provided by Jaswon and Symm (1977, pp. 41–43).

14.1.4 Lifting Aerofoil Problem

To simulate the flow around a lifting body such as an inclined aerofoil, the solution procedure described in Sect. 14.1.1 must be supplemented to provide a unique value for the lift produced by the aerofoil. The lift is related to the circulation Γ around any closed path c in the fluid which encloses the body by

$$L = \varrho U_\infty \Gamma \ , \tag{14.19}$$

where $\Gamma = \int v \cdot dc$. For the body shown in Fig. 14.1 an appropriate path to evaluate Γ would be the surface of the body.

To represent the circulation an additional surface doublet (or dipole) distribution $\mu(s)$ is introduced so that (14.2) takes the form

$$\Phi(x_k, y_k) = U_\infty x_k + \frac{1}{2\pi} \int_s \sigma(s)\,(\ln r_{jk})\,ds - \frac{1}{2\pi} \int \mu(s) \frac{\partial}{\partial n(s)} (\ln r_{jk})\,ds \ . \tag{14.20}$$

The doublet distribution $\mu(s)$ can be related to a vortex sheet (Rubbert and Saaris 1972). A linearly varying doublet panel is equivalent to a constant strength vortex sheet. In practice, $\mu(s)$ is chosen to give the appropriate value of Γ to satisfy the Kutta condition, as indicated below.

In practical implementations (Hess 1975, pp. 160–163) advantage is taken of the linear nature of (14.1) and the opportunity to superpose solutions. For an inclined aerofoil the solution after applying a conventional surface-source method (Sect. 14.1.1) would appear as shown in Fig. 14.13a. This solution is a poor representation of the real flow since it implies an infinite velocity at the sharp trailing edge and that the lift is zero.

The Kutta condition, that the velocity must be finite at the trailing edge, is invoked to fix Γ and, thereby, to determine the lift, from (14.19). The circulation Γ is generated computationally by a surface distribution of linearly varying doublet panels. This generates the purely circulatory flow shown in Fig. 14.13b. Addition of

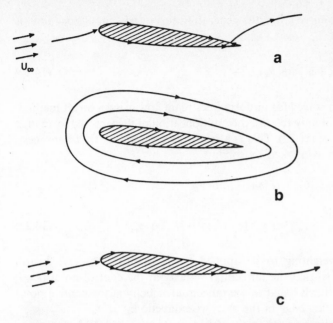

Fig. 14.13. Superposition of solutions: (**a**) non-lifting flow, (**b**) circulatory flow, (**c**) combined lifting flow

this solution and the original surface-source solution, with the doublet strengths chosen to satisfy the Kutta condition, generates the physically realistic solution shown in Fig. 14.13c. The Kutta condition is satisfied by requiring that the tangential velocities at corresponding control points adjacent to the trailing edge (Fig. 14.14) are equal.

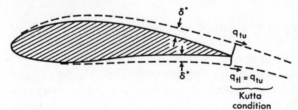

Fig. 14.14. Displacement thickness effects and the Kutta condition

The circulatory flow shown in Fig. 14.13b is generated computationally by ascribing a unit strength vortex at each control point and determining what source distribution, by solving (14.6), will produce the same flow field.

It may be recalled that $\mathbf{v}_{kj}$ is the vector of velocity components at the kth control point due to a unit source at the jth control point. If $\mathbf{v}_{kj}$ is rotated through $90°$, i.e. $(v_{kj}, -u_{kj})$, the result gives the velocity components at the kth control point due to a unit vortex at the jth control point.

Thus the total velocity induced normal to the kth control point due to the distribution of unit strength vortices is given by

$$v_n = \sum_j (v_{kj} \sin \alpha_k + u_{kj} \cos \alpha_k) , \tag{14.21}$$

and the corresponding source strengths generating this purely circulatory flow is given by the solution of

$$\underline{A}\,\sigma^c = -\sum_j (v_{kj}\sin\alpha_k + u_{kj}\cos\alpha_k)\ . \tag{14.22}$$

If (14.6) has already been solved for uniform freestream flow using a direct method, the solution of (14.22) will only require the multiplication of the factored form of $\underline{A}$ and the right-hand side of (14.22). This is a relatively economical, $O(N^2)$, process.

The Kutta condition is invoked by solving

$$\sum_j (u_{l,j}\cos\alpha_l - v_{l,j}\sin\alpha_l)(\sigma_j + \tau\sigma_j^c) + U_\infty\cos\alpha_l$$

$$= -\sum_j (u_{m,j}\cos\alpha_m - v_{m,j}\sin\alpha_m)(\sigma_j + \tau\sigma_j^c) - U_\infty\cos\alpha_m \tag{14.23}$$

for τ. This gives the weighting to be applied to the solution associated with circulatory flow (14.22) so that the tangential velocity at the control point $(k=l)$ just before the trailing edge is equal to the tangential velocity at the control point $(k=m)$ just after the trailing edge in the clockwise direction.

In terms of the parameters used in the program PANEL, (14.23) becomes

$$\tau = \frac{-\left[\sum_j (\mathrm{FT}_{l,j} + \mathrm{FT}_{m,j})\sigma_j + U_\infty(\cos\alpha_l + \cos\alpha_m) - V_\infty(\sin\alpha_l + \sin\alpha_m)\right]}{\sum_j (\mathrm{FT}_{l,j} + \mathrm{FT}_{m,j})\sigma_j^c}\ . \tag{14.24}$$

To generate more accurate pressure distributions, particularly for aerofoils (or turbine blades) producing lift, it is desirable to allow for boundary layer displacement thickness effects. This is done as follows. Given the surface pressure distribution resulting from the solution for the lifting aerofoil, Fig. 14.13c, the boundary layer solution (Sect. 15.1) is obtained and the displacement thickness distribution $\delta^*(x)$ calculated (11.67). This is added to the original geometric shape (Fig. 14.14) and the pressure distribution obtained for the augmented shape. The overall process is repeated, perhaps four or five times, until convergence is achieved. Alternative strategies for imposing the Kutta condition, which allow for the displacement thickness, are discussed by Hess (1975, p. 163).

14.1.5 Higher-Order Panel Methods and the Extension to Three Dimensions

The accuracy of the panel method depends on the number of panels, N, introduced to represent the shape. For internal flow problems (e.g. engine intakes) or the three-dimensional flows around complete aircraft or automobiles the use of constant strength source densities over flat panels forces the introduction of a very large number of panels and an excessive execution time to achieve an acceptable accuracy.

The basic method, described in Sect. 14.1.1, can be extended to allow both curved panels and source densities that vary linearly or quadratically, etc., across the panel. For a two-dimensional flow, two degrees of freedom per panel would be required to define a linear source density variation. Although there is some additional algebraic manipulation required, the execution time per degree of freedom in (14.6) is not significantly greater than for the constant strength, flat panel case. Consequently for internal flows and flows with severe reflex curvature more accurate solutions can be obtained for the same overall execution time (Hess 1975, pp. 165–168) by using higher-order panels.

The extension of the panel method to three-dimensional flows is quite straight-forward. In the case of a nonlifting body, (14.2) is replaced by

$$\Phi(x_k, y_k, z_k) = U_\infty x_k + \frac{1}{4\pi} \sum_{j=1}^{N} \sigma_j \int_j \frac{1}{r_{kj}} \, ds_j \; , \tag{14.25}$$

where the jth panel is now an area rather than a line. Equation (14.4) changes correspondingly. The main complications in three dimensions are the mechanics of defining the panel geometry and the solution of (14.6) for large values of N. This is usually carried out iteratively if $N > 1000$ (Hess 1975, p. 159).

Figure 14.15 provides a typical example of the accuracy achievable by a panel method for a complicated geometric configuration. Reviews of the method, with considerable practical detail, are provided by Hess and Smith (1967), Hess (1975) Kraus (1978) and Hess (1990).

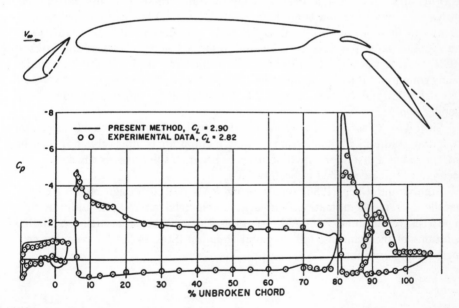

Fig. 14.15. Pressure distribution for a slotted wing (after Hess, 1975; reprinted with permission of North-Holland)

14.1.6 Panel Method for Inviscid, Compressible Flow

In this category it is useful to isolate flows for which (11.109) is valid, i.e.

$$(1 - M_\infty^2)\frac{\partial^2 \phi}{\partial x^2} + \frac{\partial^2 \phi}{\partial y^2} + \frac{\partial^2 \phi}{\partial z^2} = 0 , \qquad (14.26)$$

where ϕ is the (disturbance) velocity potential, i.e.

$$\Phi = U_\infty x + \phi , \quad \text{for example.}$$

Subsonic or supersonic flow abour slender bodies is accurately predicted by (14.26) with the boundary conditions of zero normal velocity at the body surface and the vanishing of ϕ far from the body. For supersonic flow there is a further requirement that any shock waves should be weak, since entropy is not conserved across a shock and (14.26) is based on the assumption of isentropic flow.

For subsonic inviscid flow governed by (14.26) it is possible to construct an equivalent incompressible flow using the Prandtl–Glauert transformation (Liepmann and Roshko 1957, p. 255). The equivalent incompressible flow has independent variables

$$x_i = x , \quad y_i = y\sqrt{1 - M_\infty^2} , \quad z_i = z\sqrt{1 - M_\infty^2} , \qquad (14.27)$$

and dependent variable $\Phi_i = \Phi(1 - M_\infty^2)$, and the panel method (Sect. 14.1.1) is directly applicable.

For supersonic flow a Prandtl–Glauert transformation is not introduced. Instead the panel method is applied directly to (14.26) but with zero mass flow normal to the body surface used as the body boundary condition. Kraus (1978) reviews alternative panel methods for supersonic flow and Carmichael and Erikson (1981) provide a detailed description of one particular code, PAN-AIR, that shares many common features with the panel method described in Sect. 14.1.1. PAN-AIR solves (14.26) and is therefore suitable for subsonic or supersonic flow about slender bodies.

Since (14.26) is only an approximation to the full compressible potential equation, solutions obtained from a panel method for compressible flow can be interpreted as a first approximation to more accurate solutions obtained using the techniques to be described in Sect. 14.3.3.

Tinoco and Chen (1986) report using PAN-AIR to design efficient engine nacelle airframe combinations for large commercial aircraft. In addition they report using subsonic PAN-AIR solutions as a means of partially validating solutions of the transonic full potential equation (Sect. 14.3.3) for complex geometric configurations.

14.2 Supersonic Inviscid Flow

If no restriction is placed on the slenderness of the body, subsonic inviscid flow can be computed by any of the schemes that have been developed for transonic flow (Sect. 14.3). Supersonic inviscid flow introduces the complication of shock waves, and their occurrence and accurate representation place a considerable demand on the computational algorithm. The shock wave may be moving, as in unsteady problems associated with blast waves (caused by explosions), or stationary relative to the body producing the shock wave, as with the bow shock formed by a reentry vehicle.

14.2.1 Preliminary Considerations

Supersonic flows associated with missiles, aircraft, missile engine intakes and rocket nozzles are often steady. It may be recalled (Sect. 11.6.1) that for steady inviscid flow the physical character of the flow is "elliptic" in the subsonic region and "hyperbolic" in the supersonic region. For steady supersonic flows that contain no embedded subsonic regions it is possible to construct a one-pass marching scheme in the hyperbolic direction, which is usually the approximate flow direction. Then the marching direction has the same role as time in an unsteady problem. Such marching techniques are clearly very efficient (Sect. 14.2.4).

For fully hyperbolic problems explicit schemes have been widely used in the past and will probably continue to be so where the discretisation step in the time or time-like coordinate is more restricted by the need to achieve a predetermined accuracy than by a stability limitation. For one-dimensional unsteady supersonic inviscid flow the explicit stability limit for many schemes is the generalised CFL condition (compare Sect. 9.1.2) of $(|u| + a)\Delta t/\Delta x \leq 1.0$ where a is the local sound speed.

If embedded subsonic regions are present, e.g. a blunt body problem, it is necessary to adopt a pseudo-transient formulation (Sect. 6.4) to obtain the steady solution, i.e. to march in time until the solution no longer changes. Such techniques are computationally more expensive but are generally more robust in handling instabilities associated with boundaries between subsonic and supersonic flow (i.e. sonic lines and shock waves). Since time plays the role of an iteration parameter in this case it is usual to employ implicit techniques to avoid the time step stability limitation associated with explicit techniques. The construction of appropriate implicit techniques (Sect. 14.2.8) are often based on split or approximate factor-isation algorithms (Sects. 8.2 and 9.5.1).

Traditionally, problems for which the governing equation are hyperbolic everywhere have been solved by the method of characteristics (Sect. 2.5.1 and Liepmann and Roshko 1957). However, the method of characteristics has been superseded primarily because of the difficulty of incorporating shock waves. The grid adjacent to the shock degenerates since the characteristics run together at the shock, which forms the boundary of the computational region. In addition, comparative tests (Rackich and Kutler 1972) indicate that traditional methods of characteristics are slower than competitive finite difference methods. However, some finite difference schemes, e.g. the Moretti scheme (Sect. 14.2.5), use a form of the governing equations such that knowledge of the characteristic locations can be exploited very efficiently.

Expressing the governing equations in characteristic form is very useful for determining the number and nature of the boundary conditions (Thompson 1990) and, where appropriate, for extrapolating the internal solution to the boundary along a characteristic (Rudy and Strikwerda 1981; Chakravarthy 1983). Expressing the governing equations in characteristic form is also useful (Roe 1986) in constructing special techniques for predicting the flow when strong shocks are present (Sect. 14.2.6).

For supersonic flows that contain shocks it is possible to calculate the change in flow properties across the shock from the Rankine–Hugoniot conditions, e.g. (11.110), and to link them with a computational scheme suitable for the shock-free region. Such a strategy is often referred to as a shock-fitting scheme. However, for non-simple flows, e.g. a cone at angle of attack, secondary shocks occur (Fletcher 1975) whose locations are not known, a priori. The complicated coding logic required to use a shock-fitting method for non-simple flows makes them less effective.

By casting the equations in conservation form, for example (11.116), and by using a discretised form that conserves the mass, etc., it is possible to obtain solutions that satisfy the weak form (5.6) of the governing equations. As shown by Lax and Wendroff (1960), solutions of the weak form of the governing equations automatically satisfy the Rankine–Hugoniot jump conditions across any discontinuities that may occur in the flow. Shock waves are the most common form of such discontinuities. Consequently the solution of the discretised equations automatically captures the shock behaviour, both the strength and the shock speed for unsteady flow. The finite volume method (Sect. 5.2) discretises the weak form of the governing equations directly and is therefore well-suited to the prediction of flows with embedded shocks (Vinokur 1989). The main difficulty with shock-capturing techniques is in obtaining sharply defined profiles of the flow variables through the shock without having to introduce special procedures which inevitably reduce the economy of the overall method.

14.2.2 MacCormack's Predictor–Corrector Scheme

A very effective finite-difference technique for inviscid supersonic flow, particularly as the foundation for steady-flow shock-capturing techniques, is MacCormack's explicit predictor–corrector scheme (MacCormack 1969). This scheme can be illustrated for the inviscid Burgers' equation (10.2) written in conservation form,

$$\frac{\partial u}{\partial t} + \frac{\partial F}{\partial x} = 0 \ , \tag{14.28}$$

where $F = 0.5u^2$. At the first (predictor) stage an intermediate solution u^* is calculated from

$$u_j^* = u_j^n - \frac{\Delta t}{\Delta x}(F_{j+1}^n - F_j^n) \ . \tag{14.29}$$

The corrector stage is

$$u_j^{n+1} = 0.5(u_j^n + u_j^*) - \frac{\Delta t}{2\Delta x}(F_j^* - F_{j-1}^*) \ . \tag{14.30}$$

The scheme is constructed from one-sided difference formulae at each stage but contributions to the truncation error cancel to produce a scheme which is second-order accurate in time and space. By reversing the asymmetric differencing of F in (14.29 and 30) an equivalent scheme is generated.

MacCormack's scheme is conceptually similar to the two-stage Lax–Wendroff scheme (10.11, 12). In fact for linear problems, e.g. (9.2), MacCormack's scheme reduces to the single-stage Lax–Wendroff scheme (9.16).

In order to obtain stable solutions of (14.29 and 30) for the particular choice $F = 0.5u^2$, the time-step must be restricted to $\Delta t \leq \Delta x/u$, i.e. Δt is limited by the CFL condition (Sect. 9.1.2). If $F(u)$ in (14.28) is more general the time-step restriction for the MacCormack and Lax–Wendroff schemes becomes

$$|A(u)|\frac{\Delta t}{\Delta x} \leq 1 \ , \tag{14.31}$$

where $A = dF/du$. For the vector equivalent of (14.28), e.g. (10.40), $\underline{A}$ is a matrix with elements $\partial F_j/\partial u_i$. The corresponding stability restriction is

$$|\lambda_k|\frac{\Delta t}{\Delta x} \leq 1 \ , \qquad k = 1, 2, \ldots, n \ , \tag{14.32}$$

where λ_k are the eigenvalues of $\underline{A}$.

MacCormack's scheme, and the two-stage Lax–Wendroff scheme, can be interpreted as members of the S_β^α family introduced by Lerat and Peyret (1975). The parameters α and β determine where on the discretised (x, t) grid u_j is effectively evaluated (Fig. 14.16). The MacCormack schemes correspond to $\alpha = 0$ and $\beta = 0$ or 1.

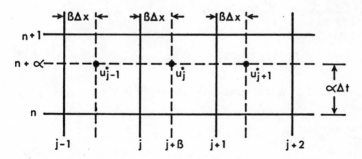

Fig. 14.16. Effective evaluation points for S_β^α family

The S_β^α family is described by Peyret and Taylor (1983, pp. 48–51). All members of the S_β^α family are conservative which is necessary for correctly predicting the shock strength and propagation speed, when used to capture shocks automatically.

The need for the equations to be expressed in conservation form and for conservative differencing to be used can be illustrated for the inviscid Burgers' equation (14.28). For initial conditions

$$u=a \quad \text{at} \quad t=0 \quad \text{and} \quad -\infty<x\leq 0 \ ,$$

$$u=b \quad \text{at} \quad t=0 \qquad \quad 0<x\leq +\infty \ ,$$

the exact solution is

$$u=a \quad \text{for} \quad x\leq U_p t$$

$$u=b \quad \text{for} \quad x>U_p t \ ,$$

where U_p is the propagation speed to be determined. If L is large enough to contain the disturbance, then from (14.28)

$$\frac{\partial}{\partial t} \int_{-L}^{L} u\,dx=[-0.5u^2]_{-L}^{L}=0.5(a^2-b^2) \ . \tag{14.33}$$

But also

$$\frac{\partial}{\partial t} \int_{-L}^{L} u\,dx=U_p[-u]_{-L}^{L}=U_p(a-b) \ , \quad \text{so that}$$

$$U_p=\frac{[0.5u^2]}{[u]}=\frac{0.5(b^2-a^2)}{b-a}=0.5(a+b) \ , \tag{14.34}$$

where $[\ ]$ denotes the evaluation of the change in the quantity across the discontinuity. If (14.28) is differenced by MacCormack's scheme (14.29, 30) and the intermediate solution is eliminated, one obtains

$$\frac{u_j^{n+1}-u_j^n}{\Delta t}+\frac{F_{j+1}^n-F_{j-1}^n}{2\Delta x}-0.5\frac{\Delta t}{\Delta x^2}\left[u_j(F_{j+1}-F_j)-u_{j-1}(F_j-F_{j-1})\right]^n$$

$$+0.25\frac{\Delta t^2}{\Delta x^3}\left[(F_{j+1}-F_j)^2-(F_j-F_{j-1})^2\right]^n=0 \ . \tag{14.35}$$

If $\partial/\partial t \int_{-L}^{L} u\,dx$ is evaluated by a mid-point rule, the result is

$$\frac{\partial}{\partial t} \int_{-L}^{L} u\,dx=\frac{\Delta x}{\Delta t}[0.5(u_1^{n+1}-u_1^n)+(u_2^{n+1}-u_2^n)+\dots$$

$$+(u_{N-1}^{n+1}-u_{N-1}^n)+0.5(u_N^{n+1}-u_N^n)] \ .$$

Substituting from (14.35) and noting the interior cancellation gives

$$\frac{\partial}{\partial t}\int_{-L}^{L} u\,dx = 0.25[-(F_0+2F_1+F_2)+(F_{N-1}+2F_N+F_{N+1})]+O(\Delta x)$$

$$= 0.5(a^2-b^2)+O(\Delta x) ,$$

i.e. (14.33) is preserved. This result is exact if $F_0=F_1=F_2$ and $F_{N-1}=F_N=F_{N+1}$.

14.2.3 SHOCK: Propagating Shock Wave Computation

The MacCormack scheme and the two-stage Lax–Wendroff scheme will be applied to the propagation of a shock wave in one-dimensional unsteady flow. The use of artificial viscosity to generate a smooth shock profile will be demonstrated.

The propagation of a shock wave in one-dimensional inviscid flow is governed by the equations

$$\frac{\partial\varrho}{\partial t}+\frac{\partial(\varrho u)}{\partial x}=0 , \tag{14.36}$$

$$\frac{\partial(\varrho u)}{\partial t}+\frac{\partial}{\partial x}(\varrho u^2+p)=0 , \tag{14.37}$$

$$\frac{\partial}{\partial t}[\varrho(e+\tfrac{1}{2}u^2)]+\frac{\partial}{\partial x}\{u[\varrho(e+\tfrac{1}{2}u^2)+p]\}=0 . \tag{14.38}$$

These equations are the conservation form of the continuity, x-momentum and energy equations respectively; (11.117) without the τ and $\dot{Q}$ terms. For an ideal gas, such as air, the specific internal energy in (14.38) can be expressed as

$$e=c_v T=\frac{p}{(\gamma-1)\varrho} , \tag{14.39}$$

where γ is specific heat ratio. With the aid of (14.39) it is clear that (14.36–38) contain three dependent variables: u, ϱ and p.

For the propagating shock problem Dirichlet boundary conditions for u, ϱ and p are required upstream and downstream of the shock. These are

$$\begin{aligned}
u=u_1 , \quad & \varrho=\varrho_1 , \quad p=p_1 \quad \text{at} \quad x=x_1 , \\
u=u_2=0 , \quad & \varrho=\varrho_2 , \quad p=p_2 \quad \text{at} \quad x=x_2 .
\end{aligned} \tag{14.40}$$

At time $t=0$, the shock is located at $x=x_0$. Therefore, appropriate initial conditions are

$$\begin{aligned}
u(x,0)=u_1 , \quad & \varrho(x,0)=\varrho_1 , \quad p(x,0)=p_1 , \quad \text{for} \quad x_1\leq x<x_0 , \\
u(x,0)=0 , \quad & \varrho(x,0)=\varrho_2 , \quad p(x,0)=p_2 , \quad \text{for} \quad x_0<x\leq x_2 .
\end{aligned} \tag{14.41}$$

Location $x = x_1$ is far upstream of the shock and location $x = x_2$ is far downstream of the shock.

It is convenient to nondimensionalise the dependent variables with respect to the conditions downstream of the shock. Noting that

$$\varrho_2 a_2^2 = \gamma p_2 ,$$
(14.42)

(14.36–38) can be written in compact nondimensional form as

$$\frac{\partial \mathbf{q}}{\partial t'} + \frac{\partial \mathbf{F}}{\partial x'} = 0 , \quad \text{where}$$
(14.43)

$$\mathbf{q} = \begin{bmatrix} \varrho' \\ \varrho' u' \\ p'/\gamma(\gamma-1) + \dfrac{\varrho}{2}(u')^2 \end{bmatrix} , \quad \mathbf{F} = \begin{bmatrix} \varrho' u' \\ \varrho'(u')^2 + p'/\gamma \\ [p'/(\gamma-1) + 0.5\varrho'(u')^2]u' \end{bmatrix}$$
(14.44)

and

$$\varrho' = \frac{\varrho}{\varrho_2} , \quad u' = \frac{u}{a_2} , \quad p' = \frac{p}{p_2} , \quad x' = \frac{x}{L} , \quad t' = a_2 \frac{t}{L} .$$

The prime denotes a nondimensional quantity.

The nondimensional boundary conditions, replacing (14.40), are

$$u'_1 = \frac{u_1}{a_2} , \quad \varrho'_1 = \frac{\varrho_1}{\varrho_2} , \quad p'_1 = \frac{p_1}{p_2} \quad \text{at} \quad x'_1 = \frac{x_1}{L} ,$$
$$u'_2 = 0 , \quad \varrho'_2 = 1 , \quad p'_2 = 1 \quad \text{at} \quad x'_2 = \frac{x_2}{L} .$$
(14.45)

The pressure ratio p_1/p_2 is the governing parameter for this problem and determines the shock strength and speed of propagation. For a given pressure ratio the boundary values u'_1 and ϱ'_1 are obtained from the Rankine–Hugoniot relations (Liepmann and Roshko 1957, p. 64)

$$u'_1 = \left(\frac{p_1}{p_2} - 1 \right) \left\{ 2 \Big/ \left[\gamma(\gamma+1)\frac{p_2}{p_1} + \gamma(\gamma-1) \right] \right\}^{1/2}$$
(14.46)

$$\varrho'_1 = \frac{1 + \dfrac{(\gamma+1)}{(\gamma-1)}\dfrac{p_2}{p_1}}{\dfrac{(\gamma+1)}{(\gamma-1)} + \dfrac{p_2}{p_1}} .$$

The shock propagation speed is needed to determine the shock location so that the accuracy of the computational solution of (14.43–45) may be assessed. The nondimensional shock propagation speed is given by the Rankine–Hugoniot relations as

$$u'_{ss} = \frac{u_{ss}}{a_2} = \left(\frac{\gamma - 1}{2\gamma} + \frac{p_2}{p_1} \frac{\gamma + 1}{2\gamma} \right)^{1/2} . \tag{14.47}$$

The primes will now be dropped. After time t, the exact solution is

$$u_{ex}(x, t) = u_1 , \quad \varrho_{ex}(x, t) = \varrho_1 \quad \text{for} \quad x_1 \leqq x \leqq x_0 + u_{ss} t ,$$

$$\tag{14.48}$$

$$u_{ex}(x, t) = 0 , \quad \varrho_{ex}(x, t) = \varrho_2 \quad \text{for} \quad x_0 + u_{ss} t \leqq x \leqq x_2 .$$

In the program SHOCK the MacCormack scheme and the two-stage Lax–Wendroff scheme are applied to (14.43) as:

i) MacCormack scheme:

$$q_j^* = q_j^n - \frac{\Delta t}{\Delta x} (F_{j+1}^n - F_j^n) , \tag{14.49}$$

$$\mathbf{q}_j^{n+1} = 0.5(\mathbf{q}_j^n + \mathbf{q}_j^*) - 0.5 \frac{\Delta t}{\Delta x} [F_j^n - F_{j-1}^n] . \tag{14.50}$$

ii) Lax–Wendroff scheme:

$$\mathbf{q}_{j+1/2}^* = 0.5(\mathbf{q}_j^n + \mathbf{q}_{j+1}^n) - 0.5 \frac{\Delta t}{\Delta x} [F_{j+1}^n - F_j^n] , \tag{14.51}$$

$$\mathbf{q}_j^{n+1} = \mathbf{q}_j^n - \frac{\Delta t}{\Delta x} [F_{j+1/2}^* - F_{j-1/2}^*] . \tag{14.52}$$

For (14.43 and 44) the eigenvalues of $\underline{A} \equiv \partial F / \partial \mathbf{q}$ are $\lambda = u, u + a, u - a$. Consequently the stability restriction (14.32) becomes $(|u| + a) \Delta t / \Delta x \leqq 1$ for both schemes.

The above formulation is implemented in program SHOCK (Fig. 14.17). The parameters used in SHOCK are described in Table 14.3. Typical shock profiles generated by program SHOCK are shown in Fig. 14.18. These solutions have been obtained with 101 spatial nodes and $\Delta x = 0.01$. The boundary conditions (14.45) are applied at $x_1 = 0$, $x_2 = 1.0$. The time step $\Delta t = 0.002$ and the shock is located at $x = 0.501$ at $t = 0$. The solutions shown in Fig. 14.18 have been obtained after 100 time steps for a pressure ratio $p_1 / p_2 = 2.5$, and a specific heat ratio $\gamma = 1.4$. For these conditions the shock propagation speed $u'_{ss} = 1.512$.

The Lax–Wendroff scheme produces a solution (Fig. 14.18) with severe oscillations just upstream of the shock. The MacCormack scheme produces a similar solution (not shown) but with oscillations of slightly reduced magnitude. The oscillations are caused, primarily, by dispersion errors (Sect. 9.2). As might be expected they get worse as the shock strength (p_1 / p_2) increases.

```
 1   C      SHOCK COMPUTES THE PROPAGATION OF A SHOCK USING
 2   C      MACCORMACK OR LAX-WENDROFF SCHEMES
 3   C      WITH ARTIFICIAL VISCOSITY OR FCT SMOOTHING
 4   C
 5          DIMENSION X(101),P(101),RH(101),U(101),T(101),UEX(101)
 6          DIMENSION Q(101,3),QD(101,3),F(101,3),DF(101,3),DL(3),DLM(3)
 7          OPEN(1,FILE='SHOCK.DAT')
 8          OPEN(6,FILE='SHOCK.OUT')
 9          READ(1,1)IME,IFCT,IPR,NX,NT,DT,GAM,PRAT
10          READ(1,2)ETA,ET1,ET2,ENL,SHST
11        1 FORMAT(5I5,3F5.3)
12        2 FORMAT(3F8.5,2F5.3)
13   C
14          NXM = NX - 1
15          ANX = NXM
16          DX = 1./ANX
17          JSST = SHST/DX
18          DTR = DT/DX
19          GMM = GAM - 1.
20          GMP = GAM + 1.
21          GMR = GMM/GMP
22          SHS = 0.5*(GMM + GMP*PRAT)/GAM
23          SHS = SQRT(SHS)
24   C
25          IF(IME .EQ. 2)WRITE(6,4)
26          IF(IME .EQ. 1)WRITE(6,3)
27          IF(IFCT.EQ. 1)WRITE(6,5)ETA,ET1,ET2
28          WRITE(6,6)NX,NT,DX,DT,ENL
29          WRITE(6,7)GAM,PRAT,SHS,SHST
30        3 FORMAT(' ONE-DIMENSIONAL SHOCK PROPAGATION,   MACCORMACK SCHEME')
31        4 FORMAT(' ONE-DIMENSIONAL SHOCK PROPAGATION,   LAX-WENDROFF'
32          1,' SCHEME')
33        5 FORMAT(' FLUX CORRECTED TRANSPORT,     ETA,ET1,ET2=',3F8.5)
34        6 FORMAT(' NX=',I3,' NT=',I3,' DX=',F5.3,' DT=',F5.3,'  ENL=',F5.3)
35        7 FORMAT(' GAM=',F5.3,'  P1/P2=',F5.2,'  SH/SP=',F5.3,' SHST=',
36          1F5.3,/)
37   C
38   C      SET INITIAL CONDITIONS
39   C
40          DUM = GMR + PRAT
41          RHD = DUM/(1. + GMR*PRAT)
42          DIM = SQRT(2.*GAM/GMP/DUM)
43          UD = (PRAT - 1.)*DIM/GAM
44          DO 8 J = 1,JSST
45          U(J) = UD
46          RH(J) = RHD
47          P(J) = PRAT
48          T(J) = P(J)/RH(J)
49        8 CONTINUE
50          JHR = JSST + 1
51          DO 9 J = JHR,NX
52          U(J) = 0.
53          RH(J) = 1.
54          P(J) = 1.
55          T(J) = 1.
56        9 CONTINUE
57          TIM = 0.
58          N = 0
59   C
60   C      SET INITIAL Q AND F
```

Fig. 14.17. Listing of program SHOCK

```
61  C
62          DO 10 J = 1,NX
63          AJ = J - 1
64          X(J) = AJ*DX
65          Q(J,1) = RH(J)
66          Q(J,2) = RH(J)*U(J)
67          Q(J,3) = P(J)/GAM/GMM + 0.5*RH(J)*U(J)*U(J)
68          F(J,1) = Q(J,2)
69          F(J,2) = P(J)/GAM + U(J)*Q(J,2)
70          F(J,3) = (P(J)/GMM + 0.5*U(J)*Q(J,2))*U(J)
71       10 CONTINUE
72          WRITE(6,21)N,TIM
73          WRITE(6,22)(X(J),J=1,NX)
74          WRITE(6,23)(U(J),J=1,NX)
75          WRITE(6,24)(RH(J),J=1,NX)
76          WRITE(6,25)(P(J),J=1,NX)
77          WRITE(6,26)(T(J),J=1,NX)
78  C
79  C       ADVANCE SOLUTION IN TIME
80  C
81          DO 30 N = 1,NT
82          AN = N
83          IF(IFCT .NE. 1)GOTO 12
84          ENU = ETA + 0.25*ET1*((U(1)+U(2))*DTR)**2
85          DO 11 K = 1,3
86       11 DF(1,K) = ENU*(Q(2,K)-Q(1,K))
87  C
88  C       OBTAIN HALF-STEP SOLUTION
89  C
90       12 DO 18 J = 2,NXM
91          DO 13 K = 1,3
92          IF(IME .EQ. 1)QD(J,K) = Q(J,K) - DTR*(F(J+1,K) - F(J,K))
93          IF(IME .EQ. 2)QD(J,K) = 0.5*(Q(J,K)+Q(J+1,K)) - 0.50*DTR*
94         1(F(J+1,K) - F(J,K))
95          IF(IFCT .EQ. 1)ENU = ETA+0.25*ET1*((U(J)+U(J+1))*DTR)**2
96          IF(IFCT .EQ. 1)DF(J,K) = ENU*(Q(J+1,K) - Q(J,K))
97       13 CONTINUE
98          F(J,1) = QD(J,2)
99          UD = QD(J,2)/QD(J,1)
100         PD = (QD(J,3) - 0.5*UD*QD(J,2))*GAM*GMM
101         F(J,2) = PD/GAM + UD*QD(J,2)
102         F(J,3) = (PD/GMM + 0.5*UD*QD(J,2))*UD
103 C
104 C       OBTAIN FULL-STEP SOLUTION
105 C
106         DO 14 K = 1,3
107         IF(IME .EQ. 1)Q(J,K) = 0.5*(Q(J,K)+QD(J,K)) - 0.5*DTR*(F(J,K)
108        1 - F(J-1,K))
109         IF(IME .EQ. 2)Q(J,K) = Q(J,K) - DTR*(F(J,K) - F(J-1,K))
110      14 CONTINUE
111 C
112 C       ARTIFICAL VISCOSITY SMOOTHING
113 C
114         IF(IFCT .EQ. 1)GOTO 18
115         IF(J .NE. 2)GOTO 16
116         DO 15 K = 1,3
117         DLM(K) = Q(2,K) - Q(1,K)
118         IF(ABS(DLM(K)) .LT. 1.0E-06)DLM(K) = 1.0E-06*SIGN(1.0,DLM(K))
119         DUC = ABS(DLM(K))
120      15 DLM(K) = DUC*DLM(K)
```

Fig. 14.17. (cont.) Listing of program SHOCK

```
121    16 DO 17 K = 1,3
122       DL(K) = Q(J+1,K) - Q(J,K)
123       IF(ABS(DL(K)) .LT.1.0E-06)DL(K)=1.0E-06*SIGN(1.0,DL(K))
124       DL(K) = ABS(DL(K))*DL(K)
125       Q(J,K) = Q(J,K) + ENL*DTR*(DL(K) - DLM(K))
126       DLM(K) = DL(K)
127    17 CONTINUE
128    18 CONTINUE
129  C
130  C     FCT SMOOTHING
131  C     OBTAIN RH,U,P AND F
132  C
133       IF(IFCT .EQ. 1)CALL FCT(NXM,ETA,ET2,DTR,Q,DF,U)
134  C
135       DO 19 J = 2,NXM
136       RH(J) = Q(J,1)
137       U(J) = Q(J,2)/Q(J,1)
138       P(J) = (Q(J,3) - 0.5*U(J)*Q(J,2))*GAM*GMM
139       T(J) = P(J)/RH(J)
140       F(J,1) = RH(J)*U(J)
141       F(J,2) = P(J)/GAM + RH(J)*U(J)*U(J)
142       F(J,3) = (P(J)/GMM + 0.5*RH(J)*U(J)*U(J))*U(J)
143    19 CONTINUE
144  C
145       TIM = AN*DT
146       IF(N .EQ. NT)GOTO 20
147       IF(IPR .EQ. 0)GOTO 30
148    20 WRITE(6,21)N,TIM
149    21 FORMAT(/,'  N=',I3,'  TIM=',E10.3)
150       WRITE(6,22)(X(J),J=1,NX)
151       WRITE(6,23)(U(J),J=1,NX)
152       WRITE(6,24)(RH(J),J=1,NX)
153       WRITE(6,25)(P(J),J=1,NX)
154       WRITE(6,26)(T(J),J=1,NX)
155    22 FORMAT('  X=',12F6.3)
156    23 FORMAT('  U=',12F6.3)
157    24 FORMAT('  RH=',12F6.3)
158    25 FORMAT('  P=',12F6.3)
159    26 FORMAT('  T=',12F6.3)
160       IF(IPR .LE. 1)GOTO 30
161       DO 27 K = 1,3
162       WRITE(6,28)(Q(J,K),J=1,NX)
163    27 WRITE(6,29)(F(J,K),J=1,NX)
164    28 FORMAT('  Q=',12F6.3)
165    29 FORMAT('  F=',12F6.3)
166    30 CONTINUE
167  C
168  C     EXACT SOLUTION, UEX
169  C
170       SHFN = SHST + SHS*TIM
171       JSST = SHFN/DX + 1.0
172       DO 31 J = 1,JSST
173       IF(J .LE. JSST)UEX(J) = U(1)
174       IF(J .GT. JSST)UEX(J) = U(NX)
175    31 CONTINUE
176       WRITE(6,32)(UEX(J),J=1,NX)
177    32 FORMAT(' UEX=',12F6.3)
178       STOP
179       END
```

Fig. 14.17. (cont.) Listing of program SHOCK

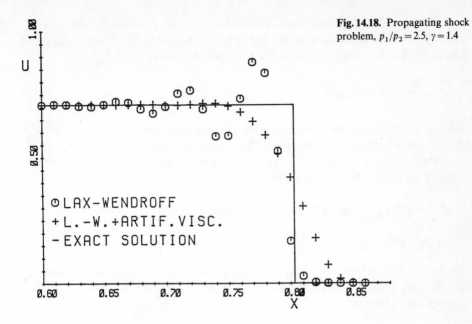

Fig. 14.18. Propagating shock problem, $p_1/p_2 = 2.5$, $\gamma = 1.4$

○ LAX-WENDROFF
+ L.-W.+ARTIF.VISC.
- EXACT SOLUTION

Table 14.3. Parameters used in program SHOCK

Parameter	Description
IME	$=1$ MacCormack scheme; $=2$ Lax–Wendroff scheme
IFCT	$=1$ flux-correct transport (FCT) smoothing, Sect. 14.2.7
IPR	$\geqq 1$ print u', ϱ', p', T'; $\geqq 2$ print Q, F
NX	number of spatial nodes
NT	number of time steps
DT	time step, $\Delta t'$
DX	spatial step, $\Delta x'$
GAM	specific heat ratio, γ
PRAT	pressure ratio, p_1/p_2
SHST	location of shock at $t' = 0$
SHS	shock speed, u_{ss}/a_2
SHFN	location of shock at $t' = NT.\Delta t'$
U, RH, P, T	u', ϱ', p', T', (14.44)
Q, F	$\mathbf{q}^n$, $\mathbf{q}^{n+1}$, $\mathbf{F}^n$, $\mathbf{F}^*$, (14.49–52)
QD	$\mathbf{q}^*$, (14.49)
ENL	artificial viscosity, v, (14.53)
DL, DLM	$\Delta\mathbf{q}^{**}_{j+1}$, $\Delta\mathbf{q}^{**}_j$, (14.54)
TIM	time, t'
UEX	exact velocity, u'_{ex}, at $t' = NT.\Delta t'$
ETA, ET1, ET2	parameters used in FCT smoothing, Table 14.4
ENU, EMU, DF	variables required for FCT smoothing, Table 14.4
FCT	subroutine containing implementation of FCT smoothing, Fig. 14.24

It is possible to substantially eliminate these oscillations by introducing artificial viscosity. It is desirable that the viscosity be equally effective for weak and strong shocks. A suitable quadratic formulation replaces (14.43) by

$$\frac{\partial \mathbf{q}}{\partial t} + \frac{\partial \mathbf{F}}{\partial x} - v \Delta x^2 [|\mathbf{q}_x| \mathbf{q}_x]_x = 0 , \qquad (14.53)$$

where v is a constant to be chosen. The artificial viscosity is introduced after a provisional solution at time level $n+1$ has been obtained.

If the solution obtained from (14.50 or 52) is interpreted as $\mathbf{q}_j^{**}$, the artificial viscosity correction is introduced as

$$\mathbf{q}_j^{n+1} = \mathbf{q}_j^{**} + v \frac{\Delta t}{\Delta x} \Delta [|\Delta \mathbf{q}_{j+1}^{**}| \Delta \mathbf{q}_{j+1}^{**}] , \quad \text{where} \qquad (14.54)$$

$$\Delta \mathbf{q}_{j+1}^{**} = \mathbf{q}_{j+1}^{**} - \mathbf{q}_j^{**} .$$

The use of artificial viscosity places a more severe stability restriction (Richtmyer and Morton, 1967, p. 336) on the time step. If the $|q_x|$ term in (14.53) is 'frozen' the stability restriction is

$$(|u'| + a)\frac{\Delta t}{\Delta x} \leqq (1+v^2)^{1/2} - v , \qquad (14.55)$$

so that as small a value of v as possible should be used for both shock profile accuracy and stability reasons.

In the program SHOCK the artificial viscosity is applied to the second and third components of (14.53). The result of using artificial viscosity ($v=1$) is seen (Fig. 14.18) to substantially alleviate the oscillations ahead of the shock at the cost of spreading the shock over more grid intervals. As might be expected from the form of (14.53 and 54), artificial viscosity has little effect on the solution away from the shock region.

For strong shocks the introduction of artificial viscosity is less satisfactory. This is illustrated in Sect. 14.2.7, where comparisons are made with the FCT algorithm for producing a sharp shock profile. Some of the coding in the program SHOCK, i.e. when IFCT$=1$, is introduced to implement the FCT algorithm. This is described in Sect. 14.2.7.

14.2.4 Inclined Cone Problem

For steady inviscid flows that are everywhere supersonic it is possible to select a marching direction for which the governing equations are hyperbolic. For the flow about a cone at angle of attack (Fig. 14.19) a cone generator (x) is a convenient marching direction. The governing equations for this problem are

$$\frac{\partial \mathbf{E}}{\partial x} + \frac{\partial \mathbf{F}}{\partial y} + \frac{\partial \mathbf{G}}{\partial \phi} + \mathbf{H} = 0 , \quad \text{where} \qquad (14.56)$$

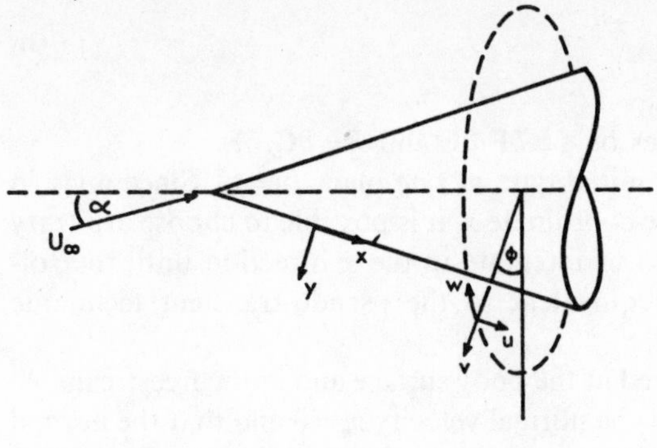

Fig. 14.19. Inclined cone geometry

$$E = r \begin{bmatrix} \varrho u \\ kp + \varrho u^2 \\ \varrho uv \\ \varrho uw \end{bmatrix} \qquad F = r \begin{bmatrix} \varrho v \\ \varrho uv \\ kp + \varrho v^2 \\ \varrho vw \end{bmatrix}$$

$$G = \begin{bmatrix} \varrho w \\ \varrho uw \\ \varrho vw \\ kp + \varrho w^2 \end{bmatrix} \qquad H = \begin{bmatrix} 0 \\ -(kp + \varrho w^2)\dfrac{\partial r}{\partial x} \\ -(kp + \varrho w^2)\dfrac{\partial r}{\partial y} \\ \varrho uw\dfrac{\partial r}{\partial x} + \varrho vw\dfrac{\partial r}{\partial y} \end{bmatrix},$$

where $k = (\gamma - 1)/2\gamma$ and γ is the specific heat ratio.

The MacCormack scheme, when applied to (14.56), is written as the following two-stage algorithm:

$$E_{j,k}^* = E_{j,k}^n - \frac{\Delta x}{\Delta y}[F_{j+1,k}^n - F_{j,k}^n]$$

$$- \frac{\Delta x}{\Delta \phi}[G_{j,k+1}^n - G_{j,k}^n] - H_{j,k}^n \Delta x \qquad (14.57)$$

and

$$E_{j,k}^{n+1} = 0.5\left[\left(E_{j,k}^n + E_{j,k}^*\right) - \frac{\Delta x}{\Delta y}(F_{j,k}^* - F_{j-1,k}^*) - \frac{\Delta x}{\Delta \phi}(G_{j,k}^* - G_{j,k-1}^*) - H_{j,k}^* \Delta x\right],$$

$$(14.58)$$

where $E_{j,k}^n = E(n\Delta x, j\Delta y, k\Delta \phi)$. The discretised equations (14.57, 58) are written as an explicit marching algorithm, exploiting the time-like nature of the x direction. However, the marching step, Δx, is expected to be limited by the following type of formula (Peyret and Taylor 1983, p. 69):

$$\left|\frac{|\lambda_{max}^A|}{\Delta y}+\frac{|\lambda_{max}^B|}{\Delta\phi}\right|\Delta x\leqq 1 \; , \tag{14.59}$$

where λ^A and λ^B are the eigenvalues of $\underline{A}\equiv\partial\mathbf{F}/\partial\mathbf{E}$ and $\underline{B}=\partial\mathbf{G}/\partial\mathbf{E}$.

The marching scheme requires initial data at one plane ($n=0$). Since inviscid conical flows are independent of the coordinate x, it is possible to choose arbitrary starting values at $x=x_0$ ($n=0$) and to integrate in the x direction until the solution no longer changes; this is equivalent to the pseudo-transient technique (Sect. 6.4).

Boundary conditions are required at the body surface and in the freestream. At the body surface it is necessary that the normal velocity $v_n=0$ and that the normal derivatives of all other variables are equal to zero. The farfield boundary is far enough away that the bow shock is within the computational domain, and automatically captured. Thus all variables in the farfield are known from the freestream Mach number. Kutler and Lomax (1971) report accurate results with 32 points in the ϕ-direction and 20 in the y-direction.

Application of the same method to the computation of space-shuttle flowfields (Kutler et al. 1973) uses the equivalent conical solution to give starting data, uses improved surface boundary conditions due to Abbett (1973) and makes the bow shock the outer boundary of the computational domain. This involves matching the Rankine–Hugoniot conditions across the shock (i.e. shock-fitting) to the values on the boundary of the computational domain so that the shock slope, and hence orientation, is determined. Secondary shocks that occur in the computational domain are captured. Typical shock locations are shown in Fig. 14.20. The BVLR method is a conceptually similar marching scheme for inviscid supersonic flow, and is described by Holt (1984).

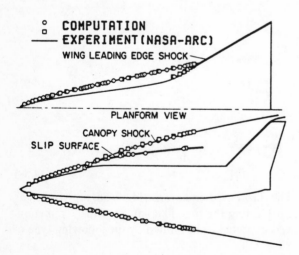

Fig. 14.20. Space-shuttle shock-wave pattern, $M_\infty=7.4$ (after Kutler et al., 1973; reprinted with permission of AIAA)

14.2.5 Moretti λ-Scheme

The simple marching scheme described in Sect. 14.2.4 is not suitable if subsonic regions occur, as with a blunt-body flow, or if the flow has substantial reflex curvature (*ABC* in Fig. 14.21). For large regions of subsonic flow, solutions can be obtained by marching the unsteady Euler equations in time until the steady state is reached. For this formulation the MacCormack scheme is possible but rather slow due to the *CFL* limitation on Δt.

The problem with shapes like that in Fig. 14.21 is that the local marching direction used with the MacCormack scheme can violate the local domain of dependence of the solution. This feature is overcome in the λ-scheme (Moretti 1979) which uses knowledge of the characteristic directions to determine which nodal values contribute to the spatial derivative discretisation.

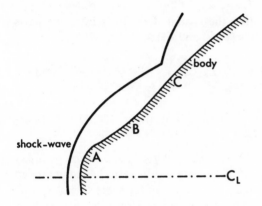

Fig. 14.21. Axisymmetric blunt body at high Mach number

The method of constructing the difference scheme can be demonstrated for one-dimensional unsteady inviscid flow which is governed by the equations

$$\frac{\partial p}{\partial t} + u\frac{\partial p}{\partial x} + \gamma\frac{\partial u}{\partial x} = 0 \quad \text{and} \tag{14.60}$$

$$\frac{\partial u}{\partial t} + \frac{a^2}{\gamma}\frac{\partial p}{\partial x} + u\frac{\partial u}{\partial x} = 0 \ , \tag{14.61}$$

where u is the velocity, p is the *logarithm* of the pressure, a is the sound-speed and γ is the specific heat ratio. Equations (14.60 and 61) are the continuity and x-momentum equations, with the density eliminated in favour of the pressure.

Characteristic (directions) $\lambda = dx/dt$ are given by

$$\lambda_1 = u - a \quad \text{and} \quad \lambda_2 = u + a \ . \tag{14.62}$$

Typical configurations of λ_1, λ_2 are shown in Fig. 14.22. This represents a well-behaved situation. Difficulties with conventional explicit schemes, like MacCormack's scheme, can arise when B and C lie on the same side of x_j. Moretti

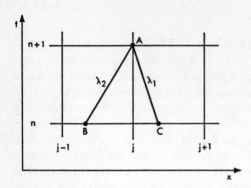

Fig. 14.22. Characteristics in the (x, t) plane

overcomes the problem by first splitting the spatial derivatives into two parts associated with the characteristics. Thus (14.60 and 61) are replaced by

$$\frac{\partial p}{\partial t} = -0.5\left(\lambda_1 \frac{\partial p}{\partial x_1} + \lambda_2 \frac{\partial p}{\partial x_2}\right) - \frac{\gamma}{2a}\left(\lambda_2 \frac{\partial u}{\partial x_2} - \lambda_1 \frac{\partial u}{\partial x_1}\right) , \tag{14.63}$$

$$\frac{\partial u}{\partial t} = -\frac{a}{2\gamma}\left(\lambda_2 \frac{\partial p}{\partial x_2} - \lambda_1 \frac{\partial p}{\partial x_1}\right) - 0.5\left(\lambda_1 \frac{\partial u}{\partial x_1} + \lambda_2 \frac{\partial u}{\partial x_2}\right) = 0 . \tag{14.64}$$

Substitution of λ_1, λ_2 from (14.62) and setting $\partial p/\partial x_1 = \partial p/\partial x_2$ and $\partial u/\partial x_1 = \partial u/\partial x_2$ in the above equations would recover (14.60 and 61), but by introducing one-sided difference formulae for $\partial p/\partial x_i$, etc., so that only points on the same side of x_j as λ_j are used, $\partial p/\partial x_1$ and $\partial p/\partial x_2$, etc., in (14.63 and 64) differ. However, the weighted evaluation, i.e. $0.5(\lambda_1 \partial p/\partial x_1 + \lambda_2 \partial p/\partial x_2)$, is a well-behaved discretisation of $u\partial p/\partial x$ whatever the values of λ_1 and λ_2.

Equations (14.63, 64) are integrated as a predictor/corrector scheme, i.e.

$$\mathbf{f}_j^* = \mathbf{f}_j^n + \left(\frac{\partial \mathbf{f}}{\partial t}\right)_j^n \Delta t \quad \text{and} \tag{14.65}$$

$$\mathbf{f}^{n+1} = 0.5\left[\mathbf{f}_j^n + \mathbf{f}_j^* + \left(\frac{\partial \mathbf{f}}{\partial t}\right)_j^* \Delta t\right] , \tag{14.66}$$

where $\mathbf{f} = (p, u)$ and $\partial \mathbf{f}/\partial t$ are obtained from (14.63 and 64) with $\partial \mathbf{f}/\partial x_i$ evaluated according to the following rules:

Predictor stage,

$$\left(\frac{\partial \mathbf{f}}{\partial x}\right)_i^n = \begin{cases} \dfrac{\mathbf{f}_{j+1}^n - \mathbf{f}_j^n}{\Delta x} , & \text{if } \lambda_i < 0 , \\[2mm] \dfrac{2\mathbf{f}_j^n - 3\mathbf{f}_{j-1}^n + \mathbf{f}_{j-2}^n}{\Delta x} , & \text{if } \lambda_i > 0 , \end{cases} \tag{14.67}$$

Corrector stage,

$$\left(\frac{\partial \mathbf{f}}{\partial x}\right)_i^* = \begin{cases} \dfrac{-2\mathbf{f}_j^* + 3\mathbf{f}_{j+1}^* - \mathbf{f}_{j+2}^*}{\varDelta x} , & \text{if } \lambda_i < 0 , \\[2ex] \dfrac{\mathbf{f}_j^* - \mathbf{f}_{j-1}^*}{\varDelta x} , & \text{if } \lambda_i > 0 . \end{cases} \tag{14.68}$$

The scheme has second-order accuracy like the MacCormack scheme. Moretti (1979) shows that an equivalent structure can be extracted in more than one spatial dimension or for steady two- or three-dimensional supersonic flow. Moretti cautions that the present scheme is not capable of accurately capturing internal shocks but is capable of handling complicated geometries like that shown in Fig. 14.21.

Dadone and Napolitano (1983, 1985) develop an implicit form of the λ-scheme for the unsteady Euler equations for isentropic compressible flows. Although the method is more efficient than the above explicit formulation, for computing steady transonic flows only weak shocks are permitted or explicit shock fitting is employed (Dadone and Moretti 1988).

Napolitano (1986) provides a very lucid review of the λ-scheme. Strengths of the method are its general efficiency and ability to incorporate appropriate boundary conditions consistent with the theory of characteristics. Its main weakness is its inability to capture shocks correctly due to its non-conservative form. However, it appears possible (Dadone and Magi 1986) to add corrective terms at the shock to accurately predict the shock strength. Consequently the modified λ-scheme is effective for steady transonic flow.

In the formulation described above it is necessary to choose an equation set, and dependent variables, so that the characteristics (14.62) appear explicitly.

The same concept can be extended to the Euler equations (14.94) written in nonconservative form

$$\frac{\partial \mathbf{q}}{\partial t} + \underline{A}\frac{\partial \mathbf{q}}{\partial x} + \underline{B}\frac{\partial \mathbf{q}}{\partial y} = 0 , \tag{14.69}$$

where $\underline{A} = \partial \mathbf{F}/\partial \mathbf{q}$ and $\underline{B} = \partial \mathbf{G}/\partial \mathbf{q}$. The terms in matrices $\underline{A}$ and $\underline{B}$ are given by (14.99). The eigenvalues, λ_A of $\underline{A}$ define the characteristic directions dx/dt, and the eigenvalues λ_B of $\underline{B}$ define the characteristic directions dy/dt. It is possible to factor $\underline{A}$ and $\underline{B}$ as

$$\underline{A} = \underline{T}\underline{\varLambda}_A^+ \underline{T}^{-1} + \underline{T}\underline{\varLambda}_A^- \underline{T}^{-1} = \underline{A}^+ + \underline{A}^- ,$$

$$\underline{B} = \underline{S}\underline{\varLambda}_B^+ \underline{S}^{-1} + \underline{S}\underline{\varLambda}_B^- \underline{S}^{-1} = \underline{B}^+ + \underline{B}^- , \tag{14.70}$$

where $\underline{T}$ and $\underline{S}$ are matrices of left eigenvectors (Isaacson and Keller 1966 p. 137) associated with $\underline{A}$ and $\underline{B}$. $\underline{\varLambda}_A^+$ and $\underline{\varLambda}_A^-$ are the diagonal matrices of positive and negative eigenvalues of $\underline{A}$, and similarly for $\underline{\varLambda}_B^+$ and $\underline{\varLambda}_B^-$. Consequently (14.69) is written as

$$\frac{\partial \mathbf{q}}{\partial t} + \underline{A}^+ \frac{\partial \mathbf{q}}{\partial x} + \underline{A}^- \frac{\partial \mathbf{q}}{\partial x} + \underline{B}^+ \frac{\partial \mathbf{q}}{\partial y} + \underline{B}^- \frac{\partial \mathbf{q}}{\partial y} = 0 , \tag{14.71}$$

where backward differencing of $\partial \mathbf{q}/\partial x$ is introduced for the term multiplied by $\underline{A}^+$ and forward differencing for the term multiplied by $\underline{A}^-$, and similarly for the $\underline{B}^+ \partial \mathbf{q}/\partial y$ and $\underline{B}^- \partial \mathbf{q}/\partial y$ terms. The current formulation (Chakravarthy et al. 1980) is seen to generalise the λ-scheme.

It may be noted that (14.71) is a nonconservative form which does not provide accurate shock capturing. However, if the contributions to the split flux vector gradients $\partial \underline{F}^+/\partial x = \underline{A}^+ \partial \mathbf{q}/\partial x$ are constructed and discretised with one-sided differences as above, the result is the flux-vector splitting technique of Steger and Warming (1981) which is capable of capturing shocks. The Steger and Warming technique can be modified to ensure that the split flux Jacobians are continuous when the eigenvalue changes sign (van Leer 1982). This produces a more robust scheme and yields sharper shocks. But flux-vector splitting schemes are less economical than the λ-type schemes.

The flux-vector splitting technique and the flux-difference splitting technique permits shocks to be captured with monotonic profiles, but without the need to add explicit artificial viscosity. However, additional procedures are required to obtain sharp, non-oscillatory shock profiles. When the additional procedures are included, explicit versions of the overall schemes are not as economical as the MacCormack scheme but can be made second-order accurate (Harten 1983) away from discontinuities. Typical schemes are described in the next section.

14.2.6 Computation of Strong Shocks

For the typical blast-wave problem (i.e. an inherently unsteady flow) in which very strong shocks occur, more physically realistic methods, such as the Godunov or Glimm schemes (Holt 1984, and Peyret and Taylor 1983), are required. The Godunov formulation can be interpreted as a finite volume method (Sect. 5.2), which assumes that each pair of contiguous grid points (x_j, x_{j+1}) is separated by a expansion fan or a shock at $x_{j+1/2}$. This permits the known form of the exact solution for such archetypical flows (the Riemann problem) to be used to estimate the fluxes, e.g. F, in the governing equations. The original Godunov scheme is only first-order accurate; shocks tend to be smeared. Van Leer (1979) describes a second-order Godunov scheme that produces sharp shocks.

Colella and Woodward (1984) introduce a higher-order extension of Gudonov-type that uses piecewise parabolic interpolation. Woodward and Colella (1984) describe the application of this method to the evolution of the interaction of two blast waves and demonstrate that the method has an impressive ability to resolve fine detail, when using a uniform grid. However, the method is about five times slower than the MacCormack method with artificial viscosity. Consequently more economical schemes that approximate only some of the physical features inherent in the Godunov schemes are of considerable interest (Roe 1981; Osher and Solomon 1982). Such methods are often referred to as approximate Riemann solvers or flux-difference splitting methods.

Here we briefly describe the flux-corrected transport schemes of Boris and Book (1973) for treating strong shocks, since such schemes can also be interpreted as an extension to simple predictor/corrector methods, like MacCormack's scheme. Subsequently, more economical extensions of the flux-correcting concept are described.

The computational representation of 'step' profiles associated with strong shocks for inviscid flow is difficult. As already indicated in Sect. 9.2, upwind differencing introduces excessive diffusion which smooths away the step. The Lax–Wendroff scheme introduces dispersion errors which appear as "ripples" on either side of the step. If density is being computed using such a scheme a negative (and physically unrealistic) value could appear.

Boris and Book developed the flux-corrected transport approach as a general technique of "predictor/corrector" type in which large diffusion is introduced in the predictor stage and an equal (almost) amount of antidiffusion is introduced in the corrector stage. However, the antidiffusion is limited so that no new maxima or minima can appear in the solution, nor can an existing extrema be accentuated. This limiting step is important because it maintains the positivity of the solution, where appropriate, and effectively allows the diffusion introduced in the predictor stage to selectively annihilate the dispersive "ripples."

The flux-corrected transport approach can be clarified by applying a typical scheme to the one-dimensional continuity equation (11.10)

$$\frac{\partial \varrho}{\partial t} + \frac{\partial}{\partial x}(\varrho u) = 0 \ . \tag{14.72}$$

For ease of exposition, the velocity u will be treated as constant; then (14.72) coincides with (9.2). For the predictor stage the following finite difference algorithm for ϱ_j^* is obtained from (14.72):

$$\varrho_j^* = \varrho_j^n - 0.5C(\varrho_{j+1}^n - \varrho_{j-1}^n) + (v + 0.5C^2)(\varrho_{j+1}^n - 2\varrho_j^n + \varrho_{j-1}^n) \ , \tag{14.73}$$

where $C = u\Delta t/\Delta x$ and v is a positive diffusion coefficient. Typically $v = 1/8$. If $v = 0$ the Lax–Wendroff scheme (9.16) is recovered.

In principle the antidiffusion corrector stage could be introduced as

$$\varrho_j^{n+1} = \varrho_j^* - \mu(\varrho_{j+1}^* - 2\varrho_j^* + \varrho_{j-1}^*) \ , \tag{14.74}$$

where $\mu = v$ would be an obvious choice. However, to suit the more general case of variable velocity and to make the scheme conservative it is desirable to introduce antidiffusive mass fluxes

$$f_{j+1/2} = \mu(\varrho_{j+1}^* - \varrho_j^*) \quad \text{and} \quad f_{j-1/2} = \mu(\varrho_j^* - \varrho_{j-1}^*) \ . \tag{14.75}$$

If cell boundaries $x_{j-1/2} = 0.5(x_{j-1} + x_j)$ and $x_{j+1/2} = 0.5(x_j + x_{j+1})$ are defined, then $f_{j-1/2}$ represents the antidiffusive mass flux across the boundary $x_{j-1/2}$; and similarly for $f_{j+1/2}$.

The crucial feature of the flux-corrected transport scheme is to replace $f_{j+1/2}$ in (14.75) with

$$f^c_{j+1/2} = \text{sgn}(\Delta \varrho_{j+1/2}) \max \{0, \min[\Delta \varrho_{j-1/2} \text{sgn}(\Delta \varrho_{j+1/2}), \mu | \Delta \varrho_{j+1/2} |,$$
$$\Delta \varrho_{j+3/2} \text{sgn}(\Delta \varrho_{j+1/2})]\} , \tag{14.76}$$

where $\Delta \varrho_{j+1/2} = \varrho^*_{j+1} - \varrho^*_j$ and $\text{sgn}\, K = K/|K|$.

An equivalent formula is used to replace $f_{j-1/2}$ in (14.75). Equation (14.76) is the quantitative means of preventing the anti-diffusive stage from introducing new maxima or minima as required above. The final stage of the algorithm replaces (14.74) with

$$\varrho^{n+1}_j = \varrho^*_j - f^c_{j+1/2} + f^c_{j-1/2} . \tag{14.77}$$

The combined scheme (14.73, 75–77) has the following restriction for stable solutions:

$$C = u \frac{\Delta t}{\Delta x} < 0.5 . \tag{14.78}$$

This is more restrictive on Δt than the MacCormack or Lax–Wendroff schemes. For the equations governing one-dimensional inviscid compressible flow (14.36–38), Woodward and Colella (1984) recommend the slightly more restrictive condition $(|u| + a)\Delta t/\Delta x < 0.4$, where a is the local sound speed.

The effect of choosing different values for $\mu = \nu$ is indicated in Fig. 14.23. A value of $\mu = \nu = 0.125$ is close to optimal in minimising both diffusion and dispersion errors. The complete absence of spurious oscillations is noteworthy.

Boris and Book (1976) recommend that ν in (14.73) and μ in (14.74) be chosen to minimise dispersion errors produced by the discretised equations. This can be done by considering the truncation error as in Sect. 9.2. The specific formulae depend on the underlying scheme to which the flux-corrected transport is applied. Boris and Book (1976) analyse such schemes as upwind differencing, Lax–Wendroff and leapfrog and consider both explicit and implicit antidiffusion steps. For (14.73–77) the following values are recommended:

$$\nu = \frac{1}{6} + \frac{C^2}{3} , \qquad \mu = \frac{1}{6} - \frac{C^2}{6} . \tag{14.79}$$

The flux-correcting concept is also applicable in multidimensions. Extensions of the method are given by Zalesak (1979, 1987) and by Book (1981, pp. 29–41). Zalesak provides an alternative interpretation of the FCT algorithm. The first step, replacing (14.73), is made with a low-order scheme guaranteed to produce a non-oscillatory solution. The antidiffusive fluxes, replacing (14.75), are constructed as the difference between a high-order discretisation of the flux and the same low-order discretisation as used in the replacement of (14.73). The limiting of the antidiffusive fluxes, equivalent to (14.76), then ensures that the solution is obtained with the high-order flux evaluation except where this would introduce spurious oscillations.

Although the flux-correcting concept is effective for generating non-oscillatory shock profiles it is difficult to place it in a suitable theoretical framework. The

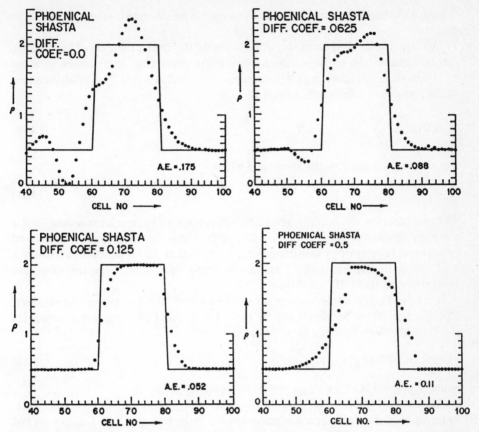

Fig. 14.23. Square-wave density comparison of flux-corrected transport schemes; $\mu = v =$ DIFF.COEF., A.E. $\equiv$ absolute error. PHEONICAL SHASTA = Lax–Wendroff for (14.72) (after Book et al., 1975; reprinted with permission of Academic Press)

theoretical development of shock-capturing schemes (Harten et al. 1983) has been guided by the requirement that numerical schemes for scalar conservation equations, like (14.28), must be monotonicity preserving if they are to converge to the physically relevant solution. For example, the numerical scheme must reject solutions associated with expansion shocks (Fig. 14.28). Schemes that select the physically relevant solution in the presence of discontinuities, shock waves or contact discontinuities, are said to be entropy-satisfying.

The concept of a monotonicity preserving solution is closely linked with the idea of preventing only spurious maxima or minima, i.e. oscillations, appearing in the solution as time is advanced. Thus if the initial data, u_j^0 are a monotonic function of x_j, solutions at a later time, u_j^n, should remain monotonic functions of x_j.

However, monotone schemes cannot be more than first-order accurate in space. Consequently such schemes are highly diffusive, shock profiles are smeared and accurate solutions cannot be obtained without unreasonable grid refinement. Thus the price of guaranteed theoretical convergence is an inaccurate solution on a finite grid.

An improvement in accuracy, without losing the strong theoretical foundation, can be achieved by replacing the monotonicity preserving requirement with the total variation diminishing (TVD) requirement (Harten 1983). The total variation of the numerical solution is defined by

$$\mathrm{TV}(u^n) = \sum_{j=-\infty}^{\infty} |u_{j+1}^n - u_j^n| \ . \tag{14.80}$$

Consequently a numerical scheme is TVD if

$$\mathrm{TV}(u^{n+1}) \leq \mathrm{TV}(u^n) \ . \tag{14.81}$$

TVD schemes do not generate spurious oscillations and can achieve second-order accuracy where the solution varies smoothly. However, they require additional restrictions to be entropy satisfying. A typical way that the higher-order accuracy is achieved is by the addition of an "antidiffusive flux" that is limited to ensure satisfaction of the TVD condition.

Clearly there is some conceptual similarity with the flux-correcting strategy. The similarity can be illustrated by considering the Lax–Wendroff scheme, i.e. (14.73) with $v=0$. This can be written as

$$\varrho_j^{n+1} = \varrho_j^n - C(\varrho_j^n - \varrho_{j-1}^n) - (f_{j+1/2} - f_{j-1/2}) \ , \tag{14.82}$$

where $f_{j+1/2} = 0.5C(1-C)(\varrho_{j+1} - \varrho_j)$ and similarly for $f_{j-1/2}$.

Equation (14.82) can be thought of as replacing (14.73 and 74). Thus the first-order upwind-differenced term replaces (14.73). If the term $(f_{j+1/2} - f_{j-1/2})$ were not present, (14.82) would be monotonicity preserving (and hence TVD). The term $(f_{j+1/2} - f_{j-1/2})$ is made up of anti-diffusive fluxes, equivalent to (14.74 and 75) with an appropriate choice of μ. However, the complete Lax–Wendroff scheme is not TVD; it produces oscillatory solutions in the presence of strong shocks. In the present interpretation the anti-diffusive fluxes inherent in the Lax–Wendroff scheme are too large and lead to the oscillatory behaviour. If they could be limited in some way the result would be a streamlined FCT algorithm. An effective way of limiting the antidiffusive fluxes is to introduce

$$f_{j+1/2}^c = \phi(r_j)[0.5C(1-C)](\varrho_{j+1} - \varrho_j) \ , \tag{14.83}$$

and a similar expression, $f_{j-1/2}^c$, to replace $f_{j-1/2}$ and $f_{j+1/2}$ in (14.82). The function $\phi(r_j)$ is the limiter. The parameter r_j measures the ratio of contiguous gradients, i.e.

$$r_j = \frac{\varrho_j - \varrho_{j-1}}{\varrho_{j+1} - \varrho_j} \ . \tag{14.84}$$

The function $\phi(r_j)$ should be chosen so that the anti-diffusive terms are as large as possible subject to the constraint that (14.82), with (14.83), is TVD. Sweby (1984) provides the following restrictions on $\phi(r)$:

$$0 < \phi(r) \le \min(2r, 2) \quad \text{for} \quad r > 0 \;,$$

$$\phi(r) = 0 \qquad\qquad \text{for} \quad r \le 0 \;. \tag{14.85}$$

The restrictions are satisfied by first- and second-order TVD schemes. To ensure second-order spatial accuracy, except at extrema $(r < 0)$, it is necessary that $\phi(1) = 1$. Roe (1986) recommends the choice

$$\phi(r) = \min(2, r) \quad \text{for} \quad r > 1 \;,$$

$$= \min(2r, 1) \quad \text{for} \quad 0 < r \le 1 \;, \tag{14.86}$$

$$= 0 \qquad\qquad \text{for} \quad r \le 0 \;.$$

Sweby (1984) shows that the flux-limiting algorithms of Van Leer (1974), Roe and Baines (1982) and Chakravarthy and Osher (1983) can be cast in the form of (14.82), with (14.83). The different algorithms correspond to different choices for $\phi(r)$.

It is clear that (14.82) with (14.83) is a one-step procedure in contrast to the two-step nature of the FCT algorithm. In addition the disposable parameters of the FCT algorithm are now concentrated into the choice of the limiter $\phi(r)$. The simple form of $\phi(r)$ and the one-step nature of the flux limiting algorithm enhances economical implementation. The one-step nature also permits efficient implicit TVD schemes to be developed (Yee et al. 1985) that are suitable for steady flow problems (Sect. 18.5), both inviscid and viscous, that involve shock waves.

For unsteady flow problems explicit TVD schemes are to be preferred since the linearisation introduced to make the implicit TVD schemes efficient, i.e. to permit exploitation of the Thomas algorithm (Sect. 6.2.2), renders the transient solution non-conservative.

It is desirable, as noted above, that the discrete form of the governing equation should be conservative, if discontinuous solutions are expected. An appropriate finite volume discretisation of (14.28) is

$$u_j^{n+1} = u_j^n - \frac{\Delta t}{\Delta x}(F_{j+1/2}^n - F_{j-1/2}^n) \;. \tag{14.87}$$

However, although this scheme is conservative in the differential sense it is desirable that it be equivalent to the integral or weak form (5.6) of the conservation law so that the jump conditions across any shock be correctly predicted. Equation (14.87) can be interpreted as a discrete integral statement if

$$u_j^n = \int_{x_{j-1/2}}^{x_{j+1/2}} u^n(x)\,dx \;,$$

i.e. the grid point value u_j is interpreted as the average value over the interval $x_{j-1/2} \le x \le x_{j+1/2}$. To be consistent with this interpretation $F_{j+1/2}$ is called a numerical flux function and is considered to be some appropriate function of local nodal values, i.e. $F_{j+1/2} = F(u_{j-k+1}, \ldots, u_{j+k})$. Equation (14.83) provides a simple example, since $(\Delta t/\Delta x)f_j = C\varrho_j$, etc.

The flux-limiting scheme (14.82, 83) can be expressed in the form of (14.87) as

$$\varrho_j^{n+1} = \varrho_j^n - \frac{\Delta t}{\Delta x}(\tilde{f}_{j+1/2}^n - \tilde{f}_{j-1/2}^n) \ , \quad \text{where} \tag{14.88}$$

$$\tilde{f}_{j+1/2} = 0.5(f_j + f_{j+1}) - 0.5\,\sigma(f_{j+1} - f_j) + 0.5\,\phi(r)(\sigma - C)(f_{j+1} - f_j) \tag{14.89}$$

and $\sigma = \operatorname{sgn} C_{j+1/2}$. This scheme allows for positive or negative u. However, more accurate solutions are obtained if (14.84) is replaced by

$$r_j = \frac{\varrho_{j+1-\sigma} - \varrho_{j-\sigma}}{\varrho_{j+1} - \varrho_j} \ , \tag{14.90}$$

i.e. the ratio of contiguous gradients is evaluated from upstream. In (14.89) the first two terms contribute to the upwind difference in (14.82). The last term is related to the limited flux (14.83). For (14.72) with constant $u, f = \varrho u$ in (14.88).

The above second-order TVD schemes can be extended to nonlinear systems of equations, like the Euler equations (14.43), by decomposing the Euler equations into characteristic form. The TVD schemes, e.g. (14.88), are applied to the individual characteristic components. The solution to the Euler equations is obtained from the summation of the contributions from the characteristic components. However, there is no theoretical guarantee, as there is for a scalar equation, that the solution will be TVD; but in practice shock profiles are non-oscillatory.

The characteristic component form of (14.43) can be written

$$\sum_{m=1}^{3} \alpha_m \left(\frac{\partial \mathbf{e}_m}{\partial t} + \frac{\partial}{\partial x}(\lambda_m \mathbf{e}_m) \right) = 0 \ , \tag{14.91}$$

where $\alpha_m = \{0.5/\gamma, \ (\gamma - 1)/\gamma, \ 0.5/\gamma\}$, the eigenvalues $\lambda_m = \{u - a, \ u, \ u + a\}$, and the corresponding characteristic vectors

$$\mathbf{e}_m = \begin{bmatrix} \varrho \\ \varrho(u-a) \\ \varrho(H-ua) \end{bmatrix} \ , \quad \begin{bmatrix} \varrho \\ \varrho u \\ 0.5\varrho u^2 \end{bmatrix} \ , \quad \begin{bmatrix} \varrho \\ \varrho(u+a) \\ \varrho(H+ua) \end{bmatrix} \ , \tag{14.92}$$

with $H = (E + p)/\varrho$.

Each component of the solution $\mathbf{q}$, has a contribution from each characteristic component, i.e. $\mathbf{q} = \sum_{m=1}^{3} \alpha_m \mathbf{e}_m$. The scalar TVD scheme, equivalent to (14.88), is applied to each characteristic component in (14.91). Thus

$$\mathbf{e}_{m,j}^{n+1} = \mathbf{e}_{m,j}^n - \frac{\Delta t}{\Delta x}(\tilde{\mathbf{F}}_{m,j+1/2}^n - \tilde{\mathbf{F}}_{m,j-1/2}^n) \ ,$$

where $\mathbf{F}_m = \lambda_m \mathbf{e}_m$.

The upwind direction will depend on $\sigma = \operatorname{sgn}(\lambda_m)$ in the equivalent of (14.89), so that different characteristic fields involve different grid point values and it is possible to use different limiters, $\phi(r)$, for different characteristic fields.

Characteristic decompositions lead to sharp shock profiles that propagate at the correct speed. Characteristic decompositions are used by Roe (1981, 1986) and Yee et al. (1985). The same type of decomposition is used with the characteristic Galerkin finite element method (Morton and Sweby 1987). If such decompositions are applied to each directional flux separately they can also provide accurate predictions in two, or more, dimensions. Thus Yee (1986) applies a flux-limiting TVD scheme to compute the solution for a shock wave passing over an inclined aerofoil. Fletcher and Morton (1986) apply a characteristic Galerkin finite element method to the problem of unsteady oblique shock reflection associated with supersonic inviscid flow over a wedge.

A more advanced description of high-accuracy TVD schemes is provided by Chakravarthy and Osher (1985) and by Yee et al. (1990).

14.2.7 FCT: Propagating Shock Wave by an FCT Algorithm

In this section an FCT algorithm similar to that described in Sect. 14.2.6 will be applied to the propagating shock problem considered in Sect. 14.2.3. The FCT algorithm can be treated, almost, as additional steps to be added to the MacCormack or Lax–Wendroff schemes used in Sect. 14.2.3. The present description will follow the phoenical Lax–Wendroff FCT scheme given by Book et al. (1975, p. 258).

The solution generated by (14.50) or (14.52) is interpreted as $\mathbf{q}^{**}$. Then the phoenical FCT algorithm consists of the following six steps:

i) Generate diffusive fluxes:

$$\mathbf{f}^{d}_{j+1/2} = v_{j+1/2}(\mathbf{q}^{n}_{j+1} - \mathbf{q}^{n}_{j})$$

ii) Generate antidiffusive fluxes:

$$\mathbf{f}^{ad}_{j+1/2} = \mu_{j+1/2}(\mathbf{q}^{**}_{j+1} - \mathbf{q}^{**}_{j})$$

iii) Diffuse the solution:

$$\mathbf{q}^{***}_{j} = \mathbf{q}^{**}_{j} + \mathbf{f}^{d}_{j+1/2} - \mathbf{f}^{d}_{j-1/2}$$

iv) Calculate first differences of $\mathbf{q}^{***}_{j}$:

$$\Delta\mathbf{q}^{***} = \mathbf{q}^{**}_{j+1} - \mathbf{q}^{***}_{j}$$

v) Limit the antidiffusive fluxes:

$$S = \text{sgn}\,\mathbf{f}^{ad}_{j+1/2}$$

$$\mathbf{f}^{cad}_{j+1/2} = S \max[0, \min\{S\Delta\mathbf{q}^{***}_{j-1/2}, |\mathbf{f}^{ad}_{j+1/2}|, S\Delta\mathbf{q}^{***}_{j+3/2}\}]$$

vi) Antidiffuse the solution:

$$q_j^{n+1} = q_j^{***} - f_{j+1/2}^{cad} + f_{j-1/2}^{cad} \ .$$

In steps i) and ii) the diffusion v and antidiffusion μ coefficients are functions of position. This follows from the generalisation of (14.79) to

$$v_{j+1/2} = \eta_0 + \eta_1 \left(u_{j+1/2} \frac{\Delta t}{\Delta x} \right)^2 , \qquad \mu_{j+1/2} = \eta_0 + \eta_2 \left(u_{j+1/2} \frac{\Delta t}{\Delta x} \right)^2 \qquad (14.93)$$

where $u_{j+1/2} = 0.5(u_j + u_{j+1})$ and, typically, $\eta_0 = 1/6$, $\eta_1 = 1/3$, $\eta_2 = -1/6$. Clearly steps i)–vi) represent a direct extension of the scalar, constant coefficient algorithm described in Sect. 14.2.6.

The diffusive fluxes, calculated at time level n, are evaluated in program SHOCK (Fig. 14.17, lines 86 and 96). All the other steps are implemented after computing q_j^{**} and are collected together in subroutine FCT (Fig. 14.24). The various parameters used in subroutine FCT are described in Table 14.4.

Table 14.4. Parameters used in subroutine FCT (and program SHOCK)

Parameter	Description
ADF	antidiffusive flux, f^{ad} and f^{cad}
DF	diffusive flux, f^d
EMU	antidiffusion coefficient, $\mu_{j+1/2}$
ENU	diffusion coefficient, $v_{j+1/2}$
ETA, ET1, ET2	η_0, η_1, η_2, (14.93)
DQ	Δq^{***}
Q	q^{**}, q^{***}, q^{n+1}

Typical results for the propagation of a strong shock are shown in Fig. 14.25. For this case 100 spatial nodes and $\Delta x = 0.01$ have been used. The shock strength is $p_1/p_2 = 5.0$ and $\gamma = 1.4$ which produce a shock propagation speed $u_{ss}' = 2.104$. Initially, the shock is located at $x' = 0.501$. After 100 steps of $\Delta t = 0.001$ the shock is located at $x' = 0.711$ and it is this solution that is shown in Fig. 14.25.

It is clear that the FCT algorithm has produced a sharp shock profile with negligible oscillation. For comparison, solutions produced with the Lax–Wendroff scheme and·artificial viscosity, $v = 1.0$ in (14.54) are also shown in Fig. 14.25. These solutions were obtained with 200 time-steps of 0.0005. This was necessary to obtain a stable solution. Although the artificial viscosity is effective in eliminating spurious oscillations, it is clear that the shock is smeared over a wide region. This problem can be reduced by going to a finer grid, but then the execution time may become excessive. If a locally fine grid is used, typically adaptively, the overall scheme must be made implicit to avoid severe stability restrictions on Δt.

A comparison of artificial viscosity, FCT and higher-order Godunov schemes is provided by Woodward and Colella (1984) for one- and two-dimensional, steady

```
 1           SUBROUTINE FCT(NXM,ETA,ET2,DTR,Q,DF,U)
 2   C
 3   C       APPLIES FLUX-CORRECTED TRANSPORT ALGORITHM TO Q(J,K)
 4   C
 5           DIMENSION Q(101,3),DQ(101,3),U(101),DF(101,3),ADF(101,3)
 6   C
 7   C       COMPUTE ANTIDIFFUSIVE FLUXES
 8   C
 9           EMU = ETA + 0.25*ET2*((U(1)+U(2))*DTR)**2
10           DO 1 K = 1,3
11         1 ADF(1,K) = EMU*(Q(2,K)-Q(1,K))
12           DO 3 J = 2,NXM
13           EMU = ETA + 0.25*ET2*((U(J)+U(J+1))*DTR)**2
14           DO 2 K = 1,3
15           ADF(J,K) = EMU*(Q(J+1,K)-Q(J,K))
16   C
17   C       DIFFUSE THE SOLUTION
18   C
19           Q(J,K) = Q(J,K) + DF(J,K) - DF(J-1,K)
20         2 DQ(J-1,K) = Q(J,K) - Q(J-1,K)
21         3 CONTINUE
22           DO 4 K = 1,3
23         4 DQ(NXM,K) = Q(NXM+1,K) - Q(NXM,K)
24   C
25   C       LIMIT ANTIDIFFUSIVE FLUXES
26   C
27           DO 6 J = 2,NXM
28           DO 5 K = 1,3
29           S = SIGN(1.0,ADF(J,K))
30           ADF(J,K) = ABS(ADF(J,K))
31           DUM = S*DQ(J-1,K)
32           ADF(J,K) = AMIN1(ADF(J,K),DUM)
33           DUM = S*DQ(J+1,K)
34           ADF(J,K) = AMIN1(ADF(J,K),DUM)
35           ADF(J,K) = AMAX1(ADF(J,K),0.)
36           ADF(J,K) = S*ADF(J,K)
37   C
38   C       ANTIDIFFUSE THE SOLUTION
39   C
40           Q(J,K) = Q(J,K) - ADF(J,K) + ADF(J-1,K)
41         5 CONTINUE
42         6 CONTINUE
43           RETURN
44           END
```

Fig. 14.24. Listing of subroutine FCT

and unsteady flows in which very strong shock waves and contact discontinuities occur. A contact discontinuity is a surface across which the density is discontinuous but velocity and pressure are continuous. In the evolution of a shock tube flow the boundary between the initially high and low pressure fluids develops as a contact discontinuity.

Woodward and Colella (1984) find that higher-order Godunov schemes produce more accurate solutions than FCT algorithms, but are more complicated to code and more expensive to run. As might be expected, FCT algorithms are more accurate than using artificial viscosity. Although not tested by Woodward and Colella it seems likely that the later flux-limiting algorithms (e.g. Yee et al. 1990), which may be interpreted as developments of the FCT concept, would produce solutions much more economically than the Godunov schemes and possibly almost as accurately.

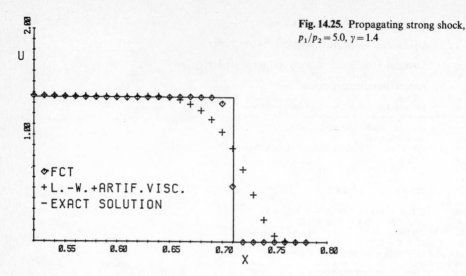

Fig. 14.25. Propagating strong shock, $p_1/p_2 = 5.0$, $\gamma = 1.4$

Various methods, from the addition of artificial viscosity to the implementation of full Godunov formulations, offer a considerable variation in execution time. For unsteady flows the accurate computation of both the shock strength and shock speed is important and greater complexity in the computational algorithm is warranted. For steady flows containing shocks accurate solutions can often be obtained with less complex algorithms.

14.2.8 Implicit Schemes for the Euler Equations

Explicit schemes for solving the Euler equations have been considered in previous sections for flows where either strong shocks were expected or where the flow was supersonic everywhere so that, for steady flow, the solution could be obtained with a single coordinate sweep. In this section implicit schemes are considered for the Euler equations governing transonic flow.

For steady transonic flow there is evidence (Rizzi and Viviand 1981) that, for certain conditions where shocks are expected, the fully conservative potential formulation, described in Sect. 14.3.3, can produce shocks in the wrong location and even multiple solutions for the same boundary conditions. Consequently where the prediction of the location and strength of a shock is crucial, e.g. a lifting aerofoil, solutions of the Euler equations are considerably more reliable.

For steady transonic flow, solutions of the Euler equations are usually obtained via a pseudo-transient formulation (Sect. 6.4). The major problem is in achieving rapid convergence of the transient process. Viviand (1981) provides a review of the problem and possible strategies.

Explicit finite volume schemes based on the MacCormack discretisation scheme are described by Rizzi and Eriksson (1982) and Lerat and Sides (1982). However, explicit schemes typically require a large number of iterations (time-steps) to reach the steady state due to CFL restrictions on the time-step for stability.

Implicit schemes, based on splitting or approximate factorisation techniques (Chap. 8) permit much larger time-steps and consequently reach the steady state in far fewer iterations. A specific comparison for the flow about a NACA-0012 aerofoil at $\alpha = 1.25°$ and $M_\infty = 0.80$ can be made. Rizzi (1981), using an explicit scheme, reports needing 4900 iterations to reach convergence on a 141×21 grid. Pulliam (1985), using an approximate factorisation (implicit) technique, obtains solutions on a 161×33 grid in close agreement with Rizzi's results in approximately 250 iterations. However, when explicit algorithms are combined with a multigrid strategy (Sect. 14.2.9) they are more competitive.

Here we develop a typical implicit algorithm for marching the Euler equations in time. In two dimensions the Euler equations can be written in conservation form as

$$\frac{\partial \mathbf{q}}{\partial t} + \frac{\partial \mathbf{F}}{\partial x} + \frac{\partial \mathbf{G}}{\partial y} = 0 \ , \quad \text{where} \tag{14.94}$$

$$\mathbf{q} \equiv \begin{bmatrix} \varrho \\ m \\ n \\ E \end{bmatrix} , \quad \mathbf{F} = \begin{bmatrix} \varrho u \\ \varrho u^2 + p \\ \varrho uv \\ u(E+p) \end{bmatrix} , \quad \mathbf{G} = \begin{bmatrix} \varrho v \\ \varrho uv \\ \varrho v^2 + p \\ v(E+p) \end{bmatrix} \tag{14.95}$$

where $m = \varrho u$, $n = \varrho v$, and for an ideal gas

$$p = (\gamma - 1)[E - 0.5\varrho(u^2 + v^2)] \ . \tag{14.96}$$

Because of the particular structure of these equations it is possible to write

$$\mathbf{F} = \underline{A}\mathbf{q} \quad \text{and} \quad \mathbf{G} = \underline{B}\mathbf{q} \ , \tag{14.97}$$

where $\underline{A}$ and $\underline{B}$ are the flux Jacobian matrices, i.e.

$$A_{ij} \equiv \frac{\partial F_i}{\partial q_j} \quad \text{and} \quad B_{ij} \equiv \frac{\partial G_i}{\partial q_j} \ . \tag{14.98}$$

Matrices $\underline{A}$ and $\underline{B}$ are

$$\underline{A} \equiv \begin{bmatrix} 0 & 1 & 0 & 0 \\ 0.5(\gamma-3)u^2 + 0.5(\gamma-1)v^2 & (3-\gamma)u & -(\gamma-1)v & (\gamma-1) \\ -uv & v & u & 0 \\ u[-\gamma E/\varrho + (\gamma-1)(u^2+v^2)] & \gamma E/\varrho - 0.5(\gamma-1)(3u^2+v^2) & -(\gamma-1)uv & \gamma u \end{bmatrix} , \tag{14.99}$$

$$
\underline{B} \equiv
\begin{bmatrix}
0 & 0 & 1 & 0 \\
-uv & v & u & 0 \\
0.5(\gamma-1)u^2 - 0.5(3-\gamma)v^2 & -(\gamma-1)u & (3-\gamma)v & (\gamma-1) \\
v[-\gamma E/\varrho + (\gamma-1)(u^2+v^2)] & -(\gamma-1)uv & \gamma E/\varrho - 0.5(\gamma-1)(u^2+3v^2) & \gamma v
\end{bmatrix} .
$$

To introduce the implicit scheme, (14.94) is discretised as

$$
(1+\gamma)\frac{\Delta \mathbf{q}^{n+1}}{\Delta t} - \gamma\frac{\Delta \mathbf{q}^{n}}{\Delta t} = -(1-\beta)(L_x\mathbf{F}^n + L_y\mathbf{G}^n) - \beta(L_x\mathbf{F}^{n+1} + L_y\mathbf{G}^{n+1}) ,
$$

$$(14.100)$$

where $\Delta \mathbf{q}^{n+1} = \mathbf{q}^{n+1} - \mathbf{q}^n$, and γ and β are parameters that will provide different marching schemes. Previously L_x and L_y were defined as central difference operators, e.g.

$$
L_x\mathbf{F}_{j,k} = \left(\frac{1}{2\Delta x}\right)(\mathbf{F}_{j+1,k} - \mathbf{F}_{j-1,k}) .
$$

Here they will be left undefined since alternative spatial discretisations may be preferred in supersonic regions of the flow.

It is assumed that the solution $\mathbf{q}$ is known at time-level n and (14.100) is to be used to advance the solution to time-level $n+1$. The broad strategy follows that of Sects. 8.2, 9.5.1 and 10.4.2. That is, a linear system of equations is developed for the correction to the solution, $\Delta \mathbf{q}^{n+1}$. The terms $\mathbf{F}^{n+1}$ and $\mathbf{G}^{n+1}$ are nonlinear functions of $\mathbf{q}^{n+1}$. These terms may be linearised by expansion about time-level n as a Taylor series, i.e.

$$
\mathbf{F}^{n+1} = \mathbf{F}^n + \underline{A}^n\frac{\partial \mathbf{q}}{\partial t}\Delta t + O(\Delta t^2)
$$

$$(14.101)$$

$$
= \mathbf{F}^n + \underline{A}^n\Delta \mathbf{q}^{n+1} + O(\Delta t^2) \quad \text{and}
$$

$$
\mathbf{G}^{n+1} = \mathbf{G}^n + \underline{B}^n\Delta \mathbf{q}^{n+1} + O(\Delta t^2) ,
$$

$$(14.102)$$

where $\underline{A}^n$ and $\underline{B}^n$ are the flux Jacobian matrices (14.99) evaluated at time level n. Substitution into (14.100) gives

$$
\left[\underline{I} + \frac{\beta\Delta t}{1+\gamma}\{L_x\underline{A}^n + L_y\underline{B}^n\}\right]\Delta \mathbf{q}^{n+1} = -\frac{\Delta t}{1+\gamma}\{L_x\mathbf{F}^n + L_y\mathbf{G}^n\} + \frac{\gamma}{1+\gamma}\Delta \mathbf{q}^n . \quad (14.103)
$$

In (14.103) the notation $\ldots L_y\underline{B}^n\}]\Delta \mathbf{q}^{n+1}$ implies $L_y(\underline{B}^n\Delta \mathbf{q}^{n+1})$.

If only the steady-state solution is of interest the choice $\gamma=0$, $\beta=1.0$ is seen to give rise to an augmented Newton's method (Sect. 6.4.1). The only difference from a conventional Newton's method is the additional diagonal term coming from the unit matrix $\underline{I}$. As $\Delta t \to \infty$, the conventional Newton's method is recovered.

However γ, β and Δt are chosen, (14.103) is computationally expensive to solve at each step, in its present form. Because of (14.101), (14.103) is only accurate to $O(\Delta t^2)$ at best. For particular choices of γ and β, e.g. the "Newton" choice $\gamma = 0$, $\beta = 1.0$, (14.103) is only accurate to $O(\Delta t)$.

The accuracy is not changed, to $O(\Delta t^2)$, if the term

$$\left\{ \frac{\beta \Delta t}{1+\gamma} \right\}^2 L_x \underline{A}^n L_y \underline{B}^n \Delta \mathbf{q}^{n+1}$$

is added to the left hand side of (14.103). The result is

$$\left[\underline{I} + \frac{\beta \Delta t}{1+\gamma} L_x \underline{A} \right] \left[\underline{I} + \frac{\beta \Delta t}{1+\gamma} L_y \underline{B} \right] \Delta \mathbf{q}^{n+1} = -\frac{\Delta t}{1+\gamma} \{ L_x \mathbf{F}^n + L_y \mathbf{G}^n \} + \frac{\gamma}{1+\gamma} \Delta \mathbf{q}^n \ . \tag{14.104}$$

Equation (14.104) provides an approximate factorisation of (14.103) which is implemented as a two-stage algorithm. The first stage is

$$\left[\underline{I} + \frac{\beta \Delta t}{1+\gamma} L_x \underline{A} \right] \Delta \mathbf{q}^* = -\frac{\Delta t}{1+\gamma} \{ L_x \mathbf{F}^n + L_y \mathbf{G}^n \} + \frac{\gamma}{1+\gamma} \Delta \mathbf{q}^n \ , \tag{14.105}$$

and the second stage,

$$\left[\underline{I} + \frac{\beta \Delta t}{1+\gamma} L_y \underline{B} \right] \Delta \mathbf{q}^{n+1} = \Delta \mathbf{q}^* \ . \tag{14.106}$$

Equation (14.105) is a 4×4 block tridiagonal system of equation associated with each grid line in the x direction. Block tridiagonal systems of equations can be solved very efficiently using an extension of the Thomas algorithm (Sect. 6.2.5). The same algorithm is applicable to (14.106), which is a 4×4 block tridiagonal system of equations associated with each grid line in the y direction.

It is possible to factorise $\underline{A}$ and $\underline{B}$ into the form

$$\underline{A} = \underline{T}_A \underline{A}_A \underline{T}_A^{-1} \quad \text{and} \quad \underline{B} = \underline{T}_B \underline{A}_B \underline{T}_B^{-1} \ , \tag{14.107}$$

where diagonal matrices $\underline{A}_A$ and $\underline{A}_B$ contain the eigenvalues of $\underline{A}$ and $\underline{B}$, i.e.

$$\text{diag.} \ \underline{A}_A \equiv \{ u, u, u+a, u-a \} \quad \text{and} \quad \text{diag} \ \underline{A}_B = \{ v, v, v+a, v-a \} \ , \tag{14.108}$$

and a is the local sound speed. Pulliam and Chaussee (1981) show that use of the factored form (14.107) permits a single 4×4 block tridiagonal system to be split into four scalar tridiagonal systems that can be solved sequentially. This produces an overall saving in execution time of about 30%. However the technique introduces an error of $O(\Delta t)$ in unsteady solutions.

Equations (14.105 and 106) are equally applicable to unsteady or steady problems. For unsteady problems the choice $\gamma = 0$, $\beta = 0.5$ produces a "Crank–Nicolson" algorithm which is accurate to $O(\Delta t^2)$. Where (14.105 and 106) are used as the basis of pseudo-transient algorithms for steady problems the choices

i) $\gamma = 0$, $\beta = 1.0$ augmented Newton, $O(\Delta t)$,

ii) $\gamma = 0.5$, $\beta = 1.0$ three-level, fully implicit, $O(\Delta t^2)$

are effective. At first sight it might appear that the augmented Newton's method with a very large Δt would be optimal for pseudo-transient calculations. However, the approximate factorisation introduces a term of $O(\Delta t^2)$ and this effectively destroys the expected quadratic convergence of Newton's method.

A von Neumann (linear) stability analysis indicates that the algorithm given by (14.105, 106) is unconditionally stable. However, nonlinear instability can still occur, particularly associated with strong shocks. If the spatial operators L_x and L_y in (14.105 and 106) are represented by central differences it is the practice to add artificial viscosity. However, this must be added to both sides of equation (14.105). The artificial viscosity may be second-order (as in Sect. 14.2.3) or fourth-order (Sect. 18.5.1).

Alternatively L_x and L_y can be constructed as upwind differences in the supersonic region. Thus

$$ L_x^- \mathbf{F}_{j,k} = \frac{1}{\Delta x} \{ \mathbf{F}_{j,k} - \mathbf{F}_{j-1,k} \} \quad \text{and} \quad L_y^- \mathbf{G}_{j,k} = \frac{1}{\Delta y} \{ \mathbf{G}_{j,k} - \mathbf{G}_{j,k-1} \} , \qquad (14.109) $$

assuming the local velocity is in the positive x and y directions. Such a discretisation is seen to introduce dissipation (Sect. 9.1) if it is interpreted as a second-order discretisation. An example for a one-dimensional shock tube problem is given by Steger and Warming (1981, Fig. 1) and is seen to produce a heavily smoothed out shock and contact discontinuity. It is possible to construct a second-order upwind differencing scheme that can be combined with the two-stage implicit algorithm (14.105, 106) in the supersonic regions. Pulliam (1985, pp. 520–523) describes such a scheme, adapted from Warming and Beam (1976), and shows that sharp shocks are generated without upstream oscillations (as in the MacCormack and Lax–Wendroff schemes, Sect. 14.2.3). However, combining with central differencing in the subsonic region requires the introduction of switching functions (as in Sect. 14.3.3) and special procedures at the interface (i.e. the sonic line and the shock).

Specification of the boundary conditions and their numerical implementation when solving the Euler equations is an important part of the overall algorithm construction. At solid boundaries the normal velocity must be set to zero to satisfy conservation of mass. Typically pressure is obtained from the normal momentum equation and density follows from holding the total enthalpy, $H = (E + p)/\varrho$, constant.

Boundaries through which flow can take place are categorised as inflow or outflow boundaries (Fig. 11.18). For internal flow problems, the classification is unambiguous. For flows external to an isolated body parts of the farfield boundary may change from being an inflow to an outflow boundary during the course of the solution.

Fortunately characteristic theory provides guidance as to the number and form of the boundary conditions (Thompson 1990). Physically information is carried along the characteristics. Therefore characteristics entering the computational domain require boundary conditions to be specified.

The one-dimensional unsteady Euler equations have eigenvalues given by $\underline{\lambda} = \{u, \ u+a, \ u-a\}^T$. The situation for one-dimensional subsonic and supersonic inflows and outflows is shown in Fig. 14.26.

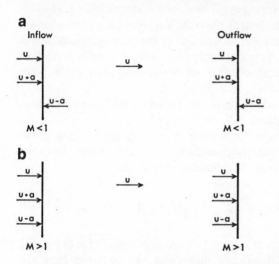

Fig. 14.26. Number of boundary conditions from characteristics theory.

Thus a subsonic inflow requires boundary conditions on two variables and requires the third variable to be computed from the boundary values and interior solution. At a subsonic outflow two characteristics point out of the computational domain and one points into the domain. Consequently one boundary condition must be set at a subsonic outflow.

For a supersonic inflow all characteristics point into the domain, therefore boundary conditions are required on all variables. Conversely at a supersonic outflow all characteristics point out of the domain so that no boundary conditions may be specified.

The above approach can be extended to multidimensions if u is interpreted as the velocity component normal to the boundary surface. This fits in conveniently with the use of generalised curvilinear coordinates (Chap. 12) since the direction normal to the boundary is usually a generalised coordinate. For an application see Steger et al. (1980).

For subsonic flow in the farfield there is some choice as to which dependent variables, or combinations, should have prescribed values. Typical boundary conditions for subsonic inflow are to specify the flow direction, entropy and total enthalpy. The density is obtained from the interior solution by using the characteristic compatibility relationship. At a subsonic outflow it is appropriate to specify the pressure, extrapolate ϱu, ϱv and E from the interior solution and obtain ϱ from (14.96).

The boundary conditions should be implemented implicitly to ensure that the overall algorithm is not subject to CFL stability restrictions on the time step. Yee (1981), Pulliam (1981), Chakravarthy (1983) and Rai and Chaussee (1984) provide appropriate implicit implementations.

The numerical formulation of the farfield boundary conditions are important in relation to the convergence of steady-state solutions in the minimum number of time steps. At any intermediate point (in time) the solution can be conceptually split into a transient part and a steady-state part. The solution algorithm, e.g. (14.105, 106), can be interpreted as seeking to destroy the transient solution in the minimum number of time steps. The transient solution consists of the propagation of waves about the computational domain. It is desirable to construct farfield boundary conditions that transmit the transient solution out of the computational domain without reflection.

An effective technique, developed by Bayliss and Turkel (1982) for external flow, is to linearise the Euler equations about uniform flow, i.e. $u = U_\infty$, $v = 0$, $p = p_\infty$ and $\varrho = \varrho_\infty$. Then a boundary condition is derived which matches the solution of the linearised equations. For subsonic outflow conditions with the flow parallel to the x-axis, Bayliss and Turkel recommend the boundary condition

$$\frac{1}{(a_\infty^2 - U_\infty^2)^{1/2}} \frac{\partial p}{\partial t} - \frac{\varrho_\infty a_\infty^2}{(a_\infty^2 - u_\infty^2)} \frac{x}{d} \frac{\partial u}{\partial t} - \varrho_\infty \frac{y}{d} \frac{\partial v}{\partial t} + \frac{1}{2d} (p - p_\infty) = 0 \ , \qquad (14.110)$$

where $d^2 = (1 - M_\infty^2) x^2 + y^2$. In application, (14.110) requires knowledge of ϱ_∞, U_∞ and a_∞. Bayliss and Turkel (1982) indicate that these can be taken from the solution at the preceding time step. Clearly as the steady state is approached (14.110) forces $p = p_\infty$, where p_∞ is the specified outflow pressure.

An alternative outflow boundary condition due to Rudy and Strikwerda (1981) is

$$\frac{\partial p}{\partial t} - \varrho_\infty a_\infty \frac{\partial u}{\partial t} + \alpha(p - p_\infty) = 0 \ , \qquad (14.111)$$

where typically $\alpha = 0.3$ for rapid convergence. Bayliss and Turkel found the use of (14.110) superior to (14.111) for most problems.

For farfield boundaries that are solid surfaces e.g. a wind-tunnel wall, one boundary condition should be applied. If the wall is parallel to the x-axis, Bayliss and Turkel recommend the following boundary conditions:

$$\frac{\partial p}{\partial t} - \varrho_\infty a_\infty \frac{\partial v}{\partial t} = 0 \ . \qquad (14.112)$$

Thompson (1987) develops nonlinear nonreflecting farfield boundary conditions by combining the characteristic compatibility relations with a condition that the amplitude of the incoming waves remains constant.

The equations in the interior can be modified to accelerate convergence to the steady state. It is convenient to write the governing equations (14.94) as

$$\underline{N}^{-1} \frac{\partial \mathbf{q}}{\partial t} + \frac{\partial \mathbf{F}}{\partial x} + \frac{\partial \mathbf{G}}{\partial y} = 0 \ , \qquad (14.113)$$

where $\underline{N}$ is a matrix constructed to enhance convergence. Viviand (1981) has reviewed pseudo-transient algorithms that fall into the class of (14.113) for transonic inviscid flow.

Turkel (1985) introduces a form for $\underline{N}$ such that if all terms in (14.113) were factored in the manner of (14.107) the resulting eigenvalues, equivalent to (14.108), would be independent of the sound speed a.

If $\mathbf{N}$ is diagonal (14.113) reduces to choosing a different time step for each grid point. Pulliam (1985) indicates the following formula is effective in compensating for a spatially varying grid:

$$\Delta t_{\text{loc}} = \frac{\Delta t_0}{1 + (J)^{1/2}} , \tag{14.114}$$

where J is the Jacobian (12.3).

Alternatively Δt_{loc} be chosen to keep an effective CFL number constant, i.e.

$$\Delta t_{\text{loc}} = k \frac{(\Delta x \, \Delta y)^{1/2}}{w + a} , \tag{14.115}$$

where $w = (u^2 + v^2)^{1/2}$ and k is $O(10)$, typically.

For steady flows with stronger shocks it is advantageous to construct flux-limited TVD schemes (Sect. 14.2.6) that are implicit and use pseudo-transient approximate factorisation algorithms conceptually similar to (14.105 and 106) to obtain the steady-state solution. Typical algorithms are described by Yang et al. (1986), Chakravarthy (1986) and Yee et al. (1990).

Implicit schemes for the Euler equations are also appropriate, with minor modifications, to the Navier–Stokes equations. Thus many of the techniques described in Chap. 18 are also applicable to the Euler equations.

14.2.9 Multigrid for Euler Equations

For steady-state solutions of the Euler equations it is possible to use multigrid techniques (Sect. 6.3.5) to accelerate the convergence. Here an algorithm due to Ni (1982) will be described which uses an explicit scheme to march the unsteady Euler equations in time until the steady state is reached. By using a multigrid strategy the effective time-step limit associated with the explicit scheme is not restrictive since marching on coarse grids allows rapid propagation of the transient solution.

The algorithm combines one-step Lax–Wendroff time differencing (10.10), which is second-order in time, with a finite volume spatial discretisation (Sect. 5.2). The algorithm will be illustrated for a uniform Cartesian grid, Fig. 14.27, but the extension to a non-uniform grid is straightforward (Ni 1982; Hall 1984). The starting point is the two-dimensional unsteady Euler equations

$$\mathbf{q}_t = -(\mathbf{F}_x + \mathbf{G}_y) , \tag{14.116}$$

where the components of $\mathbf{q}$, $\mathbf{F}$ and $\mathbf{G}$ are given by (11.117) without the τ and Q terms.

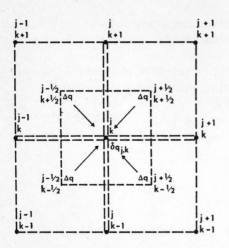

Fig. 14.27. Correspondence between control volumes and grid points

Application of the finite volume method (Sect. 5.2.1) to the control volume centered at $(j+1/2, k+1/2)$ and first-order time differencing gives the following algorithm for the correction to **q** at the centre of the control volume (Fig. 14.27):

$$\Delta\mathbf{q}_{j+1/2,\,k+1/2} = -0.5\frac{\Delta t}{\Delta x}[(\mathbf{F}_{j+1,k}+\mathbf{F}_{j+1,k+1})-(\mathbf{F}_{j,k}+\mathbf{F}_{j,k+1})]$$

$$-0.5\frac{\Delta t}{\Delta y}[(\mathbf{G}_{j,k+1}+\mathbf{G}_{j+1,k+1})-(\mathbf{G}_{j,k}+\mathbf{G}_{j+1,k})] \ . \qquad (14.117)$$

Equivalent expressions to (14.117) can be obtained for the four control volumes surrounding grid point (j, k).

When the steady state is reached both sides of (14.117) go to zero so that $\Delta\mathbf{q}_{j+1/2,k+1/2}$ is proportional to the steady-state residual associated with the $(j+1/2, k+1/2)$ control volume. This correspondence will be exploited in constructing the multigrid algorithm to be described later in this section.

The value of the gridpoint correction, $\delta\mathbf{q}_{j,k}$, can be obtained as the average of $\Delta\mathbf{q}$ for the four surrounding control volumes. However, it is also desirable to introduce second-order time-differencing. This can be achieved by introducing a one-step Lax–Wendroff scheme (10.10). Thus the time discretisation of (14.116) at node (j, k) becomes

$$\delta\mathbf{q}_{j,k} = -\Delta t(\mathbf{F}_x+\mathbf{G}_y)_{j,k} + 0.5\Delta t^2\{[\underline{A}(\mathbf{F}_x+\mathbf{G}_y)]_x + [\underline{B}(\mathbf{F}_x+\mathbf{G}_y)]_y\}_{j,k} \ , \qquad (14.118)$$

where $\underline{A}$ and $\underline{B}$ are the Jacobians, $\partial\mathbf{F}/\partial\mathbf{q}$ and $\partial\mathbf{G}/\partial\mathbf{q}$, and $\mathbf{F}_x \equiv \partial\mathbf{F}/\partial x$, etc. The first term on the right-hand side can be evaluated as an average over the surrounding control volumes using (14.117) to give

$$-\Delta t(\mathbf{F}_x+\mathbf{G}_y)_{j,k} = 0.25(\Delta\mathbf{q}_{j-1/2,k-1/2}+\Delta\mathbf{q}_{j-1/2,k+1/2}$$
$$+\Delta\mathbf{q}_{j+1/2,k+1/2}+\Delta\mathbf{q}_{j+1/2,k-1/2}] \ . \qquad (14.119)$$

Using (14.116),

$$\underline{A}(\mathbf{F}_x+\mathbf{G}_y) = -\underline{A}\mathbf{q}_t \quad \text{and} \quad \underline{B}(\mathbf{F}_x+\mathbf{G}_y) = -\underline{B}\mathbf{q}_t \ .$$

Introducing $\mathbf{q}_t = \Delta\mathbf{q}/\Delta t$ leads to

$$\Delta t[\underline{A}(\mathbf{F}_x + \mathbf{G}_y)]_x \approx -(\underline{A}\Delta\mathbf{q})_x \approx -(\Delta\mathbf{F})_x \quad \text{and}$$

$$\Delta t[\underline{B}(\mathbf{F}_x + \mathbf{G}_y)]_y \approx -(\underline{B}\Delta\mathbf{q})_y \approx -(\Delta\mathbf{G})_y , \tag{14.120}$$

where $\Delta\mathbf{F}$ and $\Delta\mathbf{G}$ denote the changes in $\mathbf{F}$ and $\mathbf{G}$ during the current time-step corresponding to the correction $\Delta\mathbf{q}$.

The terms $(\Delta\mathbf{F})_x$, $(\Delta\mathbf{G})_y$ centred at grid point (j, k) are evaluated using the finite volume method based on a control volume bounded by $(j-\tfrac{1}{2}, k-\tfrac{1}{2})$, $(j+\tfrac{1}{2}, k-\tfrac{1}{2})$, $(j+\tfrac{1}{2}, k+\tfrac{1}{2})$ and $(j-\tfrac{1}{2}, k+\tfrac{1}{2})$, as indicated in Fig. 14.27. This produces the result

$$(\Delta\mathbf{F})_x \approx [0.5(\Delta\mathbf{F}_{j+1/2, k+1/2} + \Delta\mathbf{F}_{j+1/2, k-1/2})$$

$$-0.5(\Delta\mathbf{F}_{j-1/2, k+1/2} + \Delta\mathbf{F}_{j-1/2, k-1/2})]\frac{1}{\Delta x} \tag{14.121}$$

and an equivalent result for $(\Delta\mathbf{G})_y$.

Substitution of (14.119–121) into (14.118) produces the following algorithm for the correction to $\mathbf{q}_{j,k}$:

$$\delta\mathbf{q}_{j,k} = 0.25\left\{\left[\Delta\mathbf{q} + \frac{\Delta t}{\Delta x}\Delta\mathbf{F} + \frac{\Delta t}{\Delta y}\Delta\mathbf{G}\right]_{j-1/2, k-1/2}\right.$$

$$+\left[\Delta\mathbf{q} + \frac{\Delta t}{\Delta x}\Delta\mathbf{F} - \frac{\Delta t}{\Delta y}\Delta\mathbf{G}\right]_{j-1/2, k+1/2}$$

$$+\left[\Delta\mathbf{q} - \frac{\Delta t}{\Delta x}\Delta\mathbf{F} - \frac{\Delta t}{\Delta y}\Delta\mathbf{G}\right]_{j+1/2, k+1/2}$$

$$\left.+\left[\Delta\mathbf{q} - \frac{\Delta t}{\Delta x}\Delta\mathbf{F} + \frac{\Delta t}{\Delta y}\Delta\mathbf{G}\right]_{j+1/2, k-1/2}\right\} . \tag{14.122}$$

Clearly (14.122) can be interpreted as giving the correction to $\mathbf{q}_{j,k}$ in terms of the average of the corrections occurring in adjacent control volumes. Ni (1982) refers to the individual contributions in (14.122) as distribution formulae, since one can traverse each control volume in turn and distribute its effect to its four adjacent grid points.

The basic algorithm consists of (14.117) to obtain the control volume corrections and (14.122) to obtain the grid point corrections. This algorithm is second-order accurate in time and space. However, for stability the following restriction on the time step is necessary:

$$\Delta t \leqq \min\left\{\frac{\Delta x}{(|u|+a)} , \frac{\Delta y}{(|v|+a)}\right\} . \tag{14.123}$$

It may be noted that the dependent variables are defined at the vertices of the control volume rather than at the centre of the control volume as in Sects. 5.2 and 17.2.3. It is also possible to construct a finite volume discretisation of (14.116) with grid-point values of the dependent variables defined at the centres of the control volumes (Jameson et al. 1981).

However, the identification of control volume vertices with grid points has inherent advantages (Morton and Paisley 1989). First, for a given nonuniform grid the accuracy is usually higher for the vertex grid-point form. However this advantage is restricted to steady flow (Vinokur 1989). The centred grid-point form is sensitive to the appearance of oscillatory solutions (Sect. 17.2.3). If a staggered grid is not used then it is usually necessary to include additional dissipative terms to suppress the oscillations (Jameson et al. 1981). The vertex grid-point form is much less sensitive to spurious oscillations, although additional dissipative terms are recommended (Ni 1982) if shocks are present.

Probably the greatest advantage is associated with the implementation of boundary conditions, which can be imposed directly with the vertex grid-point form since grid points coincide with the boundary, which is not the case for centred grid-point control volumes (Sect. 17.1.3).

To accelerate convergence to the steady state one would like to use larger time steps than permitted by (14.123) for a grid fine enough to achieve acceptable accuracy. The multigrid strategy (Sect. 6.3.5) achieves this by evaluating (14.122) on successively coarser grids with correspondingly larger values of Δt_{max} (14.123). Thus unwanted transient disturbances are rapidly propagated through the computational domain and expelled through the farfield boundaries, leaving the converged steady-state solution. In contrast to the multigrid algorithms described in Sect. 6.3.5, simpler multigrid algorithms to obtain the steady-state solution of the Euler equations are usually preferred. Here we briefly describe the algorithm of Ni (1982).

For an intermediate grid, the control volume correction $\Delta \mathbf{q}^m$ is not obtained from (14.117). Instead it is obtained from a restriction of the grid-point corrections $\delta \mathbf{q}^{m+1}$ on the next finer grid, i.e.

$$\Delta \mathbf{q}^m = I_{m+1}^m \delta \mathbf{q}^{m+1} ,\qquad (14.124)$$

where I_{m+1}^m is the restriction operator (Sect. 6.3.5).

Successively coarser grids are constructed by deleting alternate grid lines so that control volume centres on the coarser grid coincide with grid points on the finer grid. From the form of (14.117 and 122) it may be noted that the restriction of the corrections in (14.124) is equivalent to the restriction of the residuals in (6.85).

Given $\Delta \mathbf{q}^m$ from (14.124) the corresponding grid-point corrections $\delta \mathbf{q}^m$ are obtained from (14.122). The corrections are then either prolonged (interpolated) onto the finest grid M to give a new fine grid correction

$$\delta \mathbf{q}^M = I_m^M \delta \mathbf{q}^m \qquad (14.125)$$

or are used to give the control volume corrections on the next coarser grid, i.e. using (14.124) again with $m = m - 1$.

A subcycle of the multigrid algorithm starts with the finest grid M and progressively restricts through each coarser grid using (14.124) and marching in time using (14.122) until the mth grid is reached. Then the solution on the mth grid is interpolated back to the finest grid using (14.125). Each subcycle is a sawtooth cycle in contrast to the V-cycle illustrated in Fig. 6.21. The complete multigrid cycle consists of the solution of (14.117 and 122) on the finest (M) grid followed by a sequence of nested sawtooth cycles until the coarsest grid ($m = 1$) is reached.

Ni (1982) uses a nest of four grids ($M = 4$) with the finest grid having 65×17 points. For transonic channel flow past a bump Ni requires 900 time-steps to reach the steady state, i.e. (14.117 and 122) are used on the finest grid only. If the above multigrid algorithm is used, 130 multigrid cycles are required with a reduction in the execution time by a factor of 4, approximately. Ni also applies the above algorithm to the transonic flow around an axisymmetric nacelle and past a cascade of turbine rotor blades.

Johnson (1983) describes the extension of the Ni algorithm to utilise any of the two-step Lax–Wendroff-like explicit algorithms (Sect. 14.2.2). Davis et al. (1984) and Chima and Johnson (1985) provide extensions of a Ni-type algorithm to obtain solutions of the compressible Navier–Stokes equations.

Jameson (1983) describes a control-volume discretisation of the Euler equations based on placing grid points at the centre of the control volumes and using a four-stage Runge–Kutta scheme (Sect. 7.4) in place of the Lax–Wendroff scheme described above. Jameson's multigrid implementation is similar to Ni's except that a single sawtooth cycle extending to the coarsest grid is used and the prolongation to the finest grid is made via each intermediate grid.

Mulder (1985) combines an implicit time-marching method with a multigrid correction algorithm and flux-vector splitting. A flux-limiting procedure is used to produce sharper shocks. Hemker (1986) applies the FAS multigrid scheme (Sect. 6.3.5) with Gauss-Seidel relaxation to the steady Euler equations after introducing a second-order finite volume discretisation based on centred control-volume grid points. Rapid convergence of the multigrid algorithm is facilitated by using a first-order TVD scheme (Sect. 14.2.6) until convergence is almost achieved. The second-order flux-limiting improvement is only introduced as final convergence is approached.

Anderson et al. (1988) combine a Van Leer (1982) flux-split upwind discretisation of the Euler equations with a multigrid accelerated implicit spatial approximate factorisation to predict the transonic flow about a complete wing. Mavriplis (1990) combines an unstructured adaptive triangular mesh multigrid algorithm with an explicit five-stage time-stepping finite volume formulation to predict subsonic inviscid flow about a multi-element two-dimensional airfoil. Multigrid is implemented in a multiblock configuration with an explicit finite volume discretisation by Yu et al. (1990) to predict transonic flow about a complete wing including an engine nacelle and strut.

14.3   Transonic Inviscid Flow

The category of transonic inviscid flow is considered separately since it is amenable to special treatment via a potential equation, as long as shocks are weak. Transonic inviscid flow is characterised by the occurrence of regions of subsonic and supersonic flow (Fig. 11.15). Techniques for solving the steady potential equation will be emphasised. The extensions necessary to solve flows governed by the unsteady potential equation will be described briefly.

14.3.1 General Considerations

If the flow is irrotational, a velocity potential can be introduced (11.102) and the Euler equations can be reduced to a single partial differential equation and an auxiliary algebraic equation. In two-dimensional steady flow the governing equation in terms of the velocity potential is

$$(a^2 - u^2)\Phi_{xx} - 2uv\Phi_{xy} + (a^2 - v^2)\Phi_{yy} = 0 \ , \tag{14.126}$$

and the sound speed a is given by (11.104). If (14.126) is written in natural coordinates (s, n), that is, s is parallel to the local flow direction and n is normal to it, the result is

$$(1 - M^2)\Phi_{ss} + \Phi_{nn} = 0 \ , \tag{14.127}$$

where the local Mach number $M = q/a$, and $q^2 = u^2 + v^2$. Clearly (14.127) changes from an elliptic to a hyperbolic partial differential equation if M is locally greater than unity, i.e. if the flow is locally supersonic. As indicated in Fig. 11.15, the supersonic region is usually terminated in the flow direction by a shock wave.

Since the derivation of (14.126) is based on the assumption of irrotationality, Crocco's theorem (Liepmann and Roshko 1957 p. 193) indicates that the flow must also be isentropic for the flow around a body in a uniform freestream. However, although the Rankine–Hugoniot conditions across the shock, e.g. (11.110), require that mass, momentum and energy are conserved, there is an increase in entropy proportional to the third power of the shock strength $[\sim(M_1 - M_2)^3]$. If the normal Mach number ahead of the shock (M_1) is less than 1.1, it is acceptable to treat the change in the flow conditions through the shock as approximately isentropic, and to seek the flow solution via the velocity potential.

Equation (14.126) is usually solved in the equivalent, nondimensional conservative form (i.e. the steady continuity equation)

$$(\varrho'\Phi'_x)_x + (\varrho'\Phi'_y)_y = 0 \tag{14.128}$$

where $\Phi' = \Phi/(U_\infty L)$ and L is a characteristic length. The nondimensional density is obtained from (11.104) as

$$\varrho' = \frac{\varrho}{\varrho_\infty} = \left\{ 1 + 0.5(\gamma - 1)M_\infty^2 \left[1 - \left(\frac{u}{U_\infty}\right)^2 - \left(\frac{v}{U_\infty}\right)^2 \right] \right\}^{1/(\gamma - 1)} \ , \tag{14.129}$$

where γ is the specific heat ratio and M_∞ is the freestream Mach number.

If conservative discretisations are introduced into (14.128), the solution will satisfy the weak form, i.e. (5.6), of (14.128). Consequently "shock-like" discontinuities in the solution will be captured. However, the discontinuous solutions will conserve entropy, energy and mass, but not momentum. Thus the Rankine–Hugoniot conditions are satisfied only approximately. Jameson (1978, p. 3) notes that the momentum jump across the computed shock provides a measure of the wave drag.

Computational solutions to (14.128 and 129) can be obtained very efficiently, and working computer codes are widely used in the aircraft industry (Young et al. 1990) for the design of aircraft that operate in the transonic flow regime, where only weak shocks may be expected. For flows with stronger shocks or large regions of rotational flow, e.g. the wake of an aerofoil or turbine blade at angle of attack, it is necessary to solve the full Euler equations (11.22–24) typically by a pseudo-transient method (Sects. 14.2.8, 14.2.9). However, such methods are usually considerably slower than methods based on the solution of (14.128 and 129). The disparity in economy is even more marked for three-dimensional or unsteady flow.

Since (14.126) governs isentropic, inviscid flow it does not distinguish between a compression shock and an expansion shock (Fig. 14.28). Computational schemes must incorporate special procedures to prevent expansion shocks (physically untenable) from appearing. This can be done by using upwind differencing, or by introducing artificial viscosity, in the supersonic region. Both of these techniques provide dissipative mechanisms that block the appearance of expansion shocks.

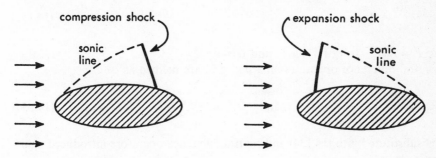

Fig. 14.28. Compression and expansion shocks

14.3.2 Transonic Small Disturbance Equation

For the flow about slender bodies it is possible to simplify (14.126) further to obtain (11.107) governing the disturbance potential ϕ. If the body profile is given by $y = \tau f(x)$, where τ is assumed small, it is possible to write (11.107) as (Cole 1975)

$$\frac{\partial}{\partial x}[K\phi_x - 0.5(\gamma + 1)\{\phi_x\}^2] + \frac{\partial^2 \phi}{\partial y^2} = 0 \ , \tag{14.130}$$

where $K = (1 - M_\infty^2)/\tau^{2/3}$ and $\phi_x \equiv \partial\phi/\partial x$. In (14.130) y has been scaled by $\tau^{1/3}$ and ϕ by $\tau^{-2/3}$.

Equation (14.130) is referred to as the transonic small disturbance equation. Consistent with (14.130) the jump conditions across the shock are

$$[\phi_y] - \frac{dy}{dx}[K\phi_x - 0.5(\gamma + 1)\{\phi_x\}^2] = 0 \ , \tag{14.131}$$

where [ ] denotes the jump in value and dy/dx is the shock slope. It is also consistent to apply the boundary condition, of $\partial\phi/\partial n = 0$ at the body surface, on the x-axis in the form

$$\frac{\partial\phi}{\partial y} = \frac{df}{dx} \quad \text{at} \quad y = 0 \ . \tag{14.132}$$

Equations (14.130–132) provide an accurate description of the flow at transonic speeds around slender bodies except in the immediate neighbourhood of blunt leading edges. This is because the velocity perturbation u is of the same order as the freestream velocity U_∞ in this region. For supersonic speeds, say $M_\infty > 1.3$, (14.130) can be simplified further since $0.5(\gamma + 1)\{\phi_x\}^2 \ll K\phi_x$. Consequently (14.130) reduces to (11.109), which is linear and amenable to solution with a panel method (e.g. PAN-AIR) as in Sect. 14.1.6.

In conservation form, which is necessary to ensure that the correct shock solution is obtained, (14.130) can be written compactly as

$$\frac{\partial F}{\partial x} + \frac{\partial G}{\partial y} = 0 \ , \tag{14.133}$$

where $F = K\phi_x - 0.5(\gamma + 1)\{\phi_x\}^2$ and $G = \partial\phi/\partial y$.

Central difference operators on a half-grid are defined as

$$P_{j,k} = \frac{[F_{j+1/2,k} - F_{j-1/2,k}]}{\Delta x} \quad \text{and} \quad Q_{j,k} = \frac{[G_{j,k+1/2} - G_{j,k-1/2}]}{\Delta y} \ . \tag{14.134}$$

If G is substituted into (14.134) and central difference operators introduced again, the result is

$$Q_{j,k} = \frac{(\phi_{j,k+1} - 2\phi_{j,k} + \phi_{j,k-1})}{\Delta y^2} \ ,$$

and similarly for $P_{j,k}$. Using (14.134) an algorithm suitable for discretising (14.133) in transonic flow can be written as

$$P_{j,k} + Q_{j,k} - \mu_{j,k}P_{j,k} + \mu_{j-1,k}P_{j-1,k} = 0 \ , \tag{14.135}$$

where

$$\mu_{j,k} = 0 \quad \text{at subsonic points, i.e. } K > (\gamma + 1)\phi_x \ ,$$

$$= 1 \quad \text{at supersonic points, i.e. } K < (\gamma + 1)\phi_x \ .$$

In the supersonic region it is clear from (14.135) that the term $\partial F/\partial x$ in (14.133) is being represented by a three-point upwind scheme (Fig. 14.29), in contrast to the centred-difference representation in subsonic regions. The difference formulae used in the supersonic region properly simulate the hyperbolic requirement of no upstream influence. If the sonic line (Fig. 11.15) or the shockwave (Fig. 11.15) lie between points $(j-1,k)$ and (j,k), (14.135) takes the form

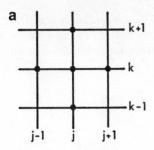 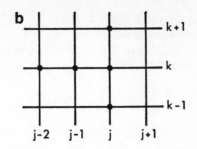

Fig. 14.29a, b. Active nodes in (14.135). **(a)** Subsonic point (j, k) **(b)** Supersonic point (j, k)

sonic line: $Q_{j,k} = 0$,

shock wave: $P_{j,k} + P_{j-1,k} + Q_{j,k} = 0$.

$$(14.136)$$

These special forms are necessary to allow the governing equations to be solved efficiently and the shock to be represented accurately. After converting (14.135) to an equivalent velocity potential form, the resulting equation can be solved iteratively. Special techniques to do this will be indicated in Sect. 14.3.5. The above formulation is essentially due to Murman (1973).

A Taylor expansion of $P_{j-1,k}$ about the node (j, k) indicates that it is consistent with $\partial F/\partial x$ but introduces an artificial viscosity proportional to Δx. Thus rather than switch to upwind formulae in the supersonic region it is possible to introduce explicit artificial viscosity in the supersonic region. This technique has also been used with the finite volume method (Caughey 1982).

14.3.3 Full Potential Equation

For steady transonic inviscid flow more accurate solutions are obtained by solving the full potential equation (11.103) written in nondimensional conservation form. In two dimensions this is given by

$$\frac{\partial}{\partial x}\left(\varrho' \frac{\partial \phi'}{\partial x}\right) + \frac{\partial}{\partial y}\left(\varrho' \frac{\partial \phi'}{\partial y}\right) = 0 , \tag{14.137}$$

where ϕ' is the disturbance potential and the density ϱ' is given by (14.129). In the rest of this section the prime is dropped for clarity. To suit (14.137) the boundary condition of zero normal velocity at the body surface, $\partial \phi/\partial n = 0$, must be applied at the surface. Far from the body the disturbance potential ϕ goes to zero.

Equation (14.137) is discretised as

$$S_{j,k} + T_{j,k} = 0 , \quad \text{where} \tag{14.138}$$

$$
S_{j,k} = \left[\left(\varrho \frac{\partial \phi}{\partial x}\right)_{j+1/2,k} - \left(\varrho \frac{\partial \phi}{\partial x}\right)_{j-1/2,k}\right]\frac{1}{\Delta x}
$$
$$
+ \left[\left(\varrho \frac{\partial \phi}{\partial y}\right)_{j,k+1/2} - \left(\varrho \frac{\partial \phi}{\partial y}\right)_{j,k-1/2}\right]\frac{1}{\Delta y} , \tag{14.139}
$$

with $(\varrho \partial \phi / \partial x)_{j+1/2,k} = 0.5(\varrho_j + \varrho_{j+1})(\phi_{j+1} - \phi_j)/\Delta x$, etc. $T_{j,k}$ provides the artificial viscosity required in the supersonic region. If the flow direction in the supersonic region is nearly parallel to the x-axis it is sufficient to introduce artificial viscosity in the x-direction only. Thus

$$T_{j,k} = \frac{[P_{j+1/2,k} - P_{j-1/2,k}]}{\Delta x} , \quad \text{where} \tag{14.140}$$

$$P_{j+1/2,k} = -(\mu_{j,k} \Delta x)[L_{xx}(\phi_{j,k} - \varepsilon\phi_{j-1,k})] \quad \text{and}$$

$$L_{xx}\phi_{j,k} = \frac{(\phi_{j-1,k} - 2\phi_{j,k} + \phi_{j+1,k})}{\Delta x^2} .$$

As before, μ is the switching function but now given by

$$\mu = \min\left[0, \varrho\left(1 - \frac{u^2}{a^2}\right)\right] .$$

Thus μ varies smoothly and is only non-zero in the supersonic region.

The parameter ε is defined by

$$\varepsilon = 1 - \lambda\Delta x ,$$

which makes (14.138) second-order accurate. The additional dissipative terms $T_{j,k}$ represent $\partial P/\partial x$ where

$$P = -\mu[(1-\varepsilon)\Delta x\phi_{xx} + \varepsilon\Delta x^2 \phi_{xxx}] ,$$

i.e. T is proportional to $\partial^2 u/\partial x^2$.

A particular problem develops with the full potential equation if the flow direction deviates too far from the x-axis. It is then possible for supersonic points to occur such that $u^2 < a^2 < u^2 + v^2$. In turn this leads to negative artificial viscosity being introduced.

The solution to this problem is to relate the additional dissipative terms, like $T_{j,k}$ in (14.138), to the local flow direction. This is most easily illustrated with reference to (14.126), which becomes (14.127) in natural coordinates with

$$\phi_{ss} = \frac{[u^2\phi_{xx} + 2uv\phi_{xy} + v^2\phi_{yy}]}{q^2} . \tag{14.141}$$

At supersonic points centred difference formulae are used to represent ϕ_{nn} but the following upwind formulae are used to represent the ϕ_{xx}, ϕ_{xy} and ϕ_{yy} contributions to ϕ_{ss}:

$$\phi_{xx} \equiv \frac{[\phi_{j,k} - 2\phi_{j-1,k} + \phi_{j-2,k}]}{\Delta x^2} ,$$

$$\phi_{xy} \equiv \frac{[\phi_{j,k} - \phi_{j-1,k} - \phi_{j,k-1} + \phi_{j-1,k-1}]}{4\Delta x\Delta y} , \tag{14.142}$$

$$\phi_{yy} \equiv \frac{[\phi_{j,k} - 2\phi_{j,k-1} + \phi_{j,k-2}]}{\Delta y^2} .$$

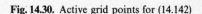

Fig. 14.30. Active grid points for (14.142)

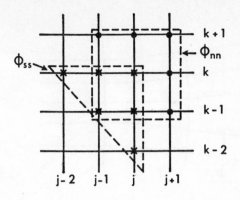

The active gridpoints are shown in Fig. 14.30. The corresponding scheme to suit (14.137) is more complicated but is provided by Jameson (1978, pp. 17–19).

Artificial viscosity can also be introduced by modifying ϱ in (14.139). Holst and Ballhaus (1979) show that (14.140) is equivalent to replacing terms like $\varrho_{j+1/2,k}$ in (14.139) by

$$\tilde{\varrho}_{j+1/2,k} = (1 - v_{j,k})\varrho_{j+1/2,k} + v_{j,k}\varrho_{j-1/2,k} \, , \tag{14.143}$$

where $v = \max\{0, [1 - a^2/q^2]\}$ and $q^2 = u^2 + v^2$. An advantage of modifying the density, as in (14.143), is that this often permits the algorithm for solving the discretised equations to be implemented more easily.

14.3.4 Transonic Inviscid Flow: Generalised Coordinates

The need to compute the flow about bodies like aerofoils and turbine blades has led to the introduction of body-fitted coordinates, Chap. 12, so that in the computational domain (Fig. 14.31) the body surface coincides with a constant value of the transform variable, e.g. $\eta = \eta_0$. Techniques for choosing an appropriate grid of points $x(\xi, \eta)$ and $y(\xi, \eta)$ are discussed in Chap. 13.

In the (ξ, η) plane (14.137) and (14.129) become

$$\frac{\partial}{\partial \xi}(\varrho^* U^c) + \frac{\partial}{\partial \eta}(\varrho^* V^c) = 0 \quad \text{and} \tag{14.144}$$

$$\varrho^* = \frac{1}{J}\left\{1 + 0.5(\gamma - 1)M_\infty^2\left[1 - \left(U^c \frac{\partial \phi}{\partial \xi} + V^c \frac{\partial \phi}{\partial \eta}\right)\right]\right\}^{1/(\gamma - 1)} , \tag{14.145}$$

where $\varrho^* = \varrho/J$ and U^c, V^c are the (contravariant) velocity components in the ξ and η directions and are related to the velocity potential by

$$U^c = A_1 \frac{\partial \phi}{\partial \xi} + A_3 \frac{\partial \phi}{\partial \eta} , \qquad V^c = A_3 \frac{\partial \phi}{\partial \xi} + A_2 \frac{\partial \phi}{\partial \eta} , \quad \text{and} \tag{14.146}$$

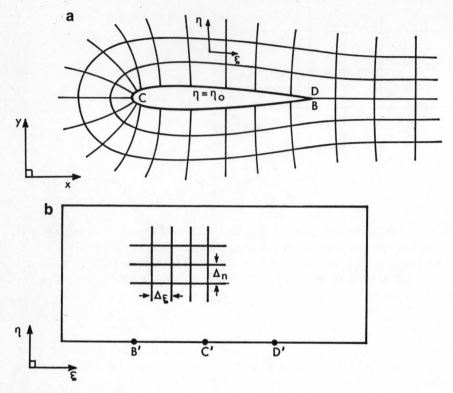

Fig. 14.31a, b. Generalised coordinates. (a) Physical domain. (b) Computational domain

$A_1 = \xi_x^2 + \zeta_y^2$, $A_2 = \eta_x^2 + \eta_y^2$ and $A_3 = \xi_x \eta_x + \xi_y \eta_y$, (12.12). J is the Jacobian (determinant) of the transformation and given by (12.3). The metric quantities, like ξ_x, are defined by the mapping (Chap. 13).

Equation (14.144) is discretised in a manner similar to (14.139), i.e.

$$\frac{[(\tilde{\varrho}^*U^c)_{j+1/2,k} - (\tilde{\rho}^*U^c)_{j-1/2,k}]}{\Delta\xi} + \frac{[(\bar{\varrho}^*V^c)_{j,k+1/2} - (\bar{\varrho}^*V^c)_{j,k-1/2}]}{\Delta\eta} = 0 \ .$$

(14.147)

In (14.147) $\tilde{\varrho}^*$ and $\bar{\varrho}^*$ are upwind evaluations of ρ^*. Thus $\tilde{\varrho}^*_{j+1/2,k}$ is given by (14.143) if U^c is positive, with ϱ^* replacing ϱ. An equivalent upwind expression, based on $\varrho^*_{j+3/2,k}$, is used if U^c is negative. Equivalent expressions to (14.143) based on the $\eta(k)$ direction are used for $\bar{\varrho}$. Upwinding of the density avoids the need for additional dissipative terms like $T_{j,k}$ in (14.138).

The densities are obtained from (14.145), which requires J, U^c and $\partial\phi/\partial\xi$, etc., to be evaluated at the half grid points like $(j+\frac{1}{2}, k)$. For example

$$U^c_{j+1/2,k} = A_1|_{j+1/2,k}\frac{(\phi_{j+1,k} - \phi_{j,k})}{\Delta\xi} + A_3|_{j+1/2,k}\frac{(\phi_{j+1/2,k+1} - \phi_{j+1/2,k-1})}{2\Delta\eta} \ ,$$

(14.148)

where

$$\phi_{j+1/2,k+1} = 0.5(\phi_{j,k+1} + \phi_{j+1,k+1}) \ .$$

Thus the ϕ solution is stored at grid points. The metric quantities, A_1–A_3 and J, are evaluated using standard second-order formulae (Sect. 12.2.1) and stored at the half grid points. The overall accuracy is sensitive to the metric quantities being evaluated without averaging (Flores et al. 1983).

14.3.5 Solution of the Algebraic Equations

The algebraic equations that arise from discretisation of equations like (14.130 and 137) are amenable to solution by a modification to the SOR technique (Sect. 6.3) called successive line over-relaxation (SLOR). The SLOR technique [e.g. (6.64, 65)] solves an implicit system for the corrections to the solution, $\Delta\phi_{j,k}^{n+1} = \phi_{j,k}^{n+1} - \phi_{j,k}^{n}$, on each line in the y direction (constant value of j) at a particular iteration level $n+1$. The tridiagonal solver, Sect. 6.2.3, is used for this purpose.

In the subsonic region the evaluation of the equations for $\Delta\phi_{j,k}^{n+1}$ requires knowledge of $\phi_{j-1,k}^{n+1}$ and $\phi_{j+1,k}^{n}$ as well as points on the jth grid line. In the supersonic region knowledge of $\phi_{j-2,k}^{n+1}$ and $\phi_{j-1,k}^{n+1}$ as well as points on the jth grid line is required. Such a scheme works well with the transonic small disturbance equation (Murman 1973) but needs to be modified for the full potential equation.

It is useful to interpret relaxation schemes in a general framework, i.e. similar to (6.51),

$$\underline{N}\Delta\boldsymbol{\phi}^{n+1} = -\omega\mathbf{R} \ , \tag{14.149}$$

where $\boldsymbol{\phi}$ is the vector of grid point values $\phi_{j,k}$. In (14.149), $\mathbf{R}$ is the residual when $\boldsymbol{\phi}$ is substituted into the discretised equations, ω is a scaling factor and $\underline{N}$ is a discretised linear operator that is economical to factorise (invert). If $\underline{N}\,\boldsymbol{\phi}^{n}$ is a close approximation to $\mathbf{R}$ then convergence to the solution will be rapid.

An equivalent time-dependent interpretation would be

$$\underline{F}\frac{\partial\boldsymbol{\phi}}{\partial t} + \omega\left(\frac{\Delta x}{\Delta t}\right)L\boldsymbol{\phi} = 0 \ , \tag{14.150}$$

where L is the operator in the steady differential equation being solved. Comparing (14.149) and (14.150) gives the equivalences

$$\mathbf{R} \equiv L\boldsymbol{\phi} \ , \qquad \Delta\boldsymbol{\phi} = \Delta t\frac{\partial\boldsymbol{\phi}}{\partial t} \ , \qquad \underline{N} = \frac{\underline{F}}{\Delta x} \ . \tag{14.151}$$

Then $\underline{N}$ should be chosen so that (14.150) represents a convergent time-dependent process.

Based on this concept, a typical relaxation scheme for (14.126), when discretised with centred difference formulae in the subsonic region and (14.142) in the supersonic region, is

$$\tau_1[\Delta\phi_{j,k} - \Delta\phi_{j-1,k}] + \tau_2[\Delta\phi_{j,k+1} - 2\Delta\phi_{j,k} + \Delta\phi_{j,k-1}] + \tau_3\Delta\phi_{j,k} = R_{j,k} \ , \tag{14.152}$$

where $\tau_1 = 1/\Delta x^2$, $\tau_2 = 1/\Delta y^2$ and $\tau_3 = (2/\omega - 1)(\tau_1 + \tau_2)$, and ω is a relaxation factor. Jameson (1978, p. 21) shows that this scheme is equivalent to solving

$$(1 - M^2)\phi_{ss} + \phi_{nn} = 2\alpha\phi_{st} + 2\beta\phi_{nt} + \gamma\phi_t \ , \tag{14.153}$$

where α, β and γ depend on τ_1, τ and τ_3. Equation (14.152) is effective in the subsonic region but in the supersonic region it is necessary, for the diagonal dominance of N in (14.152), to evaluate ϕ_{xx}, ϕ_{xy} and ϕ_{yy}. in such a way that $(1 - M^2)\phi_{ss}$ in (14.153) is replaced by

$$(1 - M^2)\phi_{ss} + 2\left(\frac{\Delta t}{q}\right)(1 - M^2)\left\{\frac{u}{\Delta x} + \frac{v}{\Delta y}\right\}\phi_{st} \ . \tag{14.154}$$

A similar approach of augmenting the equivalent time-dependent form (14.150) is effective for the conservation equation (14.137). Relaxation methods as a general class converge rapidly initially but become very slow as the equivalent steady-state solution is approached, as is noted in Sect. 6.3.5.

An alternative approach is the approximation factorisation scheme of Ballhaus et al. (1978) in which N in (14.149) is split into two factors, each of which is economical to 'invert'. The same technique applied to (14.150) is equivalent to the approximate factorisation scheme given by (8.23, 24). Here (14.149) is factorised, with $R_{j,k}$ the residual of (14.147) after substitution from (14.148). The result is

$$\{\alpha - L_\xi^+(\tilde{\varrho}^* A_1)L_\xi^-\}\{\alpha - L_\eta^+(\tilde{\varrho}^* A_2)L_\eta^-\}\Delta\phi_{j,k}^{n+1} = \alpha\omega R_{j,k} \ . \tag{14.155}$$

where

$$L_\xi^+ f_j = \frac{f_{j+1} - f_j}{\Delta\xi} \ , \qquad L_\xi^- f_j = \frac{f_j - f_{j-1}}{\Delta\xi} \ ,$$

and

$$L_\xi^+(\tilde{\varrho}^* A_1)L_\xi^- f_j = \left\{(\tilde{\varrho}^* A_1)_{j+1/2}\frac{f_{j+1} - f_j}{\Delta\xi} - (\tilde{\varrho}^* A_1)_{j-1/2}\frac{f_j - f_{j-1}}{\Delta\xi}\right\}\bigg/\Delta\xi \ .$$

Equation (14.155) is implemented in two stages:

1st stage: $\{\alpha - L_\xi^+(\tilde{\varrho}^* A_1)L_\xi^-\}\Delta\phi_{j,k}^* = \alpha\omega R_{j,k} \ , \tag{14.156}$

2nd stage: $\{\alpha - L_\eta^+(\tilde{\varrho}^* A_2)L_\eta^-\}\Delta\phi_{j,k} = \Delta\phi_{j,k}^* \ . \tag{14.157}$

For the first stage (14.156) represents a tridiagonal system of equations that can be solved along each gridline in the ξ direction in turn, using (6.29–31). For the second stage (14.157) gives a tridiagonal system that can be solved along each gridline in the η direction consecutively.

Equations (14.156 and 157) constitute the AF1 algorithm of Ballhaus et al. (1978) and is clearly of similar form to the ADI and approximate factorisation schemes for time-dependent problems described in Sect. 8.2. However, a more efficient implementation of (14.155) is available as the following two-stage algorithm (AF2):

1st stage: $\{\alpha - L_\eta^+(\bar{\varrho}^*A_2)\}\,\Delta\phi_{j,k}^* = \alpha\omega R_{j,k}$,   (14.158)

2nd stage: $\{\alpha L_\eta^- - L_\xi^+(\tilde{\varrho}^*A_1)\}\,\Delta\phi_{j,k}^{n+1} = \Delta\phi_{j,k}^*$.   (14.159)

The first stage provides a set of bidiagonal equations that are swept in the negative η direction. The second stage is in the form of a set of tridiagonal equations in the ξ direction which are solved progressively in the positive η direction. Holst (1985) discusses the practical implementation of both the AF1 and AF2 algorithms.

The parameters α and ω are chosen to accelerate convergence; ω has to be in the range $\theta < \omega \le 2$ for stability and is usually chosen to be as large as possible (say $\omega = 1.8$–1.9). The parameter α can be interpreted as $1/\Delta t$ so that, conceptually, the steady state would be reached in the smallest number of iterations by making α as small as possible. This tends to remove errors of low frequency but not those of high frequency. Therefore a better strategy is to use a sequence of values of α, say

$$\alpha_m = \alpha_1 a^{m-1} \quad \text{with} \quad \alpha_1 = \Delta y \quad \text{and} \quad a = \left(\frac{2}{\Delta y}\right)^{1/(N-1)} ,$$

where N is the number of steps in the sequence, typically $N = 11$. Different approximate factorisation schemes and analysis of optimal choices of α and ω are provided by Catherall (1982).

The above scheme can be made even more efficient by embedding it in a multigrid iteration procedure (Sect. 6.3.5).

Equation (14.147), after substitution of (14.148), can be written

$$\underline{A}^M \phi^M = 0 ,$$ (14.160)

where superscript M denotes the finest grid on which the solution is sought, as in Sect. 6.3.5. For any intermediate solution on a coarser grid, $\underline{A}^{m+1}\phi^{m+1} = R^{m+1}$, i.e. a non-zero residual occurs. Jameson (1979) uses a modified FAS algorithm (Sect. 6.3.5) in which only the residuals are restricted from the finer to the coarser grid. Equation (6.90) is replaced by

$$\underline{A}^m \phi^{m,a} = \underline{A}^m \phi^m - I_{m+1}^m R^{m+1} ,$$ (14.161)

where ϕ^m is the existing solution on the mth grid. As in the FAS algorithm $\phi^{m,a}$ is obtained by relaxation and further restriction to the coarsest grid, exact solution and relaxation and prolongation back to the mth grid as in Fig. 6.21b. The new solution on the $(m+1)$-th grid is given by

$$\phi^{m+1,a} = \phi^{m+1} + I_m^{m+1}(\phi^{m,a} - \phi^m) .$$ (14.162)

The restriction and prolongation operators I in (14.161, 162) are described in Sect. 6.3.5.

Jameson uses a modified form of the approximate factorisation scheme (14.156, 157) as a relaxation algorithm, with $R_{j,k}$ in (14.156) replaced by the right-hand side of (14.161). In Jameson's scheme the parameter α in (14.156, 157) is replaced by S where

$$S = \alpha_0 + \alpha_1 L_\xi^- + \alpha_2 L_\eta^- \;, \qquad (14.163)$$

and L_ξ^- and L_η^- are upwind operators, defined after (14.155). This modification is introduced to handle the embedded supersonic region more efficiently.

A complete V-cycle (Fig. 6.21b) consists of one relaxation using (14.156, 157 and 163) and the restriction of the residuals onto the next coarser grid until the coarsest grid is reached, on which the exact solution is obtained. Then a prolongation using (14.162) and one relaxation step are carried out on each grid until the second finest grid is reached. The finest grid solution is then obtained from (14.162). The V-cycle is repeated until (14.160) is satisfied.

A single V-cycle of the present algorithm can be compared with a complete cycle through a range of α values in the conventional approximate factorisation algorithm, e.g. (14.158 and 159).

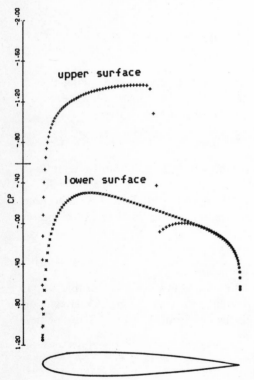

Fig. 14.32. Pressure distribution for NACA-0012 aerofoil at $= 2°$, $M_\infty = 0.75$, 192×32 grid

Jameson (1979) indicates that the multigrid implementation with six levels of grid refinement requires about a quarter of the number of operations per cycle to implement the conventional approximate factorisation scheme. In addition roughly 10 complete cycles are required to produce the converged solution (to engineering accuracy) of the flow about an inclined aerofoil with an embedded shock as indicated in Fig. 14.32. This is comparable with approximately 80–100 iterations of the conventional approximate factorisation algorithm (Holst 1985). Thus the introduction of the multigrid strategy is giving about a fourfold improvement in computational efficiency.

14.3.6 Non-isentropic Potential Formulation

Klopfer and Nixon (1984) introduce an interesting non-isentropic potential formulation that greatly improves the accuracy of (14.137) when stronger shocks occur. The formulation essentially improves the calculation of the density in (14.129) to allow for changes in entropy across the shock. Equation (14.129) is replaced by

$$\varrho/\varrho_\infty = \frac{[1+0.5(\gamma-1)M_\infty^2(1-q^2)]^{1/(\gamma-1)}}{K^{1/(\gamma-1)}} \ , \tag{14.164}$$

where $q^2 = (u^2+v^2)/U_\infty^2$. K is an entropy-related function given by

$$K = \frac{2\gamma M_{1,n}^2 - (\gamma-1)}{(\gamma+1)}\left(\frac{(\gamma-1)M_{1,n}^2+2}{(\gamma+1)M_{1,n}^2}\right)^\gamma \ , \tag{14.165}$$

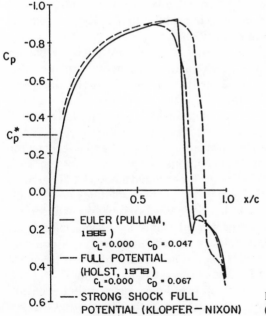

EULER (PULLIAM, 1985)
$c_L = 0.000$ $c_D = 0.047$
FULL POTENTIAL (HOLST, 1979)
$c_L = 0.000$ $c_D = 0.067$
STRONG SHOCK FULL POTENTIAL (KLOPFER − NIXON)
$c_L = 0.000$ $c_D = 0.050$

Fig. 14.33. Comparison of shock locations (after Klopfer and Nixon, 1984; reprinted with permission of AIAA)

where $M_{1,n}$ is the local Mach number based on the upstream velocity component normal to the shock. For the flow external to an aerofoil Klopfer and Nixon indicate that it is sufficiently accurate to set $M_{1,n} = u/a$. For points ahead of the shock, $K = 1$. Downstream of the shock K is constant along each streamline and these must be tracked approximately.

With the above modifications made to the full potential code of Holst (1979) a typical improvement in shock location, and hence pressure distribution, is shown in Fig. 14.33. Clearly a solution closer to that given by the Euler code (Pulliam 1985) is produced.

Since the modifications required to implement (14.164 and 165) do not greatly increase the execution time, the non-isentropic formulation of Klopfer and Nixon is a computationally efficient extension to the methods based on a velocity potential to accurately predict the flowfield when shocks of moderate strength are present. Hafez (1985) also discusses non-isentropic potential formulations.

14.3.7 Full Potential Equation, Further Comments

The general status of transonic potential flow solvers is reviewed by the papers in Habashi (1985). Finite difference, finite element and finite volume methods are all successful in obtaining accurate solutions (Rizzi and Viviand 1981) to the transonic full potential equation. Spectral methods (Hussaini and Zang 1987) are also effective if shock-fitting is employed to isolate the jump in the solution at the shock from the smooth behaviour elsewhere.

As long as the flow is irrotational except for the presence of weak shocks, accurate solutions of the transonic full potential equation can be obtained very economically. As a result these methods are being applied to relatively complicated three-dimensional geometries (Young et al. 1990).

A comparison (Flores et al. 1985) for transonic inviscid flows indicates that, for comparable accuracy, computer codes based on the full potential equation (e.g. Holst 1979) are an order of magnitude faster than the implicit Euler codes (e.g. Pulliam 1985) if the shocks are weak. However, as the shocks become stronger the Euler codes become more competitive since, in general, the isentropic full potential codes lose accuracy in predicting the shock strength and location.

A disturbing feature of conservative formulations of the full potential equation is that under certain conditions multiple solutions are generated for the same boundary conditions (Salas et al. 1983). For the flow around a NACA-0012 aerofoil at $M_\infty = 0.83$ and small angles of incidence, up to three solutions are possible leading to a variation of lift coefficient with incidence which is physically incorrect. It is conjectured (Hafez 1985) that the proper allowance for viscous effects may remove the non-uniqueness. The non-uniqueness does not occur with non-conservative formulations; however, typically mass is not conserved across the shock and drag predictions are inaccurate for non-conservative formulations.

For steady transonic flows, particularly in two-dimensions, possible problems of non-uniqueness and the difficulty of accurately predicting flows with shocks of moderate strength have caused a significant shift towards computing with the

Euler equations. However, for unsteady flows, e.g. associated with flutter, the economic advantages of working with a potential formulation are considerable, although the problem of non-uniqueness must still be overcome.

For two-dimensional unsteady potential flow the governing equations are

$$\frac{\partial \varrho}{\partial t} + \frac{\partial}{\partial x}\left(\varrho \frac{\partial \phi}{\partial x} \right) + \frac{\partial}{\partial y}\left(\varrho \frac{\partial \phi}{\partial y} \right) = 0 \tag{14.166}$$

in place of (14.128) and

$$\varrho/\varrho_\infty = \left(1 + 0.5(\gamma - 1)M_\infty^2 \left\{ 1 - \left[\frac{2}{U_\infty^2}\frac{\partial \phi}{\partial t} + \left(\frac{u}{U_\infty} \right)^2 + \left(\frac{v}{U_\infty} \right)^2 \right] \right\} \right)^{1/(\gamma - 1)} \tag{14.167}$$

in place of (14.129). Goorjian (1985) describes implicit approximate factorisation techniques that are effective for solving (14.166 and 167).

14.4 Closure

By organising the various computational techniques according to the class of flow problems being considered, it is apparent that, for restricted classes of flows, the equations reduce to relatively simple forms for which very efficient computational methods are available e.g. the panel method considered in Sect. 14.1 for (incompressible) potential flow.

For supersonic inviscid flow the major difficulty is in representing shock waves efficiently. Because of the unknown shock location, shock-capturing methods, based on conservatively discretised schemes, are preferred to shock-fitting methods. If the shock waves are very strong special techniques are required, as in Sects. 14.2.6 and 14.2.7. Supersonic inviscid flow with strong shocks requires the use of the full Euler equations.

By contrast, many transonic flows feature only weak shock waves and consequently can be solved accurately using the full potential equation in the form of (14.137 and 129). Because of the mixed subsonic, supersonic flow regimes, special computational procedures must be introduced to suit the changing character of the governing equation and to provide dissipation in the supersonic region. The relative economy of the potential flow formulation makes it particularly suitable for unsteady and three-dimensional flows involving complex geometric domains (Young et al. 1990).

For steady transonic flows the use of the potential equation permits relatively efficient techniques, e.g. multigrid, to be used to accelerate the solution to convergence; such techniques are also effective for the Euler equations. The development of acceleration techniques remains an important area of research interest for compressible inviscid flow computations based on the unsteady Euler equations. The greater reliability of computations based on the Euler equations makes this the preferred approach for steady two-dimensional inviscid transonic flows.

Finite difference and finite volume methods have been the more widely used techniques for both supersonic and transonic flow computation. However, finite element methods have been applied (e.g. Deconinck and Hirsch 1985; Ecer and Akay 1985; Jameson and Baker 1986; Young et al. 1990) with considerable success to both internal and external transonic inviscid flows. For flows with strong shocks characteristic finite element methods (Morton and Sweby 1987; Fletcher and Morton 1986; Hughes and Mallet 1985) permit efficient TVD algorithms to be constructed. Spectral methods (Hussaini and Zang 1987) are effective when used with a shock-fitting strategy; however McDonald (1989) demonstrates that an FCT pseudospectral formulation can accurately capture 'shocks' created by the inviscid Burgers equation, (10.8).

14.5 Problems

Panel Method (Sect. 14.1)

14.1 a) Use program PANEL (Fig. 14.7) to obtain the pressure distribution for the flow around a circular cylinder at $M_\infty = 0$ with 4, 8, 16 and 32 panels and compare the results with the exact pressure distribution.
   b) Repeat step a) for an ellipse with minor/major axis ratio = 0.5 and 0.2.
   c) Comment on the improvement in accuracy with increasing numbers of panels with reducing minor/major axis ratio.

14.2 Apply program PANEL to the flow about a NACA-0012 aerofoil at zero incidence. The coordinates for the NACA-0012 aerofoil are given by (13.70).
   a) Obtain solutions with 8, 16 and 32 panels for $M_\infty = 0.4$. Compare the solutions with those shown in Fig. 14.4.
   b) For 16 panels obtain solutions at $M_\infty = 0.4$ with more panels in the nose region, more in the tail region and less over the mid-section. Decide what is the best distribution for achieving the highest accuracy with a given number of elements. How does this 'optimal' distribution relate to gradients of the solution?

14.3 Replace the direct solver (subroutines FACT and SOLVE) with an iterative solver based on SOR (Sect. 6.3) and obtain solutions for 8, 16 and 32 elements representing a circle. Assume the SOR iteration has converged when the rms algebraic equation residual is less than 1×10^{-5}. From the results deduce a relationship

$$\text{NITER} = kN^P \ ,$$

where NITER is the number of iterations to convergence and N is the number of panels.

14.4 Carry out an approximate operation count (only multiplications and divisions) for program PANEL and related subroutines as a function of the number of panels N.

a) Compare the approximate operation count with that for the case where a SOR solver replaces the direct solver (subroutines FACT and SOLVE) to solve (14.6) for the source strengths σ_j.

b) Compare the approximate operation count for N panels with that for program LAGEN (Sect. 12.4.1) applied to the flow past a circle. Assume that the computational domain for LAGEN has $N \times N$ grid points external to the half-circle.

14.5 Modify program PANEL to obtain solutions about lifting aerofoils by implementing the procedure described in Sect. 14.1.4. Test the program modifications by obtaining solutions for a NACA-0012 aerofoil at $M_\infty = 0$ and $\alpha = 0, 2$ and $4°$ incidence.

Modify program PANEL to integrate the pressure distribution to evaluate the lift coefficient and compare with the theoretical result $C_L = 2\pi\alpha$.

Supersonic Inviscid Flow (Sect. 14.2)

14.6 Apply program SHOCK with the MacCormack scheme to the propagating shockwave problem with a pressure ratio $= 2.5$. Compare the results for NX $= 101$ with the Lax–Wendroff results shown in Fig. 4.18.

14.7 An alternative form of the artificial viscosity correction (Lapidus 1967) replaces (14.54) with

$$\mathbf{q}_j^{n+1} = \mathbf{q}_j^{**} + v\left(\frac{\Delta t}{\Delta x}\right) \Delta[|\Delta u_{j+1}^{**}|\Delta\mathbf{q}_{j+1}^{**}] \ . \tag{14.168}$$

Implement this form of the artificial viscosity and obtain solutions for moderate shock $(p_1/p_2 = 2.5)$ and strong shock $(p_1/p_2 = 5)$ propagation. Compare solutions with those shown in Figs. 14.18 and 14.25 with (14.168) applied to all components of $\mathbf{q}$.

14.8 Program SHOCK is to be modified to obtain the solution for the flow in a shock tube, before any reflection from the end walls has occurred.

A shock tube initially contains high pressure fluid at rest separated from low pressure fluid at rest by a diaphragm. At $t = 0$ the diaphragm is broken and the resulting flow quickly develops into a shock wave propagating into the low pressure region.

Behind the shock is a contact discontinuity which is the current location of the initial boundary between the high and low pressure regions. Across the contact discontinuity pressure and velocity are continuous but density is discontinuous.

Behind the contact discontinuity and spreading into the high pressure region is an expansion wave across which pressure, density and velocity smoothly change from the conditions of the high pressure fluid at rest. The head of the expansion wave moves into the high pressure region. The physical behaviour of shock tube flow is described by Liepmann and Roshko (1957, pp. 79–83). This example has been used to compare various computational methods by Sod (1978).

The flow in the shock tube is, to a good approximation, one-dimensional, unsteady and governed by nondimensional equations (14.43, 44). Obtain solutions for the following initial conditions:

$$u_1' = 0 \ , \quad \varrho_1' = 8.0 \ , \quad p_1' = \frac{p_1}{p_2} = 10 \quad \text{for} \quad x < 0.305 \ ,$$
$$ \tag{14.169}$$
$$u_2' = 0 \ , \quad \varrho_2' = 1.0 \ , \quad p_2' = 1.0 \quad \text{for} \quad x \geq 0.305 \ ,$$

where p_1/p_2 is the pressure ratio (PRAT in program SHOCK). Equation (14.169) also provides boundary conditions at $x = 0$ and 1.0. Obtain solutions with NX = 101 for NT = 170 and DT = 0.100, with the Lax–Wendroff scheme and artificial viscosity. The general character of the flow may be compared with numerical solutions given by Sod (1978). The exact solution to this problem is given by Liepmann and Roshko (1957, pp. 79–83). The two most difficult parts of the solution to predict accurately are the contact discontinuity and the shock.

14.9 Obtain solutions for moderate ($p_1/p_2 = 2.5$, Fig. 14.18) and strong ($p_1/p_2 = 5.00$, Fig. 14.25) shock propagation using the FCT scheme with the following choices for the diffusion and anti-diffusion parameters in (14.93):

i) $\eta_0 = 0.125$, $\eta_1 = 0$, $\eta_2 = 0$,
ii) $\eta_0 = 0.500$, $\eta_1 = 0$, $\eta_2 = 0$,
iii) $\eta_0 = 1/3$, $\eta_1 = 1/3$, $\eta_2 = -1/6$,
iv) $\eta_0 = 1/6$, $\eta_1 = 1/3$, $\eta_2 = -1/3$.

Compare the shock profiles for these various solutions.

14.10 Apply the FCT algorithm to the shock tube flow of Problem 14.8. Compare the solutions with those of the Lax–Wendroff scheme plus artificial viscosity (Problem 14.8) with particular reference to the shock and contact discontinuity profiles.

Transonic Inviscid Flow (Sect. 14.3)

14.11 The solution to the flow past a slender body defined by $y = \tau(x^2 - 1)^2$, $-1 \leq x \leq 1$ is governed by the transonic small disturbance equation (14.130) for the domain and boundary conditions shown in Fig. 14.34. Discuss the use of the finite difference discretisation given by (14.135) to obtain solutions for $\tau = 0.1$ and $M_\infty = 0.8$ and 0.9, with SOR iteration employed to obtain the converged solution to the discretised equations.

14.12 Formulate a procedure to obtain the solution to the problem shown in Fig. 14.34 using the discretised form of the full potential equation (14.129, 138, 139). Use (14.143) to provide the dissipative mechanism in the supersonic region. The boundary condition on AB, $\phi_y = f_x$, may be interpreted as a specified injection velocity profile.

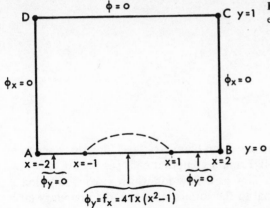

Fig. 14.34. Model problem for potential equation

Discuss the solution of the resulting discretised equation by an approximate factorisation method like (14.156, 157), but based on a pseudo-transient equivalent of (14.137), for $\tau = 0.1$ and $M_\infty = 0.8$.

14.13 Discretise Laplace's equation $\partial^2\phi/\partial x^2 + \partial^2\phi/\partial y^2 = 0$ on a rectangle $0 \leq x \leq 1, 0 \leq y \leq 1$ for boundary conditions $\phi(0, y) = \cos(0.5\pi y)$, $\phi(1, y) = \exp(\pi/2)\cos(0.5\pi y)$, $\phi(x, 0) = \exp(0.5\pi x)$, $\phi(x, 1) = 0$, using centred differences.

  i) Obtain solutions to the discretised equations using approximate factorisation, equivalent to (14.156, 157),

  ii) Consider approximate factorisation plus multigrid (14.156, 157, 161, 162), and discuss the two methods taking into account the number of iterations to convergence and the approximate operation count. The exact solution for the present problem is $\phi_{ex} = \exp(0.5\pi x)\cos(0.5\pi y)$.

15. Boundary Layer Flow

Traditionally it has been useful to consider boundary layer flow as a separate category (Table 11.4 and Sect. 11.4). From a computational perspective it is convenient to classify boundary layer flow as a flow for which viscous diffusion is significant only in directions normal to the surface on which the boundary layer occurs (Fig. 15.1) and for which the normal momentum equation can be replaced with the condition that the pressure is constant. For such flows the governing equations are non-elliptic, if the pressure solution is given. This permits very efficient single-pass marching algorithms to be introduced (in the x direction in Fig. 15.1).

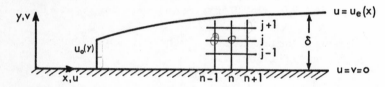

Fig. 15.1. Boundary layer flow

Boundary layer flows contain severe velocity gradients in the direction normal to the surface. This makes it desirable to introduce a transformation to render the gradients less severe in the transformed coordinates. The more effective transformations are described in Sects. 15.2 and 15.3. In addition, grids that grow geometrically in the direction normal to the surface provide an efficient way of achieving good resolution close to the surface (Sect. 15.1.2).

The equations governing three-dimensional boundary layer flow have a hyperbolic character in planes parallel to the surface on which the boundary layer develops. This introduces the complication of domains of influence and dependence (Sect. 2.2.1) in planes parallel to the surface. The domain of influence affects the allowable step-size for explicit marching schemes (Sect. 15.4).

Turbulent boundary layers are solved using the same computational techniques as laminar boundary layers. However, the increased severity of the normal velocity gradient adjacent to the surface can require the use of a very refined grid in the normal direction. This problem is avoided by adopting the Dorodnitsyn formulation (Sect. 15.3) of treating the streamwise velocity component (u) as the independent variable. An alternative way of avoiding the use of fine near-wall grids is to incorporate wall functions that effectively provide a local analytic velocity profile adjacent to the wall. Wall functions are discussed in Sect. 18.1.1.

The neglect of the streamwise diffusion and the transverse momentum equation is also appropriate to the treatment of jets, wakes and developing flows in pipes. Thus these thin-shear-layer flows can be computed efficiently with essentially the same computational techniques as are appropriate to traditional boundary layer flows.

15.1 Simple Boundary Layer Flow

The governing equations for steady laminar incompressible two-dimensional flow (Sect. 11.4.1) can be written as

$$\frac{\partial u}{\partial x} + \frac{\partial v}{\partial y} = 0 \quad \text{and} \tag{15.1}$$

$$u\frac{\partial u}{\partial x} + v\frac{\partial u}{\partial y} = u_e \frac{du_e}{dx} + v\frac{\partial^2 u}{\partial y^2} \ , \tag{15.2}$$

where the Bernoulli equation (11.49) has been used to introduce the known velocity distribution $u_e(x)$ at the outer edge of the boundary layer, Fig. 15.1. Since the equation system (15.1, 2) is mixed parabolic/hyperbolic with x having a time-like role, both initial conditions,

$$u(x_0, y) = u_0(y) \ , \tag{15.3}$$

and boundary conditions,

$$u(x, 0) = 0 \ , \quad v(x, 0) = 0 \quad \text{and} \quad u(x, \delta) = u_e(x) \tag{15.4}$$

are required.

The momentum equation (15.2) may be compared with the one-dimensional diffusion equation considered in Chap. 7, and the one-dimensional transport equation considered in Sect. 9.4. The major differences are the nonlinear nature of the convective terms and the coupling with the continuity equation through the normal velocity v. Since $u_e du_e/dx$ is known it behaves as a source term with little influence on the choice of the computational method.

Any of the schemes described in Sects. 7.2 or 9.4 are potential candidates to be applied to (15.2). Explicit schemes (Sect. 7.1) are excluded since they lead to an unacceptable restriction on the marching step size Δx for stable solutions.

Both the Crank–Nicolson scheme (Sect. 7.2.2) and the three-level fully implicit scheme (Sect. 7.2.3) are unconditionally stable and are second-order accurate (in Δt, Δx) for the diffusion equation. To achieve second-order accuracy in Δx when solving (15.2) will require a second-order treatment of the nonlinear terms $u\partial u/\partial x$

and $v \partial u/\partial y$. For the Crank–Nicolson scheme this requires iteration at each downstream location. For the three-level fully implicit scheme (Sect. 15.1.1) iteration is avoided by projecting u and v from upstream (15.6).

15.1.1 Implicit Scheme

To develop a computational algorithm, uniform-grid finite difference expressions are introduced for the various terms in (15.1 and 2) as follows:

$$\frac{\partial u}{\partial x} = \frac{(1.5u_j^{n+1} - 2u_j^n + 0.5u_j^{n-1})}{\Delta x} + O(\Delta x^2) , \quad \longrightarrow \text{ 3 points equation}$$

$$\frac{\partial u}{\partial y} = \frac{(u_{j+1}^{n+1} - u_{j-1}^{n+1})}{2\Delta y} + O(\Delta y^2) , \tag{15.5}$$

$$\frac{\partial^2 u}{\partial y^2} = \frac{(u_{j-1}^{n+1} - 2u_j^{n+1} + u_{j+1}^{n+1})}{\Delta y^2} + O(\Delta y^2) .$$

The grid identification (Fig. 15.1) and super/subscripting in the above expressions is introduced to accentuate the time-like role of the x-coordinate.

To permit a linear system of equations for u^{n+1} to be obtained, the undifferentiated velocity components, u and v, appearing on the left-hand side of (15.2) are extrapolated using

$$u_j^{n+1} = 2u_j^n - u_j^{n-1} + O(\Delta x^2) , \quad v_j^{n+1} = 2v_j^n - v_j^{n-1} + O(\Delta x^2) . \tag{15.6}$$

Substitution of the above expressions into (15.2) and rearrangement gives the following tridiagonal system of equations associated with the grid line $n+1$ across the boundary layer:

$$a_j u_{j-1}^{n+1} + b_j u_j^{n+1} + c_j u_{j+1}^{n+1} = d_j , \quad \text{where} \tag{15.7}$$

$$a_j = -\frac{\Delta x}{2\Delta y} (2v_j^n - v_j^{n-1}) - v \frac{\Delta x}{\Delta y^2} ,$$

$$b_j = 1.5(2u_j^n - u_j^{n-1}) + 2v \frac{\Delta x}{\Delta y^2} ,$$

$$c_j = \frac{\Delta x}{2\Delta y} (2v_j^n - v_j^{n-1}) - v \frac{\Delta x}{\Delta y^2} , \quad \text{and}$$

$$d_j = (2u_j^n - u_j^{n-1})(2u_j^n - 0.5u_j^{n-1}) + \Delta x \left(u_e \frac{du_e}{dx} \right)^{n+1} .$$

Equation (15.7) is not applied at $j = 1$ ($y = 0$) or $j = JMAX$ ($y = y_{max}$). For the equation formed at $j = JMAX-1$, $u_{JMAX} = u_e$, therefore d_j in (15.7) is replaced by $d_j - c_j u_e^{n+1}$ and c_j is subsequently set equal to zero. For the equation formed at

$j = 2$, $u_1 = 0$. Equation (15.7) can be solved efficiently using the Thomas algorithm (Sect. 6.2.2).

Once u_j^{n+1} is available, v_j^{n+1} is obtained from (15.1) in the discretised form

$$v_j^{n+1} = v_{j-1}^{n+1}$$

$$-0.5 \frac{\Delta y}{\Delta x} [(1.5u_j^{n+1} - 2u_j^n + 0.5u_j^{n-1}) + (1.5u_{j-1}^{n+1} - 2u_{j-1}^n + 0.5u_{j-1}^{n-1})] ,$$

$$(15.8)$$

with $v_1^{n+1} = 0$. The combination of (15.7) and (15.8) is second-order accurate in Δx, Δy, unconditionally stable (in the von Neumann sense), robust and efficient, but must be supplemented by a one-level algorithm to start the downstream march, i.e. when $n = 1$. Alternatively two levels ($n-1$ and n) of initial data (15.3) must be provided.

If a Crank–Nicolson scheme is introduced to solve (15.2) the solution u^{n-1} is not made use of. This reduces the storage requirement and necessitates only one level of initial data. However, the projection (15.6) is replaced with $u_j^{n+1} = u_j^n + O(\Delta x)$ and $v_j^{n+1} = v_j^n + O(\Delta x)$. To achieve an overall second-order accuracy in Δx an iteration is required at each downstream location. After solution of the equivalent of (15.7 and 8) the current iterative solution, u^{k+1}, v^{k+1}, is used in place of (15.6), and the equivalent of (15.7 and 8) are re-solved. At the start of the iteration, $u^k = u^n$ and $v^k = v^n$. The iteration is terminated when $u_{rms}^{k+1} = u_{rms}^k$ to some acceptable tolerance and u^{n+1}, v^{n+1} are set equal to u^{k+1}, v^{k+1}.

In practice it is more efficient not to iterate at each downstream location but to reduce the step size Δx to achieve the required accuracy, even though this will degrade the formal convergence rate.

The main problem with using a uniform grid in x and y is that special procedures must be introduced to allow for boundary layer growth and a very refined grid in y must be used to accurately predict the velocity distribution close to the wall. This problem is particularly severe for turbulent boundary layers.

15.1.2 LAMBL: Laminar Boundary Layer Flow

The implicit scheme, described in Sect. 15.1.1, is used to obtain the flow solution in the boundary layer that is produced by a uniform flow past a two-dimensional wedge (Fig. 15.2).

This problem is a member of the Falkner-Skan family (Schlicting 1968, p. 150) of boundary layer flows that produce similar velocity profiles. That is, the velocity components are a function of a single variable

$$\eta = y \left(\frac{u_e(x)}{(2-\beta)xv} \right)^{1/2}$$

$$(15.9)$$

and the governing equations (15.1, 2) can be reduced to a single equation

$$\frac{\partial^3 f}{\partial \eta^3} + f \frac{\partial^2 f}{\partial \eta^2} + \beta \left[1 - \left(\frac{\partial f}{\partial \eta} \right)^2 \right] = 0 ,$$

$$(15.10)$$

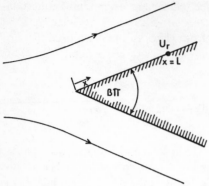

Fig. 15.2. Flow past a wedge

where $f(\eta)$ is related to the streamfunction ψ by

$$\psi = [(2-\beta)u_e \nu x]^{1/2} f(\eta) \ . \tag{15.11}$$

For the flow past a wedge, the velocity $u_e(x)$ at the outer edge of the boundary layer is given by

$$u_e = cx^{\beta/(2-\beta)} \ . \tag{15.12}$$

Accurate numerical solutions for $f(\eta)$ for various values of the wedge angle β are given by Rosenhead (1964, p. 234). Here these tabulated values will be used to provide the initial data for u and v and also to provide an 'exact' solution further downstream with which to compare the computational solution.

It is computationally efficient to introduce the following nondimensionalisation into (15.1, 2):

$$x' = \frac{x}{L} \ , \quad y' = \frac{y}{L} \, \mathrm{Re}^{1/2} \ , \quad u' = \frac{u}{U_r} \ , \quad v' = \frac{v}{U_r} \, \mathrm{Re}^{1/2} \ , \tag{15.13}$$

where the Reynolds number $\mathrm{Re} = U_r L/\nu$, and L and U_r are the characteristic length and velocity, respectively. For the wedge flow problem U_r is the value of u_e (15.12) at $x = L$.

The advantage of (15.13) is that the nondimensional y coordinate and the nondimensional normal velocity are scaled (by $\mathrm{Re}^{1/2}$) to be of the same order as x' and u', respectively. With the aid of (15.13), (15.1 and 2) can be written (dropping the prime) as

$$\frac{\partial u}{\partial x} + \frac{\partial v}{\partial y} = 0 \quad \text{and} \tag{15.14}$$

$$u \frac{\partial u}{\partial x} + v \frac{\partial u}{\partial y} = u_e \frac{du_e}{dx} + \frac{\partial^2 u}{\partial y^2} \ . \tag{15.15}$$

Initial and boundary conditions are given by (15.3 and 4) interpreted as non-dimensional equations. However, the boundary condition $u = u_e(x)$ is applied at $y = y_{max}$ where $y_{max} > \delta$, the boundary layer thickness.

For the flow past a wedge the nondimensional velocity at the outer edge of the boundary layer is given by

$$u_e = x^{\beta/(2-\beta)} . \tag{15.16}$$

To permit the use of a variable grid in the y direction, the various y-derivatives in (15.15) are discretised as in (10.30 and 32)

$$\frac{\partial u}{\partial y} = \frac{(u_j^{n+1} - u_{j-1}^{n+1})r_y + (u_{j+1}^{n+1} - u_j^{n+1})/r_y}{(1+r_y)\Delta y} + O(\Delta y^2) , \tag{15.17}$$

$$\frac{\partial^2 u}{\partial y^2} = \frac{2\left[u_{j-1}^{n+1} - \left(1+\dfrac{1}{r_y}\right)u_j^{n+1} + \dfrac{u_{j+1}^{n+1}}{r_y}\right]}{(1+r_y)\Delta y^2} + O(\Delta y) ,$$

where the grid growth ratio $r_y = (y_{j+1} - y_j)/(y_j - y_{j-1})$. The term $\partial u/\partial x$ is discretised as in (15.5).

Substituting (15.6 and 17) into (15.15) produces the tridiagonal system

$$a_j u_{j-1}^{n+1} + b_j u_j^{n+1} + c_j u_{j+1}^{n+1} = d_j , \quad \text{where} \tag{15.18}$$

$$p = (2v_j^n - v_j^{n-1})\frac{\Delta x}{(1+r_y)\Delta y} ,$$

$$q = \frac{2\Delta x}{(1+r_y)\Delta y^2} ,$$

$$a_j = -r_y p - q ,$$

$$b_j = 1.5(2u_j^n - u_j^{n-1}) + \left(r_y - \frac{1}{r_y}\right)p + \left(1 + \frac{1}{r_y}\right)q ,$$

$$c_j = \frac{p}{r_y} - \frac{q}{r_y} ,$$

$$d_j = \Delta x u_e \frac{du_e}{dx} + (2u_j^n - 0.5u_j^{n-1})(2u_j^n - u_j^{n-1}) .$$

At the wall $u_1 = 0$ and at $y = y_{max}$, $u_{JMAX} = u_e$. Equation (15.18) is repeated at the JMAX-2 interior nodes forming a tridiagonal system of equations that can be solved using the Thomas algorithm (Sect. 6.2.2) for u_j^{n+1}.

The continuity equation (15.14) is integrated across the boundary layer to give v_j^{n+1} using (15.8). The solution for the velocity distribution in the boundary layer is obtained by solving (15.18) and (15.8) sequentially at each downstream location x^{n+1}.

```
1   C      LAMBL USES AN IMPLICIT MARCHING ALGORITHM TO COMPUTE
2   C      THE SOLUTION TO A FALKNER-SKAN LAMINAR BOUNDARY LAYER (BETA = 0.5)
3   C
4          DIMENSION UP(65),U(41),UM(41),V(41),VM(41),Y(41),RHS(65)
5         1,B(5,65),UBX(41),UB(24),VB(24),YZ(24)
6          DATA UB/0.0000,0.0903,0.1756,0.2559,0.3311,0.4015,0.4669,0.5275,
7         10.5833,0.6344,0.6811,0.7614,0.8258,0.8761,0.9142,0.9422,0.9623,
8         20.9853,0.9972,0.9995,1.0000,1.0000,1.0000,1.0000/
9          DATA VB/0.,0.,-0.0003,-0.0011,-0.0027,-0.0052,-0.0089,-0.0142,
10        1-0.0211,-0.0298,-0.0406,-0.0688,-0.1065,-0.1541,-0.2114,-0.2778,
11        2-0.3521,-0.5198,-0.8008,-1.0965,-1.3954,-1.6954,-2.0954,-2.4954/
12         DATA YZ/0.0,0.1,0.2,0.3,0.4,0.5,0.6,0.7,0.8,0.9,1.0,1.2,1.4,1.6,
13        11.8,2.0,2.2,2.6,3.2,3.8,4.4,5.0,5.8,6.6/
14         OPEN(1,FILE='LAMBL.DAT')
15         OPEN(6,FILE='LAMBL.OUT')
16         READ(1,1)JMAX,NMAX,DYM,RY,XST,BETA,RE,DX
17       1 FORMAT(2I5,4F5.2,2E10.3)
18   C
19         WRITE(6,2)BETA
20       2 FORMAT(' FALKNER-SKAN SOLUTION,  BETA=',F5.2)
21         WRITE(6,3)JMAX,DYM,RY
22       3 FORMAT(' JMAX= ',I3,'  DYM= ',F5.2,'   RY= ',F5.2)
23         WRITE(6,4)NMAX,DX,XST,RE
24       4 FORMAT(' NMAX= ',I3,'  DX= ',E10.3,'  XST= ',F5.2,'  RE=',E10.3,//)
25         Y(1) = 0.
26         DY = DYM/RY
27         DO 5 J = 2,JMAX
28         DY = DY*RY
29         Y(J) = Y(J-1) + DY
30       5 CONTINUE
31         JMAP = JMAX - 1
32         AJP = JMAP
33         RYP = RY + 1.
34         BETP = BETA/(2. - BETA)
35         SQRE = SQRT(RE)
36   C
37   C      SET INITIAL VELOCITY PROFILES
38   C
39         UEST = XST**BETP
40         FALKS = SQRT((2.-BETA)*XST/UEST)
41         CALL LAG(YZ,UB,Y,UM,FALKS,JMAX)
42         CALL LAG(YZ,VB,Y,VM,FALKS,JMAX)
43         X = XST + DX
44         UE = X**BETP
45         FALK = SQRT((2.-BETA)*X/UE)
46         CALL LAG(YZ,UB,Y,U,FALK,JMAX)
47         CALL LAG(YZ,VB,Y,V,FALK,JMAX)
48   C
49         DO 6 J = 2,JMAX
50         UM(J) = UM(J)*UEST
51         U(J) = U(J)*UE
52         VM(J) = VM(J)/FALKS
53       6 V(J) = V(J)/FALK
54         UP(1) = 0.
55         U(1) = 0.
56         UM(1) = 0.
57         V(1) = 0.
58         VM(1) = 0.
59         DO 10 N = 1,NMAX
60         X = X + DX
```

Fig. 15.3. Listing of program LAMBL

```
61          UE = X**BETP
62          UEX = BETP*UE/X
63          DO 7 J = 2,JMAP
64          DY = Y(J) - Y(J-1)
65          JM = J - 1
66          P = (2.*V(J) - VM(J))*DX/RYP/DY
67          Q = 2.*DX/(RYP*DY*DY)
68          B(2,JM) = -P*RY - Q
69          B(3,JM) = 1.5*(2.*U(J) - UM(J)) + Q*RYP/RY + P*(RY-1./RY)
70          B(4,JM) =   P/RY - Q/RY
71          RHS(JM) = UE*UEX*DX + (2.0*U(J) - 0.5*UM(J))*(2.0*U(J)-UM(J))
72        7 CONTINUE
73          RHS(JM) = RHS(JM) - B(4,JM)*UE
74          B(4,JM) = 0.
75          B(2,1) = 0.
76  C
77          CALL BANFAC(B,JM,1)
78          CALL BANSOL(RHS,UP,B,JM,1)
79  C
80          UP(JMAP) = UE
81  C
82  C       OBTAIN V BY INTEGRATING CONTINUITY
83  C
84          DUM = 0.
85          SUM = 0.5*(Y(2) - Y(1))
86          DO 8 J = 2,JMAX
87          DUMH = DUM
88          VM(J) = V(J)
89          DY = Y(J) - Y(J-1)
90          DUM = 1.5*UP(J-1) - 2.*U(J) + 0.5*UM(J)
91          V(J) = V(J-1) - 0.5*(DY/DX)*(DUM + DUMH)
92          UM(J) = U(J)
93          U(J) = UP(J-1)
94          IF(J .EQ. JMAX)GOTO 8
95          SUM = SUM + 0.5*(1. - U(J)/UE)*(Y(J+1)-Y(J-1))
96        8 CONTINUE
97          DISP = SUM/SQRE
98          UYZ = (RYP*U(2) - U(3)/RYP)/RY/(Y(2)-Y(1))
99          CF = 2.*UYZ/SQRE/UE/UE
100         DUM = 0.25*X*UE*RE*(2.-BETA)
101         EXCF = 0.9278/SQRT(DUM)
102         WRITE(6,9)N,X,EXCF,CF,DISP,UE
103       9 FORMAT(' N=',I3,' X=',F4.2,' EXCF=',F9.6,'  CF=',F9.6,2X,
104        1' DISP=',F9.6,'  UE=',F6.3)
105      10 CONTINUE
106  C
107  C      COMPARE SOLUTION WITH EXACT
108  C
109         FALK = SQRT((2.-BETA)*X/UE)
110         CALL LAG(YZ,UB,Y,UBX,FALK,JMAX)
111  C
112         SUM = 0.
113         DO 11 J = 2,JMAX
114         UBX(J) = UBX(J)*UE
115      11 SUM = SUM + (U(J)-UBX(J))**2
116         RMS = SQRT(SUM/AJP)
117         WRITE(6,12)RMS
118      12 FORMAT(' RMS= ',E10.3)
119      13 CONTINUE
120         STOP
121         END
```

Fig. 15.3. (cont.) Listing of program LAMBL

Table 15.1. Parameters used in program LAMBL

Parameter	Description
JMAX	number of gridpoints in the y direction
NMAX	number of gridpoints in the x direction
DX	Δx
DY	$\Delta y = y_j - y_{j-1}$
DYM	Δy adjacent to the wall, $y_2 - y_1$
XST	x_0
X, Y	x, y
RE	Reynolds number, Re
BETA	β
UE, UEX	$u_e, du_e/dx$
UM, VM	u^{n-1}, v^{n-1}
U, V	u^n, v^n
UP	u^{n+1}
UB, VB	Falkner–Skan solution for u, v at $x = 1$
YZ	η at $x = 1$ and $u_e = 1$
LAG	interpolates Falkner–Skan velocity components to suit grid points (y_j)
B	tridiagonal matrix, with components, a_j, b_j, c_j in (15.18); factorised in BANFAC
DUM, DEM	p, q, after (15.18)
RHS	d_j, (15.18)
DISP	displacement thickness, δ^*
CF	skin friction coefficient, c_f
EXCF	exact skin friction coefficient, c_{fex}
UBX	exact u velocity solution, u_{bx}
RMS	$\|u - u_{bx}\|_{rms}$

```
1        SUBROUTINE LAG(YZ,QB,Y,Q,FALK,JMAX)
2  C
3  C     APPLIES LAGRANGE INTERPOLATION TO THE INITIAL FALKNER-SKAN
4  C     PROFILE TO OBTAIN THE F.S. PROFILE (U,V) AT DIFFERENT X
5  C
6        DIMENSION YZ(24),YB(24),QB(24),Y(41),Q(41)
7        DO 1 I = 1,24
8      1 YB(I) = YZ(I)*FALK
9        Q(1) = 0.
10       DO 6 I = 2,JMAX
11       DO 5 J = 1,23
12       IF(J .EQ. 23)GOTO 2
13       IF(Y(I) .GT. YB(J))GOTO 5
14     2 JS = J
15       IF(JS .LT. 2)JS = 2
16       Q(I) = 0.
17       DO 4 K = 1,3
18       CL = 1.
19       KK = JS - 2 + K
20       DO 3 L = 1,3
21       LL = JS - 2 + L
22       IF(LL .EQ. KK)GOTO 3
23       CL = CL*(Y(I) - YB(LL))/(YB(KK) - YB(LL))
24     3 CONTINUE
25     4 Q(I) = Q(I) + CL*QB(KK)
26       GOTO 6
27     5 CONTINUE
28     6 CONTINUE
29       RETURN
30       END
```

Fig. 15.4. Listing of program LAG

The scheme described above is coded in program LAMBL (Fig. 15.3). Since $\partial u/\partial x$ is represented by a three-level formula (15.5), two levels of data for u and v are required as initial conditions. In the program LAMBL the initial $u_0(y)$ and $v_0(y)$ profiles are given by the Falkner–Skan solutions

$$u_0(y) = u_e f(\eta) \quad \text{and} \quad v_0(y) = -(u_e/(2-\beta)/x)^{1/2}[f+(\beta-1)\eta f_\eta] \ . \tag{15.19}$$

Strictly $v_0(y)$ should be determined from the discrete form of (15.14 and 15) after substituting $u = u_0(y)$ and eliminating $\partial u/\partial x$ (Krause 1967). However, this more general procedure would not alter significantly the solution for the present problem.

Program LAMBL is written for any value of β. However, the specific data given in lines 6–13 is appropriate to $\beta = 0.5$. The variables UB and VB correspond to $f(\eta)$ and $0.5\eta f_\eta - f$ respectively and YZ is equivalent to η. To obtain the values of u_j^0 and v_j^0 it is necessary to interpolate UB and VB. This is done using Lagrange interpolation in the subroutine LAG (Fig. 15.4).

The parameters used by program LAMBL are given in Table 15.1 and typical output is indicated in Fig. 15.5. As well as producing $u(y)$, $v(y)$ at each downstream step, program LAMBL also calculates the skin friction coefficient c_f and the displacement thickness δ^*. The skin friction coefficient is calculated from

$$c_f^{n+1} = \frac{2}{\text{Re}^{1/2}u_e^2}\left[\frac{\partial u}{\partial y}\right]_{y=0}^{n+1} = -\frac{2}{\text{Re}^{1/2}u_e^2}\frac{[(1+r_y)u_2^{n+1} - u_3^{n+1}/(1+r_y)]}{[r_y(y_2-y_1)]} \ . \tag{15.20}$$

```
FALKNER-SKAN SOLUTION       BETA=  .50
JMAX=  21   DYM=   .40  RY=  1.00
NMAX=  19   DX=   .100E+00  XST=  1.00   RE=  .100E+06


N=  1 X=1.20 EXCF=  .004243  CF=  .004242  DISP=  .003343  UE= 1.063
N=  2 X=1.30 EXCF=  .004022  CF=  .004024  DISP=  .003430  UE= 1.091
N=  3 X=1.40 EXCF=  .003828  CF=  .003832  DISP=  .003512  UE= 1.119
N=  4 X=1.50 EXCF=  .003656  CF=  .003660  DISP=  .003591  UE= 1.145
N=  5 X=1.60 EXCF=  .003502  CF=  .003506  DISP=  .003668  UE= 1.170
N=  6 X=1.70 EXCF=  .003364  CF=  .003367  DISP=  .003742  UE= 1.193
N=  7 X=1.80 EXCF=  .003238  CF=  .003241  DISP=  .003813  UE= 1.216
N=  8 X=1.90 EXCF=  .003123  CF=  .003126  DISP=  .003881  UE= 1.239
N=  9 X=2.00 EXCF=  .003018  CF=  .003021  DISP=  .003947  UE= 1.260
N= 10 X=2.10 EXCF=  .002922  CF=  .002924  DISP=  .004011  UE= 1.281
N= 11 X=2.20 EXCF=  .002832  CF=  .002835  DISP=  .004072  UE= 1.301
N= 12 X=2.30 EXCF=  .002750  CF=  .002752  DISP=  .004132  UE= 1.320
N= 13 X=2.40 EXCF=  .002673  CF=  .002675  DISP=  .004191  UE= 1.339
N= 14 X=2.50 EXCF=  .002601  CF=  .002603  DISP=  .004248  UE= 1.357
N= 15 X=2.60 EXCF=  .002534  CF=  .002536  DISP=  .004303  UE= 1.375
N= 16 X=2.70 EXCF=  .002471  CF=  .002473  DISP=  .004357  UE= 1.392
N= 17 X=2.80 EXCF=  .002412  CF=  .002414  DISP=  .004409  UE= 1.409
N= 18 X=2.90 EXCF=  .002356  CF=  .002358  DISP=  .004461  UE= 1.426
N= 19 X=3.00 EXCF=  .002303  CF=  .002305  DISP=  .004511  UE= 1.442
RMS=    .674E-03
```

Fig. 15.5. Typical output from program LAMBL

This is compared with the "exact" (Falkner–Skan) skin friction coefficient

$$c_{fex} = f_{\eta\eta}(0) \left(\frac{4}{(2-\beta)} \right)^{1/2} (Re \, x \, u_e)^{-1/2} . \tag{15.21}$$

For the case $\beta = 0.5$,

$$c_{fex} = 1.5151 \, (Re \, x \, u_e)^{-1/2} . \tag{15.22}$$

At the end of the downstream march, program LAMBL calculates the exact u velocity component u_{bx} by interpolating UB and computes the rms error between u and u_{bx}. As is apparent from Fig. 15.5 the computed solution is in close agreement with the Falkner–Skan solution.

15.1.3 Keller Box Scheme

An alternative means of discretising the boundary layer equations (15.1, 2) is provided by the Keller box scheme. A feature of this method is that only first derivatives are allowed to appear. Consequently (15.2) is replaced by

$$u \frac{\partial u}{\partial x} + v \frac{\partial u}{\partial y} = u_e \frac{du_e}{dx} + \frac{\partial \tau}{\partial y} , \quad \text{where} \tag{15.23}$$

$$\tau = \mu \frac{\partial u}{\partial y} . \tag{15.24}$$

Thus an additional auxiliary variable, the shear stress τ, appears in the formulation. The discretisation is carried out within a 'box' as indicated in Fig. 15.6.

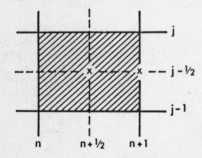

n   n+½   n+1   **Fig. 15.6.** Grid for Keller box scheme

Since only first derivatives appear in the governing equations, centred differences and two-point averages can be constructed involving only the four corner nodal values of the 'box'. For example, if w represents any of the dependent variables u, v, τ, then (Fig. 15.6)

$$[w]_{j-1/2}^{n+1} = 0.5 (w_{j-1}^{n+1} + w_j^{n+1}) \quad \text{and} \tag{15.25}$$

$$[w]_{j-1/2}^{n+1/2} = 0.5 ([w]_{j-1/2}^n + [w]_{j-1/2}^{n+1}) .$$

Derivatives are represented by

$$\left[\frac{\partial w}{\partial y}\right]_{j-1/2}^{n+1} = \frac{(w_j^{n+1} - w_{j-1}^{n+1})}{(y_j - y_{j-1})} \; ,$$

$$\left[\frac{\partial w}{\partial y}\right]_{j-1/2}^{n+1/2} = 0.5\left(\left[\frac{\partial w}{\partial y}\right]_{j-1/2}^{n} + \left[\frac{\partial w}{\partial y}\right]_{j-1/2}^{n+1}\right) , \tag{15.26}$$

$$\left[\frac{\partial w}{\partial x}\right]_{j-1/2}^{n+1/2} = 0.5\frac{([w]_{j-1/2}^{n+1} - [w]_{j-1/2}^{n})}{(x_{n+1} - x_n)} .$$

An advantage of the above discretisation is that non-uniform but rectangular grids can be used without affecting the accuracy of the discretisation.

The expressions (15.25, 26) are substituted into (15.1, 23 and 24) to give

$$\left[\frac{\partial u}{\partial x}\right]_{j-1/2}^{n+1/2} + \left[\frac{\partial v}{\partial y}\right]_{j-1/2}^{n+1/2} = 0 , \tag{15.27}$$

$$[u]_{j-1/2}^{n+1/2}\left[\frac{\partial u}{\partial x}\right]_{j-1/2}^{n+1/2} + [v]_{j-1/2}^{n+1/2}\left[\frac{\partial u}{\partial y}\right]_{j-1/2}^{n+1/2} = \left[u_e\frac{du_e}{dx}\right]_{j-1/2}^{n+1/2} + \left[\frac{\partial \tau}{\partial y}\right]_{j-1/2}^{n+1/2} \quad \text{and} \tag{15.28}$$

$$[\tau]_{j-1/2}^{n} = [v]_{j-1/2}^{n}\left[\frac{\partial u}{\partial y}\right]_{j-1/2}^{n} . \tag{15.29}$$

A Taylor series expansion about $(j-1/2, \ n+1/2)$ indicates that (15.27–29) are accurate to $O(\Delta x^2, \Delta y^2)$.

Application of (15.27–29) at each of the grid points, $j = 1, \ldots, J$, gives a system of $3J$ nonlinear coupled equations in $3J+3$ unknowns, u_j^{n+1}, v_j^{n+1} and τ_j^{n+1}. However, wall values u_0^{n+1} and v_0^{n+1} and boundary-layer edge value $u_J^{n+1} = u_e^{n+1}$ are known. Newton's method (Sect. 6.1.1) is applied to this system and produces the following system of linear equations for the corrections $\Delta \mathbf{w}^{k+1}$:

$$\underline{J}^k \Delta \mathbf{w}^{k+1} = -\mathbf{R}^k , \tag{15.30}$$

where $\Delta \mathbf{w}^{k+1} = \mathbf{w}^{k+1} - \mathbf{w}^k$, $\mathbf{w}_j \equiv (u_j, v_j, \tau_j)^T$, and k represents the iteration level. Thus, at the beginning of the iteration, $\mathbf{w}_j^0 = \mathbf{w}_j^n$, and at the end of the iteration, $\mathbf{w}_j^{n+1} = \mathbf{w}_j^{k+1}$. Typically three to four iterations are required to obtain $\mathbf{w}^{n+1}$ at each downstream step. However, due to the discretisation (15.25, 26) the Jacobian is block tridiagonal and (15.30) can be solved efficiently using the algorithm given in Sect. 6.2.5.

The Keller box scheme is described by Keller (1978) and in more detail by Cebeci and Bradshaw (1977, Chap. 7).

15.2 Complex Boundary Layer Flow

For boundary layer flows involving adverse pressure gradients (pressure increasing in the flow direction) the boundary layer thickness grows rapidly. For any type of boundary layer the velocity component in the main flow direction, u, changes rapidly across the boundary layer.

It is therefore desirable to transform the governing equations into new independent and dependent variables that are less sensitive to the above effects. If the grid is closer to uniform in the transform domain the errors introduced during discretisation (Sect. 3.1) will be correspondingly less. The general problem of using variable grids and the associated errors are discussed by Noye (1983, pp. 297–316).

15.2.1 Change of Variables

It is generally possible to transform the equations governing boundary layer flow to obtain a particular advantage. The Mangler transformation (Schlicting 1968, pp. 235–237) reduces an axisymmetric boundary layer to an equivalent two-dimensional boundary layer. The Howarth, Illingworth, Stewartson transformation (Schlicting 1968, pp. 324–328) reduces a compressible boundary layer to an equivalent incompressible boundary layer. A Blasius transformation compensates for boundary layer growth and greatly simplifies the governing equations if the problem possesses a similar solution.

The Levy–Lees transformation (Blottner 1975a), to be examined in Sect. 15.2.2, combines features of the Howarth, Mangler and Blasius transformations. The Dorodnitsyn transformation (Sect. 15.3) makes use of u as an independent variable and casts the equations in an integral form which facilitates the introduction of Galerkin finite element and spectral formulations. In the Dorodnitsyn formulation a nondimensionalised normal velocity gradient is introduced as the dependent variable. This permits accurate predictions to be obtained for the wall shear stress, and hence skin friction coefficient, (11.66).

15.2.2 Levy–Lees Transformation

The starting point for the Levy–Lees transformation are the equations governing steady laminar compressible two-dimensional (Sect. 11.6.2) or axisymmetric flow:

$$\text{cont:} \qquad \frac{\partial}{\partial x}(r_b^j \varrho u) + \frac{\partial}{\partial y}(r_b^j \varrho v) = 0 \qquad (15.31)$$

$$\text{x-mmtm:} \qquad \varrho u \frac{\partial u}{\partial x} + \varrho v \frac{\partial u}{\partial y} = -\frac{dp_e}{dx} + \frac{\partial}{\partial y}\left(\mu \frac{\partial u}{\partial y}\right) \qquad (15.32)$$

$$\text{energy:} \qquad \varrho c_p \left(u \frac{\partial T}{\partial x} + v \frac{\partial T}{\partial y}\right) = u \frac{dp_e}{dx} + \frac{\partial}{\partial y}\left[\left(\frac{\mu c_p}{Pr}\right)\frac{\partial T}{\partial y}\right] + \mu \left(\frac{\partial u}{\partial y}\right)^2 \qquad (15.33)$$

For two-dimensional flows $j = 0$. For axisymmetric flows $j = 1$ and r_b is the body radius. In (15.31–33) x is measured along the surface of the body from the tip or stagnation point and y is measured normal to the surface. In (15.32, 33) p_e is the known pressure at the outer edge of the boundary layer. For compressible flow, (15.31–33) must be supplemented by an equation of state, for example (11.1) and a viscosity temperature relationship $\mu(T)$.

For the equation set (15.31–33) the following are appropriate initial conditions:

$$u(x_0, y) = u_0(y), \qquad T(x_0, y) = T_0(y) \tag{15.34}$$

and boundary conditions:

$u(x, 0) = v(x, 0) = 0$ for no mass flow at the wall,

$$T(x, 0) = T_w(x) \quad \text{or} \quad \frac{\partial T}{\partial y}(x, 0) = -\dot{Q}_w(x) , \tag{15.35}$$

$$u(x, \delta) = u_e(x) \quad \text{and} \quad T(x, \delta) = T_e(x) .$$

The Levy–Lees transformation defines new independent variables as

$$\xi(x) = K \int_0^x (\varrho u)_{\text{ref}} u_e r_b^{2j} dx' , \qquad \eta(x, y) = u_e r_b^j \left(\frac{K}{2\xi} \right)^{1/2} \int_0^y \varrho \, dy' , \tag{15.36}$$

where K is a constant determined by the particular flow being investigated. New dependent variables are defined as

$$F = \frac{u}{u_e} , \qquad \theta = \frac{h}{h_e} = \frac{T}{T_e} \quad \text{and}$$

$$V = \frac{2\xi \left(F \dfrac{\partial \eta}{\partial x} + \dfrac{\varrho v r_b^j}{(2\xi/K)^{1/2}} \right)}{K(\varrho u)_{\text{ref}} u_e r_b^{2j}} , \tag{15.37}$$

so that (15.31–33) become

$$2\xi \frac{\partial F}{\partial \xi} + \frac{\partial V}{\partial \eta} + F = 0 , \tag{15.38}$$

$$2\xi F \frac{\partial F}{\partial \xi} + V \frac{\partial F}{\partial \eta} + \beta(F^2 - \theta) - \frac{\partial}{\partial \eta}\left(l \frac{\partial F}{\partial \eta} \right) = 0 \tag{15.39}$$

and

$$2\xi F \frac{\partial \theta}{\partial \xi} + V \frac{\partial \theta}{\partial \eta} - \alpha l \left(\frac{\partial F}{\partial \eta} \right)^2 - \frac{1}{\text{Pr}} \frac{\partial}{\partial \eta}\left(l \frac{\partial \theta}{\partial \eta} \right) = 0 , \tag{15.40}$$

where

$$\alpha = \frac{u_e^2}{c_p T_e}, \quad \beta = \left(\frac{2\xi}{u_e}\right)\frac{du_e}{d\xi} \quad \text{and} \quad l = \frac{\varrho\mu}{(\varrho\mu)_{\text{ref}}}. \tag{15.41}$$

The boundary conditions (15.35) become

$$F = V = 0 \quad \text{and} \quad \theta = \theta_w \quad \text{at} \quad \eta = 0 \quad \text{and}$$

$$F = \theta = 1 \quad \text{at} \quad \eta = \eta_e. \tag{15.42}$$

Initial conditions are obtained by setting $\xi = 0$ in (15.38–40). If terms containing ξ derivatives are set equal to zero the equations governing similar flows (e.g. the Falkner–Skan flows, Sect. 15.1.2) are obtained automatically (Blottner 1975a).

The Levy–Lees transformation has the following desirable features:
(i) Compressibility effects are suppressed; ϱ does not appear explicitly.
(ii) Axisymmetric flows are treated as equivalent two-dimensional flows.
(iii) The use of the η coordinate compensates for boundary layer growth.

15.2.3 Davis Coupled Scheme

Here we describe the Davis coupled scheme (DCS), applied to (15.38–40), for incompressible boundary layer flow. For this case $l = 1$ and $\theta = 1$ and only (15.38 and 39) need be considered. The DCS was recommended by Blottner (1975b) after comparing the accuracy and efficiency of a number of Crank–Nicolson schemes. The coupling refers to the simultaneous implicit treatment of both the continuity and x-momentum equations, in contrast to the sequential solution of (15.7 and 8).

If (15.39) is written as

$$\xi\frac{\partial F^2}{\partial \xi} = \text{RHS} = -V\frac{\partial F}{\partial \eta} - \beta(F^2 - 1) + \frac{\partial^2 F}{\partial \eta^2}, \tag{15.43}$$

then a marching algorithm is constructed from a Crank–Nicolson differencing about $(n + 1/2, j)$. That is, (15.43) is replaced by

$$0.5(\xi^n + \xi^{n+1})\frac{\{F^2\}_j^{n+1} - \{F^2\}_j^n}{\Delta\xi} = 0.5(\text{RHS}^n + \text{RHS}^{n+1}). \tag{15.44}$$

The η derivatives in (15.43) are evaluated as three-point centred difference formulae (Fig. 15.7). The nonlinear implicit terms in (15.43 and 44) are linearised using a Newton–Raphson expansion, i.e.

$$\{F^2\}^{k+1} = 2F^k F^{k+1} - \{F^2\}^k \quad \text{and} \tag{15.45}$$

$$\left\{V\frac{\partial F}{\partial \eta}\right\}^{k+1} = V^{k+1}\left(\frac{\partial F}{\partial \eta}\right)^k + V^k\left(\frac{\partial F}{\partial \eta}\right)^{k+1} - \left(V\frac{\partial F}{\partial \eta}\right)^k,$$

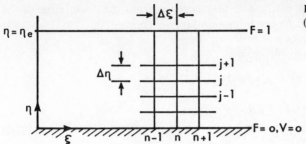

Fig. 15.7. Boundary layer grid in (ξ, η) space

where an iteration over k is carried out at each step n in the downstream direction. Convergence of the k iteration yields the solution at $n + 1$. After introducing (15.45) into (15.44) it is possible to write down a system of linear equations for F^{k+1} and V^{k+1} as

$$a_j^k F_{j-1}^{k+1} + b_j^k F_j^{k+1} + c_j^k F_{j+1}^{k+1} + g_j^k V_j^{k+1} = d_j^k \ , \quad j = 2, 3, \ldots, J - 1 \ , \quad \text{where}$$
(15.46)

$$a_j^k = -0.5(1 + 0.5 V_j^k \Delta\eta) \ ,$$

$$b_j^k = 1 + \Delta\eta^2 \left[\beta^{n+1} + \frac{(\xi^n + \xi^{n+1})}{\Delta\xi} \right] F_j^k \ ,$$

$$c_j^k = -0.5(1 - 0.5\Delta\eta V_j^k) \ ,$$

$$g_j^k = 0.25\Delta\eta(F_{j+1}^k - F_{j-1}^k) \ ,$$

$$d_j^k = -a_j^n F_{j-1}^k - c_j^n F_{j+1}^k - [1 + 0.5\Delta\eta^2 \beta^n F_j^n] F_j^n + 0.5\Delta\eta^2 [\beta^n + \beta^{n+1}]$$

$$+ 0.5\Delta\eta^2 \left[(\xi^n + \xi^{n+1}) \frac{\{(F^2)_j^n + (F^2)_j^k\}}{\Delta\xi} + V_j^k \frac{(F_{j+1}^k - F_{j-1}^k)}{(2\Delta\eta)} + \beta^k \{F^2\}_j^k \right] \ .$$

The continuity equation (15.38) is differenced as

$$2\xi^{n+1/2} \frac{[(F_j^{n+1} - F_j^n) + (F_{j-1}^{n+1} - F_{j-1}^n)]}{2\Delta\xi} + \frac{[(V_j^{n+1} - V_{j-1}^{n+1}) + (V_j^n - V_{j-1}^n)]}{2\Delta\eta}$$

$$+ 0.25[(F_j^{n+1} + F_{j-1}^{n+1}) + (F_j^n + F_{j-1}^n)] = 0 \ .$$
(15.47)

Equation (15.47) is rewritten as an algorithm for V_j^{k+1} as

$$V_j^{k+1} = V_{j-1}^{k+1} - s_j(F_{j-1}^{k+1} + F_j^{k+1}) + t_j \ , \quad j = 2, 3, \ldots, J$$
(15.48)

where $s_j = 2\Delta\eta(0.25 + \xi^{n+1/2}/\Delta\xi) \ ,$

$$t_j = -2\Delta\eta \left(0.25 - \frac{\xi^{n+1/2}}{\Delta\xi} \right) [F_j^k + F_{j-1}^k] - (V_j^k - V_{j-1}^k) \ .$$

Equations (15.46 and 48) are solved simultaneously by using a development of the tridiagonal algorithm (Sect. 6.2.2). The first sweep is from the outer edge of the boundary layer to the wall. At the boundary layer edge,

$$E_J = G_J = 0 \quad \text{and} \quad e_J = 1.0 .$$

For decreasing values, $j = (J-1), (J-2), \ldots,$

$$
\begin{aligned}
T &= b_j + c_j E_{j+1} + s_j(c_j G_{j+1} + g_j) , \\
E_j &= -[a_j - s_j(c_j G_{j+1} + g_j)]/T , \\
G_j &= -[c_j G_{j+1} + g_j]/T , \quad \text{and} \\
e_j &= [d_j - (c_j G_{j+1} + g_j)t_j - c_j e_{j+1}]/T .
\end{aligned}
\tag{15.49}
$$

At the wall $F_1 = V_1 = 0$ and for increasing values $j = 2, 3, \ldots, J$

$$F_j = E_j F_{j-1} + G_j V_{j-1} + e_j , \qquad V_j = V_{j-1} - s_j(F_{j-1} + F_j) = t_j . \tag{15.50}$$

In practice, equation systems (15.46 and 48) are solved repeatedly until solutions F_j^{k+1} and V_j^{k+1} do not differ from F_j^k and V_j^k, respectively. To start the

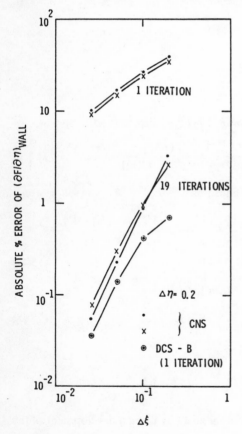

Fig. 15.8. Convergence of the DCS scheme (after Blottner, 1975b; reprinted with permission of North-Holland)

iteration $F_j^k = F_j^n$, etc., and after convergence $F^{n+1} = F^{k+1}$, etc. The major advantage of DCS is that the strong coupling between the continuity and the momentum equations permits second-order convergence in ξ with only a single iteration in k. If the equivalent scheme is solved with the continuity and momentum equations uncoupled (CNS in Fig. 15.8) 19 iterations are required to obtain second-order convergence. The convergence results shown in Fig. 15.8 correspond to a linearly retarded outer-edge velocity $u_e/U_\infty = (1 - x/L)$. The term $(\partial F/\partial \eta)_{\text{wall}}$ is proportional to the skin friction coefficient, c_f.

Blottner (1975b) indicates that DCS is computationally more efficient than other Crank–Nicolson schemes, including the Keller box scheme, for a typical laminar flow problem using a uniform grid.

15.3 Dorodnitsyn Boundary Layer Formulation

For some laminar flows and almost all turbulent flows it is necessary to use a nonuniform grid close to the wall to obtain accurate solutions. However, the use of a nonuniform grid can be avoided by treating u (in two dimensions) as an independent variable. This is a central feature of the Crocco (Blottner 1975a) and Dorodnitsyn formulations.

The Dorodnitsyn formulation converts the governing equations into integral form. This implies that a weighted residual (Chap. 5) interpretation is possible. Two interpretations will be described here, a Galerkin finite element formulation in Sect. 15.3.1 and a Galerkin spectral formulation in Sect. 15.3.3.

The Dorodnitsyn formulation will be developed for incompressible two-dimensional turbulent boundary layer flow (11.73–75). If an algebraic eddy viscosity v_T is introduced to represent the Reynolds shear stress the resulting equations can be written (in nondimensional form) as

$$\frac{\partial u}{\partial x} + \frac{\partial v}{\partial y} = 0 \quad \text{and} \tag{15.51}$$

$$u \frac{\partial u}{\partial x} + v \frac{\partial u}{\partial y} = u_e \frac{du_e}{dx} + \frac{1}{\text{Re}} \frac{\partial}{\partial y}\left[\left(1 + \frac{v_T}{v}\right)\frac{\partial u}{\partial y}\right] \tag{15.52}$$

with initial and boundary conditions given by (15.3, 4).

To implement the Dorodnitsyn formulation the following variables are introduced:

$$\xi = x \quad \text{and} \quad \eta = \text{Re}^{1/2} u_e y , \tag{15.53}$$

$$u' = \frac{u}{u_e}, \quad v' = \text{Re}^{1/2}\frac{v}{u_e} \quad \text{and} \quad w = u_e v' + \frac{\eta u'}{u_e}\frac{du_e}{d\xi} .$$

Consequently (15.51 and 52) can be written

$$\frac{\partial u'}{\partial \xi} + \frac{\partial w}{\partial \eta} = 0 \; , \tag{15.54}$$

$$u' \frac{\partial u'}{\partial \xi} + w \frac{\partial u'}{\partial \eta} = \frac{u_{e\xi}}{u_e} + \frac{\partial}{\partial \eta}\left[\left(1 + \frac{v_T}{v}\right)\frac{\partial u'}{\partial \eta}\right] \; , \tag{15.55}$$

where $u_{e\xi} \equiv du_e/d\xi$. The boundary conditions become

$$u' = w = 0 \quad \text{at} \quad \eta = 0 \quad \text{and} \quad u' = 1 \quad \text{at} \quad \eta = \infty \; .$$

A weighted sum of (15.54 and 55) is formed as

$$f_k(u') \times (15.54) + \left(\frac{df_k(u')}{du'}\right) \times (15.55) = 0 \; ,$$

where $f_k(u')$ is a weight (test) function to be chosen. The result is (after dropping the prime)

$$\frac{\partial(uf_k)}{\partial \xi} + \frac{\partial(wf_k)}{\partial \eta} = \left(\frac{u_{e\xi}}{u_e}\right)\{1 - u^2\}\frac{df_k}{du}$$

$$+ u_e \left\{\frac{df_k}{du}\right\}\frac{\partial}{\partial \eta}\left[\left(1 + \frac{v_T}{v}\right)\frac{\partial u}{\partial \eta}\right] \; . \tag{15.56}$$

An integration is made from $\eta = 0$ to $\eta = \infty$ and f_k is restricted so that $f_k(\infty) = 0$. The variable of integration is changed from η to u and the result is

$$\frac{\partial}{\partial \xi}\int_0^1 uf_k \Theta \, du = \left(\frac{u_{e\xi}}{u_e}\right)\int_0^1 \left\{\frac{df_k}{du}\right\}(1 - u^2)\Theta \, du$$

$$+ u_e \int_0^1 \left\{\frac{df_k}{du}\right\}\frac{\partial}{\partial u}\left[\left(1 + \frac{v_T}{v}\right)T\right]du \; , \tag{15.57}$$

where

$$T = \frac{1}{\Theta} = \frac{\partial u}{\partial \eta} \; . \tag{15.58}$$

Equation (15.57) is the Dorodnitsyn turbulent boundary layer formulation. In (15.57) T and Θ are the dependent variables and x and u are the independent variables.

The Dorodnitsyn formulation is attractive primarily because a uniform grid in the u direction automatically places most points adjacent to the wall where the solution is changing most rapidly (Fig. 15.9). This is particularly important for turbulent boundary layers. The uniform grid in u automatically captures downstream boundary layer growth. An additional advantage is that the normal velocity

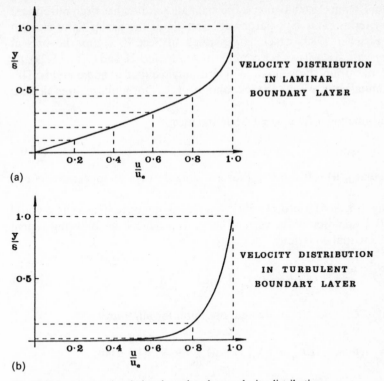

Fig. 15.9. Laminar and turbulent boundary layer velocity distributions

does not appear in (15.57), so that only one equation need be solved. The solution for v can be recovered subsequently if required. Since the wall value of T is directly proportional to the shear stress (at the wall), the Dorodnitsyn formulation provides accurate solutions for the skin friction (11.66).

15.3.1 Dorodnitsyn Finite Element Method

Here we will describe the Dorodnitsyn boundary layer formulation applied with the Galerkin finite element method (Sect. 5.3). The following approximate (trial) solutions are introduced for Θ and $(1 + v_T/v)T$ in (15.57):

$$\Theta = \sum_{j=1}^{M} N_j(u)/(1-u)\theta_j(\xi) \quad \text{and} \tag{15.59}$$

$$(1 + v_T/v)T = \sum_{j=1}^{M} (1-u)N_j(u)\{1 + v_T/v\}_j \tau_j(\xi) \ . \tag{15.60}$$

In (15.59 and 60) the factor $(1-u)$ ensures that Θ and T have the correct behaviour at the outer edge of the boundary layer. The terms $N_j(u)$ are one-dimensional interpolating functions, typically linear or quadratic (Sect. 5.3).

It is apparent, from (15.60) that the approximate solution has been introduced for a group of terms. This is a particular example of the group finite element formulation (Fletcher 1983) which is described in Sect. 10.3. For the present application the eddy viscosity v_T is a complicated function of u and η, (11.77–79). By including v_T in the group, the evaluation of v_T is only required at nodal points. This makes a substantial contribution to the economy of the Dorodnitsyn finite element formulation.

The weight function $f_k(u)$ in (15.57) has the form

$$f_k(u) = (1-u)N_k(u) , \tag{15.61}$$

which ensures that $f_k(u) = 0$ at $u = 1$, thus avoiding the explicit appearance of v in (15.57).

Substituting (15.59–61) into (15.57) produces a modified Galerkin method (Fletcher 1984). Evaluation of the various integrals produces the following system of ordinary differential equations:

$$\sum_{j=1}^{M} CC_{kj}\frac{d\theta_j}{d\xi} = \frac{u_{e\xi}}{u_e} \sum_{j=1}^{M} EF_{kj}\theta_j + u_e \sum_{j=1}^{M} AA_{kj}\left(1+\frac{v_T}{v}\right)_j \tau_j . \tag{15.62}$$

The coefficients CC_{kj}, etc., are evaluated, once and for all, from

$$CC_{kj} = \int_0^1 N_j N_k u\, du, \qquad EF_{kj} = \int_0^1 N_j\left((1-u)\frac{dN_k}{du} - N_k\right)(1+u)\,du$$

and

$$AA_{kj} = \int_0^1 \left(\frac{dN_j}{du}(1-u) - N_j\right)\left(\frac{dN_k}{du}(1-u) - N_k\right)du . \tag{15.63}$$

Although θ_j and τ_j appear separately in (15.62) the nodal values are required to satisfy $\theta_j = 1/\tau_j$.

Since $\underline{CC}$ is tridiagonal for linear elements and pentadiagonal for quadratic elements it is straightforward to construct an efficient implicit scheme as follows:

$$\sum_{j=1}^{M} CC_{kj}\Delta\theta_j^{n+1} = \Delta\xi[\beta\mathrm{RHS}^{n+1} + (1-\beta)\mathrm{RHS}^n] , \tag{15.64}$$

where

$$\mathrm{RHS} = \frac{u_{e\xi}}{u_e}\sum_{j=1}^{M} EF_{kj}\theta_j + u_e\sum_{j=1}^{M} AA_{kj}\left(1+\frac{v_T}{v}\right)_j \tau_j , \tag{15.65}$$

$\Delta\theta_j^{n+1} = \theta_j^{n+1} - \theta_j^n$ and $\beta \geq 0.6$ for stable solutions (determined empirically).

The term RHS^{n+1} is expanded about RHS^n as a Taylor series which is equivalent to the Newton–Raphson expansion (15.45) when truncated with an error $O(\Delta\xi^2)$. Consequently (15.65) is rearranged to give the following linear system of equations for $\Delta\theta_j^{n+1}$:

$$\sum_{j=1}^{M} CCC_{kj} \Delta \theta_j^{n+1} = P_k \,, \quad \text{where} \tag{15.66}$$

$$CCC_{kj} = CC_{kj} - \beta \Delta \xi \left[\left(\frac{u_{e\xi}}{u_e} \right)^{n+1} EF_{kj} - u_e^{n+1} AA_{kj} G_j \right] \,,$$

$$G_j = \left[\left(1 + \frac{v_T}{v} \right) \tau^2 - \tau \frac{\partial}{\partial \theta} \left\{ \frac{v_T}{v} \right\} \right]_j^n$$

and

$$P_k = \Delta \xi \left\{ \left[\beta \left(\frac{u_{e\xi}}{u_e} \right)^{n+1} + (1-\beta) \left(\frac{u_{e\xi}}{u_e} \right)^n \right] \sum_{j=1}^{M} EF_{kj} \theta_j^n \right.$$

$$\left. + \left[\beta u_e^{n+1} + (1-\beta) u_e^n \right] \sum_{j=1}^{M} AA_{kj} \left(1 + \frac{v_T}{v} \right)_j \tau_j^n \right\} \,.$$

Equation (15.66) can be solved efficiently using the Thomas algorithm (Sect. 6.2.2). To maintain maximum economy no iterations are made at each location ξ^n. For laminar boundary layer flow $(v_T = 0)$, (15.66) demonstrates a convergence rate of $O(\Delta u^2, \Delta x)$, Fletcher and Fleet (1984a). However, of greater practical significance is the ability of the method to generate accurate solutions on relatively coarse grids.

Typical results for the turbulent skin friction coefficient and boundary layer thickness variation in the flow direction, produced by a zero pressure gradient, are shown in Figs. 15.10 and 15.11.

Also shown are results produced by the Dorodnitsyn spectral (DOROD-SPEC) formulation (Sect. 15.3.3) and a typical finite volume code, STAN5, which is based on the GENMIX code of Patankar and Spalding (1970). It is apparent that all three

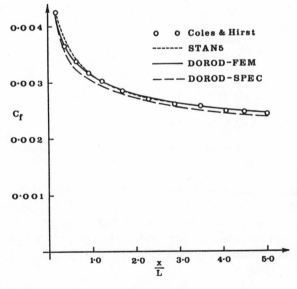

o   o   Coles & Hirst

------- STAN5

——— DOROD-FEM

— — — DOROD-SPEC

Fig. 15.10. Skin friction comparison for zero pressure gradient

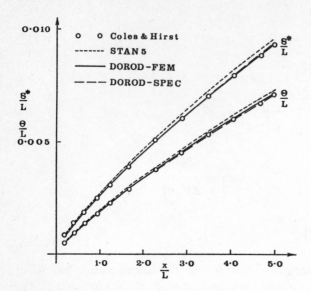

Fig. 15.11. Displacement and momentum thickness comparison for zero pressure gradient

methods are demonstrating comparable accuracy. STAN5 and GENMIX are based on a von Mises transformation (Schlicting 1968, p. 143) of the governing equations in which the streamfunction ψ replaces the normal coordinate y as an independent variable.

However, it is clear from Table 15.2 that DOROD-FEM is an order of magnitude more economical than STAN5. This benefit comes partly from the ability to represent the velocity distribution across the boundary layer with a quarter to a third of the number of nodal points and partly from requiring less steps in the downstream direction.

Table 15.2. Comparison of DOROD-FEM and STAN5

	Zero pressure gradient		Adverse pressure gradient	
	DOROD-FEM	STAN5	DOROD-FEM	STAN5
Grid pts. across b.l.	11	33–39	11	47–48
No. of steps, $\Delta x/L$	205	401	294	660
$\Delta x/L$	.0001–.071	.0004–.031	.0001–.049	.0002–.0039
Relative exec. time	1	8.99	1.55	17.70

The Dorodnitsyn finite element formulation, described above, has been applied to laminar (Fletcher and Fleet 1984a) and turbulent (Fletcher and Fleet 1984b) incompressible boundary layer flow. It has also been applied to compressible turbulent boundary layer flow (Fleet and Fletcher 1983) and to boundary layer flows with surface mass transfer (Fletcher and Fleet 1987) and internal boundary layer flows with swirl (Fletcher 1985).

15.3.2 DOROD: Turbulent Boundary Layer Flow

In this section a computer program, DOROD, is described which implements the Dorodnitsyn finite element method (Sect. 15.3.1). The structure of program DOROD|is indicated in Fig. 15.12 and a listing is provided in Fig. 15.13. Parameters used in program DOROD are described in Table 15.3.

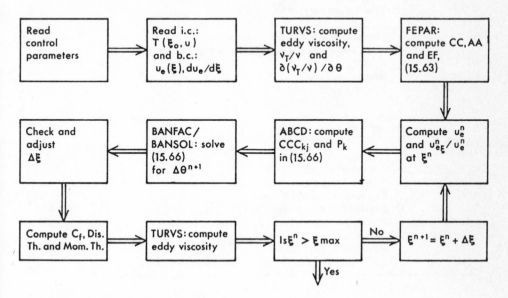

Fig. 15.12. Structure of program DOROD

An initial profile for $\tau(x_0, u_j)$ is required corresponding to u_j. This is obtained by evaluating $\partial u/\partial \eta$, (15.58), at equal intervals of y (and hence η) and interpolating to obtain τ at equal intervals of u_j.

Prior to starting the downstream integration the eddy viscosity distribution $v_T(y)/v$ is obtained from subroutine TURVS (Fig. 15.14). The parameters used in subroutine TURVS are described in Table 15.4. The eddy viscosity data evaluated in subroutine TURVS is based on the two-layer algebraic eddy viscosity formulation described in Sect. 11.4.2. That is, a mixing length prescription is applied near the wall and a Clauser formula (11.79) is used in the outer region of the boundary layer. To evaluate G_j (in subroutine ABCD), $\partial(v_T/v)/\partial\theta$ is also evaluated in subroutine TURVS.

The coefficients CC, EF and AA defined by (15.63) are evaluated in subroutine FEPAR (Fig. 15.15) from the following (interior) formulae:

$$CC_{j,j-1} = \frac{u_{j-1/2}}{6}, \quad CC_{j,j} = \frac{u_j}{3}, \quad CC_{j,j+1} = \frac{u_{j+1/2}}{6}, \tag{15.67}$$

Table 15.3. Parameters used in program DOROD

Parameter	Description
IMAX	number of steps, $\Delta\xi$
NMAX	number of grid points across the boundary layer
U, DU	u, Δu
X, DX	ξ, $\Delta\xi$
DXMI, DXMA	$\Delta\xi_{min}$, $\Delta\xi_{max}$
DXCH	$\Delta\xi$ increment for printing solution
XO, XMAX	ξ_0, ξ_{max}
XUE	ξ location at which u_e and $u_{e\xi}$ given
UEE, UEX	u_e, $u_{e\xi}$ velocity data at boundary layer edge (input)
UE, UEP	u_e^n, u_e^{n+1}
DUEDXU	$\{(du_e/d\xi)/u_e\}^n$
DUEDXP	$\{(du_e/d\xi)/u_e\}^{n+1}$
TAU	τ
DTHETA	$\Delta\theta$
RAT	$\Delta\theta_w^{n+1}/\theta_w^n$
RATCH	γ
ABC	CCC in (15.66)
D	P in (15.66)
BETA	β, controls the degree of implicitness, (15.64)
REL	Re_L, reference Reynolds number
CF	c_f, skin friction coefficient
DELTA	δ^*, displacement thickness
THKMOM	δ^{mom}, momentum thickness
SHPRTR	$H = \delta^*/\delta^{mom}$, shape factor
RTH	$u_e\delta^{mom} Re_L$, momentum thickness Reynolds number

$$EE_{j,j-1} = 0.5\frac{1-u_j^2}{\Delta u} + 0.5u_j - \frac{1+\Delta u}{6} \ ,$$

$$\tag{15.68}$$

$$EF_{j,j} = -2/3 \ , \quad EF_{j,j+1} = -0.5\frac{1-u_j^2}{\Delta u} + 0.5u_j - \frac{1-\Delta u}{6}$$

and

$$AA_{j,j-1} = -(1-u_j)\frac{1-u_{j-1}}{\Delta u^2} - \frac{1}{3} \tag{15.69}$$

$$AA_{j,j} = 2\frac{(1-u_j)^2}{\Delta u^2} + \frac{2}{3} \ , \quad AA_{j,j+1} = -(1-u_j)\frac{1-u_{j+1}}{\Delta u^2} - \frac{1}{3} \ .$$

Equations (15.67–69) are obtained by evaluating (15.63) with one-dimensional linear elements and dividing by Δu. At the wall and at the edge of the boundary layer only one element contributes and $u_j = 0$ and 1 respectively. The corresponding formulae are given in subroutine FEPAR (Fig. 15.15).

At the start of each downstream step the values of u_e^n, u_e^{n+1}, $u_{e\xi}^n/u_e^n$ and $u_{e\xi}^{n+1}/u_e^{n+1}$ are interpolated from the boundary values read in, $u_e(\xi)$ and $du_e(\xi)/d\xi$.

```
1   C       DOROD USES THE FINITE ELEMENT METHOD TO SOLVE
2   C       THE DORODNITSYN BOUNDARY LAYER FORMULATION
3   C
4           DIMENSION ABC(5,65),D(65),TAU(41),DTHETA(65)
5           DIMENSION TAUD(41),TRV(41),DTRV(41)
6           DIMENSION CC(41,5),AA(41,5),EF(41,5)
7           DIMENSION XUE(24),UEE(24),UEX(24),TITLE(8)
8           COMMON CC,AA,EF,ABC,D,TAU,TRV,DTRV
9   C
10          OPEN(1,FILE='DOROD.DAT')
11          OPEN(6,FILE='DOROD.OUT')
12          OPEN(2,FILE='DOROD.STA')
13          READ(1,1)NMAX,IMAX,BETA,AKP,APZ,PCON,ATR,REL
14        1 FORMAT(2I5,4F5.2,2E10.3)
15          READ(1,2)DX,DXMI,DXMA,DXCH,XO,XMAX,RATCH
16        2 FORMAT(7E10.3)
17          WRITE(6,3)NMAX,IMAX
18        3 FORMAT(' NMAX,IMAX = ',2I5)
19          WRITE(6,4)REL,XO,XMAX,DXCH,RATCH
20        4 FORMAT(' RE =',E10.3,' XO =',F6.3
21       1,' XMAX=',F6.3,' DXCH=',E10.3,' RATCH=',E10.3)
22          WRITE(6,5)DX,DXMI,DXMA,BETA
23        5 FORMAT(' DX=',E10.3,' DXMI=',E10.3,' DXMA=',E10.3,' BETA=',E10.3)
24          WRITE(6,6)AKP,APZ,ATR,PCON
25        6 FORMAT(' AKP=',E10.3,' APZ=',E10.3,' ATR=',E10.3,' PCON=',E10.3,/)
26          IMAP = IMAX - 1
27          IREF = 1
28          RSQ = SQRT(REL)
29   C
30   C       READ IN STARTING VELOCITY PROFILE AND EXTERNAL VELOCITY PROFILE
31   C
32          READ(2,7)TITLE
33          READ(2,8)CFST,REL,DELTA,THKMOM,NPG
34          READ(2,9)TAUD
35          READ(2,10)(XUE(N),N=1,NPG)
36          READ(2,10)(UEE(N),N=1,NPG)
37          READ(2,10)(UEX(N),N=1,NPG)
38        7 FORMAT(8A4)
39        8 FORMAT(4E10.3,I5)
40        9 FORMAT(10F8.5)
41       10 FORMAT(13F6.3)
42          UE = UEE(1)
43          DUEDXU = UEX(1)/UEE(1)
44          IDEL = 1
45          IF(IMAX .EQ. 6)IDEL = 8
46          IF(IMAX .EQ. 11)IDEL=4
47          IF(IMAX .EQ. 21)IDEL = 2
48          DO 11 I = 1,IMAX
49          IA = 1 + (I-1)*IDEL
50       11 TAU(I) = TAUD(IA)
51          WRITE(6,12)
52       12 FORMAT(' INITIAL TAU PROFILE')
53          WRITE(6,13)(TAU(I),I=1,IMAX)
54       13 FORMAT(' TAU=',8E12.5)
55          WRITE(6,14)TITLE
56       14 FORMAT(' ',8A4)
57          WRITE(6,15)IMAP
58       15 FORMAT(' ',I2,' LINEAR FINITE ELEMENTS,   TURBULENCE MODEL:',
59       1' MIXING LENGTH + VAN DRIEST DAMPING')
```

Fig. 15.13. Listing of program DOROD

```
60   C
61   C        SET INITIAL TRV
62   C
63            CALL TURVS(REL,DELTA,IMAX,UE,DUEDXU,TAU,TRV,DTRV,AKP,
64           1APZ,ATR,PCON)
65   C
66            DO 16 I = 1,IMAX
67            DTRV(I) = 0.
68        16 CONTINUE
69   C
70   C        COMPUTE CC,AA AND EF
71   C
72            CALL FEPAR(IMAX)
73   C
74            DO 18 I = 1,IMAX
75            DO 17 J = 1,5
76        17 ABC(J,I) = 0.
77        18 CONTINUE
78   C
79   C        BEGINNING OF DOWNSTREAM LOOP
80   C
81            XCH = X0
82            X = X0
83            AIM = IMAP
84            DU = 1./AIM
85            KH = 2
86            DO 27 N = 1,NMAX
87            KST = KH
88   C
89   C        COMPUTE PRESSURE GRADIENT PARAMETERS
90   C
91            DO 19 KA = KST,NPG
92            K = KA
93            IF(X .GT. XUE(K))GOTO 19
94            KH = K-1
95            IF(K+1 .GT. NPG)K = NPG-1
96            DUM = XUE(K+1) - XUE(K)
97            XI = (XUE(K-1)-XUE(K))/DUM
98            X4 = (X - XUE(K))/DUM
99            X5 = (X + DX - XUE(K))/DUM
100           DUM = UEE(K-1)-UEE(K)-XI*XI*(UEE(K+1)-UEE(K))
101           A1 = DUM/XI/(1.-XI)
102           A2 = UEE(K+1)-UEE(K)-A1
103           UE = UEE(K) + A1*X4 + A2*X4*X4
104           UEP = UEE(K) + A1*X5 + A2*X5*X5
105           DUM = UEX(K-1)-UEX(K)-XI*XI*(UEX(K+1)-UEX(K))
106           A1 = DUM/XI/(1.-XI)
107           A2 = UEX(K+1)-UEX(K)-A1
108           DUEDXU = (UEX(K)+A1*X4+A2*X4*X4)/UE
109           DUEDXP = (UEX(K) + A1*X5 + A2*X5*X5)/UE
110           GOTO 20
111       19 CONTINUE
112  C
113  C        CALCULATE TRIDIAGONAL TERMS
114  C
115       20 CALL ABCD(IMAX,DX,DUEDXU,DUEDXP,UE,UEP,BETA)
116  C
117           CALL BANFAC(ABC,IMAX,1)
118           CALL BANSOL(D,DTHETA,ABC,IMAX,1)
119  C
```

Fig. 15.13. (cont.) Listing of program DOROD

```
120  C
121  C       RAT IS THE TYPICAL FRACTIONAL CHANGE IN THETA
122  C
123          RAT = ABS(TAU(IREF)*DTHETA(IREF))
124          IF(0.5*DX .LT. DXMI)GOTO 21
125          IF(RAT .GT. RATCH)DX = 0.5*DX
126          UEP = 0.5*(UE+UEP)
127          DUEDXP = 0.5*(DUEDXU+DUEDXP)
128          IF(RAT .GT. RATCH)GOTO 20
129       21 X = X + DX
130          IF(1.5*DX .GT. DXMA)GOTO 22
131          IF(RAT .LT. 0.1*RATCH)DX = 1.5*DX
132       22 CONTINUE
133  C
134  C       EVALUATE NEW TAU ARRAY
135  C
136          DO 23 I=1,IMAX
137       23 TAU(I) = 1./(1./TAU(I)+DTHETA(I))
138          CF = 2.*TAU(1)/RSQ
139          IF(TAU(1) .LT. 0.)GOTO 25
140  C
141  C       CALCULATE DISPLACEMENT AND MOMENTUM THICKNESS
142  C
143          DELTA = 0.
144          THKMOM=0.
145          IH = IMAX/2
146          DO 24 IA = 1,IH
147          I = 2*IA - 1
148          AI = I - 1
149          UI = DU*AI
150          DELTA = DELTA + 2./TAU(I) + 4./TAU(I+1)
151          THKMOM = THKMOM + 2.*UI/TAU(I) + 4.*(UI+DU)/TAU(I+1)
152       24 CONTINUE
153          DELTA = (DELTA+1./TAU(IMAX)-1./TAU(1))*DU/3.
154          THKMOM = (THKMOM+1./TAU(IMAX))*DU/3.
155          SHPRTR = DELTA/THKMOM
156          DELTA = DELTA/UE/RSQ
157          THKMOM = THKMOM/UE/RSQ
158          RTH = REL*UEP*THKMOM
159  C
160  C       CALCULATE EDDY VISCOSITY
161  C
162          CALL TURVS(REL,DELTA,IMAX,UEP,DUEDXP,TAU,TRV,DTRV,AKP,
163         1APZ,ATR,PCON)
164  C
165          IF(X .LT. XCH)GOTO 27
166          XCH = XCH + DXCH
167       25 WRITE(6,26)N,DX,X,TAU(1),CF,DELTA,THKMOM,SHPRTR,RTH,TRV(4)
168       26 FORMAT(' N=',I3,' DX=',F5.3,' X=',F4.2,' TAU(1)=',F5.3,
169         1' CF=',F7.5,' D-TH=',F6.4,' M-TH=',F6.4,' SH=',F5.3,
170         2' RTH=',E10.3,' TRV(4)=',F5.3)
171  C
172  C       TEST FOR SEPARATION AND X .GT. XMAX
173  C
174          IF(TAU(1) .LT. 0.)GOTO 28
175          IF(X+0.001 .GE. XMAX)GOTO 28
176       27 CONTINUE
177       28 CONTINUE
178          STOP
179          END
```

Fig. 15.13. (cont.) Listing of program DOROD

```
1          SUBROUTINE TURVS(REL,DELTA,IMAX,UE,DUEDX,TAU,TRV,DTRV,AKP,
2         1APZ,ATR,PCON)
3  C
4  C       CALCULATE EDDY VISCOSITY, TRV AND DTRV
5  C
6          DIMENSION TRV(41),YP(41),TAU(41),DTRV(41),UP(41)
7  C
8  C       TRVO IS THE EDDY VISCOSITY IN THE OUTER REGION(CLAUSER FORMULATION)
9  C
10         RSQ = SQRT(REL)
11         REQ = SQRT(RSQ)
12         TRVO = ATR*UE*REL*DELTA
13         IMAP = IMAX - 1
14         AIM = IMAP
15         DU = 1./AIM
16         YP(1) = 0.
17         UP(1) = 0.
18         UT = SQRT(TAU(1))/REQ
19         PP = -DUEDX/UE/REL/UT/UT/UT
20         EN = 1. + PCON*PP
21         AP = APZ/EN
22         RUT = UT*RSQ
23         UP(2) = DU/UT
24         YP(2) = UP(2)
25         TRV(1) = 0.
26 C
27         DO 2 K = 2,IMAP
28         AK = K - 1
29         U = AK*DU
30         IF(K .EQ. 2)GO TO 1
31         UP(K) = U/UT
32         UX = 1./(1. - U + 2.*DU)
33         UY = 1./(1. - U + DU)
34         UZ = 1./(1. - U)
35         DEM = DU*(UX/TAU(K-2) + 4.*UY/TAU(K-1) + UZ/TAU(K))/3.
36         YP(K) = YP(K-2) + DEM*RUT
37       1 DUM = 1. - EXP(-YP(K)/AP)
38         EL = AKP*YP(K)*DUM
39         TRV(K) = EL*EL*TAU(K)*(1.-U)/TAU(1)
40         IF(TRV(K) .GT. TRVO)GOTO 3
41         DTRV(K) = -(1.-U)*EL*EL*TAU(K)*TAU(K)/TAU(1)
42       2 CONTINUE
43       3 DO 4 L = K,IMAX
44         DTRV(L) = 0.
45         TRV(L) = TRVO
46       4 CONTINUE
47         RETURN
48         END
```

Fig. 15.14. Listing of subroutine TURVS

Subroutine ABCD (Fig. 15.16) evaluates the terms in the tridiagonal system of equations (15.66) and subroutines BANFAC and BANSOL (Sect. 6.2.3) solve the system for the corrections $\Delta\theta^{n+1}$.

Prior to computing the new solution $\theta^{n+1} = \theta^n + \Delta\theta^{n+1}$, the marching step size $\Delta\xi$ is examined for possible alteration, depending on the size of $\Delta\theta_w^{n+1}/\theta_w^n$. If $\Delta\theta_w^{n+1}/\theta_w^n > \gamma$, $\Delta\xi$ is halved as long as this will exceed the minimum $\Delta\xi$. If $\Delta\xi$ is halved, u_e^n, etc., are recomputed leading to a new solution correction $\Delta\theta^{n+1}$. A typical value of γ is 0.02. If $\Delta\theta_w^{n+1}/\theta_w^{n+1} < 0.1\gamma$ the step size, $\Delta\xi$ is increased by 50% as long as the maximum step size is not exceeded. The step size changing

Table 15.4. Parameters used in subroutine TURVS

Parameter	Description
TRV	v_T/v
DTRV	$\dfrac{\partial}{\partial \theta}(v_T/v)$
TRVO	$(v_T/v)_{outer}$, Clauser formulation
ATR	Clauser constant, Sect. 11.4.2
AKP	κ, von Karman constant
APZ	A_0^+ basic van Driest damping parameter, Sect. 11.4.2
AP	A^+ pressure-adjusted van Driest damping parameter
PCON	pressure control parameter,
YP, UP	y^+, u^+, Sect. 18.1.1
UT	u_τ, friction velocity
EL	l, mixing length
DLTH	$\partial l/\partial \theta$

mechanism permits small step sizes when θ is changing rapidly and larger step sizes when θ is changing more slowly.

Program DOROD calculates the skin friction coefficient c_f directly from the value of τ at the wall, as

$$c_f = 2\tau(1)/\mathrm{Re}_L^{1/2} \ , \tag{15.70}$$

where the reference Reynolds number $\mathrm{Re}_L = U_\infty L/v$. The displacement δ^* and momentum δ^{mom} thicknesses are obtained by numerically integrating, by Simpson's rule, the expressions

$$\delta^* = \frac{1}{u_e \, \mathrm{Re}_L^{1/2}} \int_0^1 (1/\tau) du \ , \tag{15.71}$$

$$\delta^{mom} = \frac{1}{u_e \, \mathrm{Re}_L^{1/2}} \int_0^1 (u/\tau) du \ . \tag{15.72}$$

Program DOROD is applied to the computation of boundary layer flow with an adverse pressure gradient. Experimental results (Bradshaw 1967) are available as part of the 1968 Stanford conference (Coles and Hirst 1968) on turbulent flows. These results have been used to construct the input data (Fig. 15.17) to program DOROD. Typical tabulated output for this case is shown in Fig. 15.18. It is clear from Fig. 15.18 that the step size ($\Delta x \equiv \Delta \xi$) increases as the solution develops in x. This is consistent with the reduced rate of change of the solution. It is also clear that the step size $\Delta \xi$ changing algorithm is prevented from sustaining a premature increase in the step size $\Delta \xi$.

The variation of skin friction coefficient and the boundary layer thicknesses with downstream position are shown in Figs. 15.19 and 15.20. The results indicate that DOROD-FEM (the present method) is producing solutions of comparable accuracy to DOROD-SPEC (Sect. 15.3.3) and to a representative finite volume

```
1          SUBROUTINE FEPAR(IMAX)
2  C
3  C       COMPUTES VALUES OF CC,AA, AND EF ACROSS BOUNDARY LAYER
4  C
5          DIMENSION CC(41,5),AA(41,5),EF(41,5),ABC(5,65),D(65)
6         1,TAU(41),TRV(41),DTRV(41)
7          COMMON CC,AA,EF,ABC,D,TAU,TRV,DTRV
8          SI = 1./6.
9          TI = 1./3.
10         TTI = 2.*TI
11         IMAP = IMAX - 1
12         AIM = IMAP
13         DU = 1./AIM
14         DUS = DU*DU
15 C
16 C       AT WALL
17 C
18         CC(1,1) = 0.
19         CC(1,2) = DU/12.
20         CC(1,3) = DU/12.
21         AA(1,1) = 0.
22         AA(1,2) = 1./DUS + TI
23         AA(1,3) = -AA(1,2) + 1./DU
24         EF(1,1) = 0.
25         EF(1,2) = -0.5/DU - TI
26         EF(1,3) = -0.5/DU - SI*(1.-DU)
27 C
28 C       AT OUTER EDGE OF B.L.
29 C
30         CC(IMAX,1) = SI*(1.  - 0.5*DU)
31         CC(IMAX,2) = TI*(1.  - 0.25*DU)
32         CC(IMAX,3) = 0.
33         AA(IMAX,1) = -TI
34         AA(IMAX,2) = TI
35         AA(IMAX,3) = 0.
36         EF(IMAX,1) = TI*(1.  - 0.5*DU)
37         EF(IMAX,2) = -TI
38         EF(IMAX,3) = 0.
39 C
40 C       AT INTERIOR NODES
41 C
42         DO 1 J = 2,IMAP
43         AJ = J-1
44         U = AJ*DU
45         CC(J,1) = SI*(U - 0.5*DU)
46         CC(J,2) = TTI*U
47         CC(J,3) = SI*(U + 0.5*DU)
48         DUM = (1. - U)/DU
49         DUS = DUM*DUM
50         AA(J,1) = - DUS - DUM - TI
51         AA(J,2) = 2.*DUS + TTI
52         AA(J,3) = - DUS + DUM - TI
53         DUM = 0.5*(1.  - U*U)/DU
54         DUS = 0.5*U - SI
55         EF(J,1) = DUM + DUS - SI*DU
56         EF(J,2) = - TTI
57         EF(J,3) = - DUM + DUS + SI*DU
58 1 CONTINUE
59         RETURN
60         END
```

Fig. 15.15. Listing of subroutine FEPAR

```
1            SUBROUTINE ABCD(IMAX,DX,DUEDXU,DUEDXP,UE,UEP,BETA)
2   C
3   C        COMPUTES ABC,D  ACROSS BOUNDARY LAYER
4   C
5            DIMENSION CC(41,5),AA(41,5),EF(41,5),ABC(5,65)
6           1,D(65),TAU(41),TRV(41),DTRV(41)
7            COMMON CC,AA,EF,ABC,D,TAU,TRV,DTRV
8   C
9            DO 2 K = 1,IMAX
10           JST = 1
11           JFN = 3
12           IF(K .EQ. 1)JST = 2
13           IF(K .EQ. IMAX)JFN = 2
14           DIM = 0.
15           DO 1 J = JST,JFN
16           JP = J + 1
17           KD = K - 2 + J
18           DOM = DUEDXP*EF(K,J)
19           DEM = AA(K,J)*(1. + TRV(KD))
20           DAM = TAU(KD)*(DEM*TAU(KD)- AA(K,J)*DTRV(KD))*UEP
21           ABC(JP,K) = CC(K,J) + BETA*DX*(DAM-DOM)
22           DOM = DUEDXU*EF(K,J)*(1.-BETA) + DOM*BETA
23           DIM = DIM + DOM/TAU(KD) + DEM*TAU(KD)*(UE*(1.-BETA)+UEP*BETA)
24        1 CONTINUE
25           D(K) = DIM*DX
26        2 CONTINUE
27           RETURN
28           END
```

Fig. 15.16. Listing of subroutine ABCD

```
500    11 0.60 0.4125.0021.00 0.168E-01 1.601E+06
0.001E-00 0.001E-00 0.100E-00 0.200E-00 0.938E-00 3.810E+00 0.020E-00


BRADSHAW FLOW C:  ADVERSE P.G.
 .265E-02  .160E+07  .733E-02  .532E-02     22
1.67653 1.71531 1.75605 1.79910 1.82957 1.85014 1.83343 1.80281 1.72612 1.63548
1.53116 1.39436 1.26574 1.10231  .95764  .81205  .67268  .55842  .45748  .38151
 .31169  .23790  .18126  .09885  .07391  .05997  .04830  .03961  .03327  .02981
 .02949  .03254  .03581  .04051  .04739  .05840  .07624  .10474  .15709  .28916
 .44260
 .938 1.063 1.188 1.313 1.438 1.563 1.688 1.813 1.938 2.063 2.188 2.313 2.438
2.563 2.688 2.938 3.188 3.313 3.438 3.563 3.688 3.813
1.011  .989  .968  .949  .925  .904  .890  .872  .859  .847  .835  .822  .813
0.800 0.790 0.774 0.760 0.750 0.742 0.736 0.729 0.726
-.119 -.196 -.181 -.167 -.156 -.141 -.126 -.113 -.105 -.098 -.093 -.086 -.081
-0.076-0.072-0.067-0.064-0.061-0.059-0.057-0.056-0.055
```

Fig. 15.17. Input data to program DOROD

```
NMAX,IMAX =   500   11
RE = .160E+07 XO = .938 XMAX= 3.810 DXCH= .200E+00 RATCH= .200E-01
DX= .100E-02 DXMI= .100E-02 DXMA= .100E+00 BETA= .600E+00
AKP= .410E+00 APZ= .250E+02 ATR= .168E-01 PCON= .210E+02


INITIAL TAU PROFILE
TAU= .16765E+01 .18296E+01 .17261E+01 .12657E+01 .67268E+00 .31169E+00 .73910E-01 .33270E-01
TAU= .35810E-01 .76240E-01 .44260E+00
 BRADSHAW FLOW C:  ADVERSE P.G.
10 LINEAR FINITE ELEMENTS,  TURBULENCE MODEL: MIXING LENGTH + VAN DRIEST DAMPING
N=  1 DX= .001 X= .94 TAU(1)=1.666 CF= .00263 D-TH= .0073 M-TH= .0053 SH=1.380 RTH= .860E+04 TRV(4)= .919
N= 61 DX= .011 X=1.14 TAU(1)=1.435 CF= .00227 D-TH= .0090 M-TH= .0064 SH=1.409 RTH= .996E+04 TRV(4)=1.354
N= 79 DX= .011 X=1.34 TAU(1)=1.314 CF= .00208 D-TH= .0106 M-TH= .0074 SH=1.427 RTH= .112E+05 TRV(4)=1.631
N= 97 DX= .011 X=1.55 TAU(1)=1.229 CF= .00194 D-TH= .0123 M-TH= .0086 SH=1.441 RTH= .124E+05 TRV(4)=1.896
N=114 DX= .011 X=1.74 TAU(1)=1.179 CF= .00186 D-TH= .0139 M-TH= .0096 SH=1.448 RTH= .135E+05 TRV(4)=2.072
N=123 DX= .026 X=1.94 TAU(1)=1.138 CF= .00180 D-TH= .0154 M-TH= .0106 SH=1.453 RTH= .146E+05 TRV(4)=2.241
N=131 DX= .026 X=2.15 TAU(1)=1.107 CF= .00175 D-TH= .0170 M-TH= .0117 SH=1.456 RTH= .157E+05 TRV(4)=2.388
N=139 DX= .026 X=2.35 TAU(1)=1.082 CF= .00171 D-TH= .0187 M-TH= .0128 SH=1.459 RTH= .168E+05 TRV(4)=2.515
N=147 DX= .026 X=2.56 TAU(1)=1.062 CF= .00168 D-TH= .0204 M-TH= .0139 SH=1.460 RTH= .179E+05 TRV(4)=2.628
N=155 DX= .026 X=2.76 TAU(1)=1.043 CF= .00165 D-TH= .0220 M-TH= .0151 SH=1.462 RTH= .190E+05 TRV(4)=2.743
N=162 DX= .026 X=2.94 TAU(1)=1.026 CF= .00162 D-TH= .0235 M-TH= .0160 SH=1.463 RTH= .199E+05 TRV(4)=2.857
N=170 DX= .026 X=3.15 TAU(1)=1.005 CF= .00159 D-TH= .0251 M-TH= .0172 SH=1.466 RTH= .210E+05 TRV(4)=3.018
N=178 DX= .026 X=3.35 TAU(1)= .985 CF= .00156 D-TH= .0270 M-TH= .0184 SH=1.469 RTH= .220E+05 TRV(4)=3.172
N=186 DX= .026 X=3.56 TAU(1)= .967 CF= .00153 D-TH= .0288 M-TH= .0196 SH=1.471 RTH= .231E+05 TRV(4)=3.331
N=194 DX= .026 X=3.76 TAU(1)= .948 CF= .00150 D-TH= .0307 M-TH= .0208 SH=1.475 RTH= .242E+05 TRV(4)=3.523
N=201 DX= .026 X=3.94 TAU(1)= .930 CF= .00147 D-TH= .0320 M-TH= .0217 SH=1.478 RTH= .252E+05 TRV(4)=3.708
```

Fig. 15.18. Typical output produced by program DOROD

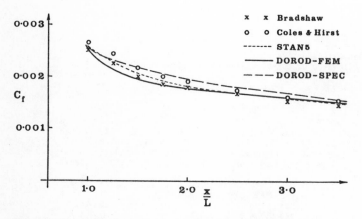

Fig. 15.19. Skin friction comparison for adverse pressure gradient

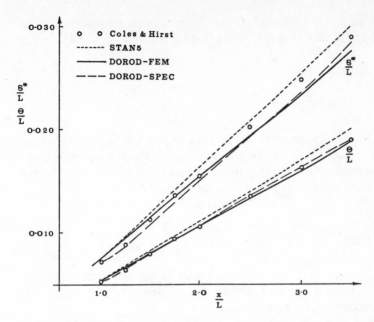

Fig. 15.20. Displacement and momentum thickness comparison for adverse pressure gradient

code, STAN5 (Reynolds 1976). Other pressure gradient cases, taken from the 1968 Stanford Conference, have been computed using program DOROD and the results are given by Fletcher and Fleet (1984b).

15.3.3 Dorodnitsyn Spectral Method

The Dorodnitsyn turbulent boundary layer formulation (15.57) can be interpreted as a Galerkin spectral method if particular choices are made for the weight function f_k and the approximate solution for Θ.

The spectral method is implemented with (15.57) written in the form

$$
\frac{\partial}{\partial \xi} \int_0^1 u f_k \Theta \, du = \frac{u_{e\xi}}{u_e} \int_0^1 \frac{df_k}{du} (1-u^2) \Theta \, du
$$
$$
- u_e \left[\left(\frac{df_k}{du} \right) \Big/ T \right]_{u=0} - u_e \int_0^1 \frac{d^2 f_k}{du^2} \left(1 + \frac{\nu_T}{\nu} \right) T \, du \ . \tag{15.73}
$$

The development of the spectral method depends on the introduction of orthonormal functions $g_j(u)$ in place of the weight functions $f_k(u)$ and in the approximate solution for Θ.

The orthonormal functions $g_j(u)$ are generated from the weight functions $f_k(u)$ by

$$
g_j(u) = \sum_{k=1}^j e_{kj} f_k(u) \ , \tag{15.74}
$$

where the coefficients e_{kj} are evaluated via the Gram-Schmidt orthonormalisation process (Isaacson and Keller 1966, p. 199) so that the functions g_j satisfy

$$\int_0^1 g_i(u)g_j(u)w(u)du = 1 \quad \text{if} \quad i = j,$$

$$= 0 \quad \text{if} \quad i \neq j. \tag{15.75}$$

The function $w(u)$ in (15.75) depends on the class of problems being solved. An appropriate form to suit (15.73) will be indicated below. The coefficients e_{kj}, and hence the orthonormal functions g_j, can be evaluated once and for all. The following approximate solution is introduced for Θ in (15.73):

$$\Theta = \frac{1}{1-u}\left[b_0 + \sum_{j=1}^{N-1} b_j g_j(u)\right], \tag{15.76}$$

where the coefficient b_0 is retained to ensure that Θ has the correct physical behaviour at the outer edge of the boundary layer. Substitution of (15.76) into (15.73) with g_k replacing f_k produces the result

$$\frac{d}{d\xi}\int_0^1 \left[b_0 + \sum_{j=1}^{N-1} b_j g_j(u)g_k(u)\right]\frac{u}{1-u}du = C_k, \quad k = 1, \ldots, N \tag{15.77}$$

where

$$C_k = \{u_{e\xi}/u_e\}\int_0^1\left(\frac{dg_k}{du}\right)(1-u^2)\Theta\,du - u_e\left[\frac{dg_k}{du}\frac{1}{\Theta}\right]_{u=0}$$

$$- u_e\int_0^1\frac{d^2 g_k}{du^2}\{1 + v_T/v\}/\Theta\,du. \tag{15.78}$$

A comparison of (15.75) and (15.77) indicates that, to exploit the orthogonal character of $g_j(u)$, the following choice is appropriate:

$$w(u) = \frac{u}{1-u}. \tag{15.79}$$

Consequently (15.77) becomes

$$V_k\frac{db_0}{d\xi} + \frac{db_k}{d\xi} = C_k, \quad k = 1, \ldots, N-1 \tag{15.80}$$

where

$$V_k = \int_0^1 g_k(u)w(u)du. \tag{15.81}$$

When $k = N$,

$$\frac{db_0}{d\xi} = C_N/V_N \; , \tag{15.82}$$

so that (15.80) can be replaced with

$$\frac{db_k}{d\xi} = C_k - C_N \frac{V_k}{V_N} \; , \qquad k = 1, \ldots, N-1 \; . \tag{15.83}$$

Equations (15.82 and 83) provide an explicit system of equations for the coefficients appearing in (15.76). This system can be integrated efficiently using the variable order, variable step-size, predictor-corrector scheme due to Gear (1971).

Typical solutions obtained with the Dorodnitsyn spectral formulation (DOROD-SPEC) are shown in Figs. 15.10, 11, 19, 20. These solutions were obtained with $N = 6$ in (15.76) and a variable marching step size $0.000015 \le \Delta\xi$ $(\equiv \Delta x) \le 0.14$. These results illustrate the ability of the Dorodnitsyn spectral method to obtain high accuracy with relatively few unknowns in the approximate solution (15.76).

The method has been applied to incompressible (Fletcher and Holt 1975) and compressible (Fletcher and Holt 1976) laminar boundary layer flow and to incompressible turbulent boundary layer flow (Yeung and Yang 1981; Fletcher and Fleet 1984b).

15.4 Three-Dimensional Boundary Layer Flow

Two major additional problems arise in computing solutions to three-dimensional boundary layers. First, although the system of governing equations has a dominant parabolic character, the occurrence of two (surface) coordinate directions to describe the boundary layer development introduces a 'hyperbolic' character to the problem. Second, since practical three-dimensional boundary layers occur on curved surfaces typically, it is necessary to introduce surface-fitted coordinate systems.

Here steady three-dimensional boundary layers will be considered. The extension of the present techniques to unsteady three-dimensional boundary layers is provided by Dwyer (1981) and the references cited therein. For steady incompressible laminar flow the three-dimensional boundary layer equations are

$$\frac{\partial u}{\partial x} + \frac{\partial v}{\partial y} + \frac{\partial w}{\partial z} = 0 \; , \tag{15.84}$$

$$u\frac{\partial u}{\partial x} + v\frac{\partial u}{\partial y} + w\frac{\partial u}{dz} = -\frac{1}{\varrho}\frac{\partial p_e}{\partial x} + v\frac{\partial^2 u}{\partial y^2} \; , \tag{15.85}$$

$$u\frac{\partial w}{\partial x} + v\frac{\partial w}{\partial y} + w\frac{\partial w}{\partial z} = -\frac{1}{\varrho}\frac{\partial p_e}{\partial z} + v\frac{\partial^2 w}{\partial y^2} \; , \tag{15.86}$$

where x and z are Cartesian coordinates locally parallel to the three-dimensional surface and y is the normal component. No-slip boundary conditions $u = v = w = 0$ are required at the body or wall surface $y = 0$. The known pressure p_e, or the equivalent velocities u_e and w_e through the Bernoulli equation (11.49), provide the boundary conditions at the boundary layer edge.

Initial conditions may be required at a single point, as for a slender inclined body of revolution, or along a stagnation line, as for a finite swept wing. In principle the distribution of all three velocity components across the boundary layer is required. These are obtained, normally, by solving a reduced form of the equations appropriate to the initial location (Blottner 1975a).

15.4.1 Subcharacteristic Behaviour

An inspection of (15.84–86) indicates that, in the x and z directions, only convection occurs. In the y direction both diffusion and convection occur. Formally (i.e. following the procedures given in Chap. 2), (15.84–86) are non-elliptic. In addition it is possible to define subcharacteristics associated with the convective operator (ignoring the normal flow component v)

$$q \frac{\partial}{\partial s} = u \frac{\partial}{\partial x} + w \frac{\partial}{\partial z} . \tag{15.87}$$

The subcharacteristic direction is given by the projection of the streamline into the plane of the surface i.e. $dz/dx = w/u$. But w/u varies across the boundary layer. Since disturbances propagate over the full y domain for a given (x, z), it is necessary to define domains of influence and dependence as shown in Fig. 15.21. For an arbitrary point P at some intermediate depth in the boundary layer the introduction of a disturbance will be felt everywhere in the downstream region bounded by the planes containing the normal coordinate through P and, on one side, the

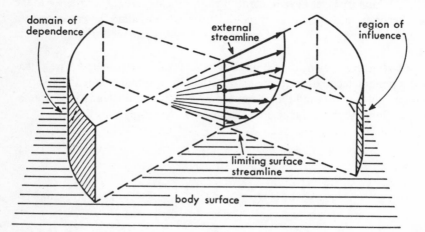

Fig. 15.21. Domains of influence and dependence

limiting streamline at the wall surface and, on the other side, the projection of the streamline at the boundary layer edge.

The hyperbolic nature of the problem, in the (x, z) plane, has important ramifications for the prescription of the initial data. If initial data, $u_0(y)$ and $w_0(y)$, are specified on a surface S_1 normal to the wall surface (Fig. 15.22), the solution can be obtained in the downstream region bounded by subcharacteristics S_2 and S_3 (and the wall and boundary layer edge). The subcharacteristic directions S_2 and S_3 are given by appropriate limiting values of the wall-surface or external flow streamlines. Attempting to compute the solution to the left of S_2 or to the right of S_3 (looking upstream) violates the Raetz-influence principle (Krause 1973) since the local solution depends on the data lying outside S_1. As indicated in Sect. 2.2.3 if additional boundary conditions, say on S_4 or S_5, are provided, the solution may be obtained anywhere in the downstream region bounded by the initial and boundary conditions.

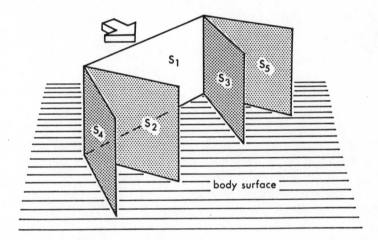

Fig. 15.22. Initial and boundary data

Computationally the hyperbolic behaviour is particularly significant if a marching algorithm is utilised that is explicit in the z direction. Just as for two-dimensional boundary layers, there is almost universal use of implicit schemes to discretise in the y-direction.

To obtain stable solutions it is necessary to obey the Courant–Friedrichs–Lewy (CFL) condition (Sect. 9.1.2) which requires that the computational domain of dependence in the (x, z) plane should include the physical domain of dependence implied by the governing partial differential equations. This effectively limits the relative size of Δx and Δz (Krause 1973).

To compare different schemes it will be assumed that x is the approximate inviscid flow direction and z is a cross-flow direction, e.g. along the span of a wing. In treating x as the primary marching direction u is presumed positive but w may be positive or negative.

A direct extension of the Crank–Nicolson scheme to three dimensions is to centre the discretisation at $(k-1/2, n+1/2)$ in the x–z plane (Fig. 15.23) and to construct two tridiagonal systems of equations from the x and z momentum equations for u and w along the $(k, n+1)$ gridline, i.e. in the y direction. The CFL condition requires that w is positive and for stability it is necessary that Δx is limited to $|w\Delta x/u\Delta z| \leqq 1$. The active nodes for the three-dimensional Crank–Nicolson scheme are shown in Fig. 15.23.

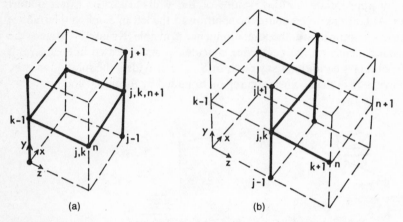

Fig. 15.23a, b. Active $(x-z)$ grid points. (a) Crank–Nicolson, (b) Krause zig-zag

The z sweep is in the direction of increasing k. If w is always negative the z sweep is made in the decreasing k direction. Thus the three-dimensional Crank–Nicolson scheme is restricted to domains where w does not change sign.

This restriction can be avoided by adopting the Krause zig-zag scheme (Krause 1973). In this scheme the discretisation is centred at $(k, n+1/2)$. However, z derivatives are evaluated using the zig-zag pattern (Fig. 15.23), e.g.

$$\frac{\partial w}{\partial z} \approx \frac{w_{j,k+1}^{n} - w_{j,k}^{n} + w_{j,k}^{n+1} - w_{j,k-1}^{n+1}}{2\Delta z}. \tag{15.88}$$

For a z sweep in the increasing k direction only the term $w_{j,k}^{n+1}$ is unknown in (15.88). Consequently it contributes to the tridiagonal system formed from the j nodes on grid line $(k, n+1)$. If w is positive there is no restriction on Δx. If w becomes negative then Δx must satisfy $|w\Delta x/u\Delta z| \leqq 1$ for stability.

The Crank–Nicolson and Krause zig-zag schemes can be thought of as semi-implicit schemes, since they are effectively explicit in the crossflow direction. The CFL restriction on Δx can be avoided by introducing a fully implicit scheme (Sect. 15.4.3), i.e. implicit in both the y and z directions.

15.4.2 Generalised Coordinates

To account for the rapid change in the velocity solution close to the surface and to allow for compressibility effects it is customary to introduce coordinate transformations. Blottner (1975a, pp. 3–35) discusses Levy–Lees type transformations (Sect. 15.2.2) and Dwyer (1981) discusses modified Blasius transformations that are also suitable for unsteady boundary layer flows.

However, prior to introducing these specialised transformations it is necessary to express the equations in surface-fitted coordinates. Here we will introduce generalized nonorthogonal coordinates (Chap. 12) on the surface (Fig. 15.24) such that ξ is approximately in the flow direction, ζ is an approximate crossflow direction and η is in the normal direction.

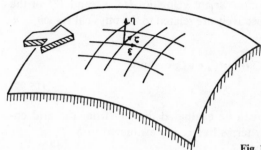

Fig. 15.24. Surface-fitted generalised coordinates

Equations (15.84–86) are rewritten in conservative vector form as

$$\frac{\partial \mathbf{E}}{\partial x} + \frac{\partial \mathbf{F}}{\partial y} + \frac{\partial \mathbf{G}}{\partial z} = 0 \ , \quad \text{where} \tag{15.89}$$

$$\mathbf{E} = \begin{bmatrix} u \\ u^2 + \dfrac{p_e}{\varrho} \\ uw \end{bmatrix} , \quad \mathbf{F} = \begin{bmatrix} v \\ uv - \tau'_{yx} \\ vw - \tau'_{yz} \end{bmatrix} , \quad \mathbf{G} = \begin{bmatrix} w \\ uw \\ w^2 + \dfrac{p_e}{\varrho} \end{bmatrix} ,$$

$$\tau'_{yx} = \frac{\tau_{yx}}{\varrho} = v \frac{\partial u}{\partial y} \quad \text{and} \quad \tau'_{yz} = \frac{\tau_{yz}}{\varrho} = v \frac{\partial w}{\partial y} \ .$$

Introducing coordinates (ξ, η, ζ) into (15.89) gives

$$\frac{\partial \hat{\mathbf{E}}}{\partial \xi} + \frac{\partial \hat{\mathbf{F}}}{\partial \eta} + \frac{\partial \hat{\mathbf{G}}}{\partial \zeta} = 0 \ , \quad \text{where} \tag{15.90}$$

$$\hat{\mathbf{E}} = J^{-1} \begin{bmatrix} U^c \\ uU^c + \xi_x \dfrac{p_e}{\varrho} \\ wU^c + \xi_z \dfrac{p_e}{\varrho} \end{bmatrix} , \qquad \hat{\mathbf{F}} = J^{-1} \begin{bmatrix} V^c \\ uV^c - \eta_y \tau'_{yx} \\ wV^c - \eta_y \tau'_{yz} \end{bmatrix} \quad \text{and}$$

$$\hat{\mathbf{G}} = J^{-1} \begin{bmatrix} W^c \\ uW^c + \zeta_x \dfrac{p_e}{\varrho} \\ wW^c + \zeta_z \dfrac{p_e}{\varrho} \end{bmatrix} . \tag{15.91}$$

In the expressions for $\hat{\mathbf{E}}$, etc., the contravariant velocities U^c, V^c and W^c in the direction of increasing ξ, η and ζ, respectively are related to the physical velocities u, v and w by

$$U^c = \xi_x u + \xi_z w , \qquad V^c = \eta_y v , \qquad W^c = \zeta_x u + \zeta_z w \quad \text{and}$$

$$J^{-1} = y_\eta (x_\xi z_\zeta - x_\zeta z_\xi) . \tag{15.92}$$

Transformation parameters like x_ξ can be evaluated directly from the grid co-ordinates (Sect. 12.2.1). Subsequently terms like ξ_x are evaluated from

$$\xi_x = J y_\eta z_\zeta , \qquad \xi_z = -J y_\eta x_\zeta , \qquad \eta_y = 1/y_\eta ,$$

$$\zeta_x = -J y_\eta z_\xi \quad \text{and} \quad \zeta_z = J y_\eta x_\xi . \tag{15.93}$$

The above generalised coordinates are the same as described in Chap. 12, except that here η is assumed to be orthogonal to the surface containing x and z, which simplifies the expressions in (15.91–93).

15.4.3 Implicit Split Marching Algorithm

An efficient marching algorithm can be constructed by writing (15.90) as

$$(1+\gamma)[\{\hat{\mathbf{E}}\}^{n+1} - \{\hat{\mathbf{E}}\}^n] - \gamma[\{\hat{\mathbf{E}}\}^n - \{\hat{\mathbf{E}}\}^{n-1}]$$

$$= \Delta\xi[\beta \text{RHS}^{n+1} + (1-\beta)\text{RHS}^n] , \tag{15.94}$$

where

$$\text{RHS} = -\frac{\partial \hat{\mathbf{F}}}{\partial \eta} - \frac{\partial \hat{\mathbf{G}}}{\partial \zeta} .$$

This is essentially the three-level scheme discussed in Sects. 8.2.3 and 9.5.1. The choice $\gamma = 0$, $\beta = 0.5$ produces a Crank–Nicolson scheme and the choice $\gamma = 0.5$,

$\beta = 1.0$ produces the 3LFI (three-level fully implicit) scheme. For problems requiring a single march in the ζ direction, as in the present situation, the disadvantage of the extra storage is usually outweighed by the greater robustness of the 3LFI scheme. The marching algorithm (15.94) relies for its effectiveness on the ξ direction coinciding approximately with the flow direction.

To implement either scheme it is necessary to construct a linear system of equations at the implicit level $n+1$. Here this is constructed by a truncated Taylor series expansion about level n. Thus

$$\{\hat{\mathbf{E}}\}^{n+1} = \{\hat{\mathbf{E}}\}^{n} + \underline{A}\Delta\hat{\mathbf{q}}^{n+1} \quad \text{where}$$

$$\underline{A} = \frac{\partial\hat{\mathbf{E}}}{\partial\hat{\mathbf{q}}} = \begin{bmatrix} \xi_x & 0 & \xi_z \\ 2\xi_x u + \xi_z w & 0 & \xi_z u \\ \xi_x w & 0 & \xi_x u + 2\xi_z w \end{bmatrix}^{n},$$

$$\{\hat{\mathbf{F}}\}^{n+1} = \{\hat{\mathbf{F}}\}^{n} + \underline{B}\Delta\hat{\mathbf{q}}^{n+1} \quad \text{where}$$

$$\underline{B} = \frac{\partial\hat{\mathbf{F}}}{\partial\hat{\mathbf{q}}} = \begin{bmatrix} 0 & \eta_y & 0 \\ V^c - v\eta_y\dfrac{\partial}{\partial\eta} & \eta_y u & 0 \\ 0 & \eta_y w & V^c - v\eta_y\dfrac{\partial}{\partial\eta} \end{bmatrix}^{n},$$

$$\{\hat{\mathbf{G}}\}^{n+1} = \{\hat{\mathbf{G}}\}^{n} + \underline{C}\Delta\hat{\mathbf{q}}^{n+1} \quad \text{where}$$

$$\underline{C} = \frac{\partial\hat{\mathbf{G}}}{\partial\hat{\mathbf{q}}} = \begin{pmatrix} \zeta_x & 0 & \zeta_z \\ 2\zeta_x u + \zeta_z w & 0 & \zeta_z u \\ \zeta_x w & 0 & 2\zeta_z w + \zeta_x u \end{pmatrix}^{n},$$

$$\hat{\mathbf{q}} \equiv J^{-1}\{u, v, w\}^T \quad \text{and} \quad \Delta\hat{\mathbf{q}}^{n+1} = \hat{\mathbf{q}}^{n+1} - \hat{\mathbf{q}}^n.$$

Substituting the above expressions into (15.94) produces the following linear systems for $\Delta\hat{\mathbf{q}}^{n+1}$:

$$\left[(1+\gamma)\underline{A} + \beta\Delta\xi\left\{\frac{\partial\underline{B}}{\partial\eta} + \frac{\partial\underline{C}}{\partial\zeta}\right\} \right]\Delta\hat{\mathbf{q}}^{n+1}$$

$$= -\Delta\xi\left\{\frac{\partial\hat{\mathbf{F}}}{\partial\eta} + \frac{\partial\hat{\mathbf{G}}}{\partial\zeta}\right\}^{n} + \gamma[\{\hat{\mathbf{E}}\}^{n} - \{\hat{\mathbf{E}}\}^{n-1}] . \tag{15.95}$$

With three-point centred difference formulae to represent $\partial/\partial\eta$ and $\partial/\partial\zeta$ the above scheme can be split or factored to $O(\Delta\xi^2)$ and implemented in two stages. For the first stage,

$$\left[\underline{A} + \Delta\xi \left\{ \frac{\beta}{(1+\gamma)} \right\} L_\zeta \underline{C} \right] \Delta\hat{\mathbf{q}}^* = - \left\{ \frac{\Delta\xi}{(1+\gamma)} \right\} \{ L_\eta \hat{\mathbf{F}} + L_\zeta \hat{\mathbf{G}} \}$$

$$+ \left\{ \frac{\gamma}{(1+\gamma)} \right\} [\{\hat{\mathbf{E}}\}^n - \{\hat{\mathbf{E}}\}^{n-1}] \ , \tag{15.96}$$

and for the second stage,

$$\left[\underline{A} + \Delta\xi \left\{ \frac{\beta}{(1+\gamma)} \right\} L_\eta \underline{B} \right] \Delta\hat{\mathbf{q}}^{n+1} = \underline{A} \Delta\hat{\mathbf{q}}^* \ . \tag{15.97}$$

During the first stage (15.96) constitutes a tridiagonal system of equations associated with each ζ gridline which can be solved using the Thomas algorithm (Sect. 6.2.2). During the second stage only terms associated with η gridlines appear implicitly. As long as the terms $\partial/\partial\eta$ in $\underline{B}$ are evaluated as three-point centred difference formulae for $L_{\eta\eta}\underline{B}'$, (15.97) also produces a tridiagonal system that can be evaluated efficiently.

Dwyer (1981) discusses a possibly more stable differencing of $\partial V^c/\partial\eta$ and Schiff and Steger (1980) discuss the implementation of schemes like (15.96 and 97). A similar scheme is considered in Sect. 16.3.1. The implementation of three-dimensional boundary layer computer codes require many ad hoc procedures depending on the problem in question. For the boundary layer flow over swept wings McLean and Randall (1979) provide insight into such pragmatic considerations.

15.5 Closure

The equations governing boundary layer flow have a dominant parabolic character. Consequently implicit marching algorithms are appropriate to obtain the downstream development of the solution. The marching algorithms should be first or second-order accurate in the marching direction and at least second-order accurate across the boundary layer. Methods that achieve higher (typically fourth-) order accuracy across the boundary layer are described by Peyret and Taylor (1983, pp. 31–37). Iteration at each downstream location is less efficient than using smaller marching steps with a non-iterative algorithm.

Boundary layer flows contain severe velocity gradients normal to the marching direction. Therefore it is advantageous to use a nonuniform grid which has the smallest grid size adjacent to the wall and a geometrically growing grid away from the wall (Sect. 15.1.2). In addition working computer codes usually operate on transformed independent and dependent variables that produce less severe gradients, and consequently more accurate discretisations, in the transform domain. Such changes of variable (Sect. 15.2) are also useful for computing compressible and axisymmetric boundary layer flows efficiently.

The Dorodnitsyn transformation (Sect. 15.3) inverts the role of the velocity component u from a dependent to an independent variable. This permits accurate solutions to be obtained with relatively few grid points across the boundary layer.

Three-dimensional boundary layer flows can be solved efficiently using the split schemes (Sect. 8.2) as long as the time-like marching direction coincides approximately with the flow direction at the outer edge of the boundary layer. This requirement can be satisfied readily when using generalised coordinates (Sect. 15.4.2).

Boundary layer calculations find practical application as a means of determining the displacement thickness contour to correct the pressure distribution calculated by inviscid algorithms (Sect. 14.1.4) and as a component of viscous, inviscid interaction algorithms (Sect. 16.3.4). Such techniques can even permit small separation bubbles to occur (Carter 1981).

Most of the methods described in this chapter have been based on finite difference discretisation. The exception is the Dorodnitsyn formulation (Sect. 15.3) which lends itself to both a finite element and spectral interpretation. However the STAN5 code referred to in Sect. 15.3 is based on a finite volume discretisation (Patankar and Spalding 1970). In addition the finite element method has been applied to the two- and three-dimensional primitive variable boundary layer equations (Baker 1983, Chap. 5).

15.6 Problems

Simple Boundary Layer Flow (Sect. 15.1)

15.1 A general two-level implicit algorithm for (15.2) can be written

$$u_j^\lambda \frac{(u_j^{n+1} - u_j^n)}{\Delta x} + v_j^\lambda [\lambda L_y u_j^{n+1} + (1-\lambda) L_y u_j^n]$$

$$= \lambda [u_e u_{ex}]^{n+1} + (1-\lambda)[u_e u_{ex}]^n + v[\lambda L_{yy} u_j^{n+1} + (1-\lambda) L_{yy} u_j^n] , \qquad (15.98)$$

where

$$u_j^\lambda = \lambda u_j^{n+1} + (1-\lambda) u_j^n , \qquad v_j^\lambda = \lambda v_j^{n+1} + (1-\lambda) v_j^n ,$$

$$L_y u_j = \frac{(u_{j+1} - u_{j-1})}{2\Delta y} , \qquad L_{yy} u_j = \frac{(u_{j-1} - 2u_j + u_{j+1})}{\Delta y^2} \quad \text{and} \quad 0 \le \lambda \le 1.$$

Show that (15.98) can be written as a tridiagonal system of equations at each iteration level k

$$a_j u_{j-1}^{k+1} + b_j u_j^{k+1} + c_j u_{j+1}^{k+1} = d_j , \quad \text{where} \qquad (15.99)$$

$$a_j = -\lambda(\gamma + \delta) , \qquad b_j = u_j^\lambda + 2\lambda\delta , \qquad c_j = \lambda(\gamma - \delta) ,$$

$$\gamma = 0.5 v_j^\lambda \frac{\Delta x}{\Delta y} , \qquad \delta = v \frac{\Delta x}{\Delta y^2} , \qquad u_j^\lambda = \lambda u_j^k + (1-\lambda) u_j^n$$

and

$$d_j = u_j^\lambda u_j^n - 0.5(1-\lambda)v_j^\lambda \left(\frac{\Delta x}{\Delta y}\right)(u_{j+1}^n - u_{j-1}^n) + \Delta x \lambda [u_e u_{ex}]^{n+1}$$

$$+ \Delta x(1-\lambda)[u_e u_{ex}]^n + (1-\lambda)v\left(\frac{\Delta x}{\Delta y^2}\right)[u_{j-1}^n - 2u_j^n + u_{j+1}^n] .$$

At each iteration v_j^{k+1} is obtained from (15.8). At the beginning of the iteration $u_j^k = u_j^n$; at the end $u_j^{n+1} = u_j^{k+1}$.

15.2 Modify program LAMBL (Fig. 15.3) to incorporate the scheme in Problem 15.1 for the Crank–Nicolson case ($\lambda = 0.5$) and the fully implicit case ($\lambda = 1.0$). Obtain solutions corresponding to Fig. 15.5 but with

(a) $JMAX = 41$ i) $DX = 0.10$, $NMAX = 20$,

ii) $DX = 0.20$, $NMAX = 10$,

iii) $DX = 0.40$, $NMAX = 5$,

b) $DX = 0.2$, $NMAX = 100$ i) $JMAX = 21$,

ii) $JMAX = 11$,

iii) $JMAX = 6$.

From results a) determine the approximate convergence rate with Δx. From results b) determine the approximate convergence rate with Δy.

15.3 Modify program LAMBL to obtain the solution to boundary layer flow over a flat plate. This corresponds to $\beta = 0$. The following data statements, replacing lines 6–13, provide the similarity (Blasius) solution which is used as a starting solution and as the "exact" solution:

DATA UB/0.0000, 0.0931, 0.1876, 0.2806, 0.3720, 0.4606, 0.5453, 0.6244, 0.6967, 0.7611, 0.8167, 0.8633, 0.9011, 0.9306, 0.9529, 0.9691, 0.9880, 0.9959, 0.9988, 0.9997, 0.9999, 1.0000/

DATA VB/0.0000, 0.0094, 0.0375, 0.0840, 0.1479, 0.2276, 0.3206, 0.4234, 0.5318, 0.6410, 0.7466, 0.8443, 0.9310, 1.0047, 1.0648, 1.1116, 1.1716, 1.2001, 1.2115, 1.2153, 1.2165, 1.2167/

DATA YZ/0.0, 0.2, 0.4, 0.6, 0.8, 1.0, 1.2, 1.4, 1.6, 1.8, 2.0, 2.2, 2.4, 2.6, 2.8, 3.0, 3.4, 3.8, 4.2, 4.6, 5.0, 5.4/

The skin friction coefficient may be compared with the "exact" value,

$$c_{fex} = 0.664(Rex)^{-1/2} .$$

15.4 Modify program LAMBL to introduce a mechanism to change the streamwise step size Δx in response to changes in the solution as in program DOROD. For $JMAX = 21$ compare the number of streamwise steps to achieve comparable accuracy to the fixed step size algorithm in the range $1 \leq x \leq 3$ for $\Delta x_{fixed} = 0.10, 0.20$.

15.5 In program LAMBL y_{max} is chosen sufficiently large as to exceed the expected boundary layer thickness at x_{max}. Devise and implement a procedure to extend y_{max} adaptively in response to the increase in boundary layer thickness. One way would be to require that $y_{max} > k\delta^*$ where k is an empirical constant.

15.6 Determine the leading terms in the truncation error of the formulae for $\partial u/\partial y$ and $\partial^2 u/\partial y^2$ in (15.17). For the $\partial^2 u/\partial y^2$ formula determine a restriction on r_y to reduce the truncation error to second order. What would be the restriction on r_y to make the following formula second order:

$$\frac{\partial u}{\partial y} \approx \frac{(u_{j+1} - u_{j-1})}{(1 + r_y)\Delta y} \ .$$

Discuss the possibility of expanding the Taylor series about a point other than y_j. Consider how the convective terms might be handled and whether a higher-order truncation error for the equation as a whole could be achieved on a nonuniform grid.

15.7 Discuss the application of the Keller box scheme, described in Sect. 15.1.3, to the problem of the boundary layer flow over a flat plate (Sect. 15.1.2 and Problem 15.3).

Complex Boundary Layer Flow (Sect. 15.2)

15.8 For the flow over a flat plate a Crank–Nicolson discretisation of (15.14, 15) gives

$$0.5\frac{(v_j^n - v_{j-1}^n)}{\Delta y} + 0.5\frac{(v_j^{n+1} - v_{j-1}^{n+1})}{\Delta y} + 0.5\frac{(u_j^{n+1} - u_j^n)}{\Delta x}$$

$$+ 0.5\frac{(u_{j-1}^{n+1} - u_{j-1}^n)}{\Delta x} = 0 \ , \tag{15.100}$$

$$0.5(u_j^n + u_j^{n+1})\frac{(u_j^{n+1} - u_j^n)}{\Delta x} + 0.25v_j^n\frac{(u_{j+1}^n - u_{j-1}^n)}{\Delta y}$$

$$+ 0.25v_j^{n+1}\frac{(u_{j+1}^{n+1} - u_{j-1}^{n+1})}{\Delta y}$$

$$= 0.5\frac{(u_{j-1}^n - 2u_j^n + u_{j+1}^n)}{\Delta y^2} + 0.5\frac{(u_{j-1}^{n+1} - 2u_j^{n+1} + u_{j+1}^{n+1})}{\Delta y^2} \ . \tag{15.101}$$

Nonlinear terms at location x^{n+1} are linearised about x^n, i.e.

$$0.5(u_j^n + u_j^{n+1})\frac{(u_j^{n+1} - u_j^n)}{\Delta x} \approx u_j^n\frac{(u_j^{n+1} - u_j^n)}{\Delta x} \quad \text{and}$$

$$v_j^{n+1}(u_{j+1}^{n+1} - u_{j-1}^{n+1}) \approx v_j^n(u_{j+1}^{n+1} - u_{j-1}^{n+1}) + v_j^{n+1}(u_{j+1}^n - u_{j-1}^n) \ .$$

Show that the resulting (15.100 and 101) can be manipulated into the form of (15.46 and 48) to permit a coupled solution to be obtained.

15.9 Implement the application of (15.49, 50) to the equations developed in Problem 15.8. Obtain solutions for flat plate flow ($\beta = 0$) and compare with those produced by program LAMBL.

Dorodnitsyn Boundary Layer Formulation (Sect. 15.3)

15.10 Obtain solutions using program DOROD corresponding to Fig. 15.18 but with RATCH $= 0.01$ and DXMA $= 0.01$ and for three cases: i) NMAX $= 6$, ii) NMAX $= 11$ and iii) NMAX $= 21$. Compare the solutions for c_f and δ^* with the tabulated experimental data given by Coles and Hirst (1968), Case 3300.

15.11 Obtain solutions using program DOROD corresponding to Fig. 15.18 for three cases: i) RATCH $= 0.01$, ii) RATCH $= 0.02$, iii) RATCH $= 0.05$. Set DXCH sufficiently small to observe the behaviour of the step size (Δx) changing mechanism. Note the effect of RATCH on the accuracy and economy of the computational solution.

15.12 Modify program DOROD to obtain the velocity distribution across the boundary layer. This is done most conveniently in terms of y^+ and u^+ (Sect. 18.1.1) where

$$y^+ = u_\tau y \mathrm{Re} \ , \qquad u^+ = u/u_\tau \ , \qquad u_\tau = \left(\frac{\tau_1}{\mathrm{Re}^{1/2}} \right)^{1/2}$$

$$\text{and} \quad y = \frac{1}{\mathrm{Re}^{1/2}} \int_0^u \frac{du'}{(1 - u')\tau(u')} \ .$$

15.13 Discuss, in broad terms, what changes would be required to introduce one-dimensional quadratic elements (Sect. 5.3.2) into program DOROD. Exact evaluation of (15.63) is possible but tedious. Numerical evaluation based on Gauss quadrature (Zienkiewicz 1977, Chap. 8) is straightforward. Experience (Fletcher and Fleet 1984a, b) indicates that quadratic elements are more efficient than linear elements for laminar boundary layers but are less effective for turbulent boundary layers.

```
 300    11 0.60 0.4125.0021.00 0.168E-01 2.167E+06
0.001E-00 0.001E-00 0.100E-00 0.200E-00 0.187E-00 5.000E+00 0.020E-00

WIEGHARDT AND TILLMANN:  Z.P.G.
 .424E-02  .217E+07  .726E-03  .498E-03    12
3.12439 3.19838 3.27596 3.35807 3.44456 3.51442 3.57650 3.60822 3.60776 3.57723
3.48747 3.39341 3.25916 3.13512 2.93587 2.76024 2.53211 2.32607 2.09397 1.87966
1.66462 1.40086 1.19528 1.02369  .87114  .66700  .54118  .45518  .40517  .38367
 .36750  .36443  .36441  .36767  .38060  .43723  .51183  .61285  .90873 1.55564
2.02850
 .187  .387  .637  .937 1.237 1.687 2.287 2.887 3.487 4.087 4.687 4.987
1.000 1.000 1.006 1.003 1.006 1.003 1.006 1.006 1.006 1.006 1.009 1.006
0.000 0.000 0.000 0.000 0.000 0.000 0.000 0.000 0.000 0.000 0.000 0.000
```

Fig. 15.25. Input data to program DOROD for flat plate turbulent layer flow

15.14 Starting data for turbulent flow over a flat plate are provided in Fig. 15.25. These results are based on the Wieghardt and Tillmann data given by Coles and Hirst (1968). Apply the program DOROD to this case and compare the predicted skin friction coefficient c_f and displacement thickness δ^* with the results tabulated by Coles and Hirst (Case 1400).

3D Boundary Layer Flow (Sect. 15.4)

15.15 Starting from (15.85 and 86) develop expressions for the coefficients a_j, b_j, c_j and d_j in

$$a_j F_{j-1,k}^{n+1} + b_j F_{j,k}^{n+1} + c_j F_{j+1,k}^{n+1} = d_j \,, \quad \text{where } F \equiv u \text{ or } w \qquad (15.102)$$

and (15.102) are the tridiagonal systems of equations associated with the grid line (k, n) in Fig. 15.23. Obtain the coefficients of a_j–d_j for the i) Crank–Nicolson scheme and ii) Krause zig-zag scheme.

15.16 Discuss the application of the Krause zig-zag scheme to the laminar boundary layer flow over a flat plate on which is placed a circular cylinder of radius a, centred at $x = x_0$ and $z = 0$. The inviscid velocity distribution ahead of the cylinder is given accurately by potential flow theory (Sect. 11.3) as

$$\frac{u_e}{U_\infty} = 1 + \left\{ \left(\frac{z}{a} \right)^2 - \left[\frac{(x-x_0)}{a} \right]^2 \right\} \Big/ \left\{ \frac{r}{a} \right\}^4 \,,$$

$$\frac{w_e}{U_\infty} = -2 \left\{ (x-x_0) \frac{z}{a^2} \right\} \Big/ \left\{ \frac{r}{a} \right\}^4 \,,$$

where $r^2 = (x-x_0)^2 + z^2$.

The adverse pressure gradient associated with the circular cylinder is sufficient to make the boundary layer separate ahead of the circular cylinder. Starting well upstream of the cylinder with a flat plate boundary layer profile at all z locations, the solution should be marched towards the cylinder until close to separation at $z = 0$. A flat plate boundary layer solution may be assumed at $z = z_{max}$. The formulation may be compared with that of Cebeci (1975) who applied the Keller box scheme to this problem.

16. Flows Governed by Reduced Navier–Stokes Equations

In this chapter categories of viscous-inviscid flow equations, intermediate between the full Navier–Stokes equations and the boundary layer equations will be considered. These intermediate categories are called reduced Navier–Stokes (RNS) equations.

For flows with thick viscous layers or with significant streamline curvature the boundary layer approximation produces inaccurate solutions primarily because it fails to account for the transverse pressure variation. However, from a computational perspective, the boundary layer equations possess the considerable virtue of being non-elliptic in the flow direction, thereby permitting accurate solutions to be obtained in a single downsteam march and, consequently, in an economical manner. The RNS equations are constructed to retain the economy of the boundary layer equations while accurately modelling physical processes that would otherwise require expensive solutions of the full Navier–Stokes equations.

As is implicit in the above description the RNS category is less sharply defined than either the full Navier–Stokes equations (Chaps. 17 and 18) or the boundary layer equations (Chap. 15). Not surprisingly other intermediate categories have been identified in the literature that share some features of the RNS equations described in this chapter.

Thus Lomax and Mehta (1984) refer to composite equations as the broad intermediate category which includes both inviscid and viscous approximations to the full Navier–Stokes equations. The composite equations include thin-layer, slender-layer and conical Navier–Stokes equations. These categories are associated with certain physical flow features, which permit particular approximations. Davis and Rubin (1980) refer to Non-Navier–Stokes viscous flow equations as defining a category essentially equivalent to composite equations.

Rubin (1981) refers to 'parabolised' Navier–Stokes equations emphasising that it is the computational advantage of being able to march parabolic equations (Chap. 7) that is the most important property of the intermediate category. This same philosophy is present in the work of Rudman and Rubin (1968), Lin and Rubin (1973), Lubard and Helliwell (1974) and Lin and Rubin (1981). More recently Rubin (1985) has referred to this intermediate category as reduced Navier–Stokes equations; it is essentially this categorisation that we follow here.

The above labelling has often been introduced in conjunction with external flow problems. For internal flows intermediate categories of equations govern 'parabolic' flows (Patankar and Spalding 1972), 'partially parabolic' flows (Pratap and Spalding 1976), 'semi-elliptic' flows (Ghia et al. 1981) and 'partially elliptic' flows

(Rhie 1985). However, just as for external flows, parabolised (Anderson 1980) and reduced (Kreskovsky and Shamroth 1978) Navier–Stokes equations, are referred to. In this chapter we will use the intermediate category, reduced Navier–Stokes equations, to describe both internal and external flows, although the precise form of the equations will often differ slightly in the two cases.

The distinguishing features of RNS equations and the flows that they govern will be discussed in Sect. 16.1. Additional approximations and formulations for internal flows, e.g. in diffusers and ducts, are considered in Sect. 16.2. For flows external to a body (or bodies) the inviscid equations (Chap. 14) provide an excellent farfield approximation. For the viscous domain, close to the body, it is useful, in considering RNS systems of equations, to distinguish between supersonic and subsonic flow (Sect. 16.3).

To complete the picture for external flows, some developments of the traditional inviscid, boundary layer splitting for handling more difficult flows will be described. Here more difficult flows include phenomena like small separation bubbles and shock/boundary layer interaction. The more advanced inviscid, boundary layer formulations are often of RNS type rather than strictly boundary layer type.

16.1 Introduction

In Chap. 1 it was noted that the rapid development of computational fluid dynamics has been caused, inter alia, by its widespread use in the design process. The design of equipment involving fluid flow (usually air or water) often involves choosing one condition at which the flow equipment, e.g. turbine, diffuser or even complete aircraft, is to operate at maximum efficiency and choosing a family of closely related "off-design" conditions, for which the reduction in performance should not be too great.

The main role of computational fluid dynamics in the design process is to determine the detailed *steady* flow behaviour at the design conditions by solving an appropriate system of governing equations and related boundary conditions. Many practical design problems involve flows with a dominant flow direction. With an appropriate Cartesian coordinate system, this implies $u \gg v, w$, if the flow is predominantly in the positive x direction. An examination of the governing equations (Sect. 16.1.1) indicates that the viscous diffusion terms in the main flow direction are much smaller than the transverse viscous diffusion terms, and can be ignored.

This last feature is a key element of the boundary layer equations (Chap. 15). If the pressure variation across the boundary layer is negligible, that is in the absence of significant streamline curvature, the local solution can be obtained very efficiently via a single downstream march.

Thus it is not surprising that the design of flow equipment is often based on a two-stage process, i.e. inviscid flow plus boundary layer correction. In the first stage the flow is assumed inviscid and non-conducting and the solution (Chap. 14) provides the pressure distribution at the body surface, typically. The second stage adjusts the velocity and temperature fields close to the body surface to take account of the viscous and thermal boundary layer behaviour (Chap. 15). The boundary layer solution provides the shear stress and heat transfer at the body surface and, with the contribution from the pressure distribution, the drag on the body. If necessary the pressure distribution can be adjusted for the displacement thickness effect (Sect. 14.1.4).

The trend in the design of flow equipment is towards higher efficiency so that the equipment can be smaller to perform a given task, leading to lower costs. This trend is exposing some of the inherent weaknesses in the traditional inviscid flow, boundary layer correction approach to the flow equipment design process. The difficulty can be appreciated with reference to aircraft design, where maximum range is closely related to the maximum lift/drag ratio of the individual wing sections (Kuchemann 1978).

The characteristic variation of lift and drag with wing section incidence is shown in Fig. 16.1. The maximum lift/drag ratio occurs at a large angle of attack and not too far from stall. At this optimal angle of attack, the viscous layer over the leeward side of the wing section is thick, particularly downstream of any shock wave that may occur. Close to the trailing edge there will be significant pressure gradients transverse to the freestream direction (Nakayama 1984) associated with the merging of the windward and leeward streams. For some off-design conditions the wing sections will operate even closer to a stall condition, with a consequent further thickening of the viscous layer on the leeward side. Solutions to the boundary layer equations are not accurate for such conditions (Horstman 1984).

Recourse to solutions of the full Navier–Stokes equations is possible, but uneconomic in the context of requiring solutions for parametric variation of the design variables: angle of attack, body geometry, Reynolds number, Mach number, etc. What is required is a category of equations that is able to model the physical processes (almost) as accurately as the full Navier–Stokes equations while retaining

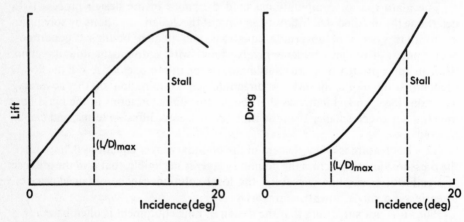

Fig. 16.1. Lift and drag variation with incidence

(almost) the economy of the boundary layer equations. This category we will call the reduced Navier–Stokes (RNS) equations.

The looseness of the categorisation implies that a number of different systems of equations can be interpreted as RNS equations. However, three features are shared by all RNS system of equations. These are:

i) There is a dominant flow direction aligned approximately with one coordinate direction.

ii) Viscous diffusion and heat conduction in the marching direction can be ignored compared with transverse viscous diffusion and heat conduction.

iii) The RNS equations reduce to the Euler equations when transverse viscous diffusion and heat conduction are negligible.

The construction of the basic RNS equations is made on an order-of-magnitude basis (Sect. 16.1.1) similar to that used to derive the boundary layer equations (Sect. 11.4). But the transverse momentum equation is retained, in approximate form, since a non-negligible transverse pressure gradient is often to be expected, on physical grounds. For external flows it is important that all terms in the Euler equations are retained exactly so that the RNS equations are applicable throughout the computational domain. In practice alternative computational techniques may be used in the inviscid region but the use of a globally applicable set of governing equations facilitates the matching of the solutions in the inviscid and viscous regions.

The requirement of an economy approaching that of the boundary layer equations implies that the solution to the viscous region should be obtained in a single downstream march. In turn this implies that the equation system should be non-elliptic in the marching direction. Consequently no downstream boundary conditions can be applied.

For internal flows and for supersonic external flows an accurate solution can usually be obtained with a single downstream march. The use of RNS equations for subsonic external flow usually requires repeated (iterative) downstream marches, particularly if small regions of reversed flow (separation bubbles) are present. For subsonic external flow the equation system has an elliptic character requiring prescription of downstream boundary conditions. However, the RNS strategy is still economical since relatively few iterations (downstream marches) are required, as long as any separated flow regions are small.

Reduced Navier–Stokes equations will be discussed in this chapter in relation to steady flow only, since this is where the main application is. However, it is possible to envisage certain classes of unsteady flow having a dominant flow direction, that could be computed more efficiently by introducing an RNS approximation.

Historically, the determination of whether a particular RNS system of equations permits a stable solution to be obtained in a single downstream march has often been approached in the empirical manner of introducing a particular approximation and examining the numerical solution. If a realistic solution is found it is assumed that the RNS system is stable. Briley and McDonald (1984) have used a characteristics analysis (Sect. 2.1.3) to test whether the RNS system of equations is elliptic with respect to the marching direction. As noted in Chap. 2 this approach requires the construction of an enlarged system of first-order partial differential equations, which may possess a singular characteristic form causing the analysis to be inconclusive.

In this chapter we prefer to use the more direct approach of a Fourier analysis (Sect. 2.1.5) applied to the governing equations. In Sect. 16.1.2 a Fourier analysis is used to determine, a priori, if the RNS system will produce a stable solution, in a single downstream march. Like the characteristic analysis (Chap. 2) the Fourier analysis determines the formal character of the system of governing equations. However, Fourier analysis has the considerable advantage of pinpointing which terms in the equations are responsible for any elliptic behaviour. It will turn out that additional approximations, often associated with the pressure, may be required to ensure non-elliptic behaviour, e.g. Sect. 16.2.2. The mechanics of applying the RNS strategy are demonstrated, in a practical way, in Sect. 16.1.4 for the prediction of the temperature profile at the entrance of a two-dimensional duct.

16.1.1 Order-of-Magnitude Analysis

In this section RNS equations will be derived from the steady two-dimensional Navier–Stokes equations for both incompressible and compressible laminar flow. The RNS equations will need to be modified for turbulent flow. However, the method of obtaining turbulent mean flow RNS equations remains essentially the same, i.e. an order-of-magnitude analysis.

The development of reduced forms of the Navier–Stokes equations (RNS) follows broadly the derivation of boundary layer equations (Cebeci and Bradshaw 1977). That is, significant viscous effects are presumed to be confined to a layer of thickness δ (Fig. 16.2), which is small compared with a characteristic length L in the flow direction.

For laminar boundary layers, δ/L is $O(\text{Re}^{-1/2})$. For a typical Reynolds number, $\text{Re} = 10^6$, this implies $\delta/L \sim 0.001$. One motivation for deriving RNS equations is to cope with viscous layers that are typically thicker than occur in boundary layers. Thus values of δ/L in the range 0.1 to 0.01 may be expected to require RNS equations. In turn, this implies that terms in the full Navier–Stokes equations of

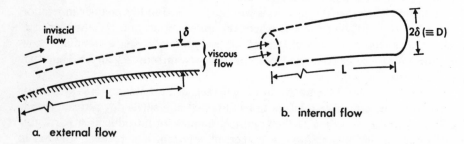

Fig. 16.2a, b. Typical viscous layer thicknesses for (**a**) external flow and (**b**) internal flow

$O((\delta/L)^2)$ may be dropped whereas terms of $O(\delta/L)$ should be retained. In contrast, the derivation of the turbulent boundary layer equations is based on discarding terms of $O(\delta/L)$.

The incompressible Navier–Stokes equations are given by (11.81). For steady laminar flow in two dimensions, the nondimensional form of (11.81) is

$$\frac{\partial u}{\partial x}+\frac{\partial v}{\partial y}=0 \ , \tag{16.1}$$

$$u\frac{\partial u}{\partial x}+v\frac{\partial u}{\partial y}+\frac{\partial p}{\partial x}=\frac{1}{\mathrm{Re}}\left(\frac{\partial^2 u}{\partial x^2}+\frac{\partial^2 u}{\partial y^2}\right) \ , \tag{16.2}$$

$$u\frac{\partial v}{\partial x}+v\frac{\partial v}{\partial y}+\frac{\partial p}{\partial y}=\frac{1}{\mathrm{Re}}\left(\frac{\partial^2 v}{\partial x^2}+\frac{\partial^2 v}{\partial y^2}\right) \ , \tag{16.3}$$

where the Reynolds number $\mathrm{Re}=\varrho U_\infty L/\mu$. Equations (16.1–3) have been non-dimensionalised as in (11.42).

To establish the relative size of terms in (16.1–3) it is assumed that $\partial/\partial x$ is $O(1)$, $\partial/\partial y$ is $O((L/\delta))$ and $\partial^2/\partial y^2$ is $O((L/\delta)^2)$ when applied to u and v. Since u is $O(1)$ and (16.1) cannot be simplified further, v is $O(\delta/L)$. All terms on the left-hand side of (16.2) are $O(1)$. On the right-hand side of (16.2), $\partial^2 u/\partial x^2$ is $O(1)$ whereas $\partial^2 u/\partial y^2$ is $O((L/\delta)^2)$; consequently $\partial^2 u/\partial x^2$ can be discarded.

In the classical boundary layer approach, $1/\mathrm{Re}$ is $O((\delta/L)^2)$. However, in a typical RNS application with Re large we expect $1/\mathrm{Re} \ll O((\delta/L)^2)$. To quantify this we assume $1/\mathrm{Re}$ is $O((\delta/L)^3)$, so that $(1/\mathrm{Re})\partial^2 u/\partial y^2$ is $O((\delta/L))$. All terms of the left-hand side of (16.3) are of $O(\delta/L)$. On the right-hand side of (16.3), $(1/\mathrm{Re})\partial^2 v/\partial x^2$ is $O((\delta/L)^4)$ whereas $(1/\mathrm{Re})\partial^2 v/\partial y^2$ is $O((\delta/L)^2)$. Thus $\partial^2 v/\partial x^2$ can be discarded. Consequently the resulting RNS equations are

$$\frac{\partial u}{\partial x}+\frac{\partial v}{\partial y}=0 \ , \tag{16.4}$$

$$u\frac{\partial u}{\partial x}+v\frac{\partial u}{\partial y}+\frac{\partial p}{\partial x}=\frac{1}{\mathrm{Re}}\frac{\partial^2 u}{\partial y^2} \ , \tag{16.5}$$

$$u\frac{\partial v}{\partial x}+v\frac{\partial v}{\partial y}+\frac{\partial p}{\partial y}=\frac{1}{\mathrm{Re}}\frac{\partial^2 v}{\partial y^2} \ . \tag{16.6}$$

In classical boundary layer analysis (16.6) is discarded and replaced by $\partial p/\partial y=0$. However, for thicker viscous layers for which the RNS equations will be required it is appropriate to retain (16.6). If significant streamline curvature occurs (16.6) should also include an additional centrifugal term. However, this term arises naturally when generalised curvilinear coordinates (Chap. 12) are used, which are usually preferred for non-simple computational domains.

According to the above order-of-magnitude analysis the term $(1/\mathrm{Re})\partial^2 v/\partial y^2$ is $O((\delta/L)^2)$ and could be ignored, as is recommended by Rubin (1984). We prefer to retain this term in the basic RNS formulation as it does not contribute to any elliptic interactions (Sect. 16.1.3) but does provide additional dissipation which may be useful computationally.

The RNS equations (16.4–6) reduce to the Euler equations when the viscous terms are negligible. This would be the case far from an isolated body in an otherwise undisturbed flow.

Appropriate boundary conditions to suit the incompressible RNS equations will depend on the mathematical character of the system (16.4–6); this is examined in Sect. 16.1.3.

The two-dimensional steady compressible laminar Navier–Stokes equations (Sect. 11.6.3) can be written in nondimensional form as

$$\frac{\partial}{\partial x}(\varrho u) + \frac{\partial}{\partial y}(\varrho v) = 0 \ , \tag{16.7}$$

$$\varrho u \frac{\partial u}{\partial x} + \varrho v \frac{\partial u}{\partial y} + \frac{\partial p}{\partial x} = \frac{1}{\mathrm{Re}} \left[\frac{\partial \tau_{xx}}{\partial x} + \frac{\partial \tau_{xy}}{\partial y} \right] \ , \tag{16.8}$$

$$\varrho u \frac{\partial v}{\partial x} + \varrho v \frac{\partial v}{\partial y} + \frac{\partial p}{\partial y} = \frac{1}{\mathrm{Re}} \left(\frac{\partial \tau_{yx}}{\partial x} + \frac{\partial \tau_{yy}}{\partial y} \right) \ , \tag{16.9}$$

$$\varrho \left(u \frac{\partial T}{\partial x} + v \frac{\partial T}{\partial y} \right) - (\gamma - 1) M_\infty^2 \left(u \frac{\partial p}{\partial x} + v \frac{\partial p}{\partial y} \right)$$

$$= \frac{1}{\mathrm{Pr\,Re}} \left[\frac{\partial}{\partial x} \left(k \frac{\partial T}{\partial x} \right) + \frac{\partial}{\partial y} \left(k \frac{\partial T}{\partial y} \right) \right] + \frac{(\gamma - 1) M_\infty^2}{\mathrm{Re}} \Phi \ , \tag{16.10}$$

$$1 + \gamma M_\infty^2 p = \varrho T \ . \tag{16.11}$$

The nondimensionalisation used to obtain (16.7–11) is the same as in Sect. 11.2.5 except that $p^{nd} = (p^d - p_\infty)/\varrho_\infty U_\infty^2$ here. In (16.8 and 9), τ_{xx}, etc., are the viscous stresses and related to the velocity gradients by (11.27), which is the same in nondimensional form. In (16.10) the viscous dissipation Φ is given by (11.39). The relative order-of-magnitude of the various terms in (16.7–11) will be considered as for (16.1–3).

The magnitude of the velocity components u and v and derivatives of u, v and p are as for incompressible flow. The terms T and $\partial T/\partial x$ are $O(1)$ as are ϱ and $\partial \varrho/\partial x$. If the wall temperature T_w is specified then $\partial T/\partial y$ and $\partial \varrho/\partial y$ are $O(L/\delta)$. However, if an adiabatic wall condition is imposed, i.e. $\partial T/\partial y|_{y=0} = 0$ and the flow is subsonic or transonic, it is more appropriate to consider average values for $\partial T/\partial y$ and $\partial \varrho/\partial y$ across the layer as being $O(1)$ or even $O(\delta/L)$. Consequently in deriving the reduced form of the compressible Navier–Stokes equations, two cases will be considered:

(i)   Specified wall temperature or supersonic flow $\left. \right\} \dfrac{\partial T}{\partial y}, \ \dfrac{\partial \varrho}{\partial y}$ are $O(L/\delta)$,

(ii) Adiabatic wall and subsonic/transonic flow $\left. \right\} \dfrac{\partial T}{\partial y}, \dfrac{\partial \varrho}{\partial y}$ are $O(1)$.

Case (i) corresponds to large temperature changes through the computational domain driven by large temperature differences between the wall and freestream or by significant compressibility due to the motion (large Mach number). Case (ii) corresponds to smaller temperature changes through the computational domain driven solely by compressibility. As the Mach number is reduced, in the absence of external heating, derivatives of ϱ and T reduce to zero.

For case (i) all terms in (16.7) are $O(1)$ and must be retained. In (16.8) all terms on the left-hand side are $O(1)$. On the right-hand side of (16.8) τ_{xx} is $O(1)$. The nondimensional viscosity behaves approximately like the nondimensional temperature. The shear stress component $\tau_{xy} = \mu(\partial u/\partial y + \partial v/\partial x)$, which is $O(L/\delta)$ from the $\partial u/\partial y$ contribution. By comparison $\partial v/\partial x$, which is $O(\delta/L)$, can be neglected. Consequently $(1/\mathrm{Re})\partial(\mu\partial u/\partial y)/\partial y$ is the only one term that need be retained on the right-hand side of (16.8).

On the left-hand side of (16.9) all terms are $O(\delta/L)$. On the right-hand side $\tau_{yx} \approx \mu\partial u/\partial y$ and parts of other terms can be amalgamated when converted to equivalent velocity derivatives. When combined with $(1/\mathrm{Re})$ many of the terms on the right-hand side of (16.9) are $O((\delta/L)^2)$.

On the left-hand side of (16.10) all terms are $O(1)$ except $v\partial p/\partial y$ which is $O((\delta/L)^2)$. However, this term is retained as it may become $O(1)$ in the inviscid region. On the right-hand side of (16.10) $\partial(k\partial T/\partial x)/\partial x/\mathrm{PrRe}$ can be neglected and the only term in the viscous dissipation that makes an $O((\delta/L))$ contribution is $\mu(\partial u/\partial y)^2/\mathrm{Re}$. Consequently, as discussed above, the following reduced Navier–Stokes equations are obtained:

$$\frac{\partial}{\partial x}(\varrho u) + \frac{\partial}{\partial y}(\varrho v) = 0 \ , \tag{16.12}$$

$$\varrho u \frac{\partial u}{\partial x} + \varrho v \frac{\partial u}{\partial y} + \frac{\partial p}{\partial x} = \frac{1}{\mathrm{Re}} \frac{\partial}{\partial y}\left(\mu \frac{\partial u}{\partial y}\right) \ , \tag{16.13}$$

$$\varrho u \frac{\partial v}{\partial x} + \varrho v \frac{\partial v}{\partial y} + \frac{\partial p}{\partial y} = \frac{1}{\mathrm{Re}}\left[\frac{4}{3}\frac{\partial}{\partial y}\left(\mu \frac{\partial v}{\partial y}\right) + \frac{\mu}{3}\frac{\partial^2 u}{\partial x \partial y}\right] \ , \tag{16.14}$$

$$\varrho\left(u\frac{\partial T}{\partial x} + v\frac{\partial T}{\partial y}\right) - (\gamma - 1)M_\infty^2\left(u\frac{\partial p}{\partial x} + v\frac{\partial p}{\partial y}\right)$$

$$= \frac{1}{\mathrm{PrRe}}\frac{\partial}{\partial y}\left(k\frac{\partial T}{\partial y}\right) + \frac{(\gamma - 1)M_\infty^2}{\mathrm{Re}}\mu\left(\frac{\partial u}{\partial y}\right)^2 \ . \tag{16.15}$$

These equations retain all the 'inviscid' terms in (16.7–11) and are appropriate when large temperature changes are expected in the computational domain.

For the second case when an adiabatic wall is used in subsonic or transonic flow the major difference is in the right-hand side of (16.14). It may be noted that $v\partial\varrho/\partial y$ in (16.7) is small when $\partial\varrho/\partial y$ is $O(1)$. Consequently if $\mathscr{D}$ in τ_{yy} is substituted from (16.7) the following equation replaces (16.14):

$$\varrho u \frac{\partial v}{\partial x} + \varrho v \frac{\partial v}{\partial y} + \frac{\partial p}{\partial y} = \frac{1}{\mathrm{Re}}\left[\frac{\partial}{\partial y}\left(\mu \frac{\partial v}{\partial y}\right) - \frac{\mu}{3}\left(\frac{\partial u}{\partial y}\right)\left(\frac{\partial \varrho}{\partial x}\right)\Big/\varrho\right] \ . \tag{16.16}$$

For the second case, even though $v\partial T/\partial y$ and $v\partial p/\partial y$ are small in (16.10) these terms are retained to recover the Euler equations in the inviscid region.

Ignoring the temperature dependence of the viscosity and the density variation the x-momentum equation is the same for compressible flow (16.13) as for incompressible flow (16.5). The terms on the right-hand sides of (16.14 and 16) are $O((\delta/L)^2)$ if $1/\text{Re}$ is assumed to be $O((\delta/L)^3)$. Consequently, as with the right-hand side of (16.6), terms on the right-hand sides of (16.14 and 16) can be dropped. However, to $O((\delta/L)^2)$, (16.14 and 16) can be put in a form equivalent to (16.6), i.e.

$$\varrho u \frac{\partial v}{\partial x} + \varrho v \frac{\partial v}{\partial y} + \frac{\partial p}{\partial y} = \left(\frac{1}{\text{Re}}\right)\frac{\partial^2 v}{\partial y^2} . \tag{16.17}$$

With the exception of the viscous dissipation term in (16.15), all dissipative terms in (16.13, 15, 17) have the same form. It is apparent from the above examples that the RNS equations differ from the full Navier–Stokes equations *only* in the dissipative terms retained.

16.1.2 Fourier Analysis for Qualitative Solution Behaviour

If the solution to the reduced Navier–Stokes equations is to be obtained by a single march in the x-direction it is necessary that the system of governing equations, e.g. (16.4–6), be non-elliptic with respect to the x-direction. This could be determined, using the techniques of Sect. 2.1.4, after constructing an equivalent first-order system of equations, in place of (16.4–6).

However, more information can be deduced about the qualitative solution behaviour by introducing a complex Fourier series representation for the dependent variables. This approach can be illustrated by a suitably simplified form of the energy equation,

$$u \frac{\partial T}{\partial x} + v \frac{\partial T}{\partial y} - \delta \frac{\partial^2 T}{\partial x^2} - \varepsilon \frac{\partial^2 T}{\partial y^2} = 0 , \tag{16.18}$$

where u and v are constants of $O(1)$ but may be positive or negative. The parameters δ and ε are positive constants, with $\delta, \varepsilon \ll 1$. Equation (16.18) is a steady two-dimensional transport equation (Sect. 9.5). A comparison with (2.1 and 2) indicates that (16.18) is elliptic. Typical boundary conditions are shown in Fig. 16.3.

It is assumed that the solution of (16.18) can be represented by a complex Fourier series, i.e.

$$T = \frac{1}{4\pi^2} \sum_{k=-\infty}^{\infty} \sum_{j=-\infty}^{\infty} \hat{T}_{kj} \exp[i(\sigma_x)_j x]\exp[i(\sigma_y)_k y] . \tag{16.19}$$

However, since (16.18) is linear, only one component of (16.19) need be considered to determine the qualitative behaviour, i.e.

$$T = \frac{\hat{T}}{4\pi^2} \exp(i\sigma_x x) \exp(i\sigma_y y) , \tag{16.20}$$

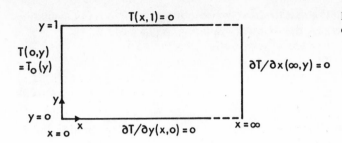

Fig. 16.3. Typical boundary conditions for (16.18)

where $\hat{T}$ can be interpreted as the Fourier transform of T. Substitution into (16.18) produces a polynomial in σ_x and σ_y

$$\delta\sigma_x^2 + iu\sigma_x + \varepsilon\sigma_y^2 + iv\sigma_y = 0 \ . \tag{16.21}$$

Equation (16.21) is sometimes referred to as the symbol of (16.18). Equation (16.21) will be used to obtain the value of σ_x for any real choice of σ_y. Substitution into (16.20) indicates the corresponding solution behaviour. Whether a particular mode present in (16.20) will contribute to the actual solution depends on the boundary conditions. The solution of (16.21) is

$$\sigma_x = -i\frac{u}{2\delta} \pm i\frac{u}{2\delta}\left(1 + 4\delta\frac{\varepsilon\sigma_y^2 + iv\sigma_y}{u^2}\right)^{1/2} \ .$$

Since $\delta \ll 1$, this can be simplified to

$$\sigma_x = -i\frac{u}{2\delta} \pm i\frac{u}{2\delta}\left(1 + 2\delta\frac{\varepsilon\sigma_y^2 + iv\sigma_y}{u^2}\right) \ .$$

Thus

$$\sigma_x = i\frac{\varepsilon\sigma_y^2}{u} - \frac{\sigma_y v}{u} \quad \text{or} \quad -i\left(\frac{u}{\delta} + \frac{\varepsilon\sigma_y^2}{u}\right) + \sigma_y\frac{v}{u} \ . \tag{16.22}$$

Given that σ_y is real, the first root indicates that the behaviour of the T solution, from (16.20), is oscillatory in x from $-\sigma_y v/u$ and has a diminishing exponential behaviour in x from $i\varepsilon\sigma_y^2/u$, if u is positive. If u is negative there is an exponential growth. The second root indicates an oscillatory solution that grows exponentially if u is positive. Thus considering both roots it is clear that T will be oscillatory in x and grow exponentially whatever the sign of u. However, since (16.18) is elliptic the boundary condition at $x = \infty$ is available to prevent the exponentially growing mode appearing in the actual solution of (16.18), subject to the boundary conditions indicated in Fig. 16.3.

The approximation that produced the reduced Navier–Stokes equations, Sect. 16.1.1, is equivalent to neglecting $\delta\partial^2 T/\partial x^2$ in (16.18). The resulting equation is parabolic in x (Sect. 2.3) so that the boundary condition at $x = \infty$ is not available. Introducing (16.20) produces the following equation in place of (16.21):

$$i u \sigma_x + \varepsilon \sigma_y^2 + i v \sigma_y = 0 \quad \text{or}$$

$$\sigma_x = i \frac{\varepsilon \sigma_y^2}{u} - \frac{\sigma_y v}{u} , \tag{16.23}$$

which coincides with the first root in (16.22). The corresponding solution for T, from (16.20), will have a diminishing oscillatory behaviour in x as long as u and ε have the same sign. If not, an exponentially growing oscillation will occur, which will preclude the use of a single march in the positive x direction to obtain the solution. The parameter ε is equivalent to the viscosity or thermal conductivity, which are always positive. Therefore for stable solutions it is necessary that u is positive. Since (16.18) with $\delta = 0$ is parabolic, positive u corresponds to information being convected in the positive time-like direction.

It is clear from the above example that the Fourier analysis of the governing equation gives quite precise information about what type of solution may be expected and which terms in the governing equation are responsible. In particular, in relation to the reduced Navier–Stokes equations, the present analysis indicates whether exponential growth in x is to be expected, which will prevent the use of a single downstream march to obtain a stable solution.

There is an obvious parallel with the use of Fourier analysis to determine the character of the solution of the discretised equations (Sect. 9.2.1). To the extent that the solution of the discrete problem converges to the solution of the continuous problem, it is expected that the results of the Fourier analysis applied directly to the governing partial differential equations will give comparable solution behaviour to the Fourier analysis of the discrete system of equations. However, where Fourier analysis is used to examine the stability of the discrete system of equations, i.e. the von Neumann stability analysis of Sect. 4.3, it is not known whether an apparent instability is a property peculiar to the discrete system of equations or a physical instability inherent in the solution of the governing equation and associated boundary conditions.

The present use of Fourier analysis identifies possible unstable growth patterns inherent in the governing partial differential equations. In principle the physical boundary conditions could be introduced into the Fourier representation so that the resulting solution, equivalent to (16.19), would constitute a valid Fourier approximation of the true solution. This is essentially the approach used to study analytically the stability of various flow phenomena, e.g. Stuart (1963) or Drazin and Reid (1981). However such a comprehensive approach goes beyond what is required to identify suitable reduced forms of the Navier–Stokes equations.

The present use of Fourier analysis may be compared with the traditional characteristic analysis of the governing partial differential equation (Chap. 2). In the traditional characteristics analysis only the highest derivatives are retained and the governing equation is reduced to characteristic form, i.e. a characteristic polynomial is obtained, e.g. (2.36). It may be noted that if all but the highest derivatives are suppressed in the present Fourier analysis, the resulting polynomial in σ_x, σ_y, e.g. (16.21) is identical with the characteristic form (also see Sect. 2.1.5). Thus the characteristic form of (16.18) can be obtained from (16.21) as

$$\delta\sigma_x^2 + \varepsilon\sigma_y^2 = 0 \quad ,$$

which has imaginary roots, so that (16.18) is classified as an elliptic partial differential equation.

However, in the present application of determining whether possible exponential growth modes are present in the governing equation, Fourier analysis is preferred to characteristics analysis for the following reasons:

i) It takes into account all terms in the governing equations, not just the highest derivatives.

ii) The contribution of specific terms in the governing equation to possible exponential growth can be identified directly.

iii) The analysis is more robust, in that the possible failure to produce a result through degenerate systems (Sect. 2.1.4) is avoided.

iv) The solution of the eigenvalue problem, e.g. (16.21) has more direct physical significance than the solution of the characteristic polynomial, e.g. (2.36).

16.1.3 Qualitative Solution Behaviour of the Reduced Navier–Stokes Equations

The various approximations to the Navier–Stokes equations can be examined, in a similar way to (16.18), after local linearization. That is u and v in the convective parts of (16.5 and 6) are assumed frozen at their local value. Thus the present analysis takes no account of any behaviour arising from nonlinear interactions. From the similarity of (16.5 and 6) with (16.18), when $\delta = 0$, it might be expected that the solution to the reduced Navier–Stokes equations will demonstrate comparable streamwise growth/decay patterns to those of (16.18) with $\delta = 0$. Whether this is, in fact, the case or not will be indicated below.

The application of Fourier analysis, as in Sect. 16.1.2, to the RNS equations, Sect. 16.1.1, implies consideration of a system of equations rather than a scalar equation, e.g. (16.18). The extension of Fourier analysis to systems of equations will be illustrated by starting with the incompressible Navier–Stokes equations for two-dimensional steady flow (16.1–3). In place of (16.20) it will be assumed that

$$u \sim \hat{u} \exp(i\sigma_x x) \exp(i\sigma_y y) \quad ,$$

$$v \sim \hat{v} \exp(i\sigma_x x) \exp(i\sigma_y y) \quad , \tag{16.24}$$

$$p \sim \hat{p} \exp(i\sigma_x x) \exp(i\sigma_y y) \quad ,$$

where $\sim$ is taken to mean that the solutions will be of the form indicated. Equations (16.24) are substituted into the frozen form of (16.1–3) producing

$$\begin{bmatrix} i\sigma_x & i\sigma_y & 0 \\ i\Lambda + (\sigma_x^2 + \sigma_y^2)/\mathrm{Re} & 0 & i\sigma_x \\ 0 & i\Lambda + (\sigma_x^2 + \sigma_y^2)/\mathrm{Re} & i\sigma_y \end{bmatrix} \begin{bmatrix} \hat{u} \\ \hat{v} \\ \hat{p} \end{bmatrix} = 0 \quad , \tag{16.25}$$

where $\Lambda \equiv u\sigma_x + v\sigma_y$.

To obtain the solution to a homogeneous system of equations like (16.25) it is necessary that $\det[\ ]=0$. For (16.25) this generates the following polynomial in σ_x:

$$(\sigma_x^2+\sigma_y^2)[i(u\sigma_x+v\sigma_y)+\frac{1}{\mathrm{Re}}(\sigma_x^2+\sigma_y^2)]=0 \ . \tag{16.26}$$

The second factor has the same form as (16.21) and has roots

$$\sigma_x=i\sigma_y^2/(u\,\mathrm{Re})-\sigma_y v/u \quad\text{and}\quad -i[u\,\mathrm{Re}+\sigma_y^2/(u\,\mathrm{Re})]+\sigma_y v/u \ .$$

Exponential growth is produced by the first root if u is negative and by the second root if u is positive. The first factor has roots $\sigma_x=\pm i\sigma_y$. The negative imaginary root will cause exponential growth in x, after substitution into (16.24).

The system of equations (16.1–3) is elliptic. This can be established by introducing auxiliary variables for the second-derivative term, and constructing an equivalent system of first-order partial differential equations. This system is analysed using the technique described in Sect. 2.1.4, leading to the characteristic polynomial (2.39).

The present Fourier analysis produces the identical polynomial if the lower-order terms, $i(u\sigma_x+v\sigma_y)$, are suppressed in (16.26). This is consistent with the classification of partial differential equations being based on the highest derivative in each independent variable.

In (16.26) it will be seen that imaginary roots still occur when $i(u\sigma_x+v\sigma_y)$ is suppressed, confirming that the system (16.1–3) is elliptic. Consequently boundary conditions are required on all boundaries, Sect. 2.4. The boundary conditions prevent the exponential growth, implied by the roots of (16.26), appearing in the solution to (16.1–3).

If the above analysis, from (16.24) onwards, is applied to the reduced Navier–Stokes equations (16.4–6) the following polynomial is obtained in place of (16.26):

$$(\sigma_x^2+\sigma_y^2)[i(u\sigma_x+v\sigma_y)+(1/\mathrm{Re})\sigma_y^2]=0 \ . \tag{16.27}$$

The neglect of the streamwise diffusion terms, $\partial^2 u/\partial x^2$ and $\partial^2 v/\partial x^2$, has a similar effect on the second factor of (16.26) as occurred for the model problem leading to (16.21). That is, the following root is obtained from the second factor in (16.27):

$$\sigma_x=i\frac{\sigma_y^2}{u\,\mathrm{Re}}-\sigma_y\frac{v}{u} \ . \tag{16.28}$$

As long as u is positive, no modes with exponential growth in x occur. Thus the reduced form of the Navier–Stokes equations is effective in suppressing the exponentially-growing mode in the convection, diffusion operator.

However, the reduced form of the Navier–Stokes equations has no impact on the first factor in (16.26) which is retained in full in (16.27). The first factor in (16.27) has a negative imaginary root and will produce exponential growth in the x direction. Suppression of the lower-order terms, $i(u\sigma_x+v\sigma_y)$, in (16.27) does not affect the imaginary roots of the first factor. This indicates that the reduced

Navier–Stokes equations (16.4–6) are mixed elliptic/parabolic, with the elliptic behaviour coming from the first factor in (16.27) and the parabolic behaviour coming from the second factor.

Any elliptic behaviour is sufficient to upset the time-like, single march strategy that motivates the interest in reduced forms of the Navier–Stokes equations. If the contributions to the first factor in (16.27) are traced back to the governing equations (16.4–6) it will be seen that the source of the elliptic behaviour is the interaction of the pressure terms in the momentum equations with the terms in the continuity equation. If there were some way of suppressing $\partial p/\partial x$ or $\partial p/\partial y$ in the governing equations, the elliptic behaviour would be avoided.

The Fourier analysis of the reduced Navier–Stokes equations for compressible flow (16.12–15) leads to a more complicated characteristic polynomial, in place of (16.27), which cannot be interpreted so precisely. An intermediate category, suitable for transonic Mach numbers, can be considered by starting from (16.12, 13, 17) and neglecting the dissipative terms in (16.15) as in Sect. 18.1.2. Consequently (16.15) can be replaced by (11.104), which can be written nondimensionally as

$$\frac{1+\gamma M_\infty^2 p}{\varrho} = \{1+0.5(\gamma-1)M_\infty^2[1-(u^2+v^2)]\} \ . \tag{16.29}$$

This equation is used to eliminate p, in favour of ϱ, u and v, from (16.13 and 17). This approximation is consistent with the observation that for transonic Mach numbers the temperature variation in the computational domain will be small if an adiabatic wall boundary condition is imposed.

In terms of ϱ, u and v the governing equations are

$$u\frac{\partial \varrho}{\partial x}+v\frac{\partial \varrho}{\partial y}+\varrho\frac{\partial u}{\partial x}+\varrho\frac{\partial v}{\partial y}=0 \ , \tag{16.30}$$

$$\frac{a^2}{\gamma}\frac{\partial \varrho}{\partial x}+\left(\frac{\varrho u}{\gamma}\right)\frac{\partial u}{\partial x}+\varrho v\frac{\partial u}{\partial y}-\frac{(\gamma-1)}{\gamma}\varrho v\frac{\partial v}{\partial x}-\frac{1}{Re}\frac{\partial^2 u}{\partial y^2}=0 \ , \tag{16.31}$$

$$\frac{a^2}{\gamma}\frac{\partial \varrho}{\partial y}-\frac{(\gamma-1)}{\gamma}\varrho u\frac{\partial u}{\partial y}+\varrho u\frac{\partial v}{\partial x}+\left(\frac{\varrho v}{\gamma}\right)\frac{\partial v}{\partial y}-\frac{1}{Re}\frac{\partial^2 v}{\partial y^2}=0 \ . \tag{16.32}$$

In the above equations the dependence of the viscosity μ on temperature is neglected and the nondimensional sound speed is defined by

$$a^2=\frac{1}{\varrho M_\infty^2}+\frac{\gamma p}{\varrho} \ . \tag{16.33}$$

If undifferentiated terms like a^2, ϱu, ϱv, etc., in (16.30–32) are frozen and a complex Fourier series introduced for ϱ, u and v, as in (16.24), the following polynomial in σ_x is obtained:

$$\frac{\varrho}{\gamma}\left(i\varrho\Lambda+\frac{\sigma_y^2}{Re}\right)\left[\Lambda^2-a^2(\sigma_x^2+\sigma_y^2)-i\frac{\gamma\sigma_y^2}{\varrho Re}(u\sigma_x+v\sigma_y)\right]=0 \ , \tag{16.34}$$

where $\Lambda=u\sigma_x+v\sigma_y$. $\tag{16.35}$

For external flow problems far from an isolated body the reduced Navier–Stokes equations coincide with the Euler equations governing inviscid flow. This situation corresponds to setting $\text{Re} = \infty$ in (16.34), which reduces to

$$\left(i\varrho^2 \frac{\Lambda}{\gamma} \right)[(u\sigma_x + v\sigma_y)^2 - a^2(\sigma_x^2 + \sigma_y^2)] = 0 .$$ (16.36)

The first factor has roots $\sigma_x/\sigma_y = -v/u$ and the second factor has roots

$$\frac{\sigma_x}{\sigma_y} = -\frac{uv}{u^2 - a^2} \pm \frac{a^2(M^2 - 1)^{1/2}}{u^2 - a^2} ,$$ (16.37)

where the local Mach number, $M = (u^2 + v^2)^{1/2}/a$. It is apparent from (16.37) that if the flow is locally subsonic, i.e. $M < 1$, a negative imaginary root will arise leading to exponential growth in the marching (x) direction. This corresponds to the inviscid equations being elliptic if $M < 1$, but hyperbolic if $M > 1$.

For the general viscous case (16.34) is directly applicable. The first factor has the same behaviour as the second factor in (16.27). That is, no exponentially growing modes are produced as long as u is positive. The second factor is quadratic in σ_x producing the following values for the roots:

$$\frac{\sigma_x}{\sigma_y} = -\frac{uv}{u^2 - a^2} + \frac{i\gamma u\sigma_y}{2\varrho\text{Re}(u^2 - a^2)}$$

$$\pm \frac{a^2}{u^2 - a^2}\left[(M^2 - 1) - \left(\frac{0.5\gamma u\sigma_y}{\varrho\text{Re}a^2} \right)^2 - \frac{i\gamma v\sigma_y}{\varrho\text{Re}a^2} \right]^{1/2} .$$ (16.38)

For large values of Re, [ ]$^{1/2}$ is dominated by the $(M^2 - 1)$ terms leading to strongly growing exponential behaviour for subsonic flow.

For flows with $u < a$, the second term on the left-hand side of (16.38) produces a weak elliptic behaviour, as does the last term in [ ]$^{1/2}$ for any flow velocity. Thus the viscous terms are producing weak exponential growth in the x direction, although of reducing magnitude as Re increases. However, this result might be modified for the equivalent turbulent form of the RNS equations.

It is seen that for both the inviscid and viscous cases, approximately subsonic flow will produce a strong exponentially growing mode in the x direction. However, if it were possible to suppress $\partial p/\partial x$ in (16.13), there would be no exponentially growing inviscid modes in x as long as u is positive.

For compressible flow with no restriction on the local Mach number the reduced Navier–Stokes equations are given by (16.12–15). The pressure can be eliminated in favour of the temperature and density via the nondimensional equation of state

$$1 + \gamma M_\infty^2 p = \varrho T .$$ (16.39)

Complex Fourier series, as in (16.24), are introduced for ϱ, u, v and T. Substitution into the governing equations generates the following polynomial in place of (16.34):

$$-\frac{\varrho}{\gamma}\left(i\varrho\Lambda+\frac{\sigma_y^2}{\text{Re}}\right)\left(i\varrho\Lambda+\frac{\gamma\sigma_y^2}{\text{PrRe}}\right)[\Lambda^2-a^2(\sigma_x^2+\sigma_y^2)]$$

$$+i\frac{\varrho^2a^2}{\text{PrRe}}\left(\frac{\gamma-1}{\gamma}\right)\Lambda\sigma_y^2(\sigma_x^2+\sigma_y^2)+\varrho^2\frac{\partial u}{\partial y}\left(\frac{\gamma-1}{\gamma\text{Re}}\right)\Lambda^2\sigma_x\sigma_y=0 \ . \tag{16.40}$$

In deriving (16.40) isolated terms containing $1/\text{Re}^2$ have been ignored as has the temperature dependence of the viscosity μ. The last term on the right-hand side contains $\partial u/\partial y$ which comes from the nonlinear dissipation term in (16.15).

For large values of Re it can be seen that the solution behaviour will be dominated by the product of the first three factors in (16.40). The first two factors produce the same behaviour as the second factor in (16.27) and will not generate exponential growth in x as long as u is positive. However, the third factor is the same as the second factor in (16.36) and leads to the same result. That is, for subsonic flow exponentially growing modes in the x direction will occur. The inclusion of the other terms in (16.40) is not expected to significantly change this conclusion.

As with the transonic flow case governed by (16.30–32), if $\partial p/\partial x$ (16.13) can be suppressed the goal of obtaining stable solutions with a single spatial march can be achieved. Of course *arbitrary* suppression of the $\partial p/\partial x$ term will lead to non-physical solutions of the resulting system of equations.

The results for the various categories of reduced Navier–Stokes equations are summarised in Table 16.1. The broad conclusion is that the neglect of streamwise dissipative terms is effective in suppressing exponential streamwise growth associated with convection diffusion interactions. However, it does not overcome the essentially elliptic behaviour associated with pressure terms for subsonic flow conditions.

Table 16.1. Dominant solution behaviour for reduced Navier–Stokes equations

Incompressible $M\sim 0$	Subsonic $0<M<1$	Supersonic $M>1$
Elliptic behaviour due to pressure/continuity interaction	Strong elliptic behaviour due to pressure. Weak elliptic behaviour from pressure/viscous interaction	Hyperbolic behaviour from inviscid terms Weak elliptic behaviour from pressure/viscous interaction

If there is some additional mechanism for "neutralising" the elliptic influence of the $\partial p/\partial x$ term in the x-momentum equation, the goal of obtaining an accurate solution in a single downstream march can be achieved throughout the range of Mach numbers. Many of the techniques described in Sects. 16.2 and 16.3 will, to a greater or lesser extent, exercise control over the influence of the $\partial p/\partial x$ term.

16.1.4 THRED: Thermal Entry Problem

In Sect. 9.5.2 the problem of a 'cold' fluid entering a 'hot' two-dimensional duct is solved using a pseudo-transient formulation to provide the steady spatial temperature distribution for a given velocity distribution. This same problem will be utilised here to illustrate the mechanics of setting up and solving an equivalent "reduced" formulation.

Using the nondimensionalisation given in (9.91), this problem is governed by the low-speed two-dimensional energy equation

$$\frac{\partial}{\partial x}(uT) + \frac{\partial}{\partial y}(vT) - \alpha_x \frac{\partial^2 T}{\partial x^2} - \alpha_y \frac{\partial^2 T}{\partial y^2} = 0 \ , \tag{16.41}$$

where $\alpha_x = 10/(\text{PrRe}^2)$ and $\alpha_y = 1.6/\text{Pr}$. The boundary conditions for (16.41) are (Fig. 9.12)

$$T(0, y) = 0 \quad \text{on} \quad x = 0 \ , \qquad \frac{\partial T}{\partial x} = 0 \quad \text{on} \quad x = x_{\max} \ ,$$

and

$$T(x, \pm 1) = 1 \quad \text{on} \quad y = \pm 1 \ . \tag{16.42}$$

Except very near the duct entrance, $x = 0$, the longitudinal temperature diffusion is very much less than the transverse temperature diffusion; consequently $\alpha_x \partial^2 T/\partial x^2$ can be dropped from (16.41). The resulting "reduced" equation is written as

$$\frac{\partial}{\partial x}(uT) = \alpha_y \frac{\partial^2 T}{\partial y^2} - \frac{\partial}{\partial y}(vT) \ . \tag{16.43}$$

Since (16.43) is parabolic in x, no boundary condition is allowed at $x = x_{\max}$. The other boundary conditions are given by (16.42).

This problem is to be solved in the computational domain $0 \le x \le 2.00$, $-1.0 \le y \le 1.0$ by marching in the positive x direction starting from the known solution at $x = 0$. Thus x has a time-like role and the temperature distribution $T(0, y)$ provides the 'initial' conditions.

The group finite element formulation (Sect. 10.3) with linear interpolation is introduced in the y direction and combined with Crank–Nicolson discretisation for the x derivatives in (16.43). The resulting algebraic equations can be written

$$M_y \frac{\Delta(uT)_j^{n+1}}{\Delta x} = 0.5\alpha_y L_{yy}(T_j^n + T_j^{n+1}) - 0.5 L_y[(vT)_j^n + (vT)_j^{n+1}] \ , \tag{16.44}$$

where n, j are the grid indices in the x and y directions, respectively. The operators L_y and L_{yy} are the three-point centred difference operators

$$L_y = \frac{\{1, 0, -1\}^T}{2\Delta y} \ , \qquad L_{yy} = \frac{\{1, -2, 1\}^T}{\Delta y^2} \ , \tag{16.45}$$

and M_y, the mass operator, is defined for both finite difference and finite element formulations:

$$M_y = \left\{ \frac{1}{6}, \frac{2}{3}, \frac{1}{6} \right\}^T , \quad \text{for the finite element formulation,}$$

$$M_y = \{0, 1, 0\}^T, \quad \text{for the finite difference formulation.} \tag{16.46}$$

It is assumed that the velocity field (u, v) is known. Consequently the following linear system of equations for $\Delta T_j^{n+1} (\equiv T_j^{n+1} - T_j^n)$ can be constructed, after appropriate linearisations about the solution T_j^n:

$$[M_y u_j^{n+1} + 0.5 \Delta x (L_y v_j^{n+1} - \alpha_y L_{yy})] \Delta T_j^{n+1}$$
$$= \Delta x \alpha_y L_{yy} T_j^n - \Delta x L_y (v_j^{n+1/2} T_j^n) - M_y T_j^n \Delta u_j^{n+1} . \tag{16.47}$$

To compare with the semi-exact solution of Brown (1960) the fully developed velocity distribution is assumed,

$$u = 1.5(1 - y^2) , \quad v = 0 . \tag{16.48}$$

For this choice of velocity distribution, which is independent of axial (x) location, (16.47) simplifies to

$$[M_y u_j + 0.5 \Delta x (L_y v_j - \alpha_y L_{yy})] \Delta T_j^{n+1} = \Delta x [\alpha_y L_{yy} T_j^n - L_y (v_j T_j^n)] . \tag{16.49}$$

Equation (16.49) is implemented in program THRED (Fig. 16.4). The main parameters appearing in program THRED are described in Table 16.2. To avoid the discontinuity in $T(0, \pm 1)$ implied by (16.42) the following 'initial' data on $x = 0$ is prescribed in program THRED:

$$T(0, y) = y^{32} . \tag{16.50}$$

The computed centreline solution $(y = 0)$ is compared with the semi-exact solution of Brown (1960). Brown obtained a separation of variables solution of (16.43 and 48) based on an exponentially decaying solution in x and an eigenvalue/eigenfunction expansion in y. The first ten terms of Brown's solution, specialised to the centre-line, are evaluated in the subroutine TEXCL (Fig. 16.5) to provide the semi-exact solution TEX.

A typical solution, with $\Delta x = 0.05$, $\Delta y = 0.20$, produced by THRED is shown in Fig. 16.6. The computed temperature solution is symmetric about $y = 0$ so that only the temperature solution corresponding to $-1 \leq y \leq 0$ is shown. The extreme right-hand column of T is the centre-line $(y = 0)$ value and may be compared with the semi-exact value TEX. For the relatively coarse-grid solution shown in Fig. 16.6 it is clear that the solution displays an oscillation close to $x \approx 0$, $y \approx -1.0$. This oscillation is associated with a rapid change in the boundary condition for T close to $(0, \pm 1.0)$. The amplitude of the oscillation reduces with increasing x. For a finer grid in x or y the oscillation does not occur. .

It may be noted that solutions, e.g. AF-FEM, produced by program THERM (Fig. 9.13) on an 11×11 grid $(\Delta x = 0.20, \Delta y = 0.20)$ do not produce a significant oscillation adjacent to $(0, \pm 1.0)$. However, program THERM is

```
 1 C     THRED SOLVES THE REDUCED FORM OF THE THERMAL ENTRY PROBLEM
 2 C     BY C.N. MARCHING
 3 C
 4       DIMENSION T(41),DT(65),U(41),V(41),R(65),B(5,65),EM(3)
 5      1,ALF(10),DYFL(10)
 6       DATA  ALF/1.6815953,5.6698573,9.6682425,13.6676614,17.6673736,
 7      121.6672053,25.6670965,29.6670210,33.6669661,37.6664327/
 8       DATA DYFL/-0.9904370,1.1791073,-1.2862487,1.3620196,-1.4213257,
 9      11.4704012,-1.5124603,1.5493860,-1.5823802,1.6122503/
10       OPEN(1,FILE='THRED.DAT')
11       OPEN(6,FILE='THRED.OUT')
12       READ(1,1)NY,NXMAX,ME,DX,DXP,XMAX,PR
13     1 FORMAT(3I5,4E10.3)
14 C
15       IF(ME .EQ. 1)WRITE(6,2)
16       IF(ME .EQ. 2)WRITE(6,3)
17     2 FORMAT(' REDUCED THERMAL ENTRY PROBLEM BY C.N.-FEM',/)
18     3 FORMAT(' REDUCED THERMAL ENTRY PROBLEM BY C.N.-FDM',/)
19       WRITE(6,4)NY,NXMAX,DX,XMAX,PR
20     4 FORMAT(' NY=',I3,' NXMAX=',I5,' DX=',E10.3,' XMAX=',F6.3,' PR=',
21      1F6.3,/)
22 C
23       NYP = NY - 1
24       NYH = NY/2 + 1
25       NYPP = NY - 2
26       ANY = NYP
27       DY = 2./ANY
28       ALY = 1.6/PR
29       CA = 0.5*DX/DY
30       CCA = ALY*DX/DY/DY
31       IF(ME .EQ. 1)EM(1) = 1./6.
32       IF(ME .EQ. 2)EM(1) = 0.
33       EM(2) = 1. - 2.*EM(1)
34       EM(3) = EM(1)
35 C
36 C     SET U,V AND T INITIAL DATA
37 C
38       DO 5 K = 1,NY
39       KM = K - 1
40       AK = KM
41       Y = -1. + AK*DY
42       U(K) = 1.5*(1. - Y*Y)
43       V(K) = 0.
44     5 T(K) = Y**32
45 C
46 C     SET UP TRIDIAGONAL COEFFICIENTS AND FACTORISE B
47 C
48       DO 6 K = 2,NYP
49       KM = K - 1
50       KP = K + 1
51       B(1,KM) = 0.
52       B(2,KM) = EM(1)*U(KM) - 0.5*CA*V(KM) - 0.5*CCA
53       B(3,KM) = EM(2)*U(K) + CCA
54       B(4,KM) = EM(3)*U(KP) + 0.5*CA*V(KP) - 0.5*CCA
55       B(5,KM) = 0.
56     6 CONTINUE
57       B(2,1) = 0.
58       B(4,KM) = 0.
59 C
```

Fig. 16.4. Listing of program THRED

```
60          CALL BANFAC(B,NYPP,1)
61 C
62          X = 0.
63          XPR = 0.
64          NCT = 0
65          SUMT = 0.
66        7 NCT = NCT + 1
67 C
68 C        GENERATE R.H.S.
69 C
70          DO 8 K = 2,NYP
71          KM = K - 1
72          KP = K + 1
73        8 R(KM) = CCA*(T(KM)-2.*T(K)+T(KP)) - CA*(V(KP)*T(KP)-V(KM)*T(KM))
74 C
75          CALL BANSOL(R,DT,B,NYPP,1)
76 C
77          DO 9 K = 2,NYP
78        9 T(K) = T(K) + DT(K-1)
79          X = X + DX
80 C
81 C        EXACT C/L SOLUTION
82 C
83          CALL TEXCL(X,TEX,PR,ALF,DYFL)
84 C
85          DMP = T(NYH) - TEX
86          IF(NCT .GT. 2)SUMT = SUMT + DMP*DMP
87          IF(X .LT. XPR)GOTO 11
88          WRITE(6,10)X,(T(K),K=1,NYH),TEX
89       10 FORMAT(' X=',F4.2,' T=',6F6.3,'  TEX=',F6.3)
90          XPR = XPR + DXP - 0.0001
91       11 IF(X .GE. XMAX)GOTO 12
92          IF(NCT .GE. NXMAX)GOTO 12
93          GOTO 7
94       12 ANCT = NCT - 2
95          RMS = SQRT(SUMT/ANCT)
96          WRITE(6,13)NCT,RMS
97       13 FORMAT(' NCT=',I5,' RMS=',E10.3)
98       14 STOP
99          END
```

Fig. 16.4. (cont.) Listing of program THRED

based on the solution of (9.90), which includes the term $\partial^2 T/\partial x^2$. This term introduces a smoothing effect and prevents an oscillation adjacent to $(0, \pm 1.0)$ from occurring.

Centre-line solutions (RED-FEM) produced by program THRED are compared in Table 16.3 with those produced by program THERM and with the semi-exact solution by Brown (1960). The rms errors calculated for the RED-FEM solutions have omitted contributions from locations $x = 0$, Δx to be consistent with the procedure adopted for program THERM (Sect. 9.5.2). With sufficient grid refinement in the x direction the RED-FEM solution can be made more accurate than the approximate factorisation finite element method (AF-FEM) solution. However, of greater significance is the economy of the RED-FEM solution, arising from the single downstream march required to obtain the solution. The RED-FEM solution with $\Delta x = 0.05$ is an order-of-magnitude more economical than the AF-FEM with $\Delta x = 0.20$ while achieving about the same solution accuracy.

Table 16.2. Parameters used in program THRED

Variable	Description
ME	=1, linear finite element method
	=2, three-point finite difference method
NY	number of points in the y direction
NXMAX	maximum number of points in the x direction
DX, DY	$\Delta x, \Delta y$
DXP	Δx increment for printing temperature solution
XMAX	downstream extent of computational domain
PR	Prandtl number, Pr
ALY	$\alpha_y = 1.6/Pr$
EM	M_y
T; U, V	temperature, velocity components in the x, y directions
B	tridiagonal matrix; left-hand side of (16.49)
R	right-hand side of (16.49)
RMS	$\| T_{c/l} - TEX \|_{rms}$
TEXCL	calculates TEX for given x and Pr
ALF, DYFL	arrays required by subroutine TEXCL

```
1
2          SUBROUTINE TEXCL(X,TEX,PR,ALF,DYFL)
3  C
4  C       FOR GIVEN X AND PR COMPUTE EXACT CENTRE-LINE
5  C       TEMPERATURE DISTRIBUTION
6  C
7          DIMENSION ALF(10),DYFL(10)
8          ZD = -3.2*X/PR/3.0
9          TB = 0.
10         DO 1 I = 1,10
11         DUM = ZD*ALF(I)*ALF(I)
12         IF(DUM .LT. -20.)GOTO 1
13         DUM = EXP(DUM)
14         CF = -2./ALF(I)/DYFL(I)
15         TB = TB + CF*DUM
16       1 CONTINUE
17         TEX = 1. - TB
18         RETURN
19         END
```

Fig. 16.5. Listing of subroutine TEXCL

```
REDUCED THERMAL ENTRY PROBLEM BY C.N.-FEM

NY= 11 NXMAX=   50 DX=  .500E-01 XMAX= 2.000 PR=  .700

X= .05 T= 1.000 1.076  .477  .188  .074  .046  TEX=  .058
X= .20 T= 1.000  .701  .767  .614  .520  .495  TEX=  .493
X= .40 T= 1.000  .868  .917  .827  .801  .793  TEX=  .786
X= .60 T= 1.000  .937  .972  .924  .918  .913  TEX=  .909
X= .80 T= 1.000  .969  .992  .966  .967  .963  TEX=  .962
X=1.00 T= 1.000  .984  .999  .985  .987  .984  TEX=  .984
X=1.20 T= 1.000  .992 1.001  .993  .995  .993  TEX=  .993
X=1.40 T= 1.000  .995 1.001  .997  .998  .997  TEX=  .997
X=1.60 T= 1.000  .997 1.001  .998  .999  .999  TEX=  .999
X=1.80 T= 1.000  .999 1.001  .999 1.000  .999  TEX=  .999
X=2.00 T= 1.000  .999 1.000 1.000 1.000 1.000  TEX= 1.000
NCT=   41 RMS=  .350E-02
```

Fig. 16.6. Typical output from program THRED

Table 16.3. Centre-line solutions for thermal entry problem, $\Delta y = 0.20$

x	0.000	0.200	0.400	0.600	0.800	1.000	1.200	1.400	1.600	1.800	2.000	rms error
(Semi-) exact	0.000	0.493	0.786	0.910	0.962	0.984	0.993	0.997	0.999	1.000	1.000	—
AF-FEM $\Delta x = 0.20$	0.000	0.462	0.794	0.910	0.963	0.984	0.994	0.997	0.999	0.999	1.000	0.003
RED-FEM, $\Delta x = 0.050$	0.000	0.495	0.793	0.913	0.963	0.984	0.993	0.997	0.999	1.000	1.000	0.0035
RED-FEM, $\Delta x = 0.010$	0.000	0.497	0.789	0.912	0.963	0.984	0.993	0.997	0.999	1.000	1.000	0.0019

16.2 Internal Flow

Many internal flows, e.g. in pipes, diffusers, ducts, engine intakes, can be described accurately by using reduced forms of the Navier–Stokes equations. However, as indicated in Table 16.1, the basic form of the RNS equations is likely to be elliptic.

For steady internal flow the total mass flow past any downstream station is constant. This property provides an additional piece of information that can be exploited to construct a *non-elliptic* RNS system of equations. Examples of this will be provided in Sects. 16.2.1 and 16.2.2. For flows in highly-curved ducts an alternative strategy, based on splitting the transverse velocity field, will be described in Sect. 16.2.3. The split velocity approach still permits an accurate viscous solution to be obtained in a single downstream march.

Flow in straight pipes and ducts can be categorised into four types (Rubin et al. 1977) depending on the distance from the entrance, as indicated in Fig. 16.7. The Reynolds number associated with the duct flow is based on the hydraulic diameter D_h and the mean velocity U_m. The four categories are:

A. Immediate entrance flow, x_A is $O(D_h/\text{Re})$.
B. Entry region flow, x_B is $O(D_h)$.
C. Fully viscous flow, x_C is $O(D_h\text{Re})$.
D. Fully developed flow, $x_D \gg D_h\text{Re}$.

At the immediate entrance to the duct the flow behaviour demonstrates very severe velocity gradients adjacent to the duct lip, under the influence of viscosity reducing the inflow velocity to zero at the duct walls. A very fine local grid and the full Navier–Stokes equations are required to compute this domain accurately. In region B boundary layers develop on the duct walls and the flow field can be determined by a coupled inviscid flow (Chap. 14) and boundary layer flow (Chap. 15) analysis. Such a formulation is provided by Rubin et al. (1977). However, region B could also be computed accurately with an RNS formulation.

Sufficiently far downstream, region C (Fig. 16.7), the "boundary layers" merge so that the flow, everywhere in the duct cross-section, must be considered viscous. That is, the magnitudes of the viscous terms in the governing equations are

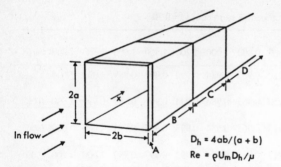

Fig. 16.7. Flow classification for flow in straight ducts

$D_h = 4ab/(a + b)$

$Re = \rho U_m D_h/\mu$

everywhere non-negligible. For this region a reduced form of the Navier–Stokes is appropriate to obtain the solution.

Even further downstream, region D, the flow behaviour becomes independent of downstream location x. In marching the reduced Navier–Stokes equations the downstream coordinate, x, has a time-like role. Consequently the solution to the reduced Navier–Stokes equations in region D is equivalent to the steady-state solution of a (pseudo-)transient formulation (Sect. 6.4).

For many internal flows the duct (or equivalent passage) terminates in regions B or C. The diffuser flows considered in Sect. 16.2.1 are in this category. For a fully developed flow, region D, the introduction of an obstruction in the interior or on the wall generates a new region C downstream of the obstruction. Further downstream a new region D occurs.

For internal flows in ducts whose centreline is curved, secondary flows are present and region C persists. A typical transverse streamline pattern (i.e. based on secondary velocity components v and w) is shown in Fig. 16.8. The duct axis is curving to the right of the local primary flow direction. For flows with mild duct axis curvature the techniques discussed in Sect. 16.2.2 are effective in making the

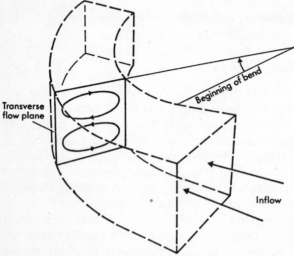

Fig. 16.8. Typical transverse flow pattern in a curved duct

governing equations non-elliptic with respect to the downstream direction. For more severe curvature the techniques described in Sect. 16.2.3 are appropriate.

As indicated in Sect. 16.1.3 the basic reduced Navier–Stokes equations are elliptic for subsonic flow due to the pressure interaction. Additional constraints or approximations to the RNS equations are required to produce a non-elliptic system. These are illustrated here for axisymmetric incompressible laminar flow in a pipe (i.e. a circular duct). The appropriate RNS equations, equivalent to (16.4–6) but in polar coordinates, are

$$\frac{\partial u}{\partial x}+\frac{\partial v}{\partial r}+\frac{v}{r}=0 \ , \tag{16.51}$$

$$u\frac{\partial u}{\partial x}+v\frac{\partial u}{\partial r}+\frac{\partial p}{\partial x}=\frac{1}{\text{Re}}\left(\frac{\partial^2 u}{\partial r^2}+\frac{1}{r}\frac{\partial u}{\partial r}\right) \tag{16.52}$$

$$u\frac{\partial v}{\partial x}+v\frac{\partial v}{\partial r}+\frac{\partial p}{\partial r}=\frac{1}{\text{Re}}\left(\frac{\partial^2 v}{\partial r^2}+\frac{1}{r}\frac{\partial v}{\partial r}-\frac{v}{r^2}\right). \tag{16.53}$$

For internal flows with small transverse velocity components the transverse variation of the pressure is small and its gradient in the marching direction (x) can be ignored in the corresponding momentum equation. Thus for axisymmetric flow the pressure can be split as

$$p=p_{c/l}(x)+p^c(x,r) \ , \tag{16.54}$$

where $p_{c/l}$ denotes the centreline pressure and p^c is a correction to account for the radial variation. Substitution into (16.52 and 53) indicates that $\partial p^c/\partial x$ is $O((\delta/L)^2)$ whereas the dominant terms in (16.52) are $O(1)$. Here δ is the thickness of the viscous layer ($=0.5D$, downstream of the merging of the viscous layers, Fig. 16.2), D is the local pipe diameter and L is a characteristic length along the axis of the pipe. Consequently $\partial p^c/\partial x$ is dropped and (16.52) becomes

$$u\frac{\partial u}{\partial x}+v\frac{\partial u}{\partial r}+\frac{dp_{c/l}}{dx}=\frac{1}{\text{Re}}\left[\frac{\partial^2 u}{\partial r^2}+\frac{1}{r}\left(\frac{\partial u}{\partial r}\right)\right] \ . \tag{16.55}$$

In the radial momentum equation (16.53) the pressure gradient is $\partial p^c/\partial r$ since $\partial p_{c/l}/\partial r=0$. Consequently the three governing equations contain four dependent variables.

However, an additional equation can be obtained by observing that, for steady flow, the overall mass flowrate $\dot{m}$ is constant. The mass flow rate is given by

$$\dot{m}=2\pi\varrho\int_0^R ur\,dr \ , \tag{16.56}$$

Thus $\partial\dot{m}/\partial x=0$, i.e.

$$\frac{\partial\dot{m}}{\partial x}=2\pi\varrho\int_0^R r\frac{\partial u}{\partial x}\,dr=0 \ , \tag{16.57}$$

which is a more useful form for combining with (16.51, 53 and 55). It may be noted that (16.57) is also obtained by integrating the continuity equation (16.51) over the pipe cross-section.

Equations (16.55 and 57) together provide u_j^{n+1} and $p_{c/l}^{n+1}$ as follows. Equation (16.55) is discretised in the x direction as

$$u^n \Delta u^{n+1}/\Delta x = J(u^{n+1/2}, v^{n+1/2}, r) - \Delta p_{c/l}^{n+1}/\Delta x \ , \tag{16.58}$$

where

$$J = \frac{1}{\mathrm{Re}} \left(\frac{\partial^2 u}{\partial r^2} + \frac{1}{r} \frac{\partial u}{\partial r} \right) - v \frac{\partial u}{\partial r} \ , \tag{16.59}$$

$$\Delta u^{n+1} = u^{n+1} - u^n \ , \quad u^{n+1/2} = 0.5(u^n + u^{n+1}) \quad \text{and} \quad v^{n+1/2} = 1.5 v^n - 0.5 v^{n-1} \ .$$

The indices n and j define the grid points in the x and r directions, respectively. With three-point centred difference expressions introduced for the r derivatives in (16.59), a linearisation of (16.58) about downstream location x^n as in Sect. 10.1.3 gives

$$\left(u^n - 0.5 \Delta x \frac{\partial J_d}{\partial u} \right) \Delta u_j^{n+1} = \Delta x \, J_d(u_j^n, v_j^{n+1/2}, r_j) - \Delta p_{c/l}^{n+1} \ , \tag{16.60}$$

where J_d is the discretised form of J. Equation (16.60) is tridiagonal and can be readily factorised into upper, $\underline{U}$, and lower, $\underline{L}$, triangular form (e.g. using BANFAC, Sect. 6.2.3). Consequently (16.60) can be written as

$$\Delta u_j^{n+1} = \Delta x \, \underline{U}^{-1} \underline{L}^{-1} J_d - \underline{U}^{-1} \underline{L}^{-1} \Delta p_{c/l}^{n+1} \ . \tag{16.61}$$

The application of (16.57) in discretised (e.g. trapezoidal) form gives an explicit expression for $\Delta p_{c/l}^{n+1}$.

$$\Delta p_{c/l}^{n+1} = \frac{\Delta x \int_d r \underline{U}^{-1} \underline{L}^{-1} J_d \, dr}{\int_d r \underline{U}^{-1} \underline{L}^{-1} \, dr} \ . \tag{16.62}$$

Consequently (16.60 and 62) constitute a modified tridiagonal system at each downstream location x^{n+1} that gives u_j^{n+1} and $p_{c/l}^{n+1}$.

The splitting of the pressure (16.54) and the introduction of the mass flow constraint (16.57) produce a system of four equations in four unknowns. Application of the Fourier analysis, described in Sect. 16.1.2, indicates that the u solution will be made up of two components, one oscillatory and one with a diminishing exponential behaviour in x. Consequently a stable marching algorithm in x will be obtained and, in fact, the system is non-elliptic. The crucial step that makes the system non-elliptic is the dropping of the term $\partial p^c/\partial x$ from the x-momentum equation to generate (16.55).

As long as this approximation is acceptable, accurate solutions can be obtained in a single march for internal flows governed by reduced forms of the Navier–Stokes equations. This will be the case for internal swirling flow, discussed in Sect. 16.2.1, and for flow in a straight duct, Sect. 16.2.2. However, for flow in highly curved ducts the transverse pressure gradients are significant, requiring an alternative means of generating a non-elliptic system of equations. How this can be done is discussed in Sect. 16.2.3.

16.2.1 Internal Swirling Flow

This problem is motivated by the experimental observation (Senoo et al. 1978) that the addition of a small amount of swirl, $w_0(r)$, to the flow entering a conical diffuser is sufficient to prevent separation of the flow at the walls for cone angles that would otherwise just produce separation ($\approx 15°$). As a result the pressure recovery through the diffuser is greater for a given energy expenditure required to overcome viscous losses.

This problem is governed by the incompressible turbulent Navier–Stokes equations (Armfield and Fletcher 1986). In a reduced form these become

$$\frac{\partial u}{\partial x} + \frac{\partial v}{\partial r} + \frac{v}{r} = 0 \ , \tag{16.63}$$

$$u\frac{\partial u}{\partial x} + v\frac{\partial u}{\partial r} + \frac{\partial p_{c/l}}{\partial x} = \frac{1}{Re}\left(\frac{\partial^2 u}{\partial r^2} + \frac{1}{r}\frac{\partial u}{\partial r}\right) - \frac{\partial \overline{(u'v')}}{\partial r} - \frac{\overline{(u'v')}}{r} \ , \tag{16.64}$$

$$\frac{\partial p^c}{\partial r} = \frac{w^2}{r} \ , \tag{16.65}$$

$$u\frac{\partial w}{\partial x} + v\frac{\partial w}{\partial r} + \frac{vw}{r} = \frac{1}{Re}\left(\frac{\partial^2 w}{\partial r^2} + \frac{1}{r}\frac{\partial w}{\partial r} - \frac{w}{r^2}\right) - \frac{\partial \overline{(v'w')}}{\partial r} - 2\frac{\overline{(v'w')}}{r} \ . \tag{16.66}$$

In obtaining the momentum equations (16.64–66) the streamwise diffusion terms are second-order in δ/L and have been neglected. In forming (16.64 and 66) some of the turbulence terms are sufficiently small that they have been dropped. In modifying the radial momentum equation (16.65) the convective and dissipative terms are second-order and have been deleted. It may be noted that the pressure splitting (16.54) has been incorporated into (16.64 and 65).

The turbulence quantities in (16.64 and 66) can be related to the mean flow by

$$\overline{u'v'} = -v_x\frac{\partial u}{\partial r} \quad \text{and} \quad \overline{v'w'} = v_\phi\left(-\frac{\partial w}{\partial r} + \frac{w}{r}\right) \ , \tag{16.67}$$

where v_x and v_ϕ are eddy viscosities which can also be related to the mean flow quantities. The specific algebraic relations are given by Armfield and Fletcher (1986).

Equations (16.63–67) and (16.57) provide a system of five equations in five unknowns, u, v, w, $p_{c/l}$, p^c, to be obtained as functions of x and r. In the actual computation, the region in the diffuser is described in spherical coordinates (Armfield and Fletcher 1986).

Application of the Fourier analysis described in Sects. 16.1.2 and 16.1.3 indicates that the system (16.63–67 and 57) is non-elliptic with respect to the x direction. Therefore 'initial' conditions are specified at one upstream location x_0, i.e. $u(x_0, r) = u_0(r)$, $w(x_0, r) = w_0(r)$. At the diffuser wall $(r = r_w)$, $u(x, r_w) = v(x, r_w) = w(x, r_w) = 0$. Along the diffuser centreline $(r = 0)$, $\partial u/\partial r = v = w = 0$.

Because of the pressure splitting (16.54), two almost independent systems can be constructed from (16.63–67). After discretisation, (16.63, 64) and the mass flow constraint (16.57) are used to obtain u_j^{n+1}, $p_{c/l}^{n+1}$ and v_j^{n+1}. Given u_j^{n+1} and v_j^{n+1}, (16.65 and 66) provide w^{n+1} and $p^{c, n+1}$.

The computational domain and the corresponding grid description are indicated in Fig. 16.9. The discretisation of (16.63–67) proceeds in two stages. First all r derivatives in (16.64 and 66) are discretised by

$$\frac{\partial \phi}{\partial r} = L_r \phi + O(\Delta r^2) = \frac{\phi_{j+1} - \phi_{j-1}}{r_{j+1} - r_{j-1}} + O(\Delta r^2) \ ,$$

$$\frac{\partial}{\partial r} v \frac{\partial \phi}{\partial r} = L_{rr}(v\phi) + O(\Delta r^2) \tag{16.68}$$

$$= 2\left[v_{j+1/2}\left(\frac{\phi_{j+1} - \phi_j}{r_{j+1} - r_j}\right) - v_{j-1/2}\left(\frac{\phi_j - \phi_{j-1}}{r_j - r_{j-1}}\right) \right] \Big/ (r_{j+1} - r_{j-1}) + O(\Delta r^2) \ ,$$

where ϕ and v are generic variables and j denotes the grid location in the radial direction. The form of discretisation indicated in (16.68) is appropriate for a nonuniform grid. In the present problem a refined grid, in the r direction, is required adjacent to the duct wall where severe radial gradients are to be expected.

Equations (16.63 and 65) will be marched in the radial direction. The required form of discretisation of the radial derivatives is indicated in (16.74 and 79).

Discretisation of the x derivatives in (16.64 and 66) is made to facilitate the construction of an efficient marching algorithm. Thus (16.64) is written in discrete form as

$$u^n \frac{\Delta u_j^{n+1}}{\Delta x^{n+1}} + \frac{\Delta p_{c/l}^{n+1}}{\Delta x^{n+1}} = J_d^{n+1/2} + O(\Delta x, \Delta r^2) \ , \tag{16.69}$$

where

$$\Delta u_j^{n+1} = u_j^{n+1} - u_j^n \ , \qquad u_j^{n+1/2} = 0.5(u_j^n + u_j^{n+1}) \ ,$$
$$v_j^{n+1/2} = (1 + 0.5 r_x)v_j^n - 0.5 r_x v_j^{n-1}$$

and

$$J_d = \frac{1}{\mathrm{Re}}\left[L_{rr}u_j + \left(\frac{1}{r_j}\right)L_r u_j \right] + L_{rr}(v_x u)_j + \left(\frac{v_x}{r}\right)_j L_r u_j - v_j L_r u_j \ . \tag{16.70}$$

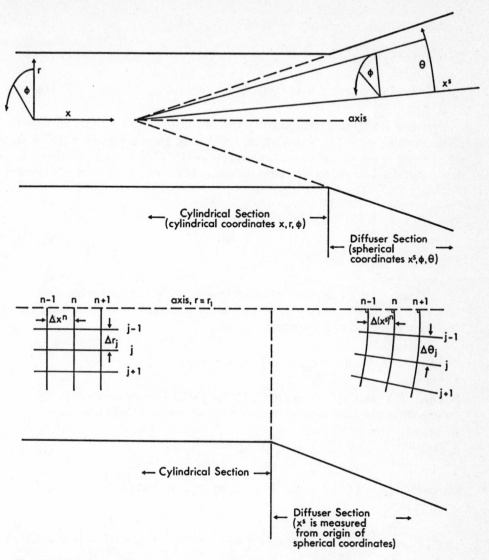

Fig. 16.9. Computational domain and grid notation for internal swirling flow

In (16.69) $\Delta p_{c/l}^{n+1} = p_{c/l}^{n+1} - p_{c/l}^{n}$ and $v_x^{n+1/2}$ is projected from upstream values like $v_j^{n+1/2}$. The grid growth ratio r_x is defined by

$$r_x = (x^{n+1} - x^n)/(x^n - x^{n-1}) = \Delta x^{n+1}/\Delta x^n \ .$$

By projecting $v^{n+1/2}$ and $v_x^{n+1/2}$ from upstream but treating $u^{n+1/2}$ implicitly, a linear, scalar system of equations for Δu_j^{n+1} can be constructed from (16.69). To achieve this, $J_d(u_j^{n+1})$ is expressed as a Taylor series expansion, as in (8.19),

$$J_d(u_j^{n+1}) = J_d(u_j^n) + \frac{\partial J_d}{\partial u} \Delta u_j^{n+1} + O(\Delta x^2) . \tag{16.71}$$

So that (16.69) becomes

$$\left(u_j^n - 0.5 \Delta x \frac{\partial J_d}{\partial u} \right) \Delta u_j^{n+1} = \Delta x \, J_d(u_j^n, v_j^{n+1/2}, r_j) - \Delta p_{c/l}^{n+1} . \tag{16.72}$$

Equation (16.72) is a tridiagonal system of equations which could be solved conventionally, e.g. as in Sect. 6.2.2, if $\Delta p_{c/l}^{n+1}$ were known. Equation (16.72) is manipulated to provide an explicit expression for $\Delta p_{c/l}^{n+1}$ by introducing the mass flow constraint as in (16.60–62). Thus the centre-line pressure adjustment is given by

$$\Delta p_{c/l}^{n+1} = \frac{\Delta x \int\limits_d r \underline{U}^{-1} \underline{L}^{-1} J_d(u_j^n, v_j^{n+1/2}, r_j) dr}{\int\limits_d r \underline{U}^{-1} \underline{L}^{-1} dr} , \tag{16.73}$$

where $\underline{U}$ and $\underline{L}$ are the upper and lower triangular factors of the left-hand side of (16.72). To be consistent with the discretisation given in (16.68), the integrals in (16.73) are evaluated by a trapezoidal formula

$$\int\limits_d^{} F dr = \sum_{j=1}^{JMAX-1} 0.5(F_j + F_{j+1})(r_{j+1} - r_j) + O(\Delta r^2) .$$

Given u_j^{n+1} and $p_{c/l}^{n+1}$ from (16.72, 73) the radial velocity component v^{n+1} is obtained from the continuity equation (16.63) written in discrete form as

$$\frac{r_{j+1/2}(v_{j+1}^{n+1/2} - v_j^{n+1/2})}{\Delta r_{j+1}} + v_{j+1/2}^{n+1/2} = -\frac{r_{j+1/2} \Delta u_{j+1/2}^{n+1}}{\Delta x} , \tag{16.74}$$

where $v_{j+1/2}^{n+1/2} = 0.5(v_j^{n+1/2} + v_{j+1}^{n+1/2})$. Thus an explicit formula for v_{j+1}^{n+1} can be obtained from

$$v_{j+1}^{n+1/2} = \left[-\frac{r_{j+1/2} \Delta u_{j+1/2}^{n+1}}{\Delta x} - v_j^{n+1/2} \left(0.5 - \frac{r_{j+1/2}}{\Delta x} \right) \right] \bigg/ \left(0.5 + \frac{r_{j+1/2}}{\Delta r_{j+1}} \right) \tag{16.75}$$

with

$$v_1^{n+1/2} = 0 \quad \text{and} \quad v_{j+1}^{n+1} = 2v_{j+1}^{n+1/2} - v_{j+1}^n .$$

Equation (16.75) provides the solution v^{n+1} via a one-dimensional radial march from the centre-line.

The circumferential momentum equation (16.66) is used to obtain w^{n+1} as follows. First, (16.66) is discretised to give

$$u_j^{n+1/2} \frac{\Delta w_j^{n+1}}{\Delta x^{n+1}} = G_d(w_j^{n+1/2}, u_j^{n+1/2}, v_j^{n+1/2}, r_j) + O(\Delta x^2, \Delta r^2) , \tag{16.76}$$

where G_d contains all the discretised radial derivatives, i.e.

$$G_d = \frac{1}{\text{Re}}\left(L_{rr}w_j + \frac{1}{r_j}L_rw_j - \frac{w_j}{r_j^2}\right) + L_{rr}(v_\phi w)_j - L_r\left(\frac{v_\phi w}{r}\right)_j$$

$$+ 2\frac{v_\phi}{r}L_rw_j - 2\left(\frac{v_\phi w}{r^2}\right)_j - v_jL_rw_j - \left(\frac{vw}{r}\right)_j . \tag{16.77}$$

Expanding w_j^{n+1} about w_j^n, as in (16.72), allows the following tridiagonal system for Δw_j^{n+1} to be constructed:

$$\left[u_j^{n+1} - 0.5\Delta x\frac{\partial G_d}{\partial w}\right]\Delta w_j^{n+1} = \Delta x\, G_d(w_j^n, u_j^{n+1/2}, v_j^{n+1/2}, r_j) + O(\Delta x^2, \Delta r^2) ,$$

$$\tag{16.78}$$

which can be solved efficiently using subroutines BANFAC and BANSOL, Sect. 6.2.3.

Finally the radial pressure correction, p^c, is obtained from (6.37) in the discrete form

$$(p^c)_{j+1}^{n+1} = (p^c)_j^{n+1} + (w_{j+1/2}^{n+1})^2\frac{r_{j+1} - r_j}{r_{j+1/2}} + O(\Delta r^2) \tag{16.79}$$

and

$$(p^c)_1^{n+1} = 0 , \quad \text{where} \quad w_{j+1/2} = 0.5(w_j + w_{j+1}) .$$

Equation (16.79) is marched from the centre-line to the duct wall to provide the pressure correction $(p^c)^{n+1}$.

The overall algorithm obtains the solution sequentially at each downstream location without iteration in the radial direction. The solution in the x direction is obtained in a single downstream march. Consequently the method is very economical. A non-uniform grid is used in both the radial and marching direction. The truncation error of the overall scheme is $O(\Delta x, \Delta r^2)$.

The flow through the conical diffuser, Fig. 16.9, uses the above formulation in the pipe region upstream of the diffuser. In the diffuser itself spherical polar coordinates are used. The equivalent form of the above algorithm in spherical polar coordinates is discussed by Armfield and Fletcher (1986).

Typical results for the axial velocity distribution are shown in Fig. 16.10. These velocity profiles are for a seven degree conical diffuser with a Reynolds number of 3.82×10^5 based on the entrance diameter. The average swirl at inlet, $w_{av}/u_{av} = 0.3$, corresponds to the experimental data of So (1964). The computational results have been obtained with 50 points in the radial direction and 150 points in the x direction. The minimum radial step size is 0.001D and occurs at the wall, where D is the upstream duct diameter. The radial step size increases by 10% in moving towards the centre-line.

The experimental data at $x/D = 0.6$, which is just inside the diffuser, has been used as starting data for the computation. Comparisons are made with the

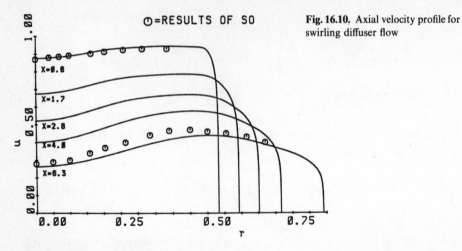

⊘=RESULTS OF SO

Fig. 16.10. Axial velocity profile for swirling diffuser flow

experimental results at $x/D = 6.3$. The turbulence has been modelled with an algebraic eddy viscosity model, as indicated above, and with a k–ε turbulence model (Sect. 11.5.2). Clearly both representations lead to close agreement with the experimental data.

The corresponding swirl velocity distribution is shown in Fig. 16.11. As with the axial velocity distribution good agreement with the experimental data of So is achieved using either turbulence model.

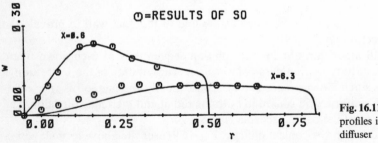

⊘=RESULTS OF SO

Fig. 16.11. Swirl velocity profiles in a conical diffuser

The broad conclusion is that the reduced Navier–Stokes equations (16.63–66) are effective in providing an accurate and economical solution to the internal swirling flow problem. However, it should be noted that too large an inlet swirl will cause a flow reversal at the axis in the diffuser.

For larger values of swirl, with or without axial flow reversal, the RNS equations given by (16.63–67) will provide accurate solutions if repeated downstream marches are made with $\partial p/\partial x$ retained in (16.64) but discretised with forward-differences (as in Sect. 16.3.3). This necessitates storing the full pressure solution from one downstream march to the next. In addition, for strongly swirling flows close to reversal, it is found necessary to construct a Poisson equation for the pressure, as in Sect. 17.1.2, and to obtain v from the radial momentum equation. If local axial flow reversal occurs it is necessary to use upwind differencing for the axial convective terms and to store the velocity field corresponding to the reverse flow region. Mean flow and Reynolds stress distributions in swirling diffuser flow are predicted by Armfield et al. (1990) using an extension of the method described here.

16.2.2 Flow in a Straight Rectangular Duct

Compared with the problem of internal swirling flow this problem has two independent variables (y, z) in the transverse direction. In developing the reduced form of the Navier–Stokes equations it will be assumed that the secondary (transverse) velocity components, v, w, are small compared with the primary (streamwise) velocity component, u. This implies that the curvature of the duct axis is small. In fact solutions will be presented in Figs. 16.13 and 16.14 for the flow in straight ducts using the formulation to be described in this section. The present formulation will be described for incompressible laminar flow; the extensions to compressible and turbulent flow being straightforward.

The flow geometry and associated grid parameters are indicated in Fig. 16.12. It is assumed that a preliminary inviscid solution is available providing an "inviscid" pressure distribution $p^{\text{inv}}(x, y, z)$. The reduced form of the nondimensional Navier–Stokes equations (16.4–6) can be written in three dimensions as

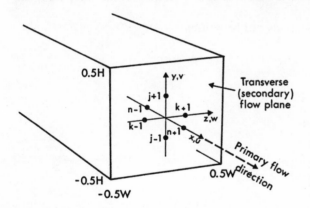

Fig. 16.12. Three-dimensional duct and associated grid definition

$$\frac{\partial u}{\partial x} + \frac{\partial v}{\partial y} + \frac{\partial w}{\partial z} = 0 \ , \tag{16.80}$$

$$u\frac{\partial u}{\partial x} = \frac{1}{\text{Re}}\left(\frac{\partial^2 u}{\partial y^2} + \frac{\partial^2 u}{\partial z^2}\right) - v\frac{\partial u}{\partial y} - w\frac{\partial u}{\partial z} - \left(\frac{\partial p^{\text{inv}}}{\partial x} + \frac{\partial p^v}{\partial x}\right) \ , \tag{16.81}$$

$$u\frac{\partial v}{\partial x} = \frac{1}{\text{Re}}\left(\frac{\partial^2 v}{\partial y^2} + \frac{\partial^2 v}{\partial z^2}\right) - v\frac{\partial v}{\partial y} - w\frac{\partial v}{\partial z} - \left(\frac{\partial p^{\text{inv}}}{\partial y} + \frac{\partial p^v}{\partial y}\right) \ , \tag{16.82}$$

$$u\frac{\partial w}{\partial x} = \frac{1}{\text{Re}}\left(\frac{\partial^2 w}{\partial y^2} + \frac{\partial^2 w}{\partial z^2}\right) - v\frac{\partial w}{\partial y} - w\frac{\partial w}{\partial z} - \left(\frac{\partial p^{\text{inv}}}{\partial z} + \frac{\partial p^v}{\partial z}\right) \ , \tag{16.83}$$

where Re is given in Fig. 16.7.

The momentum equations (16.81–83) contain a "viscous" pressure correction p^v, which is the difference between the pressure for viscous flow and the inviscid pressure p^{inv}, which is treated as a known source term. In a similar manner to that for the internal swirling flow problem, Sect. 16.2.1, the pressure correction p^v is split into two parts,

$$p^v(x, y, z) = p^v_{c/l}(x) + p^{v, c}(x, y, z) \ . \tag{16.84}$$

Substitution into (16.81–83) indicates that $\partial p^{v, c}/\partial x$ in (16.81) is of $O((\delta/L)^2)$ and can be neglected in comparison with the other terms which are of $O(1)$. In addition, for uniform inviscid flow through a straight duct, p^{inv} is constant so that its gradient is zero. Consequently for the rest of the present description, the p^{inv} terms will be dropped from (16.81–83). In addition, $p^v_{c/l}, p^{v, c}$ will be written as $p_{c/l}, p^c$.

With the above pressure splitting the system of governing equations is non-elliptic in x and suitable for a single-march solution. Since the axial coordinate (x) is time-like, initial conditions are required at the starting location x_0. Thus $u = u_0(y, z), p = p_0(y, z)$; the initial transverse velocities v_0, w_0 are adjusted to match u_0, p_0 and the interior algorithm. Boundary conditions are provided by the no-slip condition at the duct surface, e.g. $u = v = w = 0$ at $y = \pm 0.5H$. No boundary condition is required, or allowed, at the downstream boundary of the computational domain.

The three momentum equations can be written

$$u \frac{\partial \theta}{\partial x} = A\theta + \mathbf{S}_p \ , \quad \text{where} \tag{16.85}$$

$$\theta \equiv \{u, v, w\}^T \ ,$$

$$A\theta = \frac{1}{Re}\left(\frac{\partial^2 \theta}{\partial y^2} + \frac{\partial^2 \theta}{\partial z^2}\right) - v\frac{\partial \theta}{\partial y} - w\frac{\partial \theta}{\partial z} \quad \text{and}$$

$$\mathbf{S}_p \equiv \left\{-\frac{\partial p_{c/l}}{\partial x} \ , \ -\frac{\partial p^c}{\partial y}, \ -\frac{\partial p^c}{\partial z}\right\}^T \ .$$

The transverse derivatives in $A\theta$ are discretised by three-point centred difference formulae

$$L_y\theta = \frac{(\theta_{j+1, k} - \theta_{j-1, k})}{2\Delta y} + O(\Delta y^2) \ ,$$

$$L_{yy}\theta = \frac{(\theta_{j-1, k} - 2\theta_{j, k} + \theta_{j+1, k})}{\Delta y^2} + O(\Delta y^2) \ ,$$

and similarly for $L_z\theta$ and $L_{zz}\theta$. These formulae are suitable for a uniform transverse grid. If a nonuniform grid is used, (16.68) provides an appropriate discretisation.

An equivalent discrete form of (16.85) suitable for marching along the duct (in the x direction), can be written

$$u^n \frac{\Delta \theta^{n+1}}{\Delta x} = A_d^y \theta^{n+1/2} + A_d^z \theta^{n+1/2} + S_p^n , \quad \text{where} \tag{16.86}$$

$$\Delta \theta^{n+1} = \theta^{n+1} - \theta^n , \qquad \theta^{n+1/2} = 0.5(\theta^n + \theta^{n+1}) ,$$

$$A_d^y \theta^{n+1/2} = \left(\frac{1}{\text{Re}}\right) L_{yy} \theta^{n+1} - v^n L_y \theta^{n+1/2} \quad \text{and}$$

$$A_d^z \theta^{n+1/2} = \left(\frac{1}{\text{Re}}\right) L_{zz} \theta^{n+1/2} - w^n L_z \theta^{n+1/2} .$$

Equation (16.86) can be linearised using the equivalent of (16.71) giving

$$[u^n - 0.5\Delta x(A_d^y + A_d^z)]\Delta \theta^{n+1} = \Delta x(A_d^y + A_d^z)\theta^n + \Delta x S_p^n . \tag{16.87}$$

This system is linear but the structure of the left-hand side of (16.87) does not lend itself to efficient solution. However, if the term $+0.25\Delta x^2 A_d^y A_d^z \Delta \theta^{n+1}$ is added to the left-hand side of (16.87) a factorisation, as in (8.22), is possible which permits the following two-stage algorithm to be constructed:

$$(u^n - 0.5\Delta x A_d^y)\Delta \theta^* = \Delta x(A_d^y + A_d^z)\theta^n + \Delta x S_p^n \quad \text{and} \tag{16.88}$$

$$(u^n - 0.5\Delta x A_d^z)\Delta \theta^{n+1} = \Delta \theta^* . \tag{16.89}$$

Equations (16.88, 89) have the same tridiagonal structure as (9.88 and 89) and can be solved by the same techniques, i.e. subroutines BANFAC and BANSOL (Sect. 6.2.3).

Equations (16.88 and 89) have a truncation error of $(\Delta x, \Delta y^2, \Delta z^2)$ and produce a solution u^{n+1}, v^{n+1}, w^{n+1} for the velocity components, very economically. The first component, u^{n+1}, is obtained from Δu^{n+1} in conjunction with the centreline pressure correction, $\Delta p_{c/l}^{n+1}$, in the following way. The nondimensional mass flow constraint, equivalent to (16.57), can be written

$$\Delta \dot{m}^{n+1} = \iint_{d \, d} \Delta u_{j,k}^{n+1} \, dy \, dz = 0 , \tag{16.90}$$

where $\Delta u_{j,k}^{n+1}$ is obtained from the solution of (16.88 and 89). However, for arbitrary $P_{c/l}^{n+1}$, and hence arbitrary $S_p(1)$, (16.90) will not be satisfied. If $S_p(1)$ is written as

$$S_p(1) = -\frac{p_{c/l}^{n+1} - p_{c/l}^n}{\Delta x} + O(\Delta x^2) , \tag{16.91}$$

$\Delta \dot{m}^{n+1}$ is $f(p_{c/l}^{n+1})$ and an iterative algorithm can be constructed to adjust $p_{c/l}^{n+1,m}$, until (16.90) is satisfied. Thus a discrete Newton's method gives

$$p_{c/l}^{n+1,m+1} = p_{c/l}^{n+1,m} - \frac{p_{c/l}^{n+1,m} - p_{c/l}^{n+1,0}}{f(p_{c/l}^{n+1,m}) - f(p_{c/l}^{n+1,0})} f(p_{c/l}^{n+1,m}) , \tag{16.92}$$

which converges, typically, in two or three iterations of the m loop (Briley 1974). Each iteration requires a fresh $\Delta u^{n+1,m}$ from (16.88a and 89a).

After convergence of (16.92) and the solution of (16.88 and 89), the values of $p_{c/l}^{n+1}$ and $u_{j,k}^{n+1}$ are available. However, the solutions for the transverse velocity components v and w from (16.88 and 89) are considered provisional since continuity (16.80) is not satisfied. Consequently the transverse velocity components are split as

$$v^{n+1}=v^p+v^c \quad \text{and} \quad w^{n+1}=w^p+w^c , \tag{16.93}$$

where the predicted values, v^p and w^p, are obtained from (16.88, 89). The corrections, v^c and w^c, are calculated to ensure that the continuity equation (16.80) is satisfied.

As in Sect. 17.2.2 the velocity corrections are assumed to be irrotational so that a velocity potential ϕ can be introduced such that

$$v^c=\frac{\partial\phi}{\partial y} , \qquad w^c=\frac{\partial\phi}{\partial z} . \tag{16.94}$$

If (16.93 and 94) are substituted into (16.80) a Poisson equation for ϕ is generated, i.e.

$$\nabla^2\phi=-\frac{\partial u}{\partial x}-\frac{\partial v^p}{\partial y}-\frac{\partial w^p}{\partial z}=f , \tag{16.95}$$

which in discrete form becomes

$$(L_{yy}+L_{zz})\phi^{n+1}=-\left(\frac{\Delta u^{n+1}}{\Delta x}+L_y v^p+L_z w^p\right)=f_d . \tag{16.96}$$

Since the right hand side is known from the solution of (16.88, 89), (16.96) can be solved by the methods described in Sect. 6.2 or 6.3. An SOR or ADI iteration (Sect. 6.3) is constructed easily and converges rapidly if the converged solution ϕ^n is used to start the iteration.

Alternatively a direct Poisson solver (Sect. 6.2.6) can be used, if the grid is uniform. If the duct cross-section remains the same at all downstream stations the left-hand side of (16.96) need only be factorised once, assuming that sufficient storage is available. Solution of (16.96) at successive downstream locations then requires a matrix multiplication with different right-hand sides which is very economical.

In solving (16.96), homogeneous Neumann boundary conditions, i.e. $\partial\phi/\partial n=0$, are specified at the duct walls. However, $\partial\phi/\partial s$ is not zero, in general, at duct walls. Consequently although $w^c=0$ at $z=$ constant walls and $v^c=0$ at $y=$ constant walls, the no-slip velocity boundary condition is not completely satisfied. Briley (1974) recommends solving (16.88 and 89) with no-slip boundary conditions on v^p and w^p. Monitoring the errors in satisfying the no-slip boundary conditions for v^{n+1}, w^{n+1} through (16.93) then provides a test for reducing the streamwise step size Δx.

Alternatively it would be possible to force v^c, w^c to be zero at the duct walls after solving (16.96) or to adjust the wall boundary conditions on v^p, w^p to cancel $v^{c,n}$, $w^{c,n}$ values, as in Sect. 17.2.2.

In solving (16.95) it is necessary that Green's integral theorem (Gustafson 1980) be satisfied, i.e.

$$\int_A f \, dA - \int_c \frac{\partial \phi}{\partial n} \, ds = 0 \; , \tag{16.97}$$

where c is a contour enclosing an area A, with s measured along c and n the outward normal to c. Thus A is the duct cross-sectional area and c coincides with the duct walls. The discrete form (16.96) will not satisfy (16.97) identically. Consequently f_d in (16.96) is replaced by

$$f_d^c = f_d + \frac{1}{A} \left[\int_A f \, dA - \int_c \frac{\partial \phi}{\partial n} \, ds \right]_d \; . \tag{16.98}$$

A failure to satisfy (16.97) typically produces a slow divergence of the iterative scheme (Briley 1974).

The transverse pressure correction p^c in (16.84) is obtained by constructing a Poisson equation from (16.82 and 83). In discrete form this becomes

$$(L_{yy} + L_{zz})(p^c)^{n+1} = L_y F_d^y + L_z F_d^z = f_d \; , \quad \text{where} \tag{16.99}$$

$$F_d^y = -u^n \frac{(v^{n+1} - v^n)}{\Delta x} + A_d v^{n+1} \quad \text{and} \tag{16.100}$$

$$F_d^z = -u^n \frac{(w^{n+1} - w^n)}{\Delta x} + A_d w^{n+1} \; . $$

In (16.100), v^{n+1} and w^{n+1} are obtained from (16.93) and the operator A is defined after (16.85). Consequently the source term, f_d in (16.99), can be evaluated explicitly. The computational boundary for solving (16.99) is taken as the row of grid points just inside the wall. On this boundary Neumann boundary conditions on p^c are specified by

$$\frac{\partial p^c}{\partial y} = F_d^y \quad \text{or} \quad \frac{\partial p^c}{\partial z} = F_d^z \; , \tag{16.101}$$

where F_d^y and F_d^z are known from (16.100).

To satisfy Green's integral theorem, f_d in (16.99) is replaced by

$$f_d^c = f_d + \frac{1}{A} \left[\int_A f \, dA - \int_c \frac{\partial p^c}{\partial n} \, ds \right]_d \; . \tag{16.102}$$

The numerical solution of (16.99) is carried out with iterative or direct methods, as for (16.96).

The complete computational algorithm can be summarised as follows. The solution of (16.80–83) is undertaken in a single downstream march. At each downstream location x^{n+1} the following sequence is required to produce u^{n+1}, v^{n+1}, w^{n+1}, $p_{c/l}^{n+1}$ and $p^{c, n+1}$:

i) (16.88, 89 and 92) provide u^{n+1}, $p_{c/l}^{n+1}$, v^{p} and w^{p},
ii) (16.96, 94 and 93) give v^{n+1} and w^{n+1},
iii) (16.99) provides $p^{c, n+1}$.

The overall algorithm is $O(\Delta x, \Delta y^2, \Delta z^2)$ and is efficient even though some steps of the algorithm are iterative. Compared with the internal swirling flow problem, Sect. 16.2.1, the presence of two independent transverse coordinates requires a more complicated solution procedure for the transverse velocities, v and w, and pressure correction, p^c, via Poisson equations. For the internal swirling flow problem the radial velocity and transverse pressure corrections were obtained from one-dimensional radial marches.

Briley (1974) has obtained solutions for the laminar entry flow in ducts of aspect ratio 1:1 and 2:1. Typical results for the flow in a 2:1 aspect ratio duct are shown in Figs. 16.13 and 16.14. The Reynolds number, based on the mean axial velocity

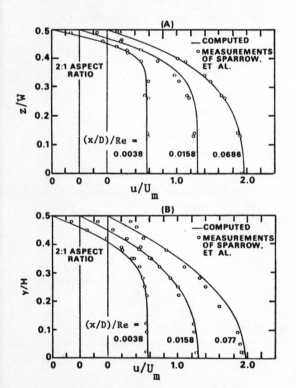

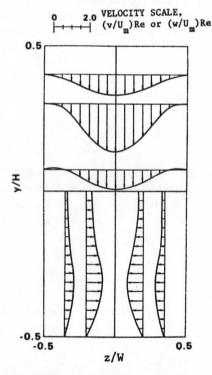

Fig. 16.13. Axial velocity distribution in a duct of aspect ratio $W/H = 2.0$ (after Briley, 1974; reprinted with permission of Academic Press)

Fig. 16.14. Secondary flow velocity profiles in a duct of aspect ratio 2:1 (after Briley, 1974; reprinted with permission of Academic Press)

U_m and the hydraulic diameter, Fig. 16.7, is $\mathrm{Re} = 1333$. The solutions are obtained on a 21×21 grid in each transverse plane and require 75 streamwise steps, typically. The transverse distribution of the axial velocity component at various downstream locations compares well with the experimental results of Sparrow et al. (1967), as indicated in Fig. 16.13. The corresponding secondary velocity components (Fig. 16.14) are seen to be rather small. Larger secondary velocity components are generated when the walls of the duct are heated differentially and buoyancy terms are included in the vertical transverse momentum equation. Briley (1974) considers this case also.

The algorithm described above has been used, in a slightly modified form, for laminar incompressible flow in straight polar ducts by Ghia et al. (1977) and in curved rectangular ducts by Ghia and Sokhey (1977). The extension to flow in curved polar ducts is described by Ghia et al. (1979).

The general problem of computing duct flows with a single downstream march is considered by Patankar and Spalding (1972), Rubin et al. (1977), Kreskovsky and Shamroth (1978), Anderson (1980) and Cooke and Dwoyer (1983), amongst others. For flows in ducts with an internal constriction, sufficient to cause streamwise separation, Ghia et al. (1981) demonstrate that reduced forms of the Navier–Stokes equations lead to accurate computational solutions as long as the single downstream march is replaced by a repeated downstream-marching iteration in which the pressure field is stored and upgraded from one iteration to the next. This iterative use of an RNS formulation is conceptually similar to that described in Sect. 16.3.3.

16.2.3 Flow in a Curved Rectangular Duct

For the flow in rectangular ducts with small curvature of the duct axis the formulation described in Sect. 16.2.2 is effective. The small curvature implies that the transverse velocity components v and w are small compared with the streamwise velocity component u. This corresponds to a relatively small variation of the pressure correction $p^{v,c}$ in (16.84), particularly in the streamwise direction. This allows the neglect of $\partial p^{v,c}/\partial x$ in the streamwise momentum equation.

For flows with large curvature of the duct axis the transverse velocity components v and w can be of the same order as the streamwise velocity component u. Consequently significant transverse pressure variation occurs and the pressure splitting introduced in Sect. 16.2.2 is inappropriate as a means of making the system of equations non-elliptic, since it will not lead to accurate solutions in a single downstream march.

An alternative non-elliptic approach, which is suitable for flows with large curvature, is provided by Briley and McDonald (1984). In this formulation an inviscid solution for the same duct geometry provides a preliminary approximate solution U^i, V^i, W^i and p^i, which is modified by the subsequent solution of the RNS equations to predict the viscous flow behaviour in the duct.

The transverse velocity is split as follows:

$$v = V^i + v_\phi + v_\psi \ , \qquad w = W^i + w_\phi + w_\psi \ , \tag{16.103}$$

or, in vector notation,

$$\mathbf{v} = \mathbf{V}^i + \mathbf{v}_\phi + \mathbf{v}_\psi \; , \tag{16.104}$$

where each $\mathbf{v}$ term has components (v, w). The potential velocity correction $\mathbf{v}_\phi \equiv \{v_\phi, w_\phi\}$ is generated primarily by the streamwise velocity gradient $\partial u / \partial x$ and is necessary to satisfy continuity (16.80), as in (16.95). The rotational velocity correction $\mathbf{v}_\psi \equiv \{v_\psi, w_\psi\}$ is generated by the streamwise vorticity Ω_x, which in Cartesian coordinates is

$$\Omega_x = \frac{\partial W^i}{\partial y} + \frac{\partial w_\psi}{\partial y} - \frac{\partial V^i}{\partial z} - \frac{\partial v_\psi}{\partial z} \; . \tag{16.105}$$

If the inviscid solution is also potential, $\partial W^i / \partial y - \partial V^i / \partial z = 0$.

The non-elliptic character depends on the recognition that potential velocity corrections $\mathbf{v}_\phi$ are small compared with the inviscid velocity field $\mathbf{V}^i$ and with the rotational velocity field $\mathbf{v}_\psi$. This follows from a consideration of the order-of-magnitude of the various terms in the governing equations, as will be indicated below.

The formulation will be described here for incompressible viscous flow in orthogonal coordinates; the corresponding compressible viscous flow formulation in orthogonal coordinates is given by Briley and McDonald (1984). The orthogonal coordinates are (ξ, η, ζ) with local velocity components u, v, w along these coordinates. The metric parameters h_1, h_2 and h_3 are defined by (12.20), i.e.

$$h_1^2 = x_\xi^2 + y_\xi^2 + z_\xi^2 \; , \quad h_2^2 = x_\eta^2 + y_\eta^2 + z_\eta^2 \; , \quad h_3^2 = x_\zeta^2 + y_\zeta^2 + z_\zeta^2 \; , \tag{16.106}$$

and are evaluated once the grid is generated as in Sect. 12.2. The equivalent Cartesian coordinate system is obtained by setting $x_\xi = y_\eta = z_\zeta = 1$ and all other transformation parameters equal to zero, so that $h_1 = h_2 = h_3 = 1$.

The transverse velocity fields $\mathbf{v}_\phi$ and $\mathbf{v}_\psi$ are linked to a velocity potential ϕ and a transverse streamfunction ψ by

$$v_\phi = \frac{1}{h_2} \frac{\partial \phi}{\partial \eta} \; , \quad w_\phi = \frac{1}{h_3} \frac{\partial \phi}{\partial \zeta} \; , \quad v_\psi = \frac{1}{h_1 h_3} \frac{\partial (h_1 \psi)}{\partial \zeta} \; , \quad w_\psi = -\frac{1}{h_1 h_2} \frac{\partial (h_1 \psi)}{\partial \eta} \; . \tag{16.107}$$

For the purposes of comparing the orders of magnitude of various terms it will be assumed that the streamwise coordinate ξ coincides with the streamlines of the inviscid solution, so that $V^i = W^i = 0$. Consequently substituting (16.107) into (16.103) gives

$$v = \frac{1}{h_2} \frac{\partial \phi}{\partial \eta} + \frac{1}{h_1 h_3} \frac{\partial (h_1 \psi)}{\partial \zeta} \; , \quad w = \frac{1}{h_3} \frac{\partial \phi}{\partial \zeta} - \frac{1}{h_1 h_2} \frac{\partial (h_1 \psi)}{\partial \eta} \; . \tag{16.108}$$

The order-of-magnitude estimates, as in Sect. 16.1.1, will assume that viscous effects are confined to a layer of nondimensional thickness δ that is small compared with a nondimensional axial length of $O(1)$. A characteristic axial length L is used for the nondimensionalisation. In the duct entry region (Fig. 16.7) δ is the boundary layer thickness; in the fully viscous and fully developed regions (Fig. 16.7) δ is half the hydraulic diameter D_h.

In (16.108), h_1, h_2 and h_3 are $O(1)$. Close to a constant η wall (Fig. 16.15),

$$u, w, \frac{\partial}{\partial \xi}, \frac{\partial}{\partial \zeta} \text{ are } O(1) \ , \quad v \text{ is } O(\delta) \quad \text{and} \quad \frac{\partial}{\partial \eta} \text{ is } O\left(\frac{1}{\delta}\right) . \tag{16.109}$$

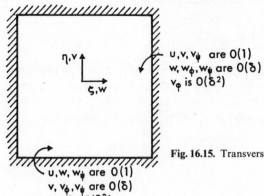

Fig. 16.15. Transverse geometry details

Close to a constant ζ wall,

$$u, v, \frac{\partial}{\partial \xi}, \frac{\partial}{\partial \eta} \text{ are } O(1) \ , \quad w \text{ is } O(\delta) \quad \text{and} \quad \frac{\partial}{\partial \zeta} \text{ is } O\left(\frac{1}{\delta}\right) . \tag{16.110}$$

To be consistent with both (16.109) and (16.110) it follows that, in (16.107), ϕ is $O(\delta^2)$ and ψ is $O(\delta)$. Corresponding to (16.109)

$$v_\phi, v_\psi \text{ are } O(\delta) \ , \quad w_\phi \text{ is } O(\delta^2) \quad \text{and} \quad w_\psi \text{ is } O(1) \ .$$

Whereas, corresponding to (16.110),

$$w_\phi, w_\psi \text{ are } O(\delta) \ , \quad v_\phi \text{ is } O(\delta^2) \quad \text{and} \quad v_\psi \text{ is } O(1) \ .$$

Since severe gradients may occur adjacent to constant ζ or constant η walls it follows that

$$\mathbf{v}_\phi \text{ is } O(\delta) \ , \quad u, \mathbf{v}_\psi \text{ are } O(1) \ . \tag{16.111}$$

Thus the transverse potential velocity corrections $\mathbf{v}_\phi$ are small compared with the streamwise velocity component u, whereas the transverse rotational velocity field $\mathbf{v}_\psi$ is of the same order as the streamwise velocity field.

The governing equations, equivalent to (16.80–83) can be written

$$\mathbf{V} \cdot \mathbf{u} = 0 \quad \text{and} \tag{16.112}$$

$$\mathbf{M} \equiv (\mathbf{u} \cdot \mathbf{V})\mathbf{u} + \mathbf{V}p - \frac{1}{\text{Re}}\mathbf{F} = 0 \ , \tag{16.113}$$

where $\mathbf{u} = (u, v, w)$, $\mathbf{M}$ is the vector of the three momentum equations and $\mathbf{F}$ is the force due to the viscous stresses.

Substitution of (16.104 and 107) into (16.112) produces a Poisson equation for the velocity potential ϕ. In Cartesian coordinates this is

$$\frac{\partial^2 \phi}{\partial y^2} + \frac{\partial^2 \phi}{\partial z^2} = -\left(\frac{\partial u}{\partial x} + \frac{\partial V^i}{\partial y} + \frac{\partial W^i}{\partial z}\right) \ . \tag{16.114}$$

As is expected from (16.107), the rotational velocity field $\mathbf{v}_\psi$ does not appear in (16.114). Equation (16.114) is equivalent to (16.95) and the same solution techniques [discussed after (16.95)] are appropriate.

The small scalar potential assumption (16.111) leads to a simplification of (16.113). The convective operator

$$\mathbf{u} \cdot \mathbf{V} = \frac{u}{h_1}\frac{\partial}{\partial \xi} + \frac{v}{h_2}\frac{\partial}{\partial \eta} + \frac{w}{h_3}\frac{\partial}{\partial \zeta} \tag{16.115}$$

is retained in full. However, $\mathbf{u}$, which it operates on, is replaced by $\mathbf{u}^a$, defined by

$$\mathbf{u}^a \equiv \{u, V^i + v_\psi, W^i + w_\psi\} \ , \tag{16.116}$$

i.e. the transverse potential velocity components are dropped, based on (16.111). The viscous force $\mathbf{F}$ can be written as

$$\mathbf{F} = \mathbf{V}^2 \mathbf{u} = -\mathbf{V} \times \mathbf{\Omega} \ , \tag{16.117}$$

where the vorticity, $\mathbf{\Omega} = \mathbf{V} \times \mathbf{u}$. The order-of-magnitude analysis permits the deletion of $\partial^2 u/\partial x^2$ from F_1 and the deletion of x derivatives in the transverse components of $\mathbf{V} \times \mathbf{\Omega}$. Thus F_2 and F_3 become

$$h_1 h_3 F_2 = -\frac{\partial(h_1 \Omega_1)}{\partial \zeta} \ , \quad h_1 h_2 F_3 = \frac{\partial(h_1 \Omega_1)}{\partial \eta} \ , \tag{16.118}$$

where the streamwise vorticity Ω_1 is given by

$$\Omega_1 = (h_2 h_3)^{-1}\left(\frac{\partial}{\partial \eta}[h_3(W^i + w_\psi)] - \frac{\partial}{\partial \zeta}[h_2(V^i + v_\psi)]\right) \ , \tag{16.119}$$

which reduces to (16.105) in Cartesian coordinates.

If (16.116, 118 and 119) are substituted into (16.113) and the Fourier analysis of Sect. 16.1.2 applied to (16.113 and 114) the following polynomial is obtained (Fletcher 1989):

$$(\sigma_y^2+\sigma_z^2)\left(u\sigma_x+v\sigma_y+w\sigma_z-i\frac{\sigma_y^2+\sigma_z^2}{Re}\right)^2=0 \; . \tag{16.120}$$

Neither factor produces any negative imaginary roots if u is positive, so that a marching solution in the x direction will be stable.

In practice, the transverse momentum equations, $M_2=0$ and $M_3=0$ in (16.113), are not solved directly. Instead they are combined as (in Cartesian coordinates)

$$\frac{\partial M_2}{\partial z}-\frac{\partial M_3}{\partial y}=0 \quad \text{and} \quad \frac{\partial M_2}{\partial y}+\frac{\partial M_3}{\partial z}=0 \; . \tag{16.121}$$

The first equation gives a transport equation for Ω_1 as

$$u\frac{\partial\Omega_1}{\partial x}+v\frac{\partial\Omega_1}{\partial y}+w\frac{\partial\Omega_1}{\partial z}-\frac{1}{Re}\left(\frac{\partial^2\Omega_1}{\partial y^2}+\frac{\partial^2\Omega_1}{\partial z^2}\right)=0 \; . \tag{16.122}$$

The second equation in (16.121) gives a Poisson equation

$$\frac{\partial^2 p'}{\partial y^2}+\frac{\partial^2 p'}{\partial z^2}=-\frac{\partial}{\partial y}(\mathbf{u}\cdot\nabla v_\psi)-\frac{\partial}{\partial z}(\mathbf{u}\cdot\nabla w_\psi) \tag{16.123}$$

for the transverse pressure correction

$$p'=p-p_{c/l} \; . \tag{16.124}$$

In solving the discrete form of (16.123), the procedure used to solve (16.99) is followed. The centre-line pressure is evaluated from the mass flow constraint (16.90) by (16.92). In the present formulation no additional approximation is introduced in evaluating the pressure field from the splitting given by (16.124).

The rotational velocity field definition (16.107) is combined with (16.119) to obtain the following Poisson equation for ψ (in Cartesian coordinates):

$$\frac{\partial^2\psi}{\partial y^2}+\frac{\partial^2\psi}{\partial z^2}=-\Omega_1-\frac{\partial V^i}{\partial z}+\frac{\partial W^i}{\partial y} \; . \tag{16.125}$$

To satisfy the no-slip conditions, $v=w=0$, using (16.103) it is necessary to couple (16.122) and (16.125) together by specifying the appropriate wall value of Ω_1. The solution of (16.114) with $\partial\phi/\partial n=0$ at the duct wall and (16.125) with $\psi=0$ at the wall ensures that the normal wall velocity is zero. At a wall of constant z (ζ in Fig. 16.15) the tangential velocity is

$$v=V^i+\frac{\partial\phi}{\partial y}+\frac{\partial\psi}{\partial z}=0 \; . \tag{16.126}$$

The vorticity at the same wall is given by

$$\Omega_1=-\frac{\partial V^i}{\partial z}-\frac{\partial^2\psi}{\partial z^2} \; . \tag{16.127}$$

For a finite difference discretisation, equivalent to that in Sect. 16.2.2, (16.126) is replaced by

$$V_{j,k}^i + \frac{\phi_{j+1,k} - \phi_{j-1,k}}{2\Delta y} + \frac{\psi_{j,k+1} - \psi_{j,k-1}}{2\Delta z} = 0 , \qquad (16.128)$$

where grid point $(j, k-1)$ lies outside the computational domain. Equation (16.127) is discretised as

$$\Omega_{1,j,k} + \frac{\partial V^i}{\partial z}\bigg|_w + \frac{\psi_{j,k-1} - 2\psi_{j,k} + \psi_{j,k+1}}{\Delta z^2} = 0 . \qquad (16.129)$$

In (16.129) $\partial V^i/\partial z|_w$ can be evaluated using a three-point one-sided finite difference expression (Chap. 3). The wall value $\psi_{j,k} = 0$. Equation (16.128) is used to eliminate $\psi_{j,k-1}$ in (16.129), i.e.

$$\Omega_{1,j,k} = -\frac{\partial V^i}{\partial z}\bigg|_w - \frac{2\psi_{j,k+1}}{\Delta z^2} - 2\left(V_{j,k}^i + \frac{\phi_{j+1,k} - \phi_{j-1,k}}{2\Delta y}\right)\frac{1}{\Delta z} . \qquad (16.130)$$

The imposition of (16.130) enforces the no-slip boundary condition on the v component at a constant z wall. An equivalent expression can be obtained to ensure that $w = 0$ at a constant y wall, and similarly for orthogonal coordinates. The ϕ values are evaluated at the known x^n station.

The various equations are discretised in the same manner as in Sect. 16.2.2. Due to the non-elliptic nature of the small potential approximation (16.111) the streamwise solution can be obtained in a single streamwise march, as in Sect. 16.2.2. At each downstream station the following sequence is followed. First (16.122 and 125) are solved as a coupled system. This produces 2×2 block tridiagonal systems of equations in place of (16.88 and 89), which are solved using the algorithm described in Sect. 6.2.5. If (16.122 and 125) were solved successively as scalar equations, the boundary condition (16.130) would place a very severe restriction on the axial step size. In discretising (16.122), the convective multipliers u, v and w are evaluated at x^n.

The Poisson equation (16.123) for the pressure correction p' is solved in the same way as (16.99). The evaluation of the right-hand side terms is made at location x^n except for v_ψ, w_ψ, which are evaluated from ψ^{n+1} via (16.107). The M_1 component of (16.113) and the mass flow constraint (16.90) are solved iteratively as in Sect. 16.2.2 to obtain u^{n+1} and $p_{c/l}^{n+1}$. Combined with the pressure correction solution, p', the complete pressure solution is obtained from (16.124). Finally the continuity equation (16.114) is solved for ϕ^{n+1}, and v_ϕ^{n+1} follows from (16.107).

Briley and McDonald (1984) provide the corresponding equations for compressible, viscous flow in orthogonal coordinates. A special fine-step iterative procedure is discussed by Briley and McDonald to allow the initial conditions to adjust to each other and the discretised form of the equations prior to marching the algorithm downstream. This helps to eliminate spurious streamwise oscillations.

Typical results for the flow in a square duct, whose axis turns through $90°$ to the right, are shown in Fig. 16.16. The radius of the turn $R/D_h = 2.3$. Six stations are

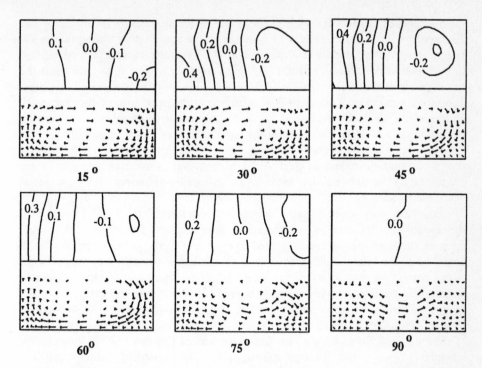

Fig. 16.16. Transverse pressure (*top*) and transverse velocity vectors (*bottom*) for flow through a 90° duct bend

shown, with 90° corresponding to the exit of the turn. The inside of the bend is to the right, the Reynolds number is $Re = 790$ (Fig. 16.7) and the viscous layer thickness at the beginning of the bend is $\delta/D_h = 0.4$.

The solution is symmetric about the z axis so that the cell of rotating fluid in the transverse plane is reflected in the horizontal line $y = 0$. This cell is seen to migrate to the inner wall as the bend is traversed. The peak transverse velocity is 0.73 and occurs at the 60° location. This strong secondary flow is associated with a significant transverse pressure gradient $\partial p/\partial z$. Such a flow as this could not be computed accurately using the pressure splitting described in Sect. 16.2.2. Such a flow could be computed using techniques described in Sect. 17.2.3. However, this requires repeated downstream marches which is much less economical than the present single-march procedure.

It may be noted that Briley and McDonald (1979) developed an earlier version of the current algorithm for a curved duct shaped like a turbine blade passage. In the 1979 algorithm the pressure field was approximated in a manner similar to that described in Sect. 16.2.2. In other respects the 1979 algorithm corresponds to that described in this section. Kreskovsky et al. (1984) and Povinelli and Anderson (1984) have used essentially the 1979 algorithm to investigate the mixer flow fields in a turbofan exhaust duct. Detailed comparisons indicate excellent agreement with experimental data.

The physical aspects of flow in curved pipes, with emphasis on laminar flow, are discussed by Berger et al. (1983). Laminar flow in a square duct undergoing a 90° bend has been computed by Humphrey et al. (1977) using a computational method similar to that described in Sect. 17.2.3. The most significant difference from the Briley and McDonald (1984) algorithm is the need to undertake repeated (iterative) downstream marches. Turbulent flow solutions in the same square duct geometry undergoing a 90° bend, and using the same computational method, are given by Humphrey et al. (1981). Calculations for laminar flow in a pipe undergoing a 180° bend are reported by Humphrey et al. (1985). The computational algorithm is similar to that described in Sect. 17.2.3 except that convective derivatives are discretised by the higher-order upwinding construction of Leonard (1979), which is described in Sect. 17.1.5.

Internal flows with strong streamline curvature have also been computed with the repeated downstream march procedure of Pratap and Spalding (1976). In this method, the pressure field is upgraded in each transverse plane essentially as in Sect. 17.2.3. But after each downstream march a one-dimensional (streamwise) global correction is made to the pressure field. The formulation has been extended to compressible flow by Moore and Moore (1979). A more recent extension based on generalised curvilinear coordinates and using a non-staggered grid is described by Rhie (1985), for compressible turbulent duct and diffuser flows.

For internal flows in general, the extension of the basic RNS formulation (Sect. 16.1) using either pressure splitting (16.52) or transverse velocity splitting (Sect. 16.2.3), permits accurate solutions to be obtained in a single downstream march, as long as the axial velocity component, u, is positive. If reversed axial flow occurs then repeated downstream marches will be required. However, it is expected (Ghia et al. 1981) that this will still be considerably more economical than solving the full Navier–Stokes equations (Chaps. 17 and 18), as long as the reversed flow region is small. Platfoot and Fletcher (1989) describe an extension of the transverse velocity splitting (Sect. 16.2.3) that includes repeated downstream marches to predict flows with regions of reversed axial flow.

16.3 External Flow

For an isolated body, the flow behaviour far from the body is closely approximated by the Euler equations (11.21). The construction of the basic RNS equations, Sect. 16.1, is constrained to ensure that the Euler equations are recovered for situations in which the viscous terms are negligible. Consequently, in principle, external flows can be computed efficiently using an RNS strategy.

As might be expected from Table 16.1, supersonic external flow is well-suited to solution via an RNS approximation. With an appropriate additional approximation to account for the subsonic surface layer (Sect. 16.3.1) the RNS equations produce an accurate solution in a single downstream march.

For subsonic (Sect. 16.3.2) and incompressible (Sect. 16.3.3) flow the RNS equations are elliptic in the far field which necessitates a repeated downstream marching strategy, even with no reversed flow. The greater efficiency with which the purely inviscid domain can be solved (Chap. 14) encourages the splitting of the computational domain into nearfield and farfield subregions. The RNS equations are used solely in the nearfield region with the farfield (inviscid) solution providing boundary conditions. A similar strategy is used in the more traditional viscous/inviscid interaction approach (Sect. 16.3.4), except that the inviscid and viscous domains overlap.

16.3.1 Supersonic Flow

From the analysis of Sect. 16.1.3, stable solutions from a single spatial march are to be expected if the flow is locally supersonic. The weak pressure, viscous interaction noted in Table 16.1 can usually be overcome by choosing a marching step that is not too small (Sect. 16.3.2). For purely inviscid flow a single spatial march algorithm is described in Sect. 14.2.4.

For viscous flow past a stationary surface, Fig. 16.17, there must always be a layer adjacent to the surface where the flow is locally subsonic since the velocity must be zero at the surface. In order to avoid exponential solution growth in the subsonic sublayer while marching parallel to the surface it is necessary to introduce an additional approximation. Typically this additional approximation is directed towards the pressure gradient in the marching direction, $\partial p / \partial x$, which appears in the x-momentum equation.

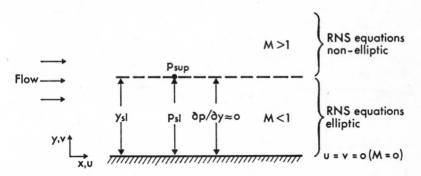

Fig.16.17. Sublayer geometry

One method for modifying the equations in the subsonic sublayer to produce a stable streamwise march is to assume that the pressure variation across the sublayer can be ignored ($\partial p / \partial y \approx 0$ in Fig. 16.17) in the governing equations (Lin and Rubin 1973). For flows that are parallel or almost parallel to the surface, the subsonic sublayer will form the inner part of the surface boundary layer. For the flow past slender bodies the boundary layer will be thin and the condition $\partial p / \partial n \approx 0$ is valid throughout the boundary layer (Sect. 11.4). Thus the use of $\partial p / \partial y \approx 0$ across the subsonic sublayer (Fig. 16.17) is consistent with boundary layer theory. In practice the condition $\partial p / \partial y = 0$ is used to obtain the pressure in the subsonic sublayer by extrapolation from the adjacent supersonic layer, i.e. $P_{sl} = P_{sup}$ (Fig. 16.17).

The above sublayer approximation will be illustrated for two-dimensional steady viscous (laminar) supersonic flow past a solid surface. The nondimensional governing equations, equivalent to (11.116), can be written

$$\frac{\partial F}{\partial x} + \frac{\partial G}{\partial y} = \frac{\partial R}{\partial x} + \frac{\partial S}{\partial y} \ , \tag{16.131}$$

where the dependent variable vector q has components

$$q \equiv \{\varrho, \varrho u, \varrho v, E\}^T \ . \tag{16.132}$$

In (16.132) E is the total energy per unit volume (11.118). The components of F and G are given by

$$F \equiv \{\varrho u, \varrho u^2 + p, \varrho u v, (E+p)u\}^T \ ,$$
$$G \equiv \{\varrho v, \varrho u v, \varrho v^2 + p, (E+p)v\}^T \ . \tag{16.133}$$

The components of R and S come from the viscous stresses (Sect. 11.6.3).

To facilitate the treatment of curved surfaces it is convenient to introduce generalised curvilinear coordinates (Chap. 12). In the present example, since the flow past slender bodies is of primary interest, it will be assumed that $\xi = \xi(x)$ and $\eta = \eta(x, y)$, where the physical orientation of ξ and η is indicated in Fig. 16.18.

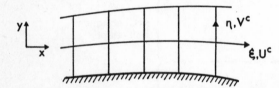

Fig. 16.18. Generalised coordinates $\xi = \xi(x)$, $\eta = \eta(x, y)$

The governing equations (16.131) become

$$\frac{\partial \hat{F}}{\partial \xi} + \frac{\partial \hat{G}}{\partial \eta} = \frac{\partial \hat{R}}{\partial \xi} + \frac{\partial \hat{S}}{\partial \eta} \ , \tag{16.134}$$

with $\hat{q} = q/J$ and the Jacobian $J = \xi_x \eta_y$. The simple form for the Jacobian follows from the restriction $\xi = \xi(x)$, which implies $\xi_y = 0$ in (12.57).

The groups $\hat{F}$ and $\hat{G}$ are given by

$$\hat{F} = \frac{\xi_x}{J} F \equiv \frac{\{\varrho U^c, \varrho u U^c + \xi_x p, \varrho v U^c, (E+p)U^c\}^T}{J} \quad \text{and} \tag{16.135}$$

$$\hat{G} = \frac{\eta_x F + \eta_y G}{J} = \frac{\{\varrho V^c, \varrho u V^c + \eta_x p, \varrho v V^c + \eta_y p, (E+p)V^c\}^T}{J} \ ,$$

where the contravariant velocity components

$$U^c = \xi_x u \quad \text{and} \quad V^c = \eta_x u + \eta_y v .$$ (16.136)

The introduction of generalised curvilinear coordinates allows the coordinates lines (constant η) to coincide with the local flow direction more closely than would be the case with Cartesian coordinates. Consequently the basic reduced Navier–Stokes approximation of discarding streamwise dissipative terms in comparison with transverse dissipative terms is equivalent to deleting $\partial \hat{\mathbf{R}}/\partial \xi$ in (16.134) and deleting ξ derivatives in $\hat{\mathbf{S}}$. This approximation coincides with the thin layer approximation (Sects. 18.1.3 and 18.4.1).

The components of $\hat{\mathbf{S}}$ are given by

$$\hat{\mathbf{S}} = \frac{1}{\text{Re}} \begin{bmatrix} 0 \\ \mu(\eta_x^2 + \eta_y^2)u_\eta + (\mu/3)[\eta_x u_\eta + \eta_y v_\eta]\eta_x \\ \mu(\eta_x^2 + \eta_y^2)v_\eta + (\mu/3)[\eta_x u_\eta + \eta_y v_\eta]\eta_y \\ (\eta_x^2 + \eta_y^2)[0.5\mu(u^2 + v^2)_\eta + (a^2)_\eta k/(\gamma - 1)/\text{Pr}] \\ + (\mu/3)(\eta_x u + n_y v)[\eta_x u_\eta + \eta_y v_\eta] \end{bmatrix} ,$$ (16.137)

where $u_\eta \equiv u/\partial \eta$, etc. The particular structure of $\hat{\mathbf{S}}$ follows partly from the deletion of ξ derivatives, partly from the simplified nature of the generalised coordinates (Fig. 16.18) and partly from the form of the nondimensionalisation. Velocities have been nondimensionalised with respect to a_∞, the freestream sound speed, density ϱ with respect to ϱ_∞, and total energy E with respect to $\varrho_\infty a_\infty^2$. Thus the Reynolds number $\text{Re} = \varrho_\infty a_\infty L/\mu_\infty$, where L is a characteristic length.

With the introduction of the basic reduced Navier–Stokes approximation, it is convenient to write (16.134) as

$$\frac{\partial \hat{\mathbf{F}}}{\partial \xi} = \frac{\partial \hat{\mathbf{S}}}{\partial \eta} - \frac{\partial \hat{\mathbf{G}}}{\partial \eta}$$ (16.138)

To invoke the sublayer approximation described earlier, it is necessary to replace $\hat{\mathbf{F}}$ (16.135a) in the sublayer with the expression

$$\mathbf{F}_{sl} = \frac{\xi_x}{J} \mathbf{F}_{sl} = \frac{\xi_x}{J} \{\varrho u, \varrho u^2 + p_{sl}, \varrho uv, u(E + p_{sl})\}^T ,$$ (16.139)

where p_{sl} is the sublayer pressure. If $u > a(1 + \varepsilon_s)$, p_{sl} is obtained from the total energy per unit volume (14.96), i.e.

$$p_{sl} = (\gamma - 1)[E - 0.5\varrho(u^2 + v^2)] .$$ (16.140)

The small parameter ε_s is introduced so that the sonic condition $u = a$ is avoided. If $u < a(1 + \varepsilon_s)$, $p_{sl} = p_{sup}$, i.e. an extrapolation is made along a constant ξ line from the adjacent supersonic region.

To obtain the solution in a single downstream (ξ) march initial values for all dependent variables are required on $\xi = \xi_0$. At the body surface $u = v = 0$, the pressure is obtained from the interior via $\partial p/\partial \eta = 0$ and T or $\partial T/\partial \eta$ are specified. Consequently ϱ follows from the equation of state. At the farfield boundary $\eta = \eta_{\max}$, the dependent variables are matched to the freestream values.

Since ξ is a time-like coordinate it is convenient to discretise (16.138) with the second-order three-level fully implicit scheme described in Sect. 7.2.3. Thus (16.138) is replaced by

$$\Delta \hat{\mathbf{F}}^{n+1} - \frac{\Delta \hat{\mathbf{F}}^n}{3} = \left(\frac{2\Delta\xi}{3}\right) \mathbf{RHS}^{n+1} \quad \text{where} \tag{16.141}$$

$\Delta \hat{\mathbf{F}}^{n+1} = \hat{\mathbf{F}}^{n+1} - \hat{\mathbf{F}}^n$, etc., and $\quad \mathbf{RHS} = L_\eta \hat{\mathbf{S}} - L_\eta \hat{\mathbf{G}}$, with

$$L_\eta \hat{\mathbf{G}} = \frac{\hat{\mathbf{G}}_{j+1} - \hat{\mathbf{G}}_{j-1}}{2\Delta\eta} \ .$$

From (16.137) it can be seen that $L_\eta \hat{\mathbf{S}}$ is of the form $L_\eta(\phi L_\eta \psi)$, which is discretised as

$$L_\eta(\phi L_\eta \psi) = \{\phi_{j+1/2}(\psi_{j+1} - \psi_j) - \phi_{j-1/2}(\psi_j - \psi_{j-1})\}\Delta\eta^2 \ , \tag{16.142}$$

where $\phi_{j+1/2} = 0.5(\phi_j + \phi_{j+1})$, etc. In (16.141 and 142) indices n and j define grid points in the ξ and η directions, respectively.

To develop an efficient algorithm for marching (16.141) downstream it is convenient to work in terms of the solution vector $\hat{\mathbf{q}}$ and to linearise about the known solution at downstream location ξ^n. A Taylor series expansion gives

$$\hat{\mathbf{F}}^{n+1} \simeq \hat{\mathbf{F}}^n + \underline{\tilde{A}}^n \Delta\hat{\mathbf{q}}^{n+1} \ , \qquad \hat{\mathbf{G}}^{n+1} \simeq \hat{\mathbf{G}}^n + \underline{\tilde{B}}^n \Delta\hat{\mathbf{q}}^{n+1} \ , \quad \text{and}$$

$$\hat{\mathbf{S}}^{n+1} \simeq \hat{\mathbf{S}}^n + \underline{\tilde{M}}^n \Delta\mathbf{q}^{n+1} \ , \quad \text{where} \quad \underline{\tilde{A}} \equiv \frac{\partial \hat{\mathbf{F}}}{\partial \hat{\mathbf{q}}} \ , \qquad \underline{\tilde{B}} = \frac{\partial \hat{\mathbf{G}}}{\partial \hat{\mathbf{q}}} \ , \qquad \underline{\tilde{M}} = \frac{\partial \hat{\mathbf{S}}}{\partial \hat{\mathbf{q}}} \ , \tag{16.143}$$

and the $\tilde{\ }$ also denotes that components of $\hat{\mathbf{q}}$ are evaluated at n whereas the metric quantities are evaluated at $n+1$.

In the sublayer, $\hat{\mathbf{F}}_{\mathrm{sl}}$ can be linearised by noting that $\hat{\mathbf{F}}_{\mathrm{sl}} = \hat{\mathbf{F}}_{\mathrm{sl}}(\hat{\mathbf{q}}, p_{\mathrm{sl}})$ when $u < a(1 + \varepsilon_\mathrm{s})$. Thus

$$\hat{\mathbf{F}}_{\mathrm{sl}}^{n+1} = \hat{\mathbf{F}}_{\mathrm{sl}}^n + \left[\frac{\partial \hat{\mathbf{F}}_{\mathrm{sl}}}{\partial \hat{\mathbf{q}}}\right]^n \Delta\hat{\mathbf{q}}^{n+1} + \left[\frac{\partial \hat{\mathbf{F}}_{\mathrm{sl}}}{\partial p_{\mathrm{sl}}}\right]^n \Delta p_{\mathrm{sl}}^{n+1} + \dots$$

$$\simeq \mathbf{F}_{\mathrm{sl}}^n + \underline{\tilde{A}}_{\mathrm{sl}}\Delta\hat{\mathbf{q}}^{n+1} + \tilde{\mathbf{F}}_p^n \ , \tag{16.144}$$

where $\tilde{\mathbf{F}}_p^n \equiv \Delta p_{\mathrm{sl}}^n \{0, 1, 0, u\}^T (\xi_x/J)^{n+1}$ and $\Delta p_{\mathrm{sl}}^{n+1}$ has been replaced by Δp_{sl}^n after extrapolation from upstream. If $u > a(1 + \varepsilon_\mathrm{s})$, $\underline{\tilde{A}}_{\mathrm{sl}} = \underline{\tilde{A}}$ and $\tilde{\mathbf{F}}_p^n = 0$.

Substituting (16.143 and 144) into (16.141) produces the following linear system for $\Delta\hat{\mathbf{q}}^{n+1}$:

$$\left[\tilde{\underline{A}}_{sl} + \frac{2\Delta\xi}{3}(L_\eta\tilde{\underline{B}} - L_\eta\underline{\tilde{M}}) \right]\Delta\hat{\mathbf{q}}^{n+1}$$

$$= \left(\frac{2\Delta\xi}{3}\right)\mathbf{RHS}^n + \frac{\Delta\hat{\mathbf{F}}_{sl}}{3} - \Delta\tilde{\underline{A}}_{sl}^n\hat{\mathbf{q}}^n - \Delta\tilde{\mathbf{F}}_p^n + \mathcal{D}\hat{\mathbf{q}}^n \ . \tag{16.145}$$

The additional terms, $\Delta\tilde{\underline{A}}_{sl}^n$ and $\Delta\tilde{\mathbf{F}}_p^n$, arise from the need to linearise $\underline{A}$ and $\mathbf{F}_p$ at both the n and $n+1$ levels. This ensures conservative differencing, which is necessary for the accurate capturing of shocks (Chap. 14 and Schiff and Steger 1980). A fourth-order dissipative term, $\mathcal{D}\hat{\mathbf{q}}^n$, is added to suppress high-frequency oscillations (Sect. 18.5.1). This term has the form

$$\mathcal{D}\hat{\mathbf{q}}^n = \varepsilon_e\tilde{\underline{A}}_{sl}^n(J^{-1})^n(\nabla_\eta\Delta_\eta)^2(J\hat{\mathbf{q}})^n \ , \tag{16.146}$$

where $\varepsilon_e < 1/8$ for stability and

$$(\nabla_\eta\Delta_\eta)^2 q_j^n = q_{j-2}^n - 4q_{j-1}^n + 6q_j^n - 4q_{j+1}^n + q_{j+2}^n \ . \tag{16.147}$$

For the region outside of the sublayer, (16.145) is applicable except that $\tilde{A}$ replaces A_{sl} and $\mathbf{F}_p = 0$.

With centred differences for L_η, (16.145) produces a 4×4 block-tridiagonal system of equations with $2 \leq j \leq J\text{MAX-1}$ at each downstream station $n+1$. The evaluation of the right-hand side of (16.145) involves only known terms at ξ locations n and $n-1$. Boundary conditions at the surface ($j=1$) and farfield ($j=J\text{MAX}$) provide appropriate values for $\Delta\hat{\mathbf{q}}_1^{n+1}$ and $\Delta\hat{\mathbf{q}}_{J\text{MAX}}^{n+1}$. Equation (16.145) can be solved efficiently using the block-tridiagonal solver described in Sect. 6.2.5.

Schiff and Steger (1980) have applied the above algorithm to the two-dimensional supersonic viscous flow past a parabolic-arc aerofoil for both laminar and turbulent flow. The solutions give good agreement with a pseudo-transient time-dependent code (Steger 1978) which also furnishes the initial data at $x/c = 0.10$ where c is the aerofoil chord.

However, if the marching step size ($\Delta\xi \equiv \Delta x$) is made too small ($\Delta x/c \leq 0.005$), the marching algorithm diverges. This behaviour is consistent with the weak elliptic behaviour associated with viscous terms (16.38). The divergent behaviour can be overcome by repeated (iterative) downstream marches with the p_{sl} distribution provided by appropriate weighted combinations of previous iterations, i.e. it is necessary to construct a convergent global iteration. Schiff and Steger find that the solution after three or four iterations is sufficiently accurate. It should be stressed that for sufficiently large Reynolds number accurate solutions are obtained in a single downstream march on a grid coarse enough to prevent divergence.

Schiff and Steger (1980) also obtain three-dimensional solutions based on the same algorithm. Typical solutions for pressure and velocity profiles on an inclined hemisphere-cylinder at $5°$ incidence are shown in Figs. 16.19 and 16.20. The marching solution is obtained in the range $3.07 \leq x/R_N \leq 40.0$, where R_N is the nose radius. Good agreement with the time-dependent code of Pulliam and Steger (1980) and the experimental results of Hsieh (1976) is apparent for both surface pressure

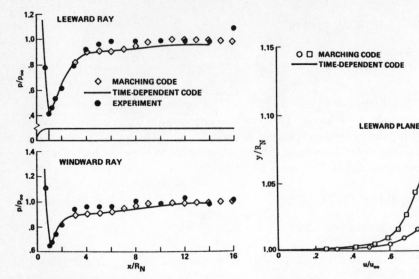

Fig. 16.19. Surface pressure distribution of an inclined hemisphere-cylinder; $M_\infty = 1.40$, $Re(R_N) = 2 \times 10^5$ (turbulent) (after Schiff and Steger, 1980; reprinted with permission of AIAA)

Fig. 16.20. Viscous layer velocity profiles on an inclined hemisphere-cylinder; $M_\infty = 1.40$, $Re_x = 1.40 \times 10^6$ (turbulent). $x/R_N = 6.98$ (after Schiff and Steger, 1980; reprinted with permission of AIAA)

and axial velocity profiles. It is noted by Schiff and Steger that the time-dependent code uses too coarse a grid in the range $9.0 \leq x/R_N \leq 14.0$ with a resulting loss of accuracy. Degani and Schiff (1986) use the algorithm with an improved algebraic eddy viscosity turbulence model to investigate leeward vortical structures for inclined pointed bodies in supersonic flow.

An alternative strategy for handling the subsonic surface layer is to weight the pressure gradient in the marching direction momentum equation with a parameter ω such that the solution does not demonstrate exponential growth in a single downstream march. A suitable expression for ω is given by (16.153). If $u > a$ the pressure gradient is handled conventionally. This technique has been used for supersonic viscous flow over delta wings by Vigneron et al. (1978), Tannehill et al. (1982) and for the supersonic viscous flow past inclined cones by Rackich et al. (1982).

The Vigneron strategy will be illustrated for the two-dimensional reduced Navier–Stokes equations governing supersonic viscous flow, i.e. (16.12, 13, 15, 17), but with $p^{nd} = p^d/\varrho_\infty U_\infty^2$ so that $a^2 = \gamma p/\varrho$ in place of (16.33). The governing equations can be written as

$$\frac{u\partial p/\partial x + v\partial p/\partial y}{p} - \frac{2}{a}\left(u\frac{\partial a}{\partial x} + v\frac{\partial a}{\partial y}\right) + \frac{\partial u}{\partial x} + \frac{\partial v}{\partial y} = 0 , \tag{16.148}$$

$$\varrho u\frac{\partial u}{\partial x} + \varrho v\frac{\partial u}{\partial y} + \omega\frac{\partial p}{\partial x} - \frac{1}{Re}\frac{\partial^2 u}{\partial y^2} = 0 , \tag{16.149}$$

$$\varrho u\frac{\partial v}{\partial x}+\varrho v\frac{\partial v}{\partial y}+\frac{\partial p}{\partial y}-\frac{1}{\text{Re}}\frac{\partial^2 v}{\partial y^2}=0 ,\tag{16.150}$$

$$p\left(\frac{\partial u}{\partial x}+\frac{\partial v}{\partial y}\right)+\omega'u\frac{\partial p}{\partial x}+\frac{v}{\gamma}\frac{\partial p}{\partial y}=\frac{2}{\gamma \text{PrRe}}\left[\left(\frac{\partial a}{\partial y}\right)^2+a\frac{\partial^2 a}{\partial y^2}\right]+\frac{(\gamma-1)}{\gamma \text{Re}}\left(\frac{\partial u}{\partial y}\right)^2 ,\tag{16.151}$$

where $\omega'=1-(\gamma-1)\omega/\gamma$. To facilitate a Fourier analysis, as in Sect. 16.1.2, density and temperature derivatives have been eliminated in favour or derivatives in the pressure, p, and sound speed, a. The dependence $\mu(T)$ and $k(T)$ has been ignored.

A Fourier analysis will be used to determine the restriction on ω such that (16.148–151), or their equivalent in terms of u, v, ϱ, p and T, will provide an accurate solution in a single spatial march. The undifferentiated terms in (16.148–151) are frozen and complex Fourier series of the form, $u\sim\hat{u}\ \exp(i\sigma_x x)\exp(i\sigma_y y)$, are introduced for u, v, p and a, as in (16.24).

The ensuing eigenvalue equation for σ_x has a somewhat simpler form if $v=0$. Thus it is assumed that locally the flow is in the x direction. This restriction facilitates the extraction of a simple analytic expression for ω_{cr}, the critical value of ω separating exponentially growing (i.e. unstable) solutions ($\omega>w_{cr}$) from stable solutions ($\omega<\omega_{cr}$).

The resulting eigenvalue equation for σ_x can be written in approximate form as

$$\frac{\varrho u\sigma_x}{\gamma a}(\sigma_x^2\{[\gamma-(\gamma-1)\omega]u^2-\omega a^2\}-a^2\sigma_y^2)$$
$$-\frac{i\sigma_y^2}{a\text{RePr}}(\sigma_x^2\{[1+0.5\gamma-0.5(\gamma-1)\omega]u^2-\omega a^2\}-a^2\sigma_y^2)=0 .\tag{16.152}$$

Far from the body surface the flow behaves as though the fluid is inviscid and the second term in (16.152) can be ignored (i.e. let $\text{Re}\rightarrow\infty$). Stable solutions obtained from a single spatial march require that σ_x should not have a negative imaginary value. From (16.152) the critical value of ω is

$$\omega_{cr,i}=\frac{\gamma M_x^2}{1+(\gamma-1)M_x^2} .\tag{16.153}$$

That is, if $\omega<\omega_{cr}$, stable solutions are obtained. For $M_x>1$, $\partial p/\partial x$ can be retained in full ($\omega=1$) without causing exponential growth in x.

If viscous effects are included we would expect from (16.38) that weak instabilities may occur. However, an approximate limit on the dominant instabilities may be obtained from (16.152) by approximating it with

$$\left(\frac{\varrho u\sigma_x}{\gamma a}-i\frac{\sigma_y^2}{a\text{RePr}}\right)[C\sigma_x^2-a^2\sigma_y^2]=0 ,\tag{16.154}$$

where $C=[0.5+0.75(\gamma-(\gamma-1)\omega)]u^2-\omega a^2$ and is the average of the two coefficients of σ_x^2 appearing in (16.152). From (16.154) no negative imaginary roots are generated as long as u and C are positive. Thus for $C > 0$

$$\omega_{\mathrm{cr,v}} = \frac{0.5 + 0.75 \gamma M_x^2}{1 + 0.75(\gamma - 1) M_x^2} .$$ (16.155)

It can be seen that $\omega_{\mathrm{cr,v}} > w_{\mathrm{cr,i}}$ so that the dominant restriction on ω when viscous terms are included is not so severe as for inviscid flow. However, the sweeping approximation required to obtain (16.155) does not exclude the possibility of weak elliptic influences due to viscous terms being present with $\omega < \omega_{\mathrm{cr,r}}$. Vigneron et al. (1978) have also obtained (16.153), but from a characteristics analysis.

If no special procedure is introduced is handle the subsonic sublayer, stable solutions in a single downstream march can be obtained as long as $\Delta x > (\Delta x)_{\min}$. This has been found empirically for supersonic flow past an inclined cone by Lin and Rubin (1973) and Lubard and Helliwell (1974). Lubard and Helliwell were able to show from a stability analysis that $(\Delta x)_{\min} \sim \Delta y_{\mathrm{sl}}$. Thus for a sufficiently thin sublayer Δy_{sl}, accurate solutions can be obtained without introducing additional approximations. Helliwell et al. (1981) provide the corresponding algorithm in generalised curvilinear coordinates.

The economy inherent in the single-march RNS strategy permits complex geometries requiring fine meshes to be considered. Thus Kaul and Chaussee (1985), using the sublayer approximation, have obtained turbulent hypersonic flow solutions about an X-24C research aircraft using a 61×30 grid in each downstream plane. Good agreement is reported with the experimental data of Neumann et al. (1978). Chaussee (1984) provides additional discussion on marching the RNS equations for supersonic flow. More recently Lawrence et al. (1989) have incorporated flux-difference splitting (Sect. 14.2.6) of the cross-flow terms to capture embedded strong shocks more accurately.

Chang and Merkle (1989) interpret the Vigneron pressure splitting as a mechanism for constructing a stable downstream-marching iterative procedure for the time-dependent full or thin-layer Navier–Stokes (Sect. 18.1.3) equations. However the use of time-dependent equations implies a local iteration at each downstream station but does provide a solution of the equations in conservation form. The rest of the pressure term can be introduced via an upstream-marching stage. Chang and Merkle show that an RNS formulation can also be based on the Steger and Warming (1981) flux-vector splitting (Sect. 14.2.5) with only the downstream-marching component retained in supersonic flow. Rubin (1988) also considers the link between RNS formulations and flux-vector splitting.

16.3.2 Subsonic Flow

The Vigneron strategy of including only part of the pressure gradient in the marching direction is used to handle, in a stable manner, the subsonic layer that occurs with a viscous supersonic flow past a solid surface (Sect. 16.3.1). The same strategy is potentially useful for completely subsonic flow.

However, an important difference arises in using an RNS formulation in external subsonic flow. The flow far from a body immersed in a uniform stream behaves as though the fluid is locally inviscid. In subsonic inviscid flow the equations are elliptic in character reflecting the physical role of pressure as a mechanism for transmitting disturbances upstream.

Consequently even with a Vigneron strategy to stabilise each spatial march in the downstream direction it will be necessary to make repeated spatial marches (in an iterative manner) to obtain an accurate solution. But if the solution produced by a single downstream march is sufficiently accurate, relatively few iterations will be required, assuming that the repeated marching iteration is stable.

For subsonic flow with no external heating, temperature changes are relatively small and accurate solutions can be obtained by replacing the energy equation with the constant total enthalpy condition (11.104). For steady two-dimensional subsonic viscous flow the RNS equations, incorporating the Vigneron approximation, become

$$u\frac{\partial \varrho}{\partial x}+v\frac{\partial \varrho}{\partial y}+\varrho\frac{\partial u}{\partial x}+\varrho\frac{\partial v}{\partial y}=0 \; , \tag{16.156}$$

$$\frac{\omega a^2}{\gamma}\frac{\partial \varrho}{\partial x}+\varrho u\frac{\partial u}{\partial x}+\varrho v\frac{\partial u}{\partial y}-\frac{\omega(\gamma-1)}{\gamma}\left(\varrho u\frac{\partial u}{\partial x}+\varrho v\frac{\partial v}{\partial x}\right)=\frac{1}{\mathrm{Re}}\frac{\partial^2 u}{\partial y^2} \tag{16.157}$$

$$\frac{a^2}{\gamma}\frac{\partial \varrho}{\partial y}-\frac{(\gamma-1)}{\gamma}\varrho u\frac{\partial u}{\partial y}+\varrho u\frac{\partial v}{\partial x}+\frac{\varrho v}{\gamma}\frac{\partial v}{\partial y}=\frac{1}{\mathrm{Re}}\frac{\partial^2 v}{\partial y^2} \; , \tag{16.158}$$

where the pressure is nondimensionalised as $p^{\mathrm{nd}}=p^{\mathrm{d}}/\varrho_\infty U_\infty^2$.

In the above equations the constant enthalpy condition has been combined with the equation of state to eliminate the pressure from the momentum equations. Equations (16.156–158) with $\omega=1$ are the same as (16.30–32). A Fourier analysis will be used here to determine the limitation on ω for a stable downstream march, just as in Sect. 16.3.1.

As in Sect. 16.1.2, complex Fourier representations are introduced for ϱ, u and v in (16.156–158). This produces the following equation for the eigenvalues σ_x:

$$\left[\left(\frac{\sigma_x}{\sigma_y}\right)^2\{u^2[\gamma-\omega(\gamma-1)]-\omega a^2\}+\frac{\sigma_x}{\sigma_y}\left\{uv[(\gamma+1)-\omega(\gamma-1)]-\frac{i\gamma u\sigma_y}{\varrho\mathrm{Re}}\right\}\right.$$

$$\left.+\left\{(v^2-a^2)-\frac{i\gamma v\sigma_y}{\varrho\mathrm{Re}}\right\}\right]\varrho\left[\varrho(u\sigma_x+v\sigma_y)-\frac{i\sigma_y^2}{\mathrm{Re}}\right]=0 \; . \tag{16.159}$$

The second factor arises from the convection diffusion operator, as in (16.34), and is associated with a stable single-march solution as long as u is positive. The Vigneron weighting parameter ω appears in the first factor of (16.159). Attention will be restricted to the inviscid case ($\mathrm{Re}\to\infty$) with $v=0$. The first factor leads to

$$\frac{\sigma_x}{\sigma_y}=\pm\frac{a}{\{u^2[\gamma-\omega(\gamma-1)]-\omega a^2\}^{1/2}} \; . \tag{16.160}$$

To avoid a negative imaginary value for σ_x it is necessary that

$$\omega \leq \frac{\gamma M_x^2}{1+(\gamma-1)M_x^2} \ . \tag{16.161}$$

This is precisely the same restriction as given by (16.153), which is to be expected since the inviscid form of (16.151) is equivalent to the constant total enthalpy condition. The use of different dependent variables to carry out the Fourier analysis does not affect the result.

If $M_x \geq 1$ (16.161) introduces no approximation, but for $M_x < 1$ (16.161) implies that the term $(1-\omega)\partial p/\partial x$ is neglected from the streamwise momentum equation. Thus for subsonic flow past an isolated body the use of (16.161) throughout most of the computational domain constitutes a much bigger approximation than that required for a subsonic surface layer in an otherwise supersonic flow.

The Fourier analysis of the governing equations (Sect. 16.1.3) has an obvious parallel with the von Neumann stability analysis (Sect. 4.3) of the discretised equations. This poses the interesting question whether there is an equivalent stability restriction to (16.161) on the marching stepsize in the algorithm formed from the discretised equivalent of (16.156–158).

For a point in the flow where $v=0$, a fully implicit discretisation of the inviscid form of (16.156–158) can be written

$$\frac{u_j^n(\varrho_j^{n+1}-\varrho_j^n)}{\Delta x} + \frac{\varrho_j^n(u_j^{n+1}-u_j^n)}{\Delta x} + \varrho_j^n L_y v_j^{n+1} = 0 \ , \tag{16.162}$$

$$\left(\frac{\omega a^2}{\gamma}\right)_j^n \frac{(\varrho_j^{n+1}-\varrho_j^n)}{\Delta x} + (\varrho u')_j^n \frac{u_j^{n+1}-u_j^n}{\Delta x} = 0 \ , \tag{16.163}$$

$$\left(\frac{a^2}{\gamma}\right)_j^n L_y \varrho_j^{n+1} - (\varrho' u)_j^n L_y u^{n+1} + (\varrho u)_j^n \frac{v_j^{n+1}-v_j^n}{\Delta x} = 0 \ , \tag{16.164}$$

where $u'=u[1-\omega(\gamma-1)/\gamma]$, $\varrho'=\varrho(\gamma-1)/\gamma$, $L_y \equiv \{1, 0, -1\}^T/2\Delta y$ and indices n and j indicate the increasing x and y directions respectively. To apply a Fourier analysis to (16.162–164) the equations are linearised by freezing the terms multiplying the difference expressions. Terms in the difference expressions are given a complex Fourier representation in y only (Sect. 4.3). Thus

$$\varrho_j^{n+1} \sim \hat{\varrho}^{n+1} e^{ij\theta} \ , \tag{16.165}$$

where $\theta = m\pi\Delta y$ as in Sect. 9.2.1. With similar representations for u and v the following matrix equation is obtained

$$\underline{A}\hat{\mathbf{q}}^{n+1} = \underline{B}\hat{\mathbf{q}}^n \ , \quad \text{where} \tag{16.166}$$

$$\hat{\mathbf{q}} \equiv \{\hat{u}, \hat{v}, \hat{\varrho}\}^T \quad \text{and}$$

$$\underline{A} \equiv \begin{bmatrix} u & \varrho & \varrho\sigma \\ \omega a^2/\gamma & \varrho u' & 0 \\ \sigma a^2/\gamma & -\sigma\varrho'u & \varrho u \end{bmatrix} , \quad \underline{B} \equiv \begin{bmatrix} u & \varrho & 0 \\ \omega a^2/\gamma & \varrho u' & 0 \\ 0 & 0 & \varrho u \end{bmatrix} .$$

In $\underline{A}$ and $\underline{B}$ all terms are evaluated at x^n, y_j, and $\sigma = i(\Delta x/\Delta y)\sin 2\theta$.

To ensure that (16.166) is stable it is necessary that all eigenvalues

$$\lambda_{\underline{A}^{-1}\underline{B}} \leq 1.0 , \quad \text{where} \tag{16.167}$$

$$\underline{A}^{-1}\underline{B} = \begin{bmatrix} 1 - \omega'\sigma^2/\beta & \dfrac{(\gamma-1)\varrho M_x^2 \omega'\sigma^2}{u\beta} & \dfrac{\gamma M_x^2 \varrho\omega'\sigma}{u\beta} \\ \dfrac{u\sigma^2\omega}{\gamma\varrho M_x^2 \beta} & 1 - \left(\dfrac{\gamma-1}{\gamma}\right)\dfrac{\omega\sigma^2}{\beta} & -\dfrac{\omega\sigma}{\beta} \\ -\dfrac{u\sigma\omega\omega_x}{\varrho\beta} & (\gamma-1)M_x^2\dfrac{\sigma}{\beta}\omega\omega_I & \dfrac{\gamma M_x^2 \omega\omega_I}{\beta} \end{bmatrix} , \tag{16.168}$$

with $\omega' = 1 - \omega(\gamma-1)/\gamma$, $\beta = \sigma^2 + \gamma M_x^2\omega\omega_I$, $\omega_I = (1/\omega_{cr} - 1/\omega)$ and $\omega_{cr} = \gamma M_x^2/[1 + (\gamma-1)M_x^2]$, i.e. from (16.161). Introducing $\tau = 1 - \lambda$ leads to the following equation for the eigenvalues of $\underline{A}^{-1}\underline{B}$,

$$\tau(\beta\tau^2 - 2\sigma^2\tau + \sigma^2) = 0 . \tag{16.169}$$

Substituting for β and introducing $\varepsilon^2 = -\sigma^2 = (\Delta x \sin 2\theta/\Delta y)^2$ gives

$$[1 + (\gamma-1)M_x^2](\omega - \omega_{cr})\tau^2 = \varepsilon^2(\tau-1)^2 . \tag{16.170}$$

The two cases $\omega < \omega_{cr}$ and $\omega > \omega_{cr}$ are of interest since they correspond to the inviscid form of (16.156–158) with $v = 0$, being non-elliptic and elliptic, respectively, in relation to the proposed marching direction, x.

For $\omega < \omega_{cr}$, (16.170) leads to

$$\mp\alpha^{1/2}(1-\lambda) = i\varepsilon\lambda , \tag{16.171}$$

where $\alpha = [1 + (\gamma-1)M_x^2](\omega_{cr} - \omega)$. Thus

$$\lambda = \frac{\mp\alpha^{1/2}(\mp\alpha^{1/2} - i\varepsilon)}{(\alpha + \varepsilon^2)} . \tag{16.172}$$

Since λ is complex it is convenient to interpret (16.167) as $\lambda\bar{\lambda} \leq 1.0$, which gives

$$\frac{\alpha}{\alpha + \varepsilon^2} \leq 1 . \tag{16.173}$$

Since α is positive and ε is real, (16.173) is true without restriction on ε and hence Δx. Thus for the non-elliptic case, $\omega < \omega_{cr}$, there is no restriction on Δx. If an explicit spatial discretisation had been used instead of (16.162–164) a stability restriction of the form $\Delta x < (\Delta x)_{max}$ would be expected, Sect. 9.1. However, such a restriction comes from the numerical scheme rather than from the character of the governing equations.

For $\omega > \omega_{cr}$, (16.170) gives

$$\mp \alpha^{1/2}(1-\lambda) = \varepsilon \lambda \, , \tag{16.174}$$

where $\alpha = [1 + (\gamma - 1)M_x^2](\omega - \omega_{cr})$. Therefore λ is given by

$$\lambda = \mp \frac{\alpha^{1/2}}{(\mp \alpha^{1/2} + \varepsilon)} \, . \tag{16.175}$$

The condition $\lambda < 1.0$ leads to $\varepsilon > \alpha^{1/2}$. The condition $\lambda > -1.0$ leads to $\varepsilon > 2\alpha^{1/2}$ or

$$\frac{\Delta x}{\Delta y} \sin \theta \cos \theta > \alpha^{1/2} \, .$$

On a finite grid, $\theta_{min} = \Delta y \pi / y_{max} \simeq \sin \theta$ and $\cos \theta \approx 1$. Thus

$$\Delta x > \alpha^{1/2} \frac{y_{max}}{\pi} \quad \text{or}$$

$$\Delta x > \{[1 + (\gamma - 1)M_x^2](\omega - \omega_{cr})\}^{1/2} \frac{y_{max}}{\pi} \, , \tag{16.176}$$

where y_{max} is the transverse extent of the computational domain. Equation (16.176) indicates that when the governing equations are elliptic there is a step-size restriction of the form $\Delta x > (\Delta x)_{min}$, where $(\Delta x)_{min}$ is proportional to the degree of ellipticity. If $(\Delta x)_{min}$ is too large then accurate solutions will not be possible with a single spatial march.

The dependence of $(\Delta x)_{min}$ on y_{max} is similar to that found by Lubard and Helliwell (1974) for the subsonic sublayer in an otherwise supersonic flow. However y_{max} in (16.176) is typically much larger than y_{sl} in Fig. 16.17. Rubin and Lin (1980) have obtained a restriction of the form $\Delta x > k y_{max}$ for incompressible flow. Rubin (1981) conjectures that $(\Delta x)_{min}$ is required to overcome the upstream influence inherent in the governing equations.

From the above analysis and the work of Rubin (1984), and references cited therein, it appears likely that obtaining solutions to elliptic equations in a single march will be subject to a stability restriction of the form $\Delta x > (\Delta x)_{min}$ and that $(\Delta x)_{min}$ is proportional to the degree of ellipticity. However, these conjectures are not yet proven in general.

As noted above it is necessary to make repeated (iterative) downstream marches to obtain an accurate solution for subsonic external flow. The Vigneron approximation can be embedded in such an iterative scheme in the following way.

Equation (16.157) is replaced by

$$\varrho u \frac{\partial u}{\partial x} + \varrho v \frac{\partial u}{\partial y} + \omega \frac{\partial p}{\partial x} - \frac{1}{Re} \frac{\partial^2 u}{\partial y^2} = -(1-\omega)\left\{\frac{\partial p}{\partial x}\right\}^*, \qquad (16.177)$$

where ω is chosen to satisfy (16.153). The term on the right-hand side of (16.177) is evaluated as an under-relaxed combination of previous iterations or is taken as the value at the outer edge of the viscous region.

The solution of a purely inviscid problem is usually much more economical (Sects. 14.1 and 14.3) than the solution of an equivalent viscous problem. Consequently it is more efficient to solve the subsonic RNS equations from the solid surface to the location where the viscous terms in the RNS equations make a negligible contribution. Outside of this domain the governing equations are treated as being purely inviscid permitting a panel method (Sect. 14.1) or full potential method (Sect. 14.3.3) to be exploited. At each stage of the iteration the inviscid problem is solved throughout the inviscid domain and the RNS equations are marched downstream once.

As long as the viscous (RNS) region is relatively thin in the transverse direction, the use of $(\partial p/\partial x)^i$ from the inner boundary of the inviscid region on the right-hand side of (16.177) is a reasonable approximation. Such an iterative scheme will be described here. The alternative, of using local values of $\partial p/\partial x$ from previous iterations will be described in Sect. 16.3.3.

The implementation of a subsonic RNS formulation can be carried out more conveniently if p appears explicitly. Thus (16.156–158) are replaced by the equivalent system of governing equations

$$\frac{\partial(\varrho u)}{\partial x} + \frac{\partial(\varrho v)}{\partial y} = 0, \qquad (16.178)$$

$$\frac{\partial}{\partial x}(\varrho u^2 + \omega p) + \frac{\partial}{\partial y}(\varrho u v) - \frac{1}{Re}\frac{\partial^2 u}{\partial y^2} = (1-\omega)\left(\frac{\partial p}{\partial x}\right)^i, \qquad (16.179)$$

$$\frac{\partial}{\partial x}(\varrho u v) + \frac{\partial}{\partial y}(\varrho v^2 + p) - \frac{1}{Re}\frac{\partial^2 v}{\partial y^2} = 0, \qquad (16.180)$$

and

$$\varrho = p \Big/ \left\{\frac{1}{\gamma M_\infty^2} + \left(\frac{\gamma-1}{2\gamma}\right)[1-(u^2+v^2)]\right\}, \qquad (16.181)$$

where all groups in (16.179 and 180) have been nondimensionalised with $\varrho_\infty^d (U_\infty^d)^2$. Thus $p_\infty^{nd} = p_\infty^d/\varrho_\infty (U_\infty^d)^2 = 1/\gamma M_\infty^2$, which leads to (16.181) having a slightly different form to (16.29).

For the present problem a change of notation is introduced. For the grid shown in Fig. 16.21 a marching algorithm (in x) is constructed to obtain the solution at x_{j+1}. A complete downstream march constitutes the $(n+1)$-th iteration. At each downstream location x_{j+1}, the transverse solution, $(u, v, p, \varrho)_{j+1,k}$, is obtained sequentially. The notation shown in Fig. 16.21 may be contrasted with the notation used in Sects. 16.2.3 and 16.3.1 where n is used to denote the downstream location.

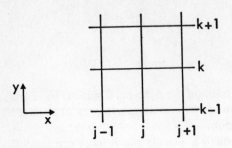

Fig. 16.21. Grid for two-dimensional subsonic RNS formulation

Equation (16.179) is used to obtain u_{j+1}^{n+1} after discretisation as

$$\frac{\{(\varrho u^2)_{j+1,k}-(\varrho u^2)_{j,k}\}}{\varDelta x}=-\frac{\{(\varrho uv)_{j+1/2,k+1}-(\varrho uv)_{j+1/2,k-1}\}}{2\varDelta y}$$

$$+\frac{\{u_{j+1/2,k-1}-2u_{j+1/2,k}+u_{j+1/2,k+1}\}}{\mathrm{Re}\varDelta y^2}$$

$$-\omega\{p_{j,k}-p_{j-1,k}\}/\varDelta x-(1-\omega)0.5\{p_{j+2,k}^{i,n}-p_{j,k}^{i,n}\}/\varDelta x$$

$$(16.182)$$

In (16.182) all terms are evaluated at x_{j+1} if no superscript is shown. Superscript i denotes the inviscid value; thus $p_j^{i,n}$ is determined from the previous inviscid iteration level at the outer edge of the viscous domain. The value of the Vigneron relaxation parameter is given by (16.181).

Terms in (16.182) evaluated at $x_{j+1/2}$ are linearised about x_j, e.g.

$$(\varrho uv)_{j+1/2,k}\approx(\varrho uv)_{j,k}+0.5\left\{\frac{\partial}{\partial u}(\varrho uv)\right\}_j\varDelta u_{j+1,k}\ ,$$

where $\varDelta u_{j+1,k}=u_{j+1,k}-u_{j,k}$. The goal of the linearisation is to construct a linear system of equations for $\varDelta u_{j+1,k}$ from (16.182). Therefore the dependence of ϱuv on ϱ and v is ignored in constructing the linear form. As a result (16.182) can be manipulated into a tridiagonal system of equations,

$$A_k\varDelta u_{j+1,k-1}^{n+1}+B_k\varDelta u_{j+1,k}^{n+1}+C_k\varDelta u_{j+1,k+1}^{n+1}=D_k\ ,\qquad(16.183)$$

where

$$A_k=-0.25\frac{\varDelta x}{\varDelta y}(\varrho v)_{j,k-1}^{n+1}-0.5\frac{\varDelta x}{\mathrm{Re}\varDelta y^2}\ ,$$

$$B_k=2(\varrho u)_{j,k}^{n+1}+\left(\frac{\varDelta x}{\mathrm{Re}\varDelta y^2}\right)\ ,$$

$$C_k=0.25\frac{\varDelta x}{\varDelta y}(\varrho v)_{j,k+1}^{n+1}-0.5\frac{\varDelta x}{\mathrm{Re}\varDelta y^2}\ ,\qquad\text{and}$$

$$D_k = -[(\varrho uv)^{n+1}_{j,k+1} - (\varrho uv)^{n+1}_{j,k-1}]0.5\frac{\varDelta x}{\varDelta y} + \left(\frac{\varDelta x}{Re\varDelta y^2}\right)(u^{n+1}_{j,k-1} - 2u^{n+1}_{j,k}$$

$$+ u^{n+1}_{j,k+1}) - \omega(p^{n+1}_{j,k} - p^{n+1}_{j-1,k}) - 0.5(1-\omega)(p^{i,n}_{j+2,k} - p^{i,n}_{j,k}) .$$

At the body surface $u_{j+1,1} = 0$, so that $\varDelta u_{j+1,1} = 0$. At the outer edge of the viscous region $(k=K)$, $\varDelta u^{n+1}_{j+1,K} = \varDelta u^{i,n}_{j+1}$, i.e. $\varDelta u^{n+1}_{j+1,K}$ is provided by the inviscid solution. Equation (16.183) is solved efficiently using the Thomas algorithm (Sect. 6.2.2) so that $u^{n+1}_{j+1,k}$ is obtained across the viscous layer at x_{j+1}.

As the second stage of the downstream march at x_{j+1}, $v^{n+1}_{j+1,k}$ is obtained by a one-dimensional sweep from the body surface to the inviscid boundary. Following Dash and Sinha (1985), (16.178, 180 and 181) are combined into a single equation of the form

$$E\frac{\partial v}{\partial y} = F\frac{\partial v}{\partial x} + G , \tag{16.184}$$

where $E = \varrho(1 - M_y^2)$, $F = \varrho M_x M_y$ and

$$G = -\left\{\varrho\frac{\partial u}{\partial x}[1 + (\gamma-1)M_x^2] + (\gamma-1)\varrho M_x M_y + \frac{\gamma u}{a^2}\left(\frac{\partial p}{\partial x}\right)^i + \frac{\gamma v}{Rea^2}\frac{\partial^2 v}{\partial y^2}\right\} ,$$

with $M_x = \dfrac{u}{a}$, $M_y = \dfrac{v}{a}$.

Equation (16.184) is discretised to $O(\varDelta x^2, \varDelta y^2)$ and manipulated to give

$$v^{n+1}_{j+1,k} = \left[F^{n+1}_{j+\frac{1}{2},k-\frac{1}{2}}\frac{\varDelta y}{\varDelta x}(v_{j+1,k-1} - v_{j,k-1} - v_{j,k})^{n+1} + 2\varDelta y G^{n+1}_{j+\frac{1}{2},k-\frac{1}{2}} \right.$$

$$\left. + E^{n+1}_{j+\frac{1}{2},k-\frac{1}{2}}(v_{j+1,k-1} + v_{j,k-1} - v_{j,k})^{n+1} \right] \Big/ \left(E^{n+1}_{j+\frac{1}{2},k-\frac{1}{2}} - \frac{\varDelta y}{\varDelta x}F^{n+1}_{j+\frac{1}{2},k-\frac{1}{2}} \right) . \tag{16.185}$$

In evaluating $E^{n+1}_{j+\frac{1}{2},k-\frac{1}{2}}$, $F^{n+1}_{j+\frac{1}{2},k-\frac{1}{2}}$ and $G^{n+1}_{j+\frac{1}{2},k-\frac{1}{2}}$, v, p, ϱ and a are projected from upstream. However, $u^{n+1}_{j+1,k}$ is available from (16.183). The value $v^{n+1}_{j+1,K}$ at the outer edge of the viscous domain provides a boundary condition for the inviscid region.

The pressure $p^{n+1}_{j+1,k-1}$ is obtained from an integration of (16.180) from the outer viscous/inviscid boundary to the wall, using

$$p^{n+1}_{j+1,k-1} = p^{n+1}_{j+1,k} + p^{n+1}_{j,k} - p^{n+1}_{j,k-1}$$

$$+ 2\varrho^{n+1}_{j+1/2,k-1/2}[u_{j+1/2,k-1/2}(v_{j+1,k-1/2} - v_{j,k-1/2})\frac{\varDelta y}{\varDelta x}$$

$$+ v_{j+1/2,k-1/2}(v_{j+1/2,k} - v_{j+1/2,k-1})]^{n+1}$$

$$- \frac{2}{Re\varDelta y}(v_{j+\frac{1}{2},k-1} - 2v_{j+\frac{1}{2},k} + v_{j+\frac{1}{2},k+1})^{n+1} . \tag{16.186}$$

Finally the density follows from (16.181).

Once a single march $(n+1)$ has been completed a new inviscid solution is obtained. Typically this would be via a velocity potential formulation (Sects. 14.1 and 14.3). The outer inviscid solution is driven by the normal velocity, $v_{j+1,K}^n$, at the outer boundary of the viscous domain. The inviscid solution provides $u_{j+1,K}^{n+1}$, at the outer boundary of the viscous domain, for the next inner solution. The interface between the viscous and inviscid domains should lie outside of the viscous region but its precise location is not critical. Since the inviscid solution can be obtained more economically than the RNS (inner) solution, it is desirable to keep the interface as close to the body as possible.

Chen and Bradshaw (1984) have used the same type of algorithm as above, but expressed in surface-orientated curvilinear coordinates, in order to obtain the viscous (turbulent) transonic flow past a two-dimensional aerofoil at small angles of attack. Dash and Sinha (1985) have used a comparable algorithm to examine turbulent subsonic, mixing layer problems. However, in both these applications $\omega = 0$, so that the axial pressure gradient in the axial momentum equations is obtained from the inviscid solution, entirely. Both applications produce accurate converged solution in relatively few (≤ 10) downstream marches.

16.3.3 Incompressible Flow

For supersonic external flow (Sect. 16.3.1) the RNS equations can be solved accurately with a single march if the sublayer approximation is applied adjacent to solid surfaces. For subsonic external flow (Sect. 16.3.2) there is an upstream influence, through the outer inviscid domain, which can be incorporated via a repeated downstream march iteration. From a consideration of the Vigneron limit on the axial pressure gradient in the downstream momentum equation (16.161) it is expected that more downstream marches may be required for convergence of the iteration as the freestream Mach number M_∞ is reduced.

In a sense, as the compressibility due to motion reduces (i.e. reducing M_∞), more upstream influence is present. This implies that for incompressible flow the solution algorithm should incorporate greater coupling between different parts of the computational domain. The coupling introduces upstream influence through the interface with the outer inviscid (elliptic) domain and through the use of forward differencing of $\partial p / \partial x$ in the downstream momentum equation. If separation of the viscous layer occurs, the reverse flow also introduces upstream influence via the convective terms.

If the computational domain contains massive separation or has no predominant flow direction the RNS strategy of a single or few downstream marches is inappropriate. The RNS equations, e.g. (16.4–6), may still be an accurate approximation but a strategy based on repeated downstream marches is no more economical typically, than the use of a pseudo-transient strategy (Sect. 6.4). In this case the strong global coupling for incompressible flow manifests itself via the need to solve a Poisson equation for the pressure (Sect. 17.1 and 17.2).

In this section the algorithm of Rubin and Reddy (1983) will be described. In this formulation a discrete Poisson equation for the pressure arises but it is coupled directly with the velocity components so as to take advantage of the RNS multiple downstream march strategy. The algorithm is effective even if small regions of reverse flow occur. In the Rubin and Reddy algorithm the RNS equations are applied throughout the computational domain. However, splitting of the domain into an inner viscous (RNS) domain and an outer inviscid domain, as in Sect. 16.3.2, creates no essential difficulty.

The incompressible RNS equations for two-dimensional laminar flow can be written in conformal (Sect. 12.1.3) coordinates as

$$\frac{\partial}{\partial \xi}(hu) + \frac{\partial}{\partial \eta}(hv) = 0 \ , \tag{16.187}$$

$$\frac{\partial}{\partial \xi}(hu^2) + \frac{\partial}{\partial \eta}(h^2 uv) + uv\frac{\partial h}{\partial \eta} - v^2\frac{\partial h}{\partial \xi} + h\frac{\partial p}{\partial \xi} = \frac{1}{Re}\frac{\partial}{\partial \eta}\left(\frac{1}{h^2}\frac{\partial}{\partial \eta}(hu)\right) \ , \tag{16.188}$$

$$\frac{\partial}{\partial \xi}(huv) + \frac{\partial}{\partial \eta}(hv^2) + uv\frac{\partial h}{\partial \xi} - u^2\frac{\partial h}{\partial \eta} + h\frac{\partial p}{\partial \eta} = 0 \ . \tag{16.189}$$

Following Rubin and Reddy no viscous term is retained in (16.189). Turbulent flow RNS equations will be comparable if Reynolds stresses are represented via an eddy viscosity construction (Sect. 11.4.2). Equations (16.187–189) are the continuity, ξ-momentum and η-momentum equations, respectively. The velocity components along ξ and η lines are denoted by u and v, here. The orthogonal metric is defined by (12.20). For conformal coordinates,

$$h_1 = h_2 = h = (x_\xi^2 + y_\xi^2)^{1/2} = (x_\eta^2 + y_\eta^2)^{1/2} \ . \tag{16.190}$$

The (ξ, η) coordinate system is constructed (Chap. 13) so that one ξ coordinate line (η_0) coincides with the body surface and other ξ coordinate lines are approximately in the local flow direction. The η coordinate lines are orthogonal to ξ coordinate lines and consequently are orthogonal to the body surface and approximately orthogonal to the local flow direction. The use of the (ξ, η) coordinate system allows the marching direction to be closely aligned with the local flow direction, which increases the accuracy of the RNS approximation.

Rubin and Reddy (1983) indicate that the neglect of the transverse viscous terms in the η-momentum equation is consistent with the neglect of the streamwise viscous terms in the ξ-momentum equation, for incompressible flow. However, the inclusion, or neglect, of the transverse viscous terms in the η-momentum equation has no effect on the character of the overall system of equations (Sect. 16.1.2).

Since (16.187–189) are to be marched in the positive ξ direction, all ξ derivatives of u and v are backward-differenced in non-separated regions. The term $\partial p/\partial \xi$ is discretised with a modified forward difference expression to introduce the required upstream influence associated with the elliptic behaviour of p. This implies that the complete p field must be stored so that it is available for the next (iterative) downstream march. In contrast the velocity field is not stored, but is computed anew on each downstream march, at least for non-separated flow.

For regions of separated flow upwind-differencing is used for the convective terms. Since all ζ-marches are made in the positive ζ direction it is necessary to store the velocity field from the previous iteration, but for the reverse flow region only.

For ease of exposition the discretised form of (16.187–189) will be presented in non-conservation form on a Cartesian grid, with x as the marching direction. The result is

$$L_x^- u_{j,k-1/2}^n + L_y^- v_{j,k}^n = 0 \ , \tag{16.191}$$

$$u_{j,k}^n L_x^- u_{j,k}^n + v_{j,k}^n L_y u_{j,k}^n + \frac{p_{j+1,k}^{n-1} - p_{j,k}^n}{\Delta x} = \frac{1}{Re} L_{yy} u_{j,k}^n \ , \tag{16.192}$$

$$u_{j,k-1/2}^n L_x^- v_{j,k-1/2}^n + v_{j,k-1/2}^n L_y v_{j,k}^n + L_y^- p_{j,k}^n = 0 \ , \tag{16.193}$$

where L_y and L_{yy} are three-point centred difference operators (Table 9.3), and L_x^-, L_y^- are one-sided operators, i.e.

$$L_x^- u_{j,k} \equiv \frac{(u_{j,k} - u_{j-1,k})}{\Delta x} \ , \qquad L_y^- v_{j,k} = \frac{(v_{j,k} - v_{j,k-1})}{\Delta y} \ . \tag{16.194}$$

All terms are evaluated at iteration level n, except $p_{j+1,k}^{n-1}$, which is from the previous iteration. It can be seen (Fig. 16.22) that the discretised continuity and y-momentum equations are centred (c, y) at (j, $k-1/2$) whereas the x-momentum equation is centred (x) at (j, k). This system has a truncation error of $O(\Delta x, \Delta y^2)$; Rubin and Reddy (1983) provide a closely related scheme that is second-order in both x and y.

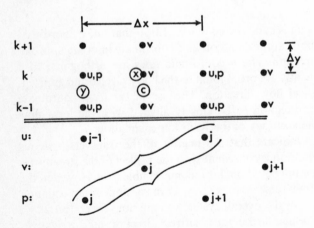

Fig. 16.22. Discretisation grid for (16.191–193)

The present algorithm has been used to compute the incompressible laminar and turbulent flow past a finite flat plate and an isolated symmetric aerofoil. Boundary conditions, Fig. 16.23, are $u = v = 0$ at $\eta = \eta_0$ (body surface); $v = 0$, $\partial u/\partial \eta = 0$ on a line of symmetry ($\eta = \eta_0$). At the outer boundary ($\eta = \eta_{max}$) $p = P_\infty$, $u = U_\infty$ and no boundary condition is required for v. At the inflow boundary ($\xi = \xi_0$), $u = U_0(y)$, $\partial v/\partial \xi = 0$ where $U_0(y)$ is a prescribed velocity distribution. For

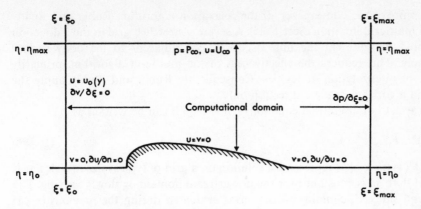

Fig. 16.23. Boundary conditions for symmetric aerofoil

the finite flat plate this is a boundary layer profile linked to $U_0 = U_\infty$ in the inviscid region. At the outflow boundary ($\xi = \xi_{max}$) the pressure or $\partial p / \partial \xi = 0$ is specified.

At each downstream station, (16.191–193) are treated as a nonlinear system of equations for $\mathbf{q}_{j,k}^n$, $k = 1, N_y$ where $\mathbf{q} \equiv \{u, v, p\}^T$. This nonlinear system is linearised and solved iteratively using the Newton–Raphson technique (Sect. 6.1.1). The Jacobian is block-tridiagonal so that each stage of the Newton–Raphson iteration can be executed efficiently using the techniques described in Sect. 6.2.5. After convergence of the Newton–Raphson iteration at the jth downstream station the process is repeated at the $(j+1)$-th station.

Each downstream march constitutes one iteration of a relaxation process (Sect. 6.3.1). Rubin and Reddy (1983) apply a von Neumann analysis (Sect. 4.3) to a linearised form (i.e. undifferentiated terms locally frozen) of (16.191–193). The maximum eigenvalue has the form

$$\lambda \sim 1 - C_1 \left(\frac{\pi \Delta x}{y_{max}} \right)^2 \left(\frac{x_{max}}{y_{max}} \right)^2 , \qquad (16.195)$$

where C_1 is a constant of order one and the extent of the computational domain is $0 \leq x \leq x_{max}$, $0 \leq y \leq y_{max}$. Since $\lambda < 1.0$ the relaxation procedure is stable but it becomes unacceptably slow as λ approaches one. This will occur for large y_{max} or small Δx or x_{max}.

The appearance of the group ($\pi \Delta x / y_{max}$) is particularly interesting. For subsonic inviscid flow it was shown (16.176) that a stable solution could be obtained in a single march if $(\pi \Delta x / y_{max}) > \alpha^{1/2}$. For the present situation of incompressible flow and $\omega = 1$, $\alpha = 1.0$. Thus although a multiple march strategy has avoided the stability restriction (16.176) any attempt to make $\pi \Delta x / y_{max} \ll 1$ produces a very uneconomic algorithm due to the slow convergence of the relaxation process. It might be expected from (16.176) that for more compressible flows the convergence of the multiple march procedure will improve by using a Vigneron-related strategy. This is confirmed by Rubin (1985).

To improve the convergence of the relaxation algorithm, Rubin and Reddy adopt a multigrid iteration (Sect. 6.3.5). A severely stretched grid in the η direction is required to represent the thin viscous region adjacent to the body surface adequately. This reduces the effectiveness of the multigrid algorithm, primarily because of interpolation difficulties. Consequently Rubin and Reddy apply the multigrid algorithm in the ξ direction only.

The system of discretised equations (16.191–193) can be written as

$$L^h Q^h = F^h \,, \tag{16.196}$$

where $\mathbf{Q}^h (\equiv \{u, v, p\}^T)$ represents a solution on a grid of typical size $h(\equiv \Delta \xi)$. It is assumed that $\Delta \xi$ is constant over the downstream domain or slowly varying. The right-hand side, F^h, contains all the terms evaluated during the previous $(n-1)$ downstream march. That is, F^h will contain p^{n-1}, and u^{n-1}, v^{n-1} in the separated region. The nth downstream march can be interpreted as a smoothing or relaxation process, as in (6.84). When a converged solution of (16.196) is obtained, $p^n = p^{n-1}$, etc. Rubin and Reddy apply the FAS multigrid formulation (Sect. 6.3.5) to solve (16.196).

Rubin and Reddy (1983) have used four levels of grids with two to three relaxation sweeps on each grid and transfers in both directions. Once the finest grid is reached only the two finest grids are used for the final convergence. For unseparated flow the use of the multigrid construction produces one to two orders of magnitude reduction in the number of iterations and execution time, even when $(\pi \Delta x / y_{max})$ is small. However, if regions of separated flow occur or if a highly nonuniform $\Delta \xi$ grid is used, Rubin and Reddy (1983) report that the multigrid technique is less effective. In a later development Himansu and Rubin (1988) find that, even with locally reversed axial flow, multigrid in the marching direction only is more efficient than multigrid applied in both axial and transverse directions.

A typical conformal grid for the flow about an aerofoil at zero angle of attack is shown in Fig. 16.24. The y-scale has been magnified to show the character of the

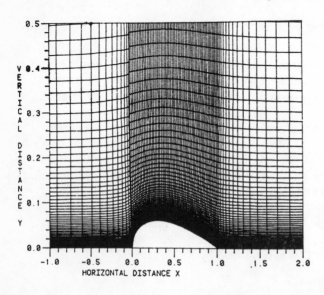

Fig. 16.24. Conformal grid for flow about a NACA-0012 aerofoil at zero incidence (after Rubin and Reddy, 1983; reprinted with permission of Pergamon Press)

grid more effectively. Solutions have been obtained for laminar flow up to
Re = 10 000. The streamlines close to the trailing edge are shown in Fig. 16.25,
demonstrating the occurrence of a small separation bubble. Solutions have also
been obtained for turbulent flow (Re = 5 × 10⁵), making use of the two-layer
algebraic eddy viscosity model described in Sect. 11.4.2. Rubin and Reddy (1983)
present results for surface pressure and skin friction distributions and indicate good
agreement with experiments.

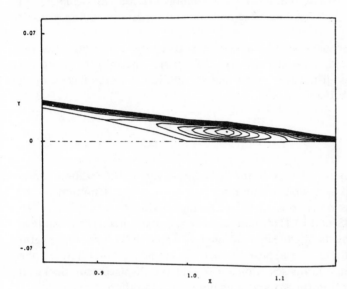

Fig. 16.25. Streamlines
near the trailing edge of
a NACA-0012 aerofoil;
Re = 10 000 (after Rubin
and Reddy, 1983; reprinted
with permission of
Pergamon Press)

Broadly the use of an RNS strategy in incompressible external flow requires
more iterations (downstream marches) to obtain an accurate solution than is the
case for high subsonic or supersonic external flow. However, for incompressible
flows with a dominant flow direction an RNS strategy is still considerably more
economical than solving the full Navier–Stokes equations by a pseudo-transient
approach, for example as in Sect. 17.2.1.

16.3.4 Viscous, Inviscid Interactions

In the preceding sections of this chapter flow problems are considered for which the
boundary layer equations are inadequate to provide accurate solutions, even
locally. The inadequacy may arise because the thickness of the viscous layer is
greater than permitted by boundary layer theory (Schlicting 1968). Alternatively,
the streamline curvature in the problem implies a significant pressure gradient
across the thickness of the viscous layer.

However, for the flow around slender bodies, such as aerofoils and turbine
blades at small angles of attack, and hence with small pressure differences between
the windward and leeward sides, the boundary layer equations with a specified
inviscid pressure distribution provide an accurate solution for the local velocity

distribution. In turn this solution can be used to compute the displacement thickness (11.67); that is, the amount by which the equivalent body contour should be displaced to permit a more accurate recalculation of the inviscid solution. The small adjustment to the inviscid pressure distribution usually leads to close agreement with experimental pressure distributions, away from the trailing-edge region and as long as reversed flow associated with separation does not occur.

Historically attempts have been made to extend the displacement thickness concept to allow for the wake, and to make the combined viscous, inviscid solution more accurate close to the trailing edge and if small separation bubbles occur (Cebeci et al. 1984).

Even when the viscous, inviscid interaction is computed conventionally the influence on the inviscid solution is incorporated more efficiently by using the displacement thickness distribution to define an equivalent normal velocity at the body surface (Lock 1981),

$$v_i(x, 0) = \frac{d}{dx} (u_e \delta^*) , \tag{16.197}$$

where u_e is the velocity at the edge of the boundary layer. This condition comes directly from the incompressible continuity equation and the definition of the displacement thickness δ^* (11.67). In the wake region (16.197) is applied as an effective source panel (Sect. 14.1.1) located on the wake centre-line. The advantage of using (16.197) in recalculating the inviscid solution is that the computational grid remains the same and (16.197) appears as a local boundary condition for the governing equations in the inviscid domain. Using the displacement thickness directly changes the grid in the inviscid domain at each iteration.

The above viscous, inviscid interaction procedure can be repeated iteratively refining the pressure distribution at the edge of the boundary layer and the boundary layer displacement thickness at each step of the iteration. This is called a direct iteration method and is described in Sect. 14.1.4.

The direct iteration method is based on the concept that the inviscid solution drives the boundary layer solution through specification of the boundary condition, $u_e(x)$ or $p_e(x)$, at the outer edge of the boundary layer and that the influence of the boundary layer solution, through the displacement thickness, on the inviscid solution is rather weak.

Close to separation the character of the mutual interaction alters dramatically. Attempts to compute a boundary layer solution up to and beyond separation fail if $p_e(x)$ is prescribed. As separation is approached the computed boundary layer normal velocity v demonstrates a singular behaviour (Goldstein 1948) which does not occur physically.

This difficulty is overcome by using a so-called inverse method. That is, the displacement thickness $\delta^*(x)$ or the skin friction coefficient $c_f(x)$ is specified and $p_e(x)$ is computed along with the boundary layer solution. When $\delta^*(x)$ is specified the definition of the displacement thickness generates $u_e(x)$, and hence $p_e(x)$, as part of the solution process, from

$$\int_0^\infty \left(1 - \frac{u}{u_e}\right) dx = \delta^{*,s}(x) \ . \tag{16.198}$$

In practice, the inverse method requires iteration of the boundary layer equations at each downstream station, x_{j+1}, until $u_{e,j+1}$ is obtained which satisfies (16.198) for the specified displacement thickness $\delta^{*,s}$. A discrete Newton's method (Sect. 6.1.1) can be used for this iteration. An auxiliary function, $f^m = \delta^{*,m} - \delta^{*,s}$, is defined. The iteration is advanced by computing $u_{e,j+1}^{m+1}$ from

$$u_{e,j+1}^{m+1} = u_{e,j+1}^m - \frac{(u_{e,j+1}^m - u_{e,j+1}^0)}{(f^m - f^0)} f^m \ , \tag{16.199}$$

with $u_{e,j+1}^0 = u_{ej}^m$. The iteration converges for $m \lesssim 10$ and is very economical. The inverse method for computing the boundary layer solution, with a reasonable choice for $\delta^{*,s}(x)$, permits a stable integration to be made through separation and into the region of reversed flow (Catherall and Mangler 1966; Klineberg and Steger 1974).

The inverse method for computing the boundary layer solution is combined with an inviscid computation, with specified $u_e(x)$, in the outer domain to produce an inverse iteration method to determine the viscous, inviscid interaction. The direct and inverse iteration methods can be compared as follows.

It is convenient to write the relationship between the outer inviscid velocity distribution $u_e(x)$ and the displacement thickness $\delta^*(x)$ in the following symbolic way:

For inviscid flow: $\qquad\qquad u_e = P[\delta^*] \ ,$

$$\tag{16.200}$$

For viscous (boundary layer) flow: $u_e = B[\delta^*] \ .$

Thus the direct iteration method can be represented as

$$u_e^{(n+1)} = P[\delta^{*(n)}] \quad \text{and} \quad \delta^{*(n+1)} = B^{-1}[u_e^{(n+1)}] \ , \tag{16.201}$$

where n indicates the iteration level of the viscous, inviscid interaction. The inverse iteration method can be represented as

$$\delta^{*(n+1)} = P^{-1}[u_e^{(n)}] \quad \text{and} \quad u_e^{(n+1)} = B[\delta^{*(n+1)}] \ . \tag{16.202}$$

The direct and inverse formulations can be combined into a semi-inverse iteration method (Le Balleur 1978; Carter 1981)

$$u_e^P = P[\delta^{*(n)}] \ , \qquad u_e^B = B[\delta^{*(n)}] \ , \quad \text{and} \tag{16.203}$$

$$\delta^{*(n+1)} = R[u_e^P, u_e^B, \delta^{*(n)}] \ ,$$

where R is an appropriate relaxation algorithm. The Le Balleur prescription of the semi-inverse method is described for transonic flow in Sect. 16.3.6. The direct, inverse and semi-inverse iteration methods are illustrated conceptually in Fig. 16.26.

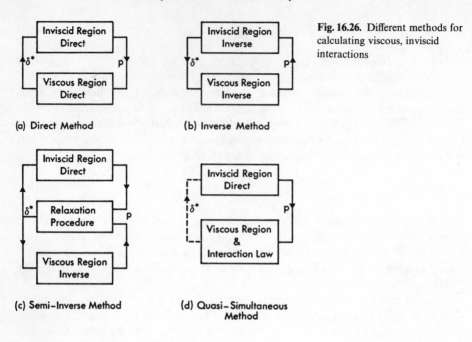

(a) Direct Method

(b) Inverse Method

(c) Semi–Inverse Method

(d) Quasi–Simultaneous Method

An alternative formulation to the semi-inverse method is the quasi-simultaneous iteration method (Veldman 1981) in which (16.200a) is replaced by

$$u_e = I[\delta^*] \ , \tag{16.204}$$

where I represents an interaction law that is an approximation to the full interaction of the inviscid flow with the viscous flow. The interaction law is constructed so that the local but not the global inviscid influence is accurately represented. Consequently the quasi-simultaneous method can be represented symbolically by

$$u_e^{(n+1)} - I[\delta^{*(n+1)}] = P[\delta^{*(n)}] - I[\delta^{*(n)}] \ ,$$
$$u_e^{(n+1)} - B[\delta^{*(n+1)}] = 0 \ . \tag{16.205}$$

Thus on convergence of the iterative process it is clear that (16.200) is satisfied. The role of $I[\delta^*]$ is to allow the most important part of the inviscid interaction to be considered simultaneously with the development of the viscous solution (Fig. 16.26).

In comparing the four methods shown in Fig. 16.26 it may be noted that the weakness of the direct method is that B^{-1} in (16.201b) is almost singular when a strong interaction occurs, i.e. near a trailing edge. Although the inverse method avoids this problem (P^{-1} is well-behaved), very slow convergence occurs because a very small underrelaxation factor is required for a stable iteration in a weak interaction region. Veldman (1984) finds that the underrelaxation factor is proportional to $\text{Re}^{-1/2}$, when Re is large.

The semi-inverse and quasi-simultaneous iteration methods achieve stronger coupling, and hence faster convergence, by simplifying one of the interactions. In the semi-inverse method the solution in the viscous flow region is represented by an integral (Cebeci and Bradshaw 1977, Chap. 5) method, so that the viscous domain solution can be interpreted as an extended boundary condition for the solution in the inviscid domain. In the quasi-simultaneous method the inviscid solution is represented in a simplified manner so that it can be considered an extended boundary condition to the boundary layer solution. More detailed comparisons of the four iterative methods, particularly in relation to the stability of the iterations, are provided by Wigton and Holt (1981) and by Veldman (1984).

16.3.5 Quasi-Simultaneous Interaction Method

The application of the quasi-simultaneous iterative method to incompressible flow is described in more detail in this section. A simplified inviscid solution is provided by thin aerofoil theory (Thwaites 1960). At the outer edge of the viscous region, the inviscid solution without interaction, $u_e^{(0)}(x)$, is given by

$$u_e^{(0)}(x) = 1 + \frac{1}{\pi} \int_0^\infty \frac{(dy_B/d\xi)}{(x-\xi)} d\xi \ , \tag{16.206}$$

where y_B is the body surface contour. Carter and Wornom (1975) model a viscous correction, $\delta u_e(x)$, to $u_e^0(x)$ as

$$\delta u_e(x) = \frac{1}{\pi} \int_{x_1}^{x_2} \frac{\{d(u_e^{(0)}\delta^*)/d\xi\}}{(x-\xi)} d\xi \ , \tag{16.207}$$

so that $u_e(x) = u_e^{(0)}(x) + \delta u_e(x)$. In (16.207) the integration limits x_1, x_2 are restricted to the region where the interaction is important. This ensures that the interaction is of a local nature.

If the integral in (16.207) is evaluated approximately by the mid-point rule the equivalent of (16.204) becomes

$$u_{e_k} = u_{e_k}^{(0)} + \sum_{j=1}^N \alpha_{kj}\delta_j^* + R_k \ , \quad \text{where} \tag{16.208}$$

$$\alpha_{kj} = -\left(\frac{u_{e_k}^{(0)}}{\pi\Delta x}\right)\Big/[(k-j)^2 - 0.25] \quad \text{and}$$

$$R_k = \frac{1}{\pi\Delta x}\left[(u_e^{(0)}\delta^*)_N(k-N-0.5) - \frac{(u_e^{(0)}\delta^*)_0}{(k-0.5)}\right] \ .$$

In (16.208) j and k correspond to grid points in the downstream (x) direction.

At each downstream march of the boundary layer equations in the quasi-simultaneous method, u_e and δ^* are coupled together by (16.208) and are obtained simultaneously with the boundary layer solution. During the nth downstream march (16.208) is implemented as a Gauss–Seidel (Sect. 6.3) process

$$u_{ek}^{(n+1)} - \alpha_{kk}\delta_k^{*(n+1)} = u_{ek}^{(0)} + \sum_{j=1}^{k-1} \alpha_{kj}\delta_j^{*(n+1)} + \sum_{j=k+1}^{N} \alpha_{kj}\delta_j^{*(n)} + R_k^{(n)} . \qquad (16.209)$$

Veldman (1981) notes that, since $\alpha_{kk}>0$, a breakdown in the boundary layer calculation at separation is avoided. The particular choice of α_{kk}, arising from the numerical integration of the Hilbert integral in (16.207), leads to relatively few ($n \sim 10$) downstream marches to achieve convergence.

The global iterative process can be described as follows:

Step A) $n=0$. A conventional inviscid solution using the panel method (Sect. 14.1.1), for a body profile $y_B(x)$ gives the basic inviscid velocity distribution $u_e^{(0)}(x)$ at the outer edge of the viscous region. The boundary layer solution is marched downstream until the strong interaction region is reached. The displacement thickness is extrapolated through the strong interaction region to give the zero-th approximation, $\delta^{*(0)}$.

Step B) $n=1, \ldots$. The boundary layer solution is marched downstream simultaneously with (16.209 and 198).

Step C). After each downstream march the following convergence criterion is checked:

$$\max |\delta_k^{*(n+1)} - \delta_k^{*(n)}| < 10^{-4} . \qquad (16.210)$$

If (16.210) is not satisfied δ^* is overrelaxed with $\omega=1.5$, typically. The iteration proceeds with Step B).

For many flows the basic inviscid solution $u_e^{(0)}(x)$ is sufficiently close to the converged inviscid solution $u_e^{(n)}$ that the inviscid solution need not be recomputed. However, if $u_e^{(n)}$ differs significantly from $u_e^{(0)}$ it would be appropriate to construct an equivalent body, $y_B + \delta^{*(n)}$, and to recompute $u_e^{(0)}(x)$ using the full inviscid domain solver. The global iterative process then restarts from Step A.

A typical result for the surface pressure distribution through the separation bubble formed in the Carter–Wornom trough (Fig. 16.27), produced by the quasi-simultaneous method is shown in Fig. 16.28. Clearly accounting for the viscous, inviscid interaction is crucial if accurate solutions are to be obtained based on the boundary layer approximation.

The solutions shown in Fig. 16.28 were obtained on a 121×81 grid covering the viscous domain and required about 10–20 iterations to satisfy (16.210) with $\omega=1.5$.

Veldman (1981) shows that the above algorithm is consistent with triple-deck theory (Stewartson 1974) which is relevant to the analysis of singular points, such as separation and the trailing edge, as the Reynolds number approaches infinity. Most significantly, triple-deck theory demonstrates that the viscous, inviscid interaction, in incompressible flow, changes from direct to inverse form as the singular point is traversed in the main flow direction. Although Veldman demon-

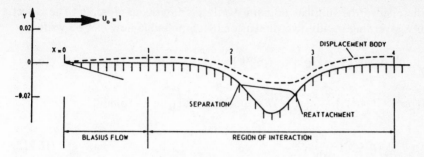

Fig. 16.27. Dented plate for calculation of separation bubbles. The *y* scale is magnified (after Veldman, 1981; reprinted with permission of AIAA)

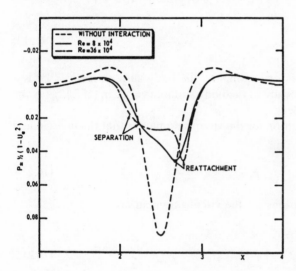

Fig. 16.28. Pressure distributions in a Carter–Wornom trough (after Veldman, 1981; reprinted with permission of AIAA)

strates the quasi-simultaneous interaction method for laminar flow, only, it is expected that the same form of algorithm would be applicable to turbulent flow or to shockwave, boundary layer interaction.

The extension to subsonic or transonic flow is straightforward if any shock waves that are present are weak, so that the inviscid domain solution can be computed from the transonic potential equation (Sect. 14.3.3). If shocks are sufficiently strong that the Euler equations are required in the inviscid domain, the interaction through the displacement thickness needs to be supplemented, as is indicated in Sect. 16.3.7.

16.3.6 Semi-Inverse Interaction Method

This method is shown schematically in Fig. 16.26 which is based on (16.203). Le Balleur's (1981) formulation utilises a defect representation in the viscous region. The defect representation of the governing equations is obtained by subtracting a reduced form of the Navier–Stokes equations from the Euler equations. This is

appropriate since the inviscid domain overlaps the viscous domain. The defect
equations governing steady two-dimensional compressible flow can be written

$$\frac{\partial}{\partial x}[(\varrho u)^i-(\varrho u)^v]+\frac{\partial}{\partial y}[(\varrho v)^i h-(\varrho v)^v h]=0 \ , \tag{16.211}$$

$$\frac{\partial}{\partial x}[(\varrho u^2)^i-(\varrho u^2)^v]+\frac{\partial}{\partial y}[(\varrho u v)^i h-(\varrho u v)^v h]+K[(\varrho u v)^i-(\varrho u v)^v]$$

$$=-\frac{\partial}{\partial x}(p^i-p^v)-\frac{\partial \tau}{\partial y} \ , \tag{16.212}$$

$$\frac{\partial}{\partial x}[(\varrho u v)^i-(\varrho u v)^v]+\frac{\partial}{\partial y}[(\varrho v^2)^i h-(\varrho v^2)^v h]-K[(\varrho u^2)^i-(\varrho u^2)^v]$$

$$=-h\frac{\partial}{\partial y}(p^i-p^v) \ , \tag{16.213}$$

where the coordinate system is orthogonal to the body surface, $h=h_1=1+Ky$,
$h_2=h_3=1$ and $K(x)$ is the curvature of the body surface ($y=0$). In (16.212) τ is the
shear stress, laminar or turbulent.

The farfield boundary condition for the viscous domain is that the inviscid and
viscous solutions coincide i.e.

$$\lim_{y\to\infty} [f^i-f^v]=0 \ , \quad f\equiv\{u, v, \varrho, p\} \ . \tag{16.214}$$

The nearfield boundary conditions for the viscous domain are

i) body surface: $u^v=v^v=0$,

ii) wake centre-line: $[u^v]=[v^v]=0$,   (16.215)

where [ ] denotes the jump.

Nearfield boundary conditions for the inviscid domain are

i) body surface: $(\varrho v)^i=\dfrac{\partial}{\partial x}\displaystyle\int_0^\infty \{(\varrho u)^i-(\varrho u)^v\}\,dy$,

ii) wake centreline: $[(\varrho v)^i]=\dfrac{\partial}{\partial x}\displaystyle\int_{-\infty}^\infty [(\varrho u)^i-(\varrho u)^v]\,dy$ and   (16.216)

$$[p^i]=-\int_{-\infty}^\infty \frac{\partial}{\partial y}(p^i-p^v)\,dy \ .$$

If $\partial(p^i-p^v)/\partial x$ in (16.212) is specified, the system (16.211–215) can be solved in a
single downstream march. Thus it is a reduced Navier–Stokes formulation nomin-
ally equivalent to (16.178–181).

However, Le Balleur (1981), perhaps guided by the additional economy associated with boundary layer integral methods (Cebeci and Bradshaw 1977),

prefers to integrate (16.211 and 212) over y and to add in empirical relations to close the system. The result is a very economical marching algorithm for the viscous region. However, the use of empirical relations implies that the overall conditions, angle of attack, body geometry profile, etc., must be restricted to those broad categories for which the empirical relations are valid. In contrast, the only restrictions in the direct solution of (16.211–216) are those introduced in establishing the reduced form of the Navier–Stokes equations, i.e. neglect of axial viscous diffusion in (16.212) and all viscous terms in (16.213).

Le Balleur et al. (1980) have used the defect formulation and the integral viscous method to establish the following coupling relationship between the inviscid and viscous domains:

$$A_1\Theta = A_2\delta*\frac{dp_e}{dx} + A_3 \ , \qquad (16.217)$$

where $\Theta \equiv \{v/(u^2+v^2)\}_{y=0}$ and A_1, A_2, A_3 and $\delta*$ are determined by the current inviscid and viscous solutions. Equation (16.217) is an equivalent statement to (16.200). In the direct mode, which is suitable for weak interactions, (16.217) provides the injection velocity, v^i, at the body surface to compute a revised inviscid solution. At separation A_1 goes to zero which necessitates either an inverse [equivalent to (16.202)] or semi-inverse [equivalent to (16.203)] approach (Fig. 16.29).

In the direct mode (Fig. 16.29) the predicted flow angle $\Theta*$ from (16.217) is under-relaxed to provide Θ^{n+1}. Close to separation and downstream a semi-inverse method is used. A crucial part of this method is the form of the relationship

$$\Theta* - \Theta^n = f\left[\left(\frac{dp}{dx}\right)^v - \left(\frac{dp}{dx}\right)^i\right] \ , \qquad (16.218)$$

where $(dp/dx)^v$ and $(dp/dx)^i$ are the pressure gradients from the current (inverse) viscous and inviscid solutions. Le Balleur (1981) applies a Fourier analysis to the system of equations underlying (16.217) to determine an explicit functional form for (16.218). This is combined with the relaxation $\Theta^{n+1} = \Theta^n + \omega(\Theta* - \Theta^n)$ to give the

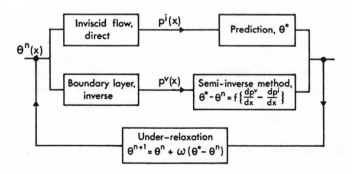

Fig. 16.29. Semi-inverse iteration method

following overall semi-inverse algorithm

$$\Theta^{n+1}-\Theta^n=\frac{\omega\beta}{\left\{\dfrac{\pi}{\Delta x}-\dfrac{\beta}{b}\right\}}\left[\left(\frac{\partial p}{\partial x}\right)^{\mathrm{v}}-\left(\frac{\partial p}{\partial x}\right)^{\mathrm{i}}\right], \tag{16.219}$$

where $\beta=(1-M_{\mathrm{s}}^2)^{1/2}$, $b=A_2\delta^*/A_1$ and $0<\omega<2$. In the above, M_{s} is the surface (inviscid) Mach number. Le Balleur (1981) provides an alternative form to (16.219) when the flow is locally supersonic. It may be noted that although A_1 in (16.217) goes to zero at separation, (16.219) is well-behaved.

Le Balleur (1981) provides the details of the above algorithm and presents results for the flow around inclined transonic aerofoils that include turbulent separation. Excellent agreement with experiments is indicated. Essentially the same approach can be used to investigate shock/boundary layer interactions (Le Balleur 1984).

16.3.7 Viscous, Inviscid Interaction Using the Euler Equations

For flows where strong shocks are expected it is preferable to solve the Euler equations in the inviscid region. This implies that additional attention must be given to the matching between the overlapping inviscid and viscous domains, Johnston and Sockol (1979) have adopted a procedure related to the interactive defect formulation (16.211–213). In the Johnston and Sockol construction the steady two-dimensional Euler equations are written as

$$\frac{\partial \mathbf{F}^{\mathrm{i}}}{\partial x}+\frac{\partial \mathbf{G}^{\mathrm{i}}}{\partial y}=0 \ . \tag{16.220}$$

The components of $\mathbf{F}^{\mathrm{i}}$ and $\mathbf{G}^{\mathrm{i}}$ are given by (14.95).

The steady two-dimensional Navier–Stokes equations, governing the flow immediately adjacent to the wall, are written as

$$\frac{\partial \mathbf{F}^{\mathrm{v}}}{\partial x}+\frac{\partial \mathbf{G}^{\mathrm{v}}}{\partial y}=0 \ . \tag{16.221}$$

The components of $\mathbf{F}^{\mathrm{v}}$ and $\mathbf{G}^{\mathrm{v}}$ in (16.221) are given in Sect. 11.6.3. By integrating (16.220 and 221) across the viscous layer δ and combining, the following condition is obtained:

$$\mathbf{G}_{y=0}^{\mathrm{i}}=\mathbf{G}_{y=0}^{\mathrm{v}}+\frac{\partial}{\partial x}\int_0^\delta (\mathbf{F}^{\mathrm{i}}-\mathbf{F}^{\mathrm{v}})\,dy \ . \tag{16.222}$$

In the viscous layer it is assumed that the Navier–Stokes solution $\mathbf{F}^{\mathrm{v}}$ is approximated by $\mathbf{F}^{\mathrm{c}}=\mathbf{F}^{\mathrm{i}}+\mathbf{F}^{\mathrm{b}}-\mathbf{F}_{y=0}^{\mathrm{i}}$ (Fig. 16.30), where $\mathbf{F}^{\mathrm{b}}$ is an economical boundary layer or reduced Navier–Stokes solution. Substituting for $\mathbf{F}^{\mathrm{c}}$ and $\mathbf{F}^{\mathrm{v}}$ in (16.222) gives

$$\mathbf{G}_{y=0}^{\mathrm{i}}=\mathbf{G}_{y=0}^{\mathrm{b}}+\frac{\partial}{\partial x}\int_0^\delta (\mathbf{F}_{y=0}^{\mathrm{i}}-\mathbf{F}^{\mathrm{b}})\,dy \ . \tag{16.223}$$

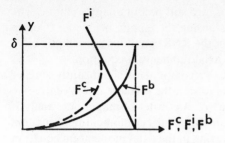

Fig. 16.30. Composite F construction

In practice, (16.223) provides boundary conditions for $G^i(1)$, $G^i(2)$ and $G^i(4)$ at $y=0$. $G^i(3)$ at $y=0$ is the surface pressure and is obtained, typically, from the normal momentum equation. It may be noted that $G^i(1)$ at $y=0$ is just the injection (normal) momentum and the relevant component of (16.223) is equivalent to (16.197). The additional boundary conditions, $G^i(2)_{y=0}$, etc., are required because the Euler equations are used in the inviscid region instead of the transonic potential equation. Le Balleur (1984) indicates that the additional boundary conditions required by the Euler equations are an advantage in the sense that more efficient coupling can be achieved between the inviscid and viscous domains. The Johnson and Sockol matching is also used by Whitfield and Thomas (1984).

In conclusion one may note that the interaction techniques, described in Sects. 16.3.4–16.3.7 have their origin in the classical idea (Lighthill 1958) of allowing the viscous domain to influence the behaviour in the inviscid domain through the displacement effect. However, the extension of this idea to cope with trailing-edge, separation and shock/boundary layer effects leads to a conceptual approach (Le Balleur 1984, pp. 446–447) very similar to the reduced Navier–Stokes philosophy discussed earlier in this chapter.

16.4 Closure

Reduced forms of the Navier–Stokes equations are useful for steady flows with a dominant flow direction if they permit the computational solution to be obtained in a single downstream march, or at worst, a few iterative downstream marches.

The primary simplification arises from an order-of-magnitude analysis which indicates that terms associated with axial (downstream) viscous diffusion or heat conduction can be dropped from the governing equations since they are much smaller than transverse diffusion or heat conduction terms. This simplification implies that the downstream direction is aligned, at least approximately, with a particular coordinate direction. This may require the introduction of generalised curvilinear coordinates (Chap. 12) to improve the alignment.

The use of Fourier analysis applied to a linearised form of the governing equations (Sect. 16.1.2) allows the character of the governing equations to be determined and establishes whether a stable solution can be generated with a single downstream march. Except for the case of inviscid supersonic flow, the basic RNS equations, i.e. with axial diffusion neglected, are still predominantly elliptic. The pressure is found to play a central role in causing the elliptic behaviour. Consequently additional approximations to render the RNS equations non-elliptic often focus on the pressure gradient term in the axial momentum equation.

For internal flows with transverse velocity components significantly smaller than the axial velocity component it is useful to split the pressure into a centre-line exponent $p_{c/l}$ and a transverse correction p^c. An order-of-magnitude analysis indicates that $\partial p^c/\partial x$ can be dropped from the axial (x) momentum equation. Consequently $p_{c/l}$, is the only pressure appearing in the axial momentum equation whereas p^c is the only pressure appearing in the transverse momentum equations. It is this establishment of separate pressures in the axial and transverse momentum equations that leads to a non-elliptic system of equations. This type of splitting is effective for axisymmetric weak swirling flow (Sect. 16.2.1), and duct flows (Sect. 16.2.2) as long as the curvature of the pipe or duct axis is small.

For ducts with significant curvature the transverse pressure gradient becomes so large that the pressure splitting, utilised in Sects. 16.2.1 and 16.2.2, is inaccurate. A splitting of the transverse velocity components, into a rotational part driven by the streamwise vorticity and an irrotational (or potential) part driven by the need to conserve mass, permits the construction of a non-elliptic system of equations, even though the pressure terms are handled without approximation. However, it should be stressed that the viscous solution, in the split transverse velocity formulation (Sect. 16.2.4), is constructed as a *correction* to a preliminary inviscid solution. For subsonic flows this preliminary inviscid solution will have an elliptic character.

For external supersonic viscous flows an accurate solution can be obtained in a single downstream march as long as the axial stepsize is not too small and as long as the subsonic region adjacent to the solid surface is rendered "non-elliptic". The extrapolation of the pressure across the subsonic sublayer from the supersonic region is an effective "non-elliptic" strategy (Sect. 16.3.1). The alternative is to introduce a Vigneron weighting on the $\partial p/\partial x$ term in the axial (x) momentum equation at grid points in the subsonic sublayer.

For external subsonic viscous flows the outer (i.e. away from an isolated body) inviscid flow is elliptic so that it is necessary to construct an iterative procedure of repeated downstream marches for solving the RNS equations. If the equations are only slightly elliptic it is necessary that the ellipticity parameter $\pi \Delta x/y_{max} > \alpha^{1/2}$, as indicated in (16.176) for a stable single march (Sect. 16.3.2). For incompressible flow $(\alpha = 1)$, the group $\pi \Delta x/y_{max}$ cannot be very much less than unity if the global iteration is to converge after a few downstream marches (Sect. 16.3.3). Multigrid techniques (Sect. 6.3.5) are effective in accelerating the iteration.

Most of the techniques considered in this chapter have been described in the context of laminar flow. However, they extend to turbulent flow without altering the non-elliptic nature of the various RNS formulations, as long as the turbulence is modelled via an eddy viscosity construction. Although the qualitative analysis (Sect. 16.1.2) mitigates against the computation of reversed flow with an RNS formulation, the empirical evidence (Sect. 16.3.3) suggests that the RNS equations are adequate if an appropriate iterative algorithm is used. However, if the separated flow region is a significant proportion of the computational domain, the RNS strategy of repeated downstream marches may offer no advantage over a more conventional (Sect. 6.4) treatment of the full Navier–Stokes equations.

Since the equations governing inviscid external flow can be computed more economically than the RNS equations it is usually more efficient to split the domain into a (RNS) region adjacent to the body and an outer inviscid domain. In the RNS formulation described in Sect. 16.3.2 these two regions do not overlap. However, in the traditional inviscid flow, boundary layer splitting the two domains do overlap. For computing strong inviscid/viscous interactions (Sect. 16.3.4) the traditional approach can be modified to account for small regions of reversed flow and the rapid axial variation in skin friction and pressure that occurs at a sharp aerofoil trailing edge. As the traditional approach is extended to handle more complicated flows, Sects. 16.3.5–16.3.7, the resulting formulations resemble more closely the RNS formulation described in Sect. 16.3.2.

All the algorithms described in this chapter have been developed via finite difference discretisation. Other means of discretisation, e.g. the finite element method (Baker 1983, Chap. 7), are used with the RNS equations, but it is customary to use finite difference discretisation in the time-like direction.

16.5 Problems

Introduction to RNS Equations (Sect. 16.1)

16.1 The time-averaged governing equations for incompressible turbulent flow are given by (11.92–94). For a steady time-averaged turbulent flow near the entrance of a two-dimensional duct parallel to the x-axis use an order-of-magnitude analysis to determine a reduced form of the governing equations, equivalent to (16.4–6). Assume that all Reynolds stresses are $O(\delta/L)$.

16.2 In orthogonal coordinates the incompressible Navier–Stokes equations are

$$\frac{\partial}{\partial x}(h_2 u) + \frac{\partial}{\partial y}(h_1 v) = 0 \ ,$$

$$\frac{u}{h_1}\frac{\partial u}{\partial x} + \frac{v}{h_2}\frac{\partial u}{\partial y} + uv K_{12} - v^2 K_{21} + \frac{1}{h_1}\frac{\partial p}{\partial x} - \frac{1}{\mathrm{Re}}\nabla^2 u = 0 \ ,$$

$$\frac{u}{h_1}\frac{\partial v}{\partial x} + \frac{v}{h_2}\frac{\partial v}{\partial y} - u^2 K_{12} + uv K_{21} + \frac{1}{h_2}\frac{\partial p}{\partial y} - \frac{1}{\mathrm{Re}}\nabla^2 v = 0 \ ,$$

where

$$\nabla^2 u \equiv \frac{\partial}{\partial x}\left(\frac{h_2}{h_1}\frac{\partial u}{\partial x}\right) + \frac{\partial}{\partial y}\left(\frac{h_1}{h_2}\frac{\partial u}{\partial y}\right) \ , \qquad K_{12} = \frac{1}{h_1 h_2}\frac{\partial h_1}{\partial y} \ ,$$

$$K_{21} = \frac{1}{h_1 h_2}\frac{\partial h_2}{\partial x} \ .$$

Assuming that h_1, h_2, and K_{12} and K_{21} are $O(1)$ and that $u \gg v$, apply an order-of-magnitude analysis to deduce the following reduced form:

$$\frac{\partial}{\partial x}(h_2 u) + \frac{\partial}{\partial y}(h_1 v) = 0 \; ,$$

$$\frac{u}{h_1}\frac{\partial u}{\partial x} + \frac{v}{h_2}\frac{\partial u}{\partial y} + uvK_{12} + \frac{1}{h_1}\frac{\partial p}{\partial x} - \frac{1}{Re}\frac{\partial}{\partial y}\left\{\frac{h_1}{h_2}\frac{\partial u}{\partial y}\right\} = 0 \; ,$$

$$\frac{u}{h_1}\frac{\partial v}{\partial x} + \frac{v}{h_2}\frac{\partial v}{\partial y} - u^2 K_{12} + uvK_{21} + \frac{1}{h_2}\frac{\partial p}{\partial y} = 0 \; .$$

16.3   Sketch and discuss the form of the solution T with x given by (16.20) for the following cases:

i) $u, v \approx 1$; ε, δ small,
ii) $u \approx 1$; v small; ε, δ small,
iii) u small; $v \approx 1$; ε, δ small,
iv) u small; $v \approx 1$; ε small, δ large,
v) u small; $v \approx 1$, $\varepsilon \approx 1$, $\delta = 0$.

16.4   For incompressible laminar flow past a flat plate aligned parallel to the flow, $\partial p/\partial x \approx 0$ away from the leading edge. Carry out a Fourier analysis of:

i) (16.1–3) with $\partial p/\partial x$ deleted from (16.2),
ii) (16.4–6) with $\partial p/\partial x$ deleted from (16.5).

Discuss the expected character of the solutions in relation to whether the system of equations are elliptic or non-elliptic.

16.5   It is sometimes proposed that the term $\varepsilon \partial^2 \varrho/\partial y^2$ is added to the right-hand side of (16.30) as an aid to stabilising the computation. Here ε is a small positive constant chosen empirically. Apply a Fourier analysis to (16.30–32) with such a modification and determine whether the solution is expected to behave differently from that indicated after (16.37 and 38).

16.6   Obtain solutions using program THRED with ME$=2$, for the cases

i) $\Delta x = 0.20$, $\Delta y = 0.20$,
ii) $\Delta x = 0.05$, $\Delta y = 0.20$,
iii) $\Delta x = 0.05$, $\Delta y = 0.10$,
iv) $\Delta x = 0.01$, $\Delta y = 0.20$.

Compare the accuracy of the centre-line solutions with those produced by RED-FEM and ADIFEM (Table 16.3). Compare the computational efficiency (Sect. 4.5) via an approximate operation count and/or direct measurement of the CPU-time.

Internal Flow (Sect. 16.2)

16.7 Introduce the pressure splitting (16.54) and show that $\partial p^c/\partial x$ in (16.52) is $O((\delta/L)^2)$. It will be helpful to consider the magnitude of $\partial p^c/\partial r$ in (16.53) first.

16.8 After introducing the pressure splitting (16.54) and deleting $\partial p^c/\partial x$ from (16.52) use the Fourier analysis of Sect. 16.1.2 to show that the modified system (16.51–53) is expected to produce a stable computational solution in a single downstream march.

16.9 Show that (16.60) can be obtained from the discretised form of (16.55).

16.10 Apply the Fourier analysis of Sect. 16.1.2 to the system of equations (16.80–83) with a constant p^{inv} and show that exponential solution growth in the x direction is to be expected. Introduce the viscous pressure splitting (16.84), drop $\partial p^c/\partial x$ from (16.81) and show that the resulting system is expected to produce a stable solution from a single march in the x direction.

16.11 Show that replacing f_d in (16.96) with f_d^c from (16.98) satisfies (16.97).

16.12 Use the Fourier analysis of Sect. 16.1.2 to derive (16.120) from the system of equations (16.113, 114, 116, 118 and 119). Obtain the equivalent polynomial to (16.120) if (16.116) is *not* introduced in (16.113) and comment on the expected stability of a single-march solution.

16.13 Obtain an expression for $\Omega_{1,j,k}$, equivalent to (16.130) to ensure that $w=0$ at a constant y wall.

External Flow (Sect. 16.3)

16.14 Consider the sublayer approximation in Cartesian coordinates with the constant total enthalpy condition (16.181). The governing equations are

$$u\frac{\partial\varrho}{\partial x}+v\frac{\partial\varrho}{\partial y}+\varrho\frac{\partial u}{\partial x}+\varrho\frac{\partial v}{\partial y}=0\;,$$

$$\varrho u\frac{\partial u}{\partial x}+\varrho v\frac{\partial u}{\partial y}+\frac{\partial p}{\partial x}-\frac{1}{Re}\frac{\partial^2 u}{\partial y^2}=0\;,$$

$$\varrho u\frac{\partial v}{\partial x}+\varrho v\frac{\partial v}{\partial y}-\frac{1}{Re}\frac{\partial^2 v}{\partial y^2}=0\;,$$

plus (16.181).
Apply a Fourier analysis to derive the characteristic polynomial

$$\varrho\left(i\varrho\Lambda+\frac{\sigma_y^2}{Re}\right)\left[(a^2-u^2)\sigma_x^2-(\gamma+1)uv\sigma_x\sigma_y-\gamma v^2\sigma_y^2+i\left(\frac{\gamma\Lambda}{\varrho Re}\right)\sigma_y^2\right]=0\;,$$

where $\Lambda=u\sigma_x+v\sigma_y$. Show that the first factor does not produce a growing exponential solution if u is positive. Show that, if the last term in the second factor can be neglected, stable solutions in the x direction will be obtained.

16.15 For subsonic inviscid flow a Vigneron weighting applied to $\partial p/\partial y$ gives the following governing equations in place of (16.156–158):

$$u\frac{\partial\varrho}{\partial x}+v\frac{\partial\varrho}{\partial y}+\varrho\frac{\partial u}{\partial x}+\varrho\frac{\partial v}{\partial y}=0 \ ,$$

$$\frac{a^2}{\gamma}\frac{\partial\varrho}{\partial x}+\frac{\varrho u}{\gamma}\frac{\partial u}{\partial x}+\varrho v\frac{\partial u}{\partial y}-\frac{(\gamma-1)}{\gamma}\varrho v\frac{\partial v}{\partial x}=0 \ ,$$

$$\omega\frac{a^2}{\gamma}\frac{\partial\varrho}{\partial y}+\varrho u\frac{\partial v}{\partial x}+\varrho v\frac{\partial v}{\partial y}-\omega\frac{(\gamma-1)}{\gamma}\left(\varrho u\frac{\partial u}{\partial y}+\varrho v\frac{\partial v}{\partial y}\right)=0 \ .$$

Utilise a Fourier analysis of these equations to show that the following characteristic polynomial is obtained:

$$\frac{\varrho^2}{\gamma}(u\sigma_x+v\sigma_y)(\sigma_x^2(u^2-a^2)+\sigma_x\sigma_y uv[(\gamma+1)-(\gamma-1)\omega]$$

$$+\sigma_y^2\{v^2[\gamma-(\gamma-1)\omega]-a^2\omega\})=0 \ .$$

Hence show that if $v\approx0$, there is no choice of ω that will avoid exponential growth in x if $u<a$.

16.16 The reduced incompressible Navier–Stokes equations can be solved in a time-like iterative manner if they are written

$$\frac{\partial u}{\partial x}+\frac{\partial v}{\partial y}=0 \ ,$$

$$u\frac{\partial u}{\partial x}+v\frac{\partial u}{\partial y}+\frac{\partial p}{\partial x}+\alpha\frac{\partial^2 p}{\partial x\partial t}-\frac{1}{Re}\frac{\partial^2 u}{\partial y^2}=0 \ ,$$

$$u\frac{\partial v}{\partial x}+v\frac{\partial v}{\partial y}+\frac{\partial p}{\partial y}-\frac{1}{Re}\frac{\partial^2 v}{\partial y^2}=0 \ ,$$

where $\alpha\partial^2 p/\partial x\partial t$ is introduced to stabilise the time-like pressure iteration. Apply the Fourier analysis to show that the time-like iteration will be stable if α is positive.

16.17 Show that (16.184) can be derived from (16.178, 180 and 181).

16.18 Carry out Taylor series expansions of (16.191) about $(j, k-1/2)$, (16.192) about (j, k) and (16.193) about $(j, k-1/2)$ and show that all three equations have truncation errors of $O(\Delta x, \Delta y^2)$.

16.19 Derive (16.197) from the incompressible continuity equation and the definition of the displacement thickness (11.67). What would be the equivalent expression for compressible flow?

16.20 Derive (16.223) from (16.220 and 221) and obtain explicit expressions for $G_{y=0}^i$, $i=1, 2, 4$.

17. Incompressible Viscous Flow

In this chapter no assumption is made about the relative magnitude of the velocity components, consequently, reduced forms of the Navier–Stokes equations (Chap. 16) are not available. Instead the full Navier–Stokes equations must be considered; however, it will be assumed that the flow is incompressible.

The techniques developed in this chapter will be applicable to problems where there is no obviously dominant flow direction, e.g. room ventilation. In addition, for many flow geometries regions of reversed flow will occur. If these regions are large or occur in conjunction with flow unsteadiness, e.g. flow past a bluff body, the (repeated) marching techniques described in Chap. 16 are not suitable.

The restriction to incompressible flow introduces the computational difficulty that the continuity equation (11.13) contains only velocity components, and there is no obvious link with the pressure as there is for compressible flow through the density. Two broad approaches to computing incompressible flow are available.

First the primitive variables, (u, v, p) in two-dimensions, are used and special procedures are introduced to handle the continuity equation. The extension to three spatial dimensions creates no additional difficulty. Appropriate computational techniques utilizing primitive variable formulations for unsteady flow are considered in Sect. 17.1. Alternative primitive variable formulations better suited to steady flow are described in Sect. 17.2.

In two dimensions the explicit treatment of the continuity equation can be avoided by introducing the stream function. In addition, construction of a transport equation for the vorticity leads to the stream function vorticity formulation which is described in Sect. 17.3. The extension of this formulation to three dimensions is not straightforward, since a three-dimensional stream function is not available. Instead it is necessary to consider the vorticity, vector potential formulation (Sect. 17.4.1).

Most flows of practical interest will be turbulent, unless the Reynolds number is very small. The effects of turbulence on incompressible viscous flow are usually accounted for either by an algebraic eddy viscosity representation (Sect. 11.4.2) or a k–ε turbulence model (Sect. 11.5.2). From a computational perspective the use of an algebraic eddy viscosity construction causes little change to the corresponding algorithm used for laminar flow. The differential equations for k and ε (11.95 and 96) are structurally similar to the momentum equations and are usually discretised in the same way. Thus computational algorithms that are effective for incompressible laminar viscous flow are also effective, with minor adjustments, for incompressible turbulent flow. Consequently, no explicit attention will be given to the computation of turbulent flow in this chapter.

17.1 Primitive Variables: Unsteady Flow

In two dimensions the governing equations for unsteady incompressible laminar flow are

$$\frac{\partial u}{\partial x}+\frac{\partial v}{\partial y}=0 \; , \tag{17.1}$$

$$\frac{\partial u}{\partial t}+\frac{\partial u^2}{\partial x}+\frac{\partial}{\partial y}(uv)+\frac{\partial p}{\partial x}=\frac{1}{Re}\left\{\frac{\partial^2 u}{\partial x^2}+\frac{\partial^2 u}{\partial y^2}\right\} \; , \tag{17.2}$$

$$\frac{\partial v}{\partial t}+\frac{\partial}{\partial x}(uv)+\frac{\partial v^2}{\partial y}+\frac{\partial p}{\partial y}=\frac{1}{Re}\left\{\frac{\partial^2 v}{\partial x^2}+\frac{\partial^2 v}{\partial y^2}\right\} \; . \tag{17.3}$$

Equations (17.1–3) are written in nondimensional form with the density absorbed into the Reynolds number, Re. By making use of (17.1), the left-hand sides of (17.2 and 3) have been written in conservation form, as in (11.116).

For unsteady flow, initial conditions $u=u_0(x, y)$, $v=v_0(x, y)$ are required that satisfy (17.1). Boundary conditions at a solid surface require no relative motion between the fluid and solid surface, which fixes the velocity components. No boundary conditions are required for the pressure at a solid surface. If velocity components are specified on all boundaries of the computational domain, e.g. as in the driven-cavity problem (Sect. 17.3.1), it is necessary to ensure that the following global constraint is satisfied:

$$\int_c \mathbf{v}\cdot\mathbf{n}\; ds=0 \; , \tag{17.4}$$

where c is the boundary of the computational domain. Equation (17.4) is the global equivalent of (17.1), as can be seen from comparing (11.7) and (11.10) with constant density ϱ.

If the computational domain includes open boundaries, as in the backward-facing step problem, Sect. 17.3.3, the number of boundary conditions required on open boundaries can be obtained from Table 11.5. At an inflow boundary two boundary conditions are required for two-dimensional flow, typically specification of one velocity component and pressure. At an outflow boundary zero normal derivatives for the velocity components would be appropriate with no boundary condition on the pressure. In some applications it is useful to specify both velocity components at inflow and pressure and zero normal derivative of the streamwise velocity component at outflow. Since pressure appears only in derivative form in the governing equations it is appropriate to specify the magnitude of the pressure at one reference point.

It should be stressed that the above boundary condition specification is made to ensure that the combination of the governing equations and boundary conditions constitutes a well-posed problem with a well-behaved solution. However, additional boundary conditions may be required to ensure that the discretised equations, which are actually solved, lead to a well-behaved computational solution.

The methods described in this section will be based, primarily, on finite difference discretisations and on the solution of a Poisson equation to determine the pressure behaviour (Sect. 17.1.2). To obtain accurate solutions without excessive grid refinement it is often necessary to discretise the convective terms more accurately (Sect. 17.1.5). Many of the solution strategies described in this section in a finite difference context are also applicable with other means of discretisation, e.g. spectral methods (Sect. 17.1.6).

17.1.1 Staggered Grid

Computational solutions of (17.1–3) are often obtained on a staggered grid. This implies that different dependent variables are evaluated at different grid points. Peyret and Taylor (1983, p. 155) compare various staggered grids for the treatment of the pressure. Haltiner and Williams (1980, p. 226) discuss the ability of different staggered grid configurations to represent different Fourier modes (as in Sect. 9.2.1) of the solution to the shallow water equations, which are analogous to the Euler equations (Chap. 14). The preferred staggered grid configuration is that shown in Fig. 17.1.

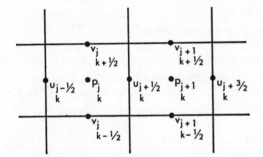

Fig. 17.1. Staggered grid

It can be seen that pressures are defined at the centre of each cell and that velocity components are defined at the cell faces. Such an arrangement makes the grid suitable for a control volume discretisation (Sect. 5.2); this connection will be exploited in Sect. 17.2.3. Discretisation of (17.1) on the staggered grid shown in Fig. 17.1 gives

$$\frac{u_{j+1/2,k}-u_{j-1/2,k}}{\Delta x}+\frac{v_{j,k+1/2}-v_{j,k-1/2}}{\Delta y}=0 , \tag{17.5}$$

which can be rewritten as

$$u_{j+1/2,k}\,\Delta y+v_{j,k+1/2}\,\Delta x-u_{j-1/2,k}\,\Delta y-v_{j,k-1/2}\,\Delta x=0 . \tag{17.6}$$

Equation (17.6) is a discrete form of (17.4), i.e. (17.6 and 5) conserve mass on the smallest grid size. In addition a Taylor series expansion about the cell centre indicates that (17.5) has a truncation error of $O(\Delta x^2, \Delta y^2)$, even though only four grid points are involved.

In discretising (17.2) finite difference expressions centred at grid point $(j+1/2, k)$ are used. This allows $\partial p/\partial x$ to be discretised as $(p_{j+1,k}-p_{j,k})/\Delta x$ which is a second-order discretisation about grid point $(j+1/2, k)$. Similarly (17.3) is discretised with finite difference expressions centred at grid point $(j, k+1/2)$ and $\partial p/\partial y$ is represented as $(p_{j,k+1}-p_{j,k})/\Delta y$.

The use of the staggered grid permits coupling of the u, v and p solutions at adjacent grid points. This in turn prevents the appearance of oscillatory solutions, particularly for p, that can occur if centred differences are used to discretise all derivatives on a non-staggered grid. The oscillatory solution is a manifestation of two separate pressure solutions associated with alternate grid points, which the use of centred differences on a non-staggered grid permits. The oscillatory behaviour is usually worse at high Reynolds number where the dissipative terms, which do introduce adjacent grid point coupling for u and v, are small. Clearly, from (17.1–3), there are no dissipative terms for p.

An additional advantage of a staggered grid is that the Poisson equation for the pressure (17.13) automatically satisfies the discrete form of the integral boundary condition (17.4). This avoids additional adjustments to the right-hand side of the Poisson equation, as is required in (16.98).

The use of staggered grids has some disadvantages. Computer programs based on staggered grids tend to be harder to interpret because it is desirable to associate a cluster of dependent variables with corresponding storage locations. Thus arrays storing u, v and p might associate storage location (j, k) with $u_{j+1/2,k}$, and $v_{j,k+1/2}$ and $p_{j,k}$ in Fig. 17.1. Generally boundary conditions are more difficult to impose consistently with a staggered grid, since at least one dependent variable, u or v, will not be defined on a particular boundary. If the grid is non-rectangular, and generalised coordinates (Chap. 12) are used, the incorporation of a staggered grid is more complicated.

The staggered grid shown in Fig. 17.1 is used in the MAC method (Sect. 17.1.2). In discretising (17.1–3) the following finite difference expressions are utilised:

$$\left[\frac{\partial u}{\partial t}\right]_{j+1/2,k} = \frac{(u^{n+1}_{j+1/2,k}-u^{n}_{j+1/2,k})}{\Delta t}+O(\Delta t) \ ,$$

$$\left[\frac{\partial u^2}{\partial x}\right]_{j+1/2,k} = \frac{(u^2_{j+1,k}-u^2_{j,k})}{\Delta x}+O(\Delta x^2) \ ,$$

$$\left[\frac{\partial (uv)}{\partial y}\right]_{j+1/2,k} = \frac{[(uv)_{j+1/2,k+1/2}-(uv)_{j+1/2,k-1/2}]}{\Delta y}+O(\Delta y^2) \ , \qquad (17.7)$$

$$\left[\frac{\partial^2 u}{\partial x^2}\right]_{j+1/2,k} = \frac{(u_{j+3/2,k}-2u_{j+1/2,k}+u_{j-1/2,k})}{\Delta x^2}+O(\Delta x^2) \ ,$$

$$\left[\frac{\partial^2 u}{\partial y^2}\right]_{j+1/2,k} = \frac{(u_{j+1/2,k-1}-2u_{j+1/2,k}+u_{j+1/2,k+1})}{\Delta y^2}+O(\Delta y^2) \ ,$$

$$\left[\frac{\partial p}{\partial x}\right]_{j+1/2,k} = \frac{(p_{j+1,k}-p_{j,k})}{\Delta x}+O(\Delta x^2) \ .$$

In the above expressions terms like $u_{j+1,k}$ appear, which are not defined in Fig. 17.1. To evaluate such terms averaging is employed, i.e.

$$u_{j+1,k} = 0.5(u_{j+1/2,k} + u_{j+3/2,k}) \ .$$

Similarly $(uv)_{j+1/2,k+1/2}$ is evaluated as

$$(uv)_{j+1/2,k+1/2} = [(u_{j+1/2,k} + u_{j+1/2,k+1})/2][(v_{j+1,k+1/2} + v_{j,k+1/2})/2] \ .$$

17.1.2 MAC Formulation

One of the earliest, and most widely used, methods for solving (17.1–3) is the Marker and Cell (MAC) method due to Harlow and Welch (1965). The method is characterised by the use of a staggered grid (Sect. 17.1.1) and the solution of a Poisson equation for the pressure at every time-step. Although the original form of the MAC method has certain weaknesses, the use of a staggered grid and a Poisson equation for the pressure has been retained in many modern methods derived from the MAC method.

The method was developed initially for unsteady problems involving free surfaces. To allow the surface location to be determined as a function of time, markers (massless particles) are introduced into the flow. The markers are convected by the velocity field but play no role in determining the velocity or pressure fields. They will not be discussed further here. The impressive ability of the MAC formulation to give qualitatively correct simulations of complicated free-surface flows is illustrated in Fig. 17.2, which shows the time history of a drop falling into a stationary fluid.

In the MAC formulation the discretisations (17.7) allow the following explicit algorithm to be generated from (17.2 and 3):

$$u_{j+1/2,k}^{n+1} = F_{j+1/2,k}^{n} - \frac{\Delta t}{\Delta x}[p_{j+1,k}^{n+1} - p_{j,k}^{n+1}] \ , \tag{17.8}$$

where

$$F_{j+1/2,k}^{n} = u_{j+1/2,k}^{n} + \Delta t \left[\frac{\{u_{j+3/2,k} - 2u_{j+1/2,k} + u_{j-1/2,k}\}}{Re\Delta x^2} \right.$$
$$+ \frac{\{u_{j+1/2,k-1} - 2u_{j+1/2,k} + u_{j+1/2,k+1}\}}{Re\Delta y^2} - \frac{\{u_{j+1,k}^2 - u_{j,k}^2\}}{\Delta x}$$
$$\left. - \frac{\{(uv)_{j+1/2,k+1/2} - (uv)_{j+1/2,k-1/2}\}}{\Delta y} \right]^{n} \ . \tag{17.9}$$

Similarly the discretised form of (17.3) can be written as

$$v_{j,k+1/2}^{n+1} = G_{j,k+1/2}^{n} - \frac{\Delta t}{\Delta y}(p_{j,k+1}^{n+1} - p_{j,k}^{n+1}) \ , \tag{17.10}$$

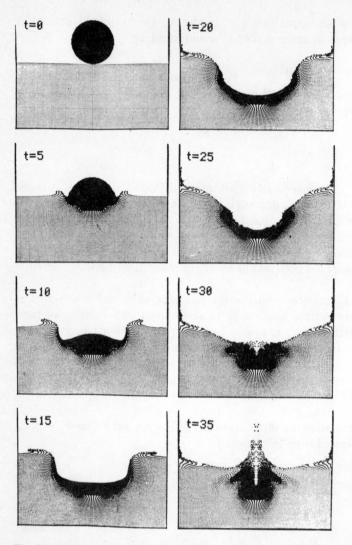

Fig. 17.2. Falling drop problem (after Harlow and Shannon, 1967; reprinted with permission of the American Association for the Advancement of Science)

where

$$G^n_{j,k+1/2} = v^n_{j,k+1/2} + \Delta t \left[\frac{\{v_{j+1,k+1/2} - 2v_{j,k+1/2} + v_{j-1,k+1/2}\}}{Re\Delta x^2} \right.$$

$$+ \frac{\{v_{j,k+3/2} - 2v_{j,k+1/2} + v_{j,k-1/2}\}}{Re\Delta y^2} - \frac{\{(uv)_{j+1/2,k+1/2} - (uv)_{j-1/2,k+1/2}\}}{\Delta x}$$

$$\left. - \frac{\{v^2_{j,k+1} - v^2_{j,k}\}}{\Delta y} \right]^n .$$

(17.11)

In (17.8 and 10) p appears implicitly; however, p^{n+1} is obtained before (17.8 and 10) are used, as follows. The continuity equation (17.1) is discretised as

$$D_{j,k}^{n+1} = \frac{(u_{j+1/2,k}^{n+1} - u_{j-1/2,k}^{n+1})}{\Delta x} + \frac{(v_{j,k+1/2}^{n+1} - v_{j,k-1/2}^{n+1})}{\Delta y} = 0 \; , \tag{17.12}$$

where $D_{j,k}$ is the dilatation for cell (j, k).

Substitution for $u_{j+1/2,k}^{n+1}$, etc., from (17.8 and 10) allows (17.12) to be rewritten as a discrete Poisson equation for the pressure, i.e.

$$\left[\frac{(p_{j-1,k} - 2p_{j,k} + p_{j+1,k})}{\Delta x^2} + \frac{(p_{j,k-1} - 2p_{j,k} + p_{j,k+1})}{\Delta y^2} \right]^{n+1}$$

$$= \frac{1}{\Delta t} \left[\frac{\{F_{j+1/2,k}^{n} - F_{j-1/2,k}^{n}\}}{\Delta x} + \frac{\{G_{j,k+1/2}^{n} - G_{j,k-1/2}^{n}\}}{\Delta y} \right] \; . \tag{17.13}$$

If the various terms on the right-hand side of (17.13) are substituted from (17.9 and 11) the result can be written

$$\text{RHS}_{(17.13)} = \frac{D_{j,k}^{n}}{\Delta t} - [L_{xx} u_{j,k}^2 + 2L_{xy}(uv)_{j,k} + L_{yy} v_{j,k}^2$$

$$- (1/\text{Re}\{L_{xx} + L_{yy}\} D_{j,k})]^{n} \; , \quad \text{where} \tag{17.14}$$

$$L_{xx} u_{j,k}^2 = (u_{j-1,k}^2 - 2u_{j,k}^2 + u_{j+1,k}^2)/\Delta x^2 \quad \text{and}$$

$$L_{xy}(uv)_{j,k} = \{(uv)_{j+1/2,k+1/2} - (uv)_{j-1/2,k+1/2} - (uv)_{j+1/2,k-1/2}$$

$$+ (uv)_{j-1/2,k-1/2}\}/\Delta x \Delta y \; .$$

In (17.14) $D_{j,k}^{n}/\Delta t$ may be interpreted as a discretisation of $-\partial D/\partial t|_{j,k}$ with $D_{j,k}^{n+1} = 0$. Thus the converged pressure solution resulting from (17.13) is such as to cause the discrete form of the continuity equation to be satisfied at time level $n+1$.

Equation (17.13) is solved at each time-step, either using the iterative techniques described in Sect. 6.3 or the direct Poisson solvers described in Sect. 6.2.6. Once a solution for p^{n+1} has been obtained from (17.13), substitution into (17.8 and 10) permits $u_{j+1/2,k}^{n+1}$ and $v_{j,k+1/2}^{n+1}$ to be computed.

Since (17.8 and 10) are explicit algorithms for u^{n+1} and v^{n+1} there is a restriction on the maximum time step for a stable solution (Peyret and Taylor 1983, p. 148),

$$0.25(|u| + |v|)^2 \, \Delta t \, \text{Re} \leq 1 \quad \text{and}$$

$$\tag{17.15}$$

$$\Delta t/(\text{Re} \, \Delta x^2) \leq 0.25 \; , \quad \text{assuming that } \Delta x = \Delta y \; .$$

Solution of (17.13) requires boundary conditions on p (Dirichlet) or, preferably, normal derivatives of p (Neumann) on all boundaries. For the flow over a backward-facing step (Fig. 17.14), it would be appropriate to impose Dirichlet boundary conditions on AF and AB and Neumann (normal) boundary conditions on the walls *FE*, *ED* and *DC*. Typically the discretised form of the normal momentum equation is used to provide Neumann boundary conditions. For boundaries like *FE*, where the primary flow is parallel to the surface, the boundary layer assumption $\partial p/\partial n = 0$ can provide an appropriate Neumann boundary condition for p, if Re is large.

For internal flow problems Neumann (normal) boundary conditions are often prescribed for the pressure on all boundaries. In this case it is necessary that an additional global boundary condition be satisfied (as in Sect. 16.2.2). That is,

$$\iint \left(\frac{\partial^2 p}{\partial x^2} + \frac{\partial^2 p}{\partial y^2} \right) dx\, dy = \int_c \frac{\partial p}{\partial n} ds \ , \tag{17.16}$$

where the integral over c is made along the boundary of the computational domain. The left-hand side of (17.16) is evaluated in discrete form from the right-hand side of (17.13).

If the discrete form of (17.16) is applied with the MAC method to an internal cell, e.g. cell j, k in Fig. 17.1, it is satisfied exactly if $\partial p/\partial n$ on the right-hand side of (17.16) is evaluated from the normal momentum equation. If (17.16) is applied over the complete computational domain it is necessary that mass is conserved globally, i.e. (17.4) is satisfied, and that $\partial p/\partial n$ is evaluated on the boundary from the momentum equations in a manner consistent with the interior discretisation, or as $\partial p/\partial n = 0$, if appropriate.

Failure to satisfy the discrete form of (17.16) causes either very slow convergence of the solution of (17.13) or even divergence. Even when (17.16) is satisfied the occurrence of Neumann boundary conditions for the pressure leads to a slower convergence of the iteration than if all boundary conditions are of Dirichlet type.

However the use of Neumann boundary conditions for the pressure equation is recommended since the resulting solution is more likely to also be a solution of (17.1) to (17.3). An interesting discussion of the "correct" boundary conditions to be used with Poisson pressure equations is provided by Gresho and Sani (1987).

17.1.3 Implementation of Boundary Conditions

The grid is arranged so that boundaries pass through velocity points but not pressure points. For example, Fig. 17.3 shows the corner of a computational domain for which it is assumed that *BC* is a solid wall and *AB* is an inflow boundary.

Clearly $v_{1, 1/2} = v_{2, 1/2} = \ldots = 0$, since *BC* is a solid wall. Evaluation of (17.9) at node $(\frac{3}{2}, 1)$ requires knowledge of $u_{3/2, 0}$. This can be obtained from the wall value,

$$u_{3/2, 1/2} = 0 = 0.5(u_{3/2, 1} + u_{3/2, 0}) \quad \text{or} \quad u_{3/2, 0} = -u_{3/2, 1} \ .$$

On *AB*, u and v are given. The component u is used directly but $v_{1/2, k}$ is used to give a value to $v_{0, k}$. Thus in the evaluation of (17.11) at node $(1, \frac{3}{2})$, $v_{0, 3/2}$ is given by

$$v_{0, 3/2} = 2 v_{1/2, 3/2} - v_{1, 3/2} \ .$$

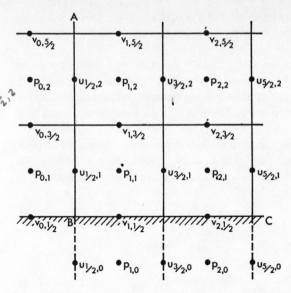

Fig. 17.3. Typical boundary orientation in relation to the staggered grid

If AB were an outflow boundary, with u positive, typical boundary conditions would be

$$\left.\frac{\partial u}{\partial x}\right|_{AB}=0 \ , \qquad \left.\frac{\partial v}{\partial x}\right|_{AB}=0 \ . \tag{17.17}$$

If (17.9) is evaluated on AB at node $(\frac{1}{2},2)$, (17.17) would be used to set $u_{3/2,2}=u_{-1/2,2}$. Similarly in evaluating (17.11) at node $(0,\frac{3}{2})$, (17.17) would be used to set $v_{1,3/2}=v_{0,3/2}$.

The evaluation of the Poisson equation for the pressure (17.13) requires values of the pressure outside of the domain. When (17.13) is evaluated centred at node $(2,1)$ values of $p_{2,0}$ and $v_{2,-1/2}$ are required. $p_{2,0}$ is obtained by evaluating (17.3) centred at the wall, i.e. $\partial p/\partial y=(\partial^2 v/\partial y^2)/\mathrm{Re}$, since v at the boundary is not a function of time. In discretised form this becomes

$$\frac{p_{2,1}-p_{2,0}}{\Delta y}=\frac{1}{\mathrm{Re}}\frac{v_{2,3/2}-2v_{2,1/2}+v_{2,-1/2}}{\Delta y^2} \ .$$

To satisfy (17.1) at the wall, $\partial v/\partial y=0$ so that $v_{2,-1/2}=v_{2,3/2}$ and

$$p_{2,0}=p_{2,1}-\frac{2v_{2,3/2}}{\mathrm{Re}\,\Delta y}$$

Harlow and Welch (1965) and Viecelli (1971) discuss appropriate boundary condition implementation at free surfaces.

17.1.4 Developments of the MAC Method

In the MAC method the pressure solution has the auxiliary task of satisfying continuity. A simplified Marker and Cell (SMAC) method has been developed by

Amsden and Harlow (1970) in which a second Poisson equation for an auxiliary velocity potential is solved to satisfy continuity more directly. A similar concept is discussed in Section 17.2.2.

In the original MAC formulation it is necessary to evaluate pressures outside the domain using (17.2) or (17.3), when Neumann boundary conditions are required for p. Easton (1972) showed that a homogeneous Neumann boundary condition ($\partial p/\partial n = 0$) could be used instead, which is both more economical and easier to implement.

In the current notation the justification for a homogeneous Neumann boundary condition for p will be established for the node $(1, 2)$ adjacent to boundary AB in Fig. 17.3. The discrete Poisson equation for the pressure, centred at node $(1, 2)$ can be written

$$
\left[\frac{-p_{1,2}+p_{2,2}}{\Delta x^2} + \frac{p_{1,1}-2p_{1,2}+p_{1,3}}{\Delta y^2} \right]^{n+1} = \frac{(p_{1,2}-p_{0,2})^{n+1}}{\Delta x^2}
$$
$$
+ \frac{1}{\Delta t}\left[\frac{F_{3/2,2}^n - F_{1/2,2}^n}{\Delta x} + \frac{G_{1,5/2}^n - G_{1,3/2}^n}{\Delta y} \right].
\tag{17.18}
$$

The Neumann boundary condition for the pressure centred at node $(\frac{1}{2}, 2)$ can be obtained from (17.8) as

$$
\frac{(p_{1,2}-p_{0,2})^{n+1}}{\Delta x} = \frac{F_{1/2,2}^n - u_{1/2,2}^{n+1}}{\Delta t}.
\tag{17.19}
$$

If (17.19) is used to eliminate $(p_{1,2}-p_{0,2})$ from (17.18) the result is found to be independent of $F_{1/2,2}^n$ and, hence, independent of any u or v values, outside of the domain, that appear in (17.9). Since the solution is independent of $F_{1/2,2}^n$ the following substitution can be made in (17.18), $F_{1/2,2}^n = u_{1/2,2}^{n+1}$, and from (17.19), $p_{0,2} = p_{1,2}$. Thus a homogeneous Neumann boundary condition is introduced for the pressure. The term $u_{1/2,2}^{n+1}$ is available as a boundary condition.

It should be emphasized that the permissible use of $\partial p/\partial n = 0$ is specific to the present particular discretisation. As a boundary condition for the pressure Poisson equation the use of the normal momentum equation, (17.2) or (17.3), to evaluate $\partial p/\partial n$ reduces to $\partial p/\partial n = 0$ for a uniform flow in the far-field. However adjacent to a solid surface or if the flow at the boundary is not uniform the use of the normal momentum equation to evaluate $\partial p/\partial n$ is preferred to forcing $\partial p/\partial n = 0$.

The present description broadly follows Peyret and Taylor (1983, p. 164) who also bring out important similarities between the MAC method and the projection method (17.22–24).

For many time-dependent problems the time-step restriction (17.15) required by the explicit treatment of (17.8 and 10) is unnecessarily restrictive. The extension of the MAC formulation to allow (17.2 and 3) to be marched in time with implicit approximate factorisations of the velocity terms is provided by Deville (1974) for very low Reynolds number flows and by Ghia et al. (1979) for high Reynolds number flows. A general discussion is provided by Peyret and Taylor (1983, pp. 164–166).

The discrete equations to advance the velocity field (17.8, 10) can be written in symbolic vector form as

$$\mathbf{u}^{n+1} = \mathbf{F}^n - \varDelta t \nabla_d p^{n+1} \; , \tag{17.20}$$

where $\mathbf{F} \equiv (F, G)^T$ and ∇_d denotes the discrete gradient operator.

The Poisson equation for the pressure (17.13) can be written as

$$\nabla_d^2 p^{n+1} = \frac{1}{\varDelta t} \nabla_d \cdot \mathbf{F}^q \; . \tag{17.21}$$

Equations (17.20 and 21) provide a concise description of the MAC method.

Alternative MAC-like methods are available. The projection method is closely related to the MAC method and was proposed independently by Chorin (1968) and Temam (1969). In the present notation the projection method splits (17.20) into two stages

$$\mathbf{u}^* = \mathbf{F}^n \quad \text{and} \tag{17.22}$$

$$\mathbf{u}^{n+1} = \mathbf{u}^* - \varDelta t \nabla_d p^{n+1} \; . \tag{17.23}$$

Substituting (17.23) into (17.12), written as $\nabla_d \mathbf{u}^{n+1} = 0$, gives

$$\nabla_d^2 p^{n+1} = \frac{1}{\varDelta t} \nabla_d \cdot \mathbf{u}^* \; , \tag{17.24}$$

to ensure that $\mathbf{u}^{n+1}$ satisfies (17.1) as well as (17.2 and 3). In the projection method (17.22) is solved for $\mathbf{u}^*$, (17.24) for p^{n+1} and (17.23) for $\mathbf{u}^{n+1}$. Originally the projection method was formulated on a non-staggered grid. However, Peyret and Taylor (1983, p. 161) recommend that the projection method be used with a staggered (MAC) grid (Fig. 17.1). It can be seen from (17.22 and 24) that the projection method coincides with the MAC method in the interior. However, the treatment of the boundary condition is slightly different. Peyret and Taylor show that for the projection method the solution is independent of the prescription of $\mathbf{u}^*$ on the boundary. This is essentially equivalent to the cancellation of $F_{1/2,2}^n$ from (17.18) after substituting (17.19).

Goda (1979) has used the projection method to obtain the steady viscous flow in two and three-dimensional driven cavities. To avoid the explicit time-step restriction (17.15) Goda replaces (17.22) with

$$\begin{aligned} \left[I + \varDelta t \left(u L_x - \frac{1}{\mathrm{Re}} L_{xx} \right) \right] \mathbf{u}' &= \mathbf{u}^n \; , \\[4pt] \left[I + \varDelta t \left(v L_y - \frac{1}{\mathrm{Re}} L_{yy} \right) \right] \mathbf{u}'' &= \mathbf{u}' \; , \\[4pt] \left[I + \varDelta t \left(w L_z - \frac{1}{\mathrm{Re}} L_{zz} \right) \right] \mathbf{u}^* &= \mathbf{u}'' \; . \end{aligned} \tag{17.25}$$

Equation (17.25) is an approximate factorisation, similar to those discussed in Sects. 8.2 and 9.5, that leads to a sequence of tridiagonal systems of equations if L_x and L_{xx}, etc., are three-point finite difference operators. Goda indicates the need to apply a CFL-type restriction of the form $\Delta t \leq \Delta x / |u_{max}|$ for stable solutions.

Another variant of the basic MAC formulation (17.20 and 21) is that due to Hirt and Cook (1972). In the present notation a preliminary velocity field is obtained from (17.2 and 3) as

$$\mathbf{u}^* = \mathbf{F}^n - \Delta t \, \nabla_d p^n \ .$$ (17.26)

A pressure correction, $\delta p = p^{n+1} - p^n$, is evaluated from

$$\nabla_d^2 \delta p = \frac{1}{\Delta t} \nabla_d \mathbf{u}^* \ .$$ (17.27)

The pressure correction is used to ensure that $\mathbf{u}^{n+1}$ satisfies continuity, i.e.

$$\mathbf{u}^{n+1} = \mathbf{u}^* - \frac{1}{\Delta t} \nabla_d \delta p \ .$$ (17.28)

Finally the new pressure is given by

$$p^{n+1} = p^n + \delta p \ .$$ (17.29)

Boundary conditions are implemented as in the MAC formulation (Sect. 17.1.3). Hirt and Cook have used the above formulation to examine incompressible (laminar) viscous flow past three-dimensional structures.

The Hirt and Cook formulation has been used to examine unsteady turbulent flow in a ventilated three-dimensional room by Sakamoto and Matuo (1980) and by Kato and Murakami (1986) with a k–ε turbulence model (Sect. 11.5.2). A comparison of the experimental and simulated flow behaviour in room model is shown in Fig. 17.4. These results were obtained using a $20(x) \times 24(y) \times 15(z)$ grid, which was the coarsest grid that gave a reasonable prediction of the main recirculating flow and secondary flows. The flow is driven by an inlet in the roof. The ensuing jet strikes the floor and causes a recirculation up the walls in the symmetry planes, Fig. 17.4a and d. At floor level the outflow pattern, Fig. 17.4c, is well predicted. At roof level, Fig. 17.4e, a return flow induced by the inlet jet is apparent.

The description of the MAC-family of methods given above has assumed that computational boundaries coincide with Cartesian coordinate lines. For computational domains of arbitrary shape it is possible to introduce boundary-fitted curvilinear coordinates (Chap. 12), transform the governing equations to be functions of the curvilinear coordinates and to apply the MAC formulation in the uniform computational domain.

Patel and Briggs (1983) applied the original explicit MAC formulation, in curvilinear coordinates, to the problem of unsteady two-dimensional laminar natural convection. A staggered grid is used with the contravariant velocities (12.65) defined at each cell face and the pressure defined at each cell centre. Because

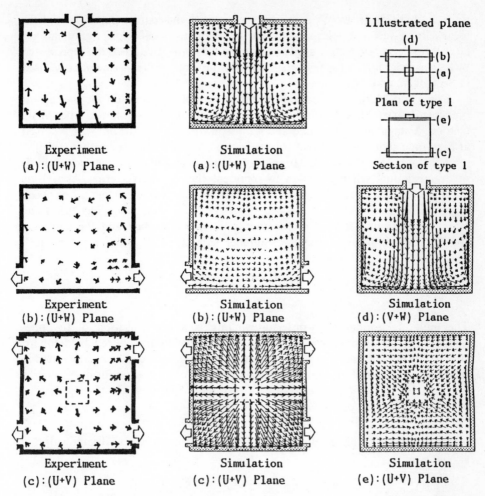

Fig. 17.4. Room ventilation simulation (after Kato and Murakami, 1986; reprinted with permission of Japan Soc. of Comp. Fluid Dynamics)

pressure derivatives p_ξ and p_η appear in the equivalent of (17.13), Patel and Briggs consider two overlapping grids to evaluate p_ξ and p_η using adjacent pressure values. However, this means that the system of Poisson equations, equivalent to (17.13), is twice as large as when a Cartesian grid is used.

17.1.5 Higher-Order Upwinding Differencing

The difference formulae (17.7) used in the original MAC formulation are centred difference expressions. For convection-dominated flows the use of centred difference formulae for the convective terms leads to solutions with severe non-physical oscillations (Sect. 9.3.1). Attempts have been made to stabilize the flow solution by discretising the convective terms with two-point upwind difference expressions (Sect. 9.3.1) or a weighted average of centred and upwind difference expressions

(Hirt et al. 1975). However, solutions are generally inaccurate particularly if the local velocity vector is oblique to the local grid and the local velocity gradients are large. More accurate solutions are obtained if the convective terms are represented by higher-order upwind schemes, such as the four-point upwind schemes discussed in Sects. 9.3.2 and 9.4.3.

Davis and Moore (1982) use a primitive variable formulation for the incompressible Navier–Stokes equations that is conceptually similar to the MAC method. A staggered grid is used and a Poisson equation for the pressure is solved at each time step using a direct solver (Sect. 6.2.6) due to Swartztrauber (1974).

A feature of the Davis and Moore formulation is that a multidimensional third-order upwinding scheme is used to represent the convective terms. As well as providing an accurate representation, the third-order upwind scheme is free of cell Reynolds number limitations associated with centred differencing (Sect. 9.3.1) The marching algorithm for the momentum equations is explicit and the empirical stability restriction at high Reynolds number is that the CFL condition is satisfied, i.e.

$$\frac{u\Delta t}{\Delta x} \leqq 1.0 \quad \text{and} \quad \frac{v\Delta t}{\Delta y} \leqq 1.0 \ .$$

The multidimensional third-order upwinding is a generalisation of the one-dimensional quadratic upstream interpolation schemes introduced by Leonard (1979). The one-dimensional third-order upwinding scheme can be be illustrated in relation to the transport equation (9.56) written in conservation form,

$$\frac{\partial T}{\partial t} + \frac{\partial(uT)}{\partial x} - \alpha\frac{\partial^2 T}{\partial x^2} = 0 \ , \tag{17.30}$$

where u varies through the domain but is known. A conservative discretisation of (17.30) is

$$\frac{T_j^{n+1} - T_j^n}{\Delta t} + \frac{(uT)_{j+1/2}^n - (uT)_{j-1/2}^n}{\Delta x} - \alpha\left\{\frac{T_{j-1}^n - 2T_j^n + T_{j+1}^n}{\Delta x^2}\right\} = 0 \ . \tag{17.31}$$

The velocity field u is defined on the uniform staggered grid (as for the MAC scheme, Fig. 17.1). However, $T_{j+1/2}$ and $T_{j-1/2}$ are related to the T field as

$$T_{j-1/2} = 0.5(T_j + T_{j-1}) - \frac{q}{3}(T_{j-2} - 2T_{j-1} + T_j) \quad \text{and} \tag{17.32}$$

$$T_{j+1/2} = 0.5(T_j + T_{j+1}) - \frac{q}{3}(T_{j-1} - 2T_j + T_{j+1}) \ . \tag{17.33}$$

The above discretisation of $\partial(uT)/\partial x$ is equivalent to the four point upwind scheme (9.71, 72); the parameter q can be chosen to increase accuracy or to alter the dissipation, dispersion characteristics (Sect. 9.4.3).

The choice $q = 0.375$ corresponds to the QUICK scheme (Leonard 1979), which is effective for steady or quasi-steady flows. The QUICK scheme has been widely used with SIMPLE-type algorithms (Sect. 17.2.3). However, for unsteady problems it is desirable to choose q so that temporal as well as spatial errors are reduced (as in Sect. 9.4.3).

The one-dimensional scheme QUICKEST introduced by Leonard (1979) resembles a modified Lax–Wendroff scheme (Table 9.3) if u is a constant $(C=u\Delta t/\Delta x)$ in (17.30), i.e.

$$T_j^{n+1}-T_j^n+0.5C(T_{j+1}^n-T_{j-1}^n)+\left(\alpha\frac{\Delta t}{\Delta x^2}+0.5C^2\right)(T_{j-1}^n-2T_j^n+T_{j+1}^n)$$

$$=C\left(\frac{1-C^2}{6}-\alpha\frac{\Delta t}{\Delta x^2}\right)[(T_{j-1}^n-2T_j^n+T_{j+1}^n)-(T_{j-2}^n-2T_{j-1}^n+T_j^n)] , \quad (17.34)$$

where the additional terms on the right-hand side of (17.34) are introduced to give an overall truncation error of $O(\Delta t^2, \Delta x^2)$. The convective terms are discretised to $O(\Delta x^3)$ and the time derivative is discretised to $O(\Delta t^3)$ in the limit $\alpha\to 0$.

Davis and Moore generalise (17.34) to two dimensions and allow for a variable velocity field. Thus, for the two-dimensional transport equation,

$$\frac{\partial T}{\partial t}+\frac{\partial(uT)}{\partial x}+\frac{\partial(vT)}{\partial y}-\alpha\left(\frac{\partial^2 T}{\partial x^2}+\frac{\partial^2 T}{\partial y^2}\right)=0 . \quad (17.35)$$

Davis and Moore obtain the following explicit algorithm

$$T_{j,k}^{n+1}=T_{j,k}^n-C_{j+1/2}\left[0.5(T_{j,k}+T_{j+1,k})-0.5C_{j+1/2}(T_{j+1,k}-T_{j,k})\right.$$

$$\left.-\left(\frac{1-C_{j+1/2}^2}{6}-\gamma_x\right)(T_{j-1,k}-2T_{j,k}+T_{j+1,k})\right]^n$$

$$+C_{j-1/2}\left[0.5(T_{j-1,k}+T_{j,k})-0.5C_{j-1/2}(T_{j,k}-T_{j-1,k})\right.$$

$$\left.-\left(\frac{1-C_{j+1/2}^2}{6}-\gamma_x\right)(T_{j-2,k}-2T_{j-1,k}+T_{j,k})\right]^n$$

$$-C_{k+1/2}\left[0.5(T_{j,k}+T_{j,k+1})-0.5C_{k+1/2}(T_{j,k+1}-T_{j,k})\right.$$

$$\left.-\left(\frac{1-C_{k+1/2}^2}{6}-\gamma_y\right)(T_{j,k-1}-2T_{j,k}+T_{j,k+1})\right]^n$$

$$+C_{k-1/2}\left[0.5(T_{j,k-1}+T_{j,k})-0.5C_{k-1/2}(T_{j,k}-T_{j,k-1})\right.$$

$$\left.-\left(\frac{1-C_{k-1/2}^2}{6}-\gamma_y\right)(T_{j,k-2}-2T_{j,k-1}+T_{j,k})\right]^n$$

$$+\gamma_x(T_{j-1,k}-2T_{j,k}+T_{j+1,k})+\gamma_y(T_{j,k-1}-2T_{j,k}+T_{j,k+1}) , \quad (17.36)$$

where the Courant numbers are given by

$$C_{j+1/2} = \frac{\Delta t u_{j+1/2}}{\Delta x} \; , \qquad C_{j-1/2} = \frac{\Delta t u_{j-1/2}}{\Delta x} \; ,$$

$$C_{k+1/2} = \frac{\Delta t v_{k+1/2}}{\Delta y} \; , \qquad C_{k-1/2} = \frac{\Delta t v_{k-1/2}}{\Delta y} \; , \quad \text{and}$$

$$\gamma_x = \alpha \frac{\Delta t}{\Delta x^2} \; , \qquad \gamma_y = \alpha \frac{\Delta t}{\Delta y^2} \; .$$

In forming (17.36) some $O(\Delta t^2)$ spatial cross-derivatives have been omitted to create a simpler algorithm. Formally this reduces the temporal accuracy of (17.36) to $O(\Delta t)$. Consequently Davis and Moore (1982) employ small time-steps to minimise the error associated with the neglected $O(\Delta t^2)$ terms.

For unsteady laminar flow past a rectangular obstacle (Fig. 17.5) Davis and Moore have used an algorithm, equivalent to (17.36) on a uniform grid, to explicitly march the discrete form of the momentum equation (17.2, 3) in time. In their 1982 formulation, Davis and Moore solve the Poisson equation for the pressure (17.13) by an SOR technique (Sect. 6.3.1).

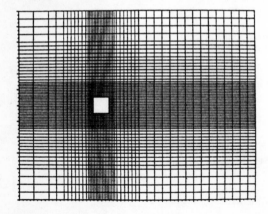

Fig. 17.5. Grid for flow past an obstacle, 51×62 grid

Subsequently (Davis et al. 1984), a direct solver based on Swartztrauber's (1974) method is used to solve (17.13) at each time step. This improves the computational efficiency considerably and permits continuity to be satisfied at each time-step to much greater precision than is economically possible with SOR iteration. A typical solution showing an intermediate development of a vortex street is indicated in Fig. 17.6.

The Davis and Moore algorithm (1982) has been used to examine the detailed structure of flow in a driven cavity at $Re = 10^4$ by Takemoto et al. (1985). More recently, Takemoto et al. (1986) have modified the algorithm in a number of important ways. First, they have expressed the governing equations in generalized curvilinear coordinates (Chap. 12). Second, they have integrated the momentum equations (17.2, 3) using a fractional step method, in which the convective terms are handled explicitly and the viscous terms are handled implicitly. A grid with

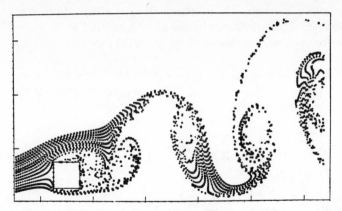

Fig. 17.6. Vortex shedding for Re = 1000 with the incident flow inclined at 15° to the horizontal (after Davis and Moore, 1982; reprinted with permission of Cambridge University Press)

pressure and velocity components defined at the same nodes is used in preference to a staggered grid. The revised algorithm is used to examine the unsteady flow over a sinusoidal bump in a channel at Re = 10^4.

It is interesting that though the flow is unsteady the QUICK algorithm, essentially (17.32 and 33) with $q = 0.375$, is used in preference to the QUICKEST algorithm, equivalent to (17.34). It is conjectured that the relative simplicity and economy of the QUICK algorithm is preferred. Takemoto and Nakamura (1986) apply the same algorithm to the three-dimensional driven cavity problem. This flow is also computed by Perng and Street (1989) using modified QUICK differencing in conjunction with a SIMPLE-type solution algorithm (Sect. 17.2.3) and an incomplete Choleski conjugate gradient (ICCG, Sect. 6.3.4) method for the pressure Poisson equation.

17.1.6 Spectral Methods

The spectral method (Sect. 5.6) is more accurate per nodal unknown than any of the local methods, e.g. finite difference or finite element methods (Fletcher 1984, Chap. 6). For problems with sufficiently smooth solutions and benign (e.g. periodic) boundary conditions the spectral method demonstrates very considerable computational efficiency (Sect. 4.5). However, for flow problems with more difficult boundary conditions the very high accuracy per nodal unknown may not be achieved and indeed there may be difficulty in obtaining stable solutions (Gottlieb and Orszag 1977).

The important role of the boundary conditions, in relation to the formulation of effective spectral methods, has led to the increased use of pseudospectral formulations based on Chebyshev polynomials (Sect. 5.6.3). Since the nodal-point solution is often obtained directly with modern pseudospectral formulations, the satisfaction of boundary conditions can also be handled more explicitly.

The Chebyshev pseudospectral matrix (CPSM) approach is described briefly in Sect. 5.6.3. The crucial feature of this method is that explicit discretisation formulae for higher-derivatives, e.g. (5.162 and 165), are obtained based on the assumption that the function behaviour is accurately represented by a Chebyshev series spanning the global domain. The application of the CPSM method to unsteady incompressible viscous flow will be described here, based on the formulation of Ku et al. (1987a, b).

The nondimensional governing equations are given by (17.1–3). One-dimensional Chebyshev approximate solutions, like (5.150), are used to generate the following discretisations of spatial derivatives:

$$
\left[\frac{\partial u}{\partial x}\right]_{j,k} = \sum_{l=1}^{N_x+1} \hat{G}_{j,l}^{x(1)} u_{l,k} \;, \qquad
\left[\frac{\partial u}{\partial y}\right]_{j,k} = \sum_{m=1}^{N_y+1} \hat{G}_{k,m}^{y(1)} u_{j,m}
$$

$$(17.37)$$

$$
\left[\frac{\partial^2 u}{\partial x^2}\right]_{j,k} = \sum_{l=1}^{N_x+1} \hat{G}_{j,l}^{x(2)} u_{l,k} \;, \qquad
\left[\frac{\partial^2 u}{\partial y^2}\right]_{j,k} = \sum_{m=1}^{N_y+1} \hat{G}_{k,m}^{y(2)} u_{j,m} \;,
$$

where the components of $\hat{G}^{x(1)}$, etc., correspond to the components of $\hat{G}^{(1)}$, etc., in (5.161). Equivalent formulae to (17.37) are available for derivatives of v and p.

Time derivatives are discretised using finite difference expressions, and the projection method (17.22–24) is used to advance the solution in time. The first component of $\mathbf{F}^n$ in (17.22) is interpreted as

$$
F_1^n = u_j^n + \Delta t \left[\frac{1}{\mathrm{Re}}\left(\frac{\partial^2 u}{\partial x^2} + \frac{\partial^2 u}{\partial y^2}\right) - \frac{\partial u^2}{\partial x} - \frac{\partial uv}{\partial y} \right]_j^n \;,
$$

$$(17.38)$$

where the spatial discretisation utilises (17.37). Similarly F_2^n is based on the CPSM discretisation of the equivalent terms in (17.3).

After $\mathbf{u}^*$ has been obtained from (17.22), p^{n+1} follows from (17.24) with a particular treatment of the boundary conditions. Using (17.37), (17.24) can be discretised as

$$
\sum_{r=2}^{N_x} \hat{G}_{j,r}^{x(1)} \sum_{l=1}^{N_x+1} \hat{G}_{r,l}^{x(1)} p_{l,k} + \sum_{s=2}^{N_y} \hat{G}_{k,s}^{y(1)} \sum_{m=1}^{N_y+1} \hat{G}_{s,m}^{y(1)} p_{j,m} = \mathrm{RHS}^I + \mathrm{RHS}^B \;,
$$

$$(17.39)$$

where

$$
\mathrm{RHS}^I = \frac{1}{\Delta t}\left(\sum_{r=2}^{N_x} \hat{G}_{j,r}^{x(1)} u_{r,k}^* + \sum_{s=2}^{N_y} \hat{G}_{k,s}^{y(1)} v_{j,s}^* \right) \;,
$$

and

$$
\mathrm{RHS}^B = \hat{G}_{j,1}^{x(1)}\left(\frac{1}{\Delta t} u_{1,k}^* - \sum_{l=1}^{N_x+1} G_{1,l}^{x(1)} p_{l,k} \right) + \hat{G}_{j,N_x+1}^{x(1)}\left(\frac{1}{\Delta t} u_{N_x+1,k}^* \right.
$$

$$
- \sum_{l=1}^{N_x+1} \hat{G}_{N_x+1,l}^{x(1)} p_{l,k} \right) + \hat{G}_{k,1}^{y(1)}\left(\frac{1}{\Delta t} v_{j,1}^* - \sum_{m=1}^{N_y+1} G_{1,m}^{y(1)} p_{j,m} \right)
$$

$$
+ \hat{G}_{k,N_y+1}^{y(1)}\left(\frac{1}{\Delta t} v_{j,N_y+1}^* - \sum_{m=1}^{N_y+1} \hat{G}_{N_y+1,m}^{y(1)} p_{j,m} \right) \;.
$$

The summations in (17.39) have been split up to isolate some of the boundary values. Using the CPSM form of (17.23), RHS^B can be replaced with

$$\text{RHS}^B = \frac{\hat{G}_{j,1}^{x(1)} u_{1,k}^{n+1}}{\Delta t} + \frac{\hat{G}_{j,N_x+1}^{x(1)} u_{N_x+1,k}^{n+1}}{\Delta t} + \frac{\hat{G}_{k,1}^{y(1)} v_{j,1}^{n+1}}{\Delta t} + \frac{\hat{G}_{k,N_y+1}^{y(1)} v_{j,N_y+1}^{n+1}}{\Delta t} . \tag{17.40}$$

The boundary velocities appearing in (17.40) are all prescribed boundary values. Equation (17.39), with (17.40), is applied at all interior nodes. Consequently values of u^*, v^* on the boundary are not required. However, boundary values of p appear in (17.39) so boundary conditions for p are required. These are constructed from (17.23) and (17.1). For $j=1$ and N_x+1, this gives

$$\left.\frac{\partial^2 p}{\partial x^2}\right|_j^n = \frac{1}{\Delta t}\left(\left.\frac{\partial v}{\partial y}\right|_j^{n+1} + \left.\frac{\partial u^*}{\partial x}\right|_j\right) , \tag{17.41}$$

and for $k=1$ and N_y+1,

$$\left.\frac{\partial^2 p}{\partial y^2}\right|_k^n = \frac{1}{\Delta t}\left(\left.\frac{\partial u}{\partial x}\right|_k^{n+1} + \left.\frac{\partial v^*}{\partial y}\right|_k\right) . \tag{17.42}$$

These expressions appear to involve u^* and v^* on the boundary. However, in applying the CPSM method the same simplification that led to (17.40) permits substitution of appropriate values of u^{n+1} and v^{n+1}. Thus the CPSM form of (17.41 and 42) is

i) for $j=1$ and N_x+1,

$$\sum_{r=2}^{N_x} \hat{G}_{j,r}^{x(1)} \sum_{l=1}^{N_x+1} \hat{G}_{r,l}^{x(1)} p_{l,k} = \frac{1}{\Delta t}\left(\sum_{s=1}^{N_y+1} \hat{G}_{k,s}^{y(1)} v_{j,s}^{n+1} + \sum_{r=2}^{N_x} \hat{G}_{j,r}^{x(1)} u_{r,k}^*\right.$$
$$\left. + \hat{G}_{j,1}^{x(1)} u_{i,k}^{n+1} + \hat{G}_{j,N_x+1}^{x(1)} u_{N_x+1,k}^{n+1}\right) , \tag{17.43}$$

ii) for $k=1$ and N_y+1,

$$\sum_{s=2}^{N_y} \hat{G}_{k,s}^{y(1)} \sum_{m=1}^{N_y+1} \hat{G}_{s,m}^{y(1)} p_{j,m} = \frac{1}{\Delta t}\left(\sum_{r=1}^{N_x+1} \hat{G}_{j,r}^{x(1)} u_{r,k}^{n+1} + \sum_{s=2}^{N_y} \hat{G}_{k,s}^{y(1)} v_{j,s}^*\right.$$
$$\left. + \hat{G}_{k,1}^{y(1)} v_{j,1}^{n+1} + \hat{G}_{k,N_y+1}^{y(1)} v_{j,N_y+1}^{n+1}\right) . \tag{17.44}$$

The combination of (17.39) at interior nodes and (17.43) or (17.44) at boundary nodes produces a dense linear system of equations for the pressure p^{n+1}. Because this system is linear it can be factorised once, e.g. using the subroutine FACT (Sect. 6.2.1), at the first time-step. Subsequent time-steps require back-substitution with different right-hand sides to obtain p^{n+1}. Subroutine SOLVE (Sect. 6.2.1) performs each back-substitution in $O(N^2)$ operations. At each time step, once p^{n+1} is available, u^{n+1}, v^{n+1} follow from the discrete form of (17.23) based on the equivalent of (17.37).

Since the implementation of (17.22) in the CPSM method is explicit there is a restriction on the time-step of the form

$$\Delta t \leq \left(|u|N_x^2 + |v|N_y^2 + \frac{N_x^4}{Re} + \frac{N_y^4}{Re} \right)^{-1} . \tag{17.45}$$

Clearly an increase in the number of grid points causes a severe restriction on the time-step. For the problem of a driven cavity, Ku et al. (1987b) report accurate solutions for a Reynolds number of 10^3 with $N_x = N_y = 31$ but require 15 000 to 30 000 time steps to reach the steady state.

As well as using (17.43 and 44) as boundary conditions for the pressure solution, Ku et al. (1987a) have also tested the more conventional specification of the pressure gradient through the momentum equations (17.2, 3) at the boundary. For a thermally driven cavity the two boundary condition specifications produce similar solutions at a Rayleigh number of 10^4. However, at a Rayleigh number of 10^6 the use of (17.43 and 44), which effectively enforces continuity at the boundaries, produces more accurate solutions.

Ku et al. (1987b) extend the formulation to a three-dimensional driven cavity and obtain solutions for Reynolds numbers up to 1000 and a grid of $31 \times 31 \times 16$ points (the problem is symmetric about $z = 0.5$). The major computational development is the use of an eigenfunction expansion to reduce the three-dimensional Poisson equation to a sequence of one-dimensional problems. This greatly reduces the storage that would be necessary to factorise the three-dimensional Poisson equation directly.

Typical steady-state solutions obtained by Ku et al. (1987b) indicate (Fig. 17.7) that the character of the flow is significantly different in a three-dimensional driven cavity to that in a two-dimensional driven cavity. The differences are less for lower Reynolds numbers. The moving lid of the driven cavity is located at $y = 1$ and is of extent $0 \leq x \leq 1$, $0 \leq z \leq 1$.

Ku et al. (1987a) also consider a CPSM implementation of the traditional MAC method (Sect. 17.1.2) but without a staggered grid. The global coupling implicit in the CPSM discretisation of pressure derivatives blocks the appearance of intertwined but uncoupled pressure solutions associated with the use of centred differences on a non-staggered grid (Sect. 17.1.1). Ku et al. (1987a) note that when boundary values of pressure gradient are eliminated in favour of $\partial u / \partial t$ and $\partial v / \partial t$, through the momentum equations, the resulting time-marching algorithm is stable. This appears to overcome the instability produced by the explicit pseudospectral method with pressure gradient boundary conditions satisfied by the steady momentum equations (Moin and Kim 1980).

The main weakness of the CPSM strategy for steady or slowly varying flows is the severe restriction on the time-step (17.45). If (17.22) were treated implicitly a dense matrix factorisation and solution would be required at every time step, due

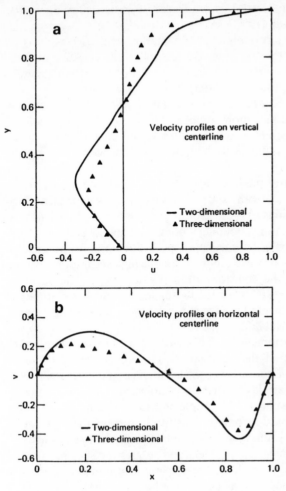

Fig. 17.7a,b. Comparisons of 2D and 3D velocity distributions for Re = 1000. (a) Velocity profiles on vertical centreline, $x = 0.5$, $z = 0.5$. (b) Velocity profiles on horizontal centreline, $y = 0.5$, $z = 0.5$ (after Ku et al., 1987b; reprinted with permission of Academic Press)

to the nonlinear nature of (17.22). For realistic problems in two and three dimensions the resulting scheme would be very uneconomical.

Gottlieb et al. (1984) review time stepping methods for incompressible viscous flow and suggest that the equivalent of (17.22) can be split into two steps. One step, involving the convective terms, is handled explicitly. The second step involving the viscous terms can be handled implicitly. Since this system is linear the CPSM strategy would only require a matrix factorisation at the first step. Such a splitting has been used with the pseudospectral method by Moin and Kim (1980) and with a mixed spectral/pseudospectral method by Orszag and Kells (1980). Orszag and Kells (1980) handle the convective terms partially implicitly by approximately linearising the convective terms at each time-step. However, Moin and Kim (1980) and Orszag and Kells (1980) only use Chebyshev polynomials in one direction; periodic boundary conditions permit the use of Fourier series in the other two directions. In addition, both of these formulations use FFTs instead of the matrix operations favoured by Ku et al. (1987a, b).

Spectral methods are used to examine fundamental instabilities which trigger transition from laminar to turbulent flow. Orszag and Kells (1980) show that transition occurs at a Reynolds number of the order of 1000 for plane Poiseuille and plane Couette flows. Orszag and Patera (1983) have shown that three-dimensional disturbances are necessary for transition in shear flows. For transition research high temporal resolution is more important than high spatial resolution. However, in the direct simulation of turbulence (Orszag and Patera 1981; Brachet et al. 1983) high spatial resolution and high temporal resolution is equally important.

Most applications of spectral methods have been to problems with computational boundaries that coincide with constant values of the independent variables. For computational domains of more arbitrary shape it is possible to transform the governing equations using boundary-fitted generalised curvilinear coordinates (Chap. 12) and to apply the particular spectral method in the uniform generalised coordinate domain. However, to maintain the high accuracy of spectral methods it is necessary that the transformation parameters, ξ_x, etc., also be evaluated using spectral methods (Orszag 1980). For finite difference methods of second-order accuracy it is sufficient to use second-order formulae to determine the transformation parameters (Sect. 12.2).

The use of Chebyshev polynomials to define the collocation sampling points (5.152) produces a refined grid at the boundary of the domain but a relatively coarse grid in the interior. For viscous flow problems this is particularly well-suited to resolving the narrow boundary layers that occur adjacent to solid surfaces for large values of the Reynolds number. For problems with severe gradients in the interior, e.g. viscous compressible flow or a moving front problem, the distribution of sampling points produced by Chebyshev polynomials may be less efficient.

One effective way to overcome this problem and to give the spectral method greater geometric flexibility is to break up the single global domain into a few, $O(10)$, subdomains. In each subdomain a spectral method is applied. For the flow over a backward-facing step (Patera 1984), the global domain (Fig. 17.14) is broken up into seven subdomains. In each subdomain an interpolation with six to seven Chebyshev polynomials in each direction is sufficient to produce accurate solutions.

The use of separate spectral expansions in each subdomain raises the additional problem of enforcing interdomain solution continuity. For the incompressible Navier–Stokes equations (17.1–3) continuity of the pressure, velocity components and first derivatives of the velocity components normal to the subdomain boundary is required. These conditions are imposed directly by Ku and Hatziavramidis (1985) in a velocity, vorticity formulation for pipe entrance flow.

However, Patera (1984) avoids having to impose continuity on velocity derivatives explicitly. Patera develops a spectral element formulation within which Lagrange interpolation of the nodal unknowns is undertaken based on Chebyshev polynomials and collocation points. A modified projection method, equivalent to (17.22–24), advances the solution from time level n to $n+1$. An intermediate velocity solution is obtained by explicitly advancing the convective terms in the momentum equation. This intermediate velocity solution provides the source term for the pressure Poisson equation, equivalent to (17.24). The solution for the pressure is used in the equivalent of (17.23) to further advance the velocity solution. The remaining step is to advance the velocity solution due to the viscous terms. This step is given an equivalent variational formulation and advanced in time with a Crank–Nicolson (implicit) construction. The use of the variational formulation avoids the need for continuity of velocity derivatives at inter-element boundaries. An alternative means of handling continuity of the velocity derivatives is provided by the global flux balance method of Macaraeg and Streett (1986).

Ku et al. (1989) extend the CPSM method, described earlier, to the three-dimensional flow over a backward-facing step with a domain decomposition strategy based on spectral elements and inter-element continuity of the velocity components and pressure only. They also demonstrate the efficiency of the algorithm on an Alliant FX/8 mini-supercomputer with up to eight parallel processors.

For simple geometries, like the backward-facing step, the spectral formulation in each subdomain can be expressed in terms of the physical coordinates. However, for more general global domains or if internal severe gradients are to be resolved accurately the use of distorted subdomains is appropriate. Korczak and Patera (1986) describe such a modification based on the isoparametric construction (Sect. 5.5.3). Macaraeg and Streett (1986) make use of a generalised curvilinear coordinate mapping in each distorted subdomains to achieve the same end result.

In conclusion it may be noted that spectral methods are still at a less well-developed stage than finite difference methods. Their main advantage is the high spatial accuracy that can be achieved with relatively few terms in the approximate solution, or equivalently relatively few collocation points (Hussaini and Zang 1987). For time-dependent viscous flow in which small time-steps are required to achieve sufficient accuracy, spectral methods are already competitive with finite difference methods, particularly in regular domains. For irregular domains and if the time dependence is not a limiting factor for accuracy then spectral methods are usually marched in time, or iterated to the steady state, with considerably less economy than purely local methods, like the finite difference and finite element methods. It appears, at present, that the overall computational efficiency of spectral methods for incompressible viscous flow is not as great as the best finite difference and finite element methods in complicated geometric domain. However, this could possibly change for computers with parallel-processing architectures (Ortega and Voigt 1985; Korczak and Patera 1986; Macaraeg and Streett 1986).

Zang et al. (1989) review relatively recent applications of the spectral method and conclude that active areas for the future are likely to be improved iterative convergence techniques, spectral domain decomposition methods and spectral implementations on massively parallel computers.

17.2 Primitive Variables: Steady Flow

Any of the techniques described in Sect. 17.1 are applicable to steady flow by marching in time until the solution no longer changes. In addition if the transient solution is of no consequence the pseudotransient formulation (Sect. 6.4) is available to implement more effective algorithms for obtaining the steady flow behaviour. The pseudotransient philosophy is exploited in the artificial compressibility construction (Sect. 17.2.1) and in the practical use of the auxiliary potential function (Sect. 17.2.2). Although the SIMPLE method (Sect. 17.2.3) was developed originally to solve the steady Navier–Stokes equations directly, it is found to be more effective if cast in a pseudotransient form. Finite element methods (Sect. 17.2.4) are suitable for either unsteady or steady flow. However, to determine steady flow behaviour they are usually applied directly to the steady Navier–Stokes equations.

17.2.1 Artificial Compressibility

In this method the solution to the steady Navier–Stokes equations are sought by applying a pseudotransient formulation (Sect. 6.4) to the unsteady momentum equations (17.2, 3) with the continuity equation (17.1) replaced by

$$\frac{\partial p}{\partial t} + a^2 \left(\frac{\partial u}{\partial x} + \frac{\partial v}{\partial y} \right) = 0 \ . \tag{17.46}$$

In the limit that $t \to \infty$, (17.46) coincides with (17.1). The transient solution of (17.46, 17.2 and 3) is not physically meaningful but the steady-state solution is. Equation (17.46) resembles the compressible continuity equation (11.10) and hence the name given to the method by Chorin (1967).

The parameter a can be interpreted as an artificial sound speed, such that $p = a^2 \varrho$. However, in practice ϱ does not appear explicitly and a and Δt have the role of relaxation parameters. Limitations on Δt will typically be determined by the stability of the computational algorithm. However, limitations will be necessary on the value of a as indicated by (17.53–55).

Since a pseudotransient formulation is used to march (17.46, 2 and 3) in time, appropriate boundary conditions for a specific problem are the same as given in Sect. 17.1.

In the original formulation, Chorin (1967) used leapfrog time-differencing (Sect. 9.1.3) and Dufort–Frankel spatial differencing (Sect. 7.1.2) with pressure and velocity components specified at the same grid points. Peyret and Taylor (1983) recommend the use of a staggered grid (Sect. 17.1.1) and show that if explicit time differencing is used, as in Sect. 17.1.2, the pseudotransient artificial compressibility construction can be interpreted as an iterative procedure to solve the discrete Poisson equation for the pressure (17.13) at $t = \infty$.

Here we describe an implicit pseudotransient algorithm to solve (17.46, 2 and 3) based on the approximate factorisation procedure (Sect. 8.2.2).

Equations (17.46, 2 and 3) can be written collectively as

$$\frac{\partial \mathbf{q}}{\partial t}+\frac{\partial \mathbf{F}}{\partial x}+\frac{\partial \mathbf{G}}{\partial y}-\frac{1}{\mathrm{Re}}\underline{D}\nabla^2 \mathbf{q}=0 \ , \quad \text{where} \tag{17.47}$$

$$\mathbf{q}\equiv\begin{bmatrix} p \\ u \\ v \end{bmatrix} , \quad \mathbf{F}\equiv\begin{bmatrix} a^2 u \\ u^2+p \\ uv \end{bmatrix} , \quad \mathbf{G}\equiv\begin{bmatrix} a^2 v \\ uv \\ v^2+p \end{bmatrix} , \quad \underline{D}\equiv\begin{bmatrix} 0 & 0 & 0 \\ 0 & 1 & 0 \\ 0 & 0 & 1 \end{bmatrix} .$$

As in (14.98), Jacobians $\underline{A}\equiv\partial\mathbf{F}/\partial\mathbf{q}$, $\underline{B}\equiv\partial\mathbf{G}/\partial\mathbf{q}$ are defined. For the present situation

$$\underline{A}\equiv\begin{bmatrix} 0 & a^2 & 0 \\ 1 & 2u & 0 \\ 0 & v & u \end{bmatrix} \quad \text{and} \quad \underline{B}\equiv\begin{bmatrix} 0 & 0 & a^2 \\ 0 & v & u \\ 1 & 0 & 2v \end{bmatrix} . \tag{17.48}$$

However, in contrast to (14.97), the following relationships arise:

$$\mathbf{F}=\underline{A}\mathbf{q}-u\underline{D}\mathbf{q} \quad \text{and} \quad \mathbf{G}=\underline{B}\mathbf{q}-v\underline{D}\mathbf{q} \ .$$

Equation (17.47) is discretised with p, u and v specified at the same grid points and trapezoidal (Crank–Nicolson) time differencing

$$\frac{\Delta\mathbf{q}^{n+1}}{\Delta t}+0.5L_x(\mathbf{F}^n+\mathbf{F}^{n+1})+0.5L_y(\mathbf{G}^n+\mathbf{G}^{n+1})$$

$$-\frac{0.5\underline{D}}{\mathrm{Re}}(L_{xx}+L_{yy})(\mathbf{q}^n+\mathbf{q}^{n+1})=0 \ , \tag{17.49}$$

where $\Delta\mathbf{q}^{n+1}=\mathbf{q}^{n+1}-\mathbf{q}^n$ and L_x, L_{xx}, etc., are three-point centred finite difference operators. For example $L_{xx}\mathbf{q}^n=(\mathbf{q}^n_{j-1,k}-2\mathbf{q}^n_{j,k}+\mathbf{q}^n_{j+1,k})/\Delta x^2$.

As in Sect. 8.2.2, (17.49) is to be manipulated to give a linear system in $\Delta\mathbf{q}^{n+1}$. The terms $\mathbf{F}^{n+1}$, $\mathbf{G}^{n+1}$ and $\mathbf{q}^{n+1}$ are expanded as Taylor series about the nth time level to give

$$\mathbf{F}^{n+1}\approx\mathbf{F}^n+\underline{A}^n\,\Delta\mathbf{q}^{n+1} \ , \quad \mathbf{G}^{n+1}\approx\mathbf{G}^n+\underline{B}^n\Delta\mathbf{q}^{n+1} \ , \quad \mathbf{q}^{n+1}\approx\mathbf{q}^n+\Delta\mathbf{q}^{n+1} \ .$$

Consequently (17.49) can be rewritten as

$$\left\{\underline{I}+0.5\Delta t\left[L_x\underline{A}^n+L_y\underline{B}^n-\frac{\underline{D}}{\mathrm{Re}}(L_{xx}+L_{yy})\right]\right\}\Delta\mathbf{q}^{n+1}=\mathrm{RHS} \ , \tag{17.50}$$

where

$$\mathrm{RHS}=\Delta t[(\underline{D}/\mathrm{Re})(L_{xx}+L_{yy})\mathbf{q}^n-L_x\mathbf{F}-L_y\mathbf{G}] \ .$$

The left-hand side is approximately factorised (as in Sect. 8.2.2) and solved as a two-stage algorithm, with additional artificial dissipation,

$$\left[\underline{I} + 0.5\Delta t\left(L_x\underline{A}^n - \frac{D}{Re}L_{xx}\right) + \varepsilon_i\Delta x^2 L_{xx}\right]\Delta\mathbf{q}^*$$
$$= \text{RHS} - \varepsilon_e[(\nabla_x\Delta_x)^2 + (\nabla_y\Delta_y)^2]\mathbf{q}^n \tag{17.51}$$

and

$$\left[\underline{I} + 0.5\Delta t\left(L_y\underline{B}^n - \frac{D}{Re}L_{yy}\right) + \varepsilon_i\Delta y^2 L_{yy}\right]\Delta\mathbf{q}^{n+1} = \Delta\mathbf{q}^* \ ,$$

where

$$(\nabla_x\Delta_x)^2\mathbf{q}^n \equiv q^n_{j-2,k} - 4q^n_{j-1,k} + 6q^n_{j,k} - 4q^n_{j+1,k} + q^n_{j+2,k} \ .$$

Clearly each stage of the algorithm leads to block tridiagonal systems of equations that can be solved efficiently (Sect. 6.2.5). The explicit fourth-order smoothing on the right-hand side of (17.51a) is introduced to overcome any nonlinear instability, since trapezoidal time differencing is neutrally stable. The implicit second-order smoothing introduced on the left-hand side of (17.51) acts to balance the explicit smoothing during the transient to avoid slowing down the convergence to the steady state. The parameters ε_e and ε_i are chosen so that the artificial dissipation is negligible compared with the physical dissipation, here determined by the value of $1/Re$. The use of artificial dissipation is discussed further in Sect. 18.5.

For non-simple geometries it is appropriate to use generalised coordinates (Chap. 12). The corresponding algorithm can be deduced from Sect. 18.4 and is used by Steger and Kutler (1977) to examine vortex wakes and by Kwak et al. (1986a) to investigate flow in engine manifolds.

It may be noted that the artificial sound speed, a, appears in the expressions for $\underline{A}$ and $\underline{B}$ (17.48). The eigenvalues of $\underline{A}$ and $\underline{B}$ are

$$\lambda_A = u, u, u \pm (a^2 + u^2)^{1/2} \ , \qquad \lambda_B = v, v, v \pm (a^2 + v^2)^{1/2} \ . \tag{17.52}$$

The transient solution can be thought of as being decomposed into different modes of the form $\exp(-\lambda_A t)$, etc. Thus for large values of a^2, different modes will decay at significantly different rates and the system of equations (17.47) is said to be stiff (Sect. 7.4). If an explicit algorithm were used to march (17.47) in time the stiffness associated with large values of a^2 would show up as a severe stability restriction on Δt. This is avoided by using the implicit algorithm (17.51). However, Steger and Kutler (1977) recommend that

$$a^2 < \frac{1}{\Delta t} \tag{17.53}$$

to maintain first-order time accuracy in (17.51). If a^2 is made too small the continuity equation (17.1) will not be satisfied sufficiently accurately with a destabilising effect on the transient solution. Kwak et al. (1986a) provide the following lower bounds for a^2. For laminar flow

$$a^2 > \left[1 + \left(\frac{4}{\text{Re}}\right)\left(\frac{x_{\text{ref}}}{x_\delta}\right)^2\left(\frac{x_L}{x_{\text{ref}}}\right)\right]^2 - 1 \; , \tag{17.54}$$

and for turbulent flow,

$$a^2 > \left[1 + \left(\frac{1}{\text{Re}_t}\right)\left(\frac{x_{\text{ref}}}{x_\delta}\right)\left(\frac{x_L}{x_{\text{ref}}}\right)\right]^2 - 1 \; , \tag{17.55}$$

where Re_t is the Reynolds number based on the turbulent eddy viscosity and x_δ and x_L are the characteristic lengths that the vorticity and pressure waves have to propagate during a given time span. For a duct flow, x_L is the total duct length and $2x_\delta$ is the distance between the duct walls. For the example of flow past a circular cylinder at $\text{Re} = 40$ and using $\Delta t = 0.1$, Kwak et al. recommend $0.1 < a^2 < 10$. Results are presented which suggest that the rate of convergence is not overly sensitive to the choice of a^2 within this range.

The Kwak et al. algorithm has been used (Kwak et al. 1986b) to compute the horseshoe vortex pattern produced by the incompressible flow around a post of circular cross-section on a flat plate at a Reynolds number (based on the cylinder diameter) of 1000 (Fig. 17.8). The particle paths shown in Fig. 17.8 were obtained with approximately 100 000 grid points and the solution required approximately 800 iterations to reach convergence. The particle paths show the presence of a

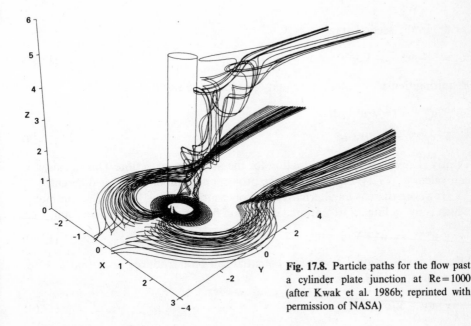

Fig. 17.8. Particle paths for the flow past a cylinder plate junction at $\text{Re} = 1000$ (after Kwak et al. 1986b; reprinted with permission of NASA)

strong primary horseshoe vortex structure and also filaments of secondary vorticity being convected normal to the flat plate (increasing z) before being swept downstream (increasing y).

Rogers and Kwak (1990) extend the method to compute unsteady incompressible flows by fully converging (17.46) at each time step. They also introduce flux-vector splitting (Sect. 14.2.5) and use the method to predict the development of a von Karman vortex street behind a two-dimensional circular cylinder.

17.2.2 Auxiliary Potential Function

This method is described in the context of a pseudotransient formulation (Sect. 6.4) to obtain the steady-state solution. It is convenient to write (17.1–3) in vector form. Discretising in time produces the following representation for the momentum equations:

$$\frac{(\mathbf{u}^{n+1}-\mathbf{u}^n)}{\Delta t}+\nabla\cdot(\mathbf{u}^n\mathbf{u}^{n+1})+\nabla p^n+\nabla\delta p^{n+1}-\left(\frac{1}{\text{Re}}\right)\nabla^2\mathbf{u}^{n+1}=0 \ , \tag{17.56}$$

where

$$p^{n+1}=p^n+\delta p^{n+1} \ .$$

To advance the solution, (17.56) is split into two parts. First an estimate of the solution, $\mathbf{u}^{n+1}$, is obtained by solving

$$\mathbf{u}^*+\Delta t\,\nabla\cdot(\mathbf{u}^n\mathbf{u}^*)-\left(\frac{\Delta t}{\text{Re}}\right)\nabla^2\mathbf{u}^*=\mathbf{u}^n-\Delta t\,\nabla p^n \tag{17.57}$$

for $\mathbf{u}^*$. In the second stage

$$\mathbf{u}^{n+1}=\mathbf{u}^*-\Delta t\,\nabla\delta p^{n+1} \ . \tag{17.58}$$

Requiring that $\mathbf{u}^{n+1}$ satisfy continuity produces the result

$$\nabla\cdot\mathbf{u}^{n+1}=0=\nabla\cdot\mathbf{u}^*-\Delta t\,\nabla^2\delta p^{n+1} \ , \quad \text{or}$$

$$\nabla^2\delta p^{n+1}=(1/\Delta t)\nabla\cdot\mathbf{u}^* \ . \tag{17.59}$$

Clearly (17.59) is a Poisson equation for the pressure correction. This equation is equivalent to (17.24) that arose in the projection method, but $\mathbf{u}^*$ is different.

However, there is an alternative way of ensuring that $\mathbf{u}^{n+1}$ satisfies continuity, which is by adding an irrotational correction field $\mathbf{u}^c$ and enforcing continuity, i.e.

$$\nabla\cdot\mathbf{u}^{n+1}=\nabla\cdot\mathbf{u}^*+\nabla\cdot\mathbf{u}^c=0 \ . \tag{17.60}$$

Since $\mathbf{u}^c$ is irrotational a velocity potential can be introduced such that $\mathbf{u}^c=\nabla\phi$. Consequently (17.60) becomes

$$\nabla^2\phi=-\nabla\cdot\mathbf{u}^* \ . \tag{17.61}$$

Equation (17.61) is a Poisson equation for ϕ which is equivalent to (17.59). If equivalent boundary conditions are used

$$\delta p^{n+1} = -\frac{\phi}{\Delta t} \; . \tag{17.62}$$

Thus once (17.61) has been solved for ϕ, p^{n+1} follows directly from (17.62). Boundary conditions for (17.61) are usually specified values of $\partial\phi/\partial n$ subject to the global constraint, equivalent to (17.4),

$$\int_c \frac{\partial\phi}{\partial n} ds = \int \nabla^2\phi \; dA = -\int \nabla\cdot\mathbf{u}^* \; dA \; . \tag{17.63}$$

In practice $\partial\phi/\partial n$ is set to zero on solid surfaces but the values on open boundaries are adjusted so that (17.63) is satisfied. The ϕ solution will not necessarily lead to $\partial\phi/\partial s = 0$ at the boundary. Therefore it is recommended that when solving (17.57) for $\mathbf{u}^*$ the following boundary condition at a solid surface is applied:

$$\mathbf{u}^* = -\left[\frac{\partial\phi}{\partial s}\right]^n \; . \tag{17.64}$$

To $O(\Delta t)$ the imposition of (17.64) will ensure that $\mathbf{u}_s^{n+1} = 0$ where the direction s is tangential to the boundary. Thus the complete no-slip boundary condition are satisfied at the $(n+1)$-th time level.

For an unsteady problem (17.61) would be solved exactly at every time-step. However, for steady incompressible viscous flow it is more efficient to replace (17.61) with

$$\frac{\partial\phi}{\partial t} - \nabla^2\phi + \nabla\cdot\mathbf{u}^* = 0 \; , \tag{17.65}$$

which can be marched forward in time in conjunction with (17.57). Usually a larger time step is used in (17.65) or it is advanced three or four time-steps for each time step that (17.57) is advanced. For small values of time (17.61) will be only approximately satisfied and it is necessary to under-relax the pressure correction, i.e. (17.62) is replaced with

$$p^{n+1} = p^n - \beta\frac{\phi}{\Delta t} \; . \tag{17.66}$$

The concept of splitting the velocity field so that one part has zero curl, i.e. can be obtained from a velocity potential as in (17.61), underlies Chorin's (1967) projection method (Sect. 17.1.4). The first explicit use of an auxiliary potential function was in the SMAC method of Amsden and Harlow (1970). More recently the method has been used by Dodge (1977), Cazalbou et al. (1983) and Kim and Moin (1985) with finite difference discretisation and by Ku et al. (1986b) in conjunction with a pseudospectral method. Briley (1974), Ghia and Sokhey (1977),

Yashchin et al. (1984) and Briley and McDonald (1984) have used an auxiliary potential to obtain the transverse velocity field in modelling duct flows (Sect. 16.2.2). Khosla and Rubin (1983) have used a conceptually similar idea in constructing a composite velocity procedure to obtain the flow behaviour in thick viscous layers.

The present algorithm is used in conjunction with multigrid and an explicit approximate-factorisation relaxation step by Fletcher and Bain (1991) to develop an efficient algorithm for implementation on parallel processors.

Essentially the same algorithm has been incorporated into a general-purpose finite volume code, RANSTAD, suitable for predicting incompressible and compressible turbulent flows. First derivatives are modelled with an extension of the higher-order upwind scheme described in Sect. 9.3.2 and the inclusion of limiters (14.86) controls the solution in the vicinity of shocks. Turbulence is modelled with an algebraic Reynolds stress model. For compressible flow Eq. (17.61) is replaced by a transport equation for ϕ, which is solved efficiently using a strongly implicit procedure (Sect. 6.3.3). Cho et al. (1991) demonstrate the effectiveness of RANSTAD for turbulent flows about wing-body junctions.

17.2.3 SIMPLE Formulations

This family of algorithms is based on a finite volume (Sect. 5.2) discretisation on a staggered grid (Sect. 17.1.1) of the governing equations. The method was introduced by Patankar and Spalding (1972) and is described in detail by Patankar (1980). The acronym, SIMPLE, stands for Semi-Implicit Method for Pressure-Linked Equations and describes the iterative procedure by which the solution to the discretised equations is obtained. The iterative procedure will be interpreted as a pseudotransient treatment of the unsteady governing equations (17.1–3) in discrete form to obtain the steady-state solution. An important link with the auxiliary potential function method (Sect. 17.2.2) will be indicated.

On a staggered grid different control volumes are used, Fig. 17.9, to discretise different equations. In addition the grid notation associated with particular dependent variables is staggered, Fig. 17.9. Thus the physical locations of $p_{j+1/2, k}$ and $u_{j, k}$ are the same, as are the physical locations of $p_{j, k+1/2}$ and $v_{j, k}$. The discretisation indicated below corresponds to a uniform grid. The more general case of a nonuniform grid can be obtained from Sect. 5.2 or Patankar (1980).

For the control volume shown in Fig. 17.9a the application of the finite volume method (Sect. 5.2) to the continuity equation (17.1) produces the discrete equation

$$(u_{j, k}^{n+1} - u_{j-1, k}^{n+1})\Delta y + (v_{j, k}^{n+1} - v_{j, k-1}^{n+1})\Delta x = 0 \ . \tag{17.67}$$

Application of the finite volume method to the x-momentum equation (17.2) using the control volume shown in Fig. 17.9b leads to the discrete equation

$$\left(\frac{\Delta x \Delta y}{\Delta t}\right)(u_{j, k}^{n+1} - u_{j, k}^{n}) + (F_{j+1/2, k}^{(1)} - F_{j-1/2, k}^{(1)})\Delta y + (G_{j, k+1/2}^{(1)} - G_{j, k-1/2}^{(1)})\Delta x$$

$$+ (p_{j+1, k}^{n+1} - p_{j, k}^{n+1})\Delta y = 0 \ , \tag{17.68}$$

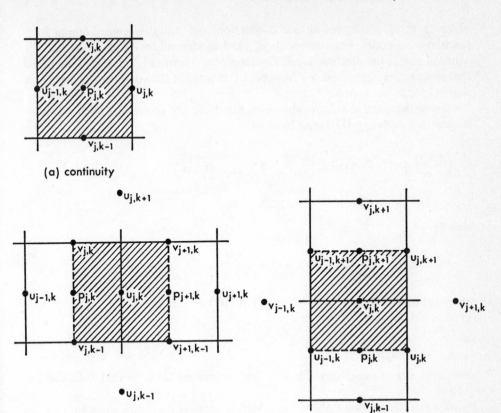

Fig. 17.9. Control volumes used in SIMPLE formulation

where

$$F^{(1)} = u^2 - \frac{1}{\text{Re}} \frac{\partial u}{\partial x} \quad \text{and} \quad G^{(1)} = uv - \frac{1}{\text{Re}} \frac{\partial u}{\partial y} .$$

Thus

$$F^{(1)}_{j+1/2, k} = 0.25(u_{j,k} + u_{j+1,k})^2 - \frac{1}{\text{Re}} \frac{u_{j+1,k} - u_{j,k}}{\Delta x} \quad \text{and}$$

$$G^{(1)}_{j,k+1/2} = 0.25(v_{j,k} + v_{j+1,k})(u_{j,k} + u_{j,k+1}) - \frac{1}{\text{Re}} \frac{u_{j,k+1} - u_{j,k}}{\Delta y} .$$

Consequently (17.68) can be written as

$$\left(\frac{\Delta x \Delta y}{\Delta t} + a^u_{j,k} \right) u^{n+1}_{j,k} + \sum a^u_{nb} u^{n+1}_{nb} + b^u + \Delta y (p^{n+1}_{j+1,k} - p^{n+1}_{j,k}) = 0 , \qquad (17.69)$$

where $\sum a_{nb}^u u_{nb}^{n+1}$ denotes all the convection and diffusion contributions from neighbouring nodes. The coefficients $a_{j,k}^u$ and a_{nb}^u depend on the grid sizes and the solution u, v at the nth time level. The term $b^u = -\Delta x \Delta y \, u_{j,k}^n / \Delta t$. It may be noted that some terms in $F^{(1)}$ and $G^{(1)}$ have been evaluated at the nth time-level to ensure that (17.69) is linear in u^{n+1}.

Using the control volume shown in Fig. 17.9c the discretised form of the y-momentum equation (17.3) can be written

$$\left(\frac{\Delta x \Delta y}{\Delta t}\right)(v_{j,k}^{n+1} - v_{j,k}^n) + (F_{j+1/2,k}^{(2)} - F_{j-1/2,k}^{(2)})\Delta y + (G_{j,k+1/2}^{(2)} - G_{j,k-1/2}^{(2)})\Delta x$$
$$+ (p_{j,k+1}^{n+1} - p_{j,k}^{n+1})\Delta x = 0 \ , \tag{17.70}$$

where

$$F^{(2)} = uv - \frac{1}{Re}\frac{\partial v}{\partial x} \quad \text{and} \quad G^{(2)} = v^2 - \frac{1}{Re}\frac{\partial v}{\partial y} \ .$$

Substituting for $F^{(2)}$ and $G^{(2)}$ allows (17.70) to be written

$$\left(\frac{\Delta x \Delta y}{\Delta t} + a_{j,k}^v\right)v_{j,k}^{n+1} + \sum a_{nb}^v v_{nb}^{n+1} + b^v + \Delta x(p_{j,k+1}^{n+1} - p_{j,k}^{n+1}) = 0 \ , \tag{17.71}$$

where the various coefficients have a similar interpretation to that indicated for (17.69).

At any intermediate stage of the SIMPLE iterative procedure the solution is to be advanced from the nth time level to the $(n+1)$-th time level. The velocity solution is advanced in two stages. First the momentum equations (17.69 and 71) are solved to obtain an approximation, $\mathbf{u}^*$, of $\mathbf{u}^{n+1}$ that does not satisfy continuity. Using the approximate velocity solution $\mathbf{u}^*$, a pressure correction δp is sought which will both give $p^{n+1} = p^n + \delta p$ and provide a velocity correction, $\mathbf{u}^c$, such that $\mathbf{u}^{n+1} = \mathbf{u}^* + \mathbf{u}^c$, where $\mathbf{u}^{n+1}$ satisfies continuity in the form of (17.67).

To obtain $\mathbf{u}^*$, (17.69 and 71) are approximated as

$$\left(\frac{\Delta x \Delta y}{\Delta t} + a_{j,k}^u\right)u_{j,k}^* + \sum a_{nb}^u u_{nb}^* = -b^u - \Delta y(p_{j+1,k}^n - p_{j,k}^n) \ , \tag{17.72}$$

$$\left(\frac{\Delta x \Delta y}{\Delta t} + a_{j,k}^v\right)v_{j,k}^* + \sum a_{nb}^v v_{nb}^* = -b^v - \Delta x(p_{j,k+1}^n - p_{j,k}^n) \tag{17.73}$$

Patankar (1980) recommends writing (17.72 and 73) as scalar tridiagonal systems along each x grid line (k constant) and solving them using the Thomas algorithm (Sect. 6.2.2). Subsequently (17.72 and 73) are written as scalar tridiagonal systems along each y grid line (j constant) and solved using the Thomas algorithm. This is conceptually similar to, although not identical with, the ADI procedure discussed in Sect. 8.2.1.

To obtain equations for the subsequent velocity correction $\mathbf{u}^c$, (17.72) is subtracted from (17.69) to give

$$\left(\frac{\Delta x \Delta y}{\Delta t} + a^u_{j,k}\right)u^c_{j,k} = -\sum a^u_{nb}\, u^c_{nb} - \Delta y(\delta p_{j+1,k} - \delta p_{j,k}) \ , \tag{17.74}$$

and (17.73) is subtracted from (17.71) to give a corresponding equation for v^c. However, to make the link between $\mathbf{u}^c$ and δp as explicit as possible, the SIMPLE algorithm approximates (17.74) as

$$u^c_{j,k} = d_{j,k}(\delta p_{j,k} - \delta p_{j+1,k}) \ , \quad \text{where} \tag{17.75}$$

$$d_{j,k} = E\Delta y/\{(1+E)a^u_{j,k}\} \quad \text{and} \quad E = \Delta t\, a^u_{j,k}/\Delta x \Delta y \ . \tag{17.76}$$

An equivalent expression can be obtained to link $v^c_{j,k}$ with $(\delta p_{j,k} - \delta p_{j,k+1})$. Substitution of $\mathbf{u}^{n+1}_{j,k} = \mathbf{u}^*_{j,k} + \mathbf{u}^c_{j,k}$ into (17.67) and use of (17.75), etc., produces the following explicit algorithm for $\delta p_{j,k}$:

$$a^p_{j,k}\, \delta p_{j,k} = \sum a^p_{nb}\, \delta p_{nb} + b^p \ , \tag{17.77}$$

where $b^p = -(u^*_{j,k} - u^*_{j-1,k})\Delta y - (v^*_{j,k} - v^*_{j,k-1})\Delta x$. Equation (17.77) is a disguised discrete Poisson equation that can be written symbolically as

$$\nabla^2_d \delta p = \frac{1}{\Delta t}\nabla_d \cdot \mathbf{u}^* \ , \tag{17.78}$$

which is equivalent to (17.59). It may also be noted that (17.75) is equivalent to

$$\mathbf{u}^c = -\frac{1}{\Delta t}\nabla_d \delta p \ . \tag{17.79}$$

Comparing (17.79) and (17.62) indicates that δp is an effective velocity potential and the velocity correction, $\mathbf{u}^c$, is irrotational. The complete SIMPLE algorithm can be summarised as follows:

1) $\mathbf{u}^*$ is obtained from (17.72 and 73),
2) δp is obtained from (17.77),
3) $\mathbf{u}^c$ is obtained from (17.75) and equivalent form for v^c,
4) p^{n+1} is obtained from $p^{n+1} = p^n + \alpha_p \delta p$, where α_p is a relaxation parameter.

The SIMPLE algorithm contains two relaxation parameters α_p and $E(\equiv \Delta t)$. Solving the steady momentum equation is equivalent to setting $E = \infty$. In this case it is recommended that $\alpha_p = 0.075$ to achieve a stable convergence. A more rapid convergence is found, empirically, if $E = 1$ and $\alpha_p = 0.8$ (Patankar 1980). Raithby and Schneider (1979) have made a systematic study of SIMPLE-type algorithms and conclude that a more efficient algorithm is obtained if $E \approx 4$ and

$$\alpha_p = \frac{1}{1+E} \ . \tag{17.80}$$

Van Doormaal and Raithby (1984) call (17.80) a consistent SIMPLE algorithm, or SIMPLEC as an acronym. However, Van Doormaal and Raithby give an alternative interpretation of SIMPLEC. It is argued that the approximation inherent in passing from (17.74) to (17.75) causes an increase in the number of iterations to convergence, although it does improve the economy of each iteration.

A closer approximation to (17.74) is obtained by subtracting $\Sigma\, a_{nb}^u u_{j,k}^c$ from both sides and dropping $\Sigma\, a_{nb}^u (u_{nb}^c - u_{j,k}^c)$ from the right-hand side to give

$$u_{j,k}^c = d_{j,k}'(\delta p_{j,k} - \delta p_{j+1,k}) \,, \tag{17.81}$$

in place of (17.75), where

$$d_{j,k}' = E \varDelta y / [(1+E)a_{j,k}^u - E \,\Sigma\, a_{nb}^u] \,.$$

If the correction u^c is slowly varying in space only a small error is introduced in dropping $\Sigma\, a_{nb}^u (u_{nb}^c - u_{j,k}^c)$, but (17.81) retains the significant economy of being explicit. In forming the Poisson equation for δp, (17.81) is used instead of (17.75) and similarly when computing $u_{j,k}^c$. However, when obtaining p^{n+1}, as in step 4) of the SIMPLE algorithm, no underrelaxation is required, i.e. $\alpha_p = 1$, if the SIMPLEC option (17.81) is introduced. A conceptually similar modification to SIMPLE is discussed by Connell and Stow (1986).

Application of the original SIMPLE algorithm to a range of problems suggests that δp is an effective mechanism for adjusting the velocity field but is often less effective in rapidly converging the pressure field. Patankar (1980) introduced a revised algorithm, SIMPLER, to improve the situation. The SIMPLER algorithm consists of the following steps:

1) A velocity field $\hat{\mathbf{u}}$ is computed from (17.72 and 73) with the pressure terms deleted from the right-hand sides.
2) Equation (17.77) then becomes a Poisson equation for p^{n+1} rather than δp with $\hat{\mathbf{u}}$ replacing the $\mathbf{u}^*$ terms in b.
3) The p^{n+1} (solution from step 2) replaces p^n in (17.72 and 73), which are solved as in SIMPLE to give $\mathbf{u}^*$.
4) Equation (17.77) is solved for δp and it is used to provide $\mathbf{u}^{n+1} = \mathbf{u}^* + \mathbf{u}^c$ but *no* further adjustment is made to p^{n+1} from step 2.

Clearly SIMPLER involves solving two Poisson steps and two momentum steps per iteration cycle. Although more expensive per iteration than SIMPLE, convergence is reached in sufficiently few iterations that SIMPLER is typically 50% more efficient. It may be noted that steps 1) and 2) of SIMPLER correspond to the projection method (17.22 and 24).

Van Doormaal and Raithby (1984) have compared SIMPLE, SIMPLEC and SIMPLER for a recirculating flow and flow over a backward-facing step. In obtaining solutions, (17.77) is repeated v times at each iteration until $\|r_p\|^v \leq \gamma_p \|r_p\|^0$, where $\|r_p\|$ is the rms residual of (17.77), i.e.

$$r_p = \sum a_{nb}^p\, \delta p_{nb} + b^p - a_{j,k}^p\, \delta p_{j,k} \quad \text{and} \quad \|r_p\| = \left[\sum_j \sum_k r_p^2 \right]^{1/2} \,.$$

Optimal values of γ_p range from 0.05 to 0.25. A comparison of the computational effort (CPU-secs) to reach convergence is shown in Fig. 17.10. Clearly both SIMPLEC and SIMPLER are more efficient than SIMPLE, with SIMPLEC to be preferred. However, the optimal choice of E, and to a lesser extent γ_p, is problem dependent.

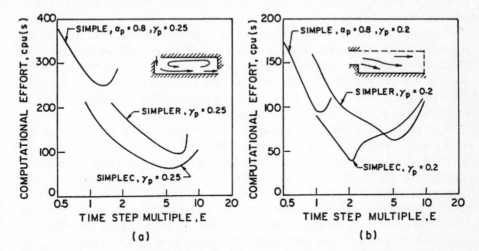

Fig. 17.10a, b. Comparison of SIMPLE, SIMPLEC and SIMPLER (after van Doormaal and Raithby, 1984; reprinted with permission of Hemisphere Publishing Co.)

SIMPLE-type algorithms on staggered grids have also been used with generalised (body-fitted) coordinates (Chap. 12). Raithby et al. (1986) have used the SIMPLEC algorithm with orthogonal generalised coordinates. It is found that formulating and discretising the problem at the stress level, as in (11.26) produces a more efficient code. Appropriate laminar or turbulent stress strain relations are introduced subsequently in an appropriate discrete form. However, if first-order upwind discretisations are introduced at this stage, the overall solution accuracy is often reduced. If higher-order discretisations are used more grid points are coupled and the solution algorithm is less economical.

Shyy et al. (1985) have applied the SIMPLE algorithm on a staggered grid with non-orthogonal generalised coordinates. The QUICK differencing scheme (Sect. 17.1.5) and a three-point second-order upwind scheme [$q=1.5$ in (9.53)] for the convective terms are compared. Two-dimensional turbulent flow in a kidney-shaped channel provides the test problem and 31×26 and 56×36 grids are used. Although this problem has no exact solution the second-order upwind scheme is generally preferred as it is more robust and more efficient without producing obviously less accurate solutions. The QUICK scheme [$q=0.375$ in (9.53)] is divergent on a very distorted mesh and generally requires more iterations when it is convergent.

The lack of robustness of the QUICK scheme has also been reported by Pollard and Siu (1982) and Patel and Markatos (1986) in applying the SIMPLE algorithm on a Cartesian grid. In relation to (9.53) it can be seen that reducing q produces a scheme closer to the three-point centred difference formula ($q = 0$). Thus the lack of robustness with the QUICK scheme ($q = 0.375$) compared with the second-order upwind scheme ($q = 1.5$) is not unexpected.

Phillips and Schmidt (1985) have combined a SIMPLE algorithm on a staggered grid with QUICK differencing of the convective terms. A multigrid procedure (Sect. 6.3.5) is employed to accelerate the convergence to the steady state. Phillips and Schmidt consider the driven cavity problem at Re = 400 and natural convection in a vertical cavity (de Vahl Davis and Jones 1983) at Ra = 10^6. The multigrid procedure is used with different grid refinements in different parts of the domain. Typically the finest grids ($h = 1/32$) are introduced adjacent to the walls but a less fine grid ($h = 1/16$) is employed in the interior. The coarsest grid ($h = 1/4$) in the multigrid procedure is used throughout the domain.

Hortmann et al. (1990) also consider natural convection in a vertical cavity and for Ra = 10^6 obtain results on grids up to 640 × 640 using a V-cycle full multigrid procedure (Sect. 6.3.5) based on SIP (Sect. 6.3.3) relaxation. Accuracies up to 0.01% are claimed. As the grid gets finer the relative efficiency of using multigrid over single-grid iteration increases; from approximately three times faster on a 40 × 40 grid to approximately 100 times faster on a 320 × 320 grid. Leschziner and Dimitriadis (1989) develop a finite volume procedure based on the SIMPLE algorithm to predict three-dimensional turbulent flow in and around a duct junction representative of IC-engine inlet manifolds. A feature of this formulation is that the overall computational domain is broken down into two subdomains, one corresponding to the main duct and one to the side duct. An overlap region, one control volume wide, is common to both subdomains. Where the flow is unidirectional (upstream of the junction) a repeated downstream marching iterative procedure is utilised. Downstream of the junction in the main duct some reversed axial flow occurs and this region is iterated without assuming a preferred direction for the iterative modification of the local solution. Good agreement is obtained with experimental data for both mean velocity and pressure distributions.

17.2.4 Finite Element Formulation

Most of the algorithms described above substitute a Poisson equation for pressure or potential for the solution of the continuity equation (17.1). In contrast the traditional finite element method operates on (17.1) directly. This method will be described first; subsequently the penalty finite element method will be described. This method solves a combination of the continuity and momentum equations for the velocity components, and the pressure does not appear explicitly.

A Galerkin finite element formulation (Sects. 5.3–5.5) is applied to the steady two-dimensional incompressible Navier–Stokes equations, i.e. (17.1–3) without the time dependent terms.

The starting point is to interpolate the velocity and pressure fields as in (5.58),

$$u = \sum_l u_l \phi_l^u(\xi, \eta) \ , \qquad v = \sum_l v_l \phi_l^v(\xi, \eta) \ , \qquad p = \sum_l p_l \phi_l^p(\xi, \eta) \ . \tag{17.82}$$

The interpolation is assumed to take place within quadrilateral elements with (ξ, η) being local element-based coordinates. Appropriate orders of interpolation for the different dependent variables will be indicated below.

Substitution of (17.82) into the governing equations produces non-zero residuals. Formation of the Galerkin weighted integral (5.5 and 10) gives

$$\iint \left(\frac{\partial u}{\partial x} + \frac{\partial v}{\partial y} \right) \phi_m^p \, dx \, dy = 0 \ , \tag{17.83}$$

$$\iint \left\{ \frac{\partial}{\partial x}(u^2 + p) + \frac{\partial}{\partial y}(uv) \right\} \phi_m^u \, dx \, dy$$

$$+ \frac{1}{\text{Re}} \int \left[\left(\frac{\partial u}{\partial x} \right) \left(\frac{\partial \phi_m^u}{\partial x} \right) + \left(\frac{\partial u}{\partial y} \right) \left(\frac{\partial \phi_m^u}{\partial y} \right) \right] dx \, dy$$

$$= \frac{1}{\text{Re}} \int \left[\left(\frac{\partial u}{\partial x} \right) \phi_m^u \right]_{x_L}^{x_R} dy + \frac{1}{\text{Re}} \int \left[\left(\frac{\partial u}{\partial y} \right) \phi_m \right]_{y_B}^{y_T} dx \ , \tag{17.84}$$

$$\iint \left[\frac{\partial}{\partial x}(uv) + \frac{\partial}{\partial y}(v^2 + p) \right] \phi_m^v \, dx \, dy$$

$$+ \frac{1}{\text{Re}} \int \left[\left(\frac{\partial v}{\partial x} \right) \left(\frac{\partial \phi_m^v}{\partial x} \right) + \left(\frac{\partial v}{\partial y} \right) \left(\frac{\partial \phi_m^v}{\partial y} \right) \right] dx \, dy$$

$$= \frac{1}{\text{Re}} \int \left[\left(\frac{\partial v}{\partial x} \right) \phi_m^v \right]_{x_L}^{x_R} dy + \frac{1}{\text{Re}} \int \left[\left(\frac{\partial v}{\partial y} \right) \phi_m^v \right]_{y_B}^{y_T} dx \ . \tag{17.85}$$

In (17.84 and 85) an integration by parts has been made of the weighted viscous terms; for ease of exposition it is assumed that the computational domain is rectangular with $x_L \leq x \leq x_R$ and $y_B \leq y \leq y_T$. Substitution of (17.82) into (17.83–85) produces a system of equations that can be written

$$\underline{A}\mathbf{q} = \mathbf{R}^c \ , \tag{17.86a}$$

$$\underline{S}(\mathbf{q})\mathbf{q} + \underline{B}\mathbf{p} = \mathbf{R}^m \ , \tag{17.86b}$$

where $\mathbf{q}$ is the vector containing all the unknown nodal values of both velocity components and $\mathbf{R}^m$ and $\mathbf{R}^c$ are known vectors arising from the Dirichlet boundary conditions.

Equation (17.86) constitutes a nonlinear global system of equations. Typically the solution is obtained iteratively using Newton's method (Sect. 6.1). At each stage of the iteration a sparse linear system of equations must be solved. This is usually carried out with a sparse Gauss elimination procedure (Sect. 6.2) based on the frontal method (Hood 1976).

The major problem with the above finite element formulation for the incompressible Navier–Stokes equations is the choice of the approximating and weight functions ϕ^u, ϕ^v and ϕ^p in (17.82–85). It might be expected that equal-order interpolation of u, v and p in (17.82) would be appropriate. However, this can cause (17.86) to become singular, because the discrete continuity equation (17.83) is applied at too many nodes. This was found empirically by Hood and Taylor (1974). Even if (17.86) is mathematically well-behaved the resulting solution demonstrates a highly oscillatory pressure field; however, the velocity field is usually well-behaved.

It may be recalled that a similar situation arises with the use of a centred difference representation for the pressure derivatives in the momentum equations if the velocity components and pressure are defined at the same grid points (Sect. 17.1.1). This was the major reason for introducing a staggered grid.

Taylor and Hood overcame the oscillatory pressure problem by introducing mixed interpolation, biquadratic interpolation for the velocity components u and v and bilinear interpolation for the pressure. This produces smooth solutions and has been widely used subsequently. In a sense this formulation is inefficient in that the overall accuracy is second-order from the bilinear pressure interpolation whereas the economy is governed by the use of biquadratic interpolation for the velocity components. This leads to relatively dense matrices $\underline{S}$, $\underline{A}$ and $\underline{B}$ in (17.86), which implies a high operation count in the subsequent iterative algorithm (Fletcher 1984, pp. 95–97).

The computational efficiency is improved somewhat by using linear interpolation of the velocity components and a constant value for the pressure in each element. However, this combination can produce oscillation in the pressure field for certain computational domains and choices of boundary conditions. The general problem of oscillatory pressure solutions occurring for various combinations of interpolation has been studied by Sani et al. (1981).

Broadly the problem can be reduced, or eliminated completely, by carefully matching additional computational boundary conditions to the choice of interpolation used. It is always necessary to use pressure interpolation that is of lower order than the velocity interpolation. If this still permits oscillatory pressure solutions to occur it is possible to filter or smooth the pressure solution to obtain useful information. It is shown theoretically by Sani et al., and confirmed numerically, that the velocity solution is well-behaved even if the pressure solution is oscillatory.

It is emphasised that the necessity of using lower-order interpolation for the pressure than that for the velocity stems directly from the use of (7.1) in (17.83). Schneider et al. (1978) demonstrate that if an auxiliary potential formulation (Sect. 17.2.2) is constructed in conjunction with a Galerkin finite element discretisation, equal-order interpolation for velocity and pressure is well-behaved. This is also the experience with a group finite element formulation (Fletcher 1982; Srinivas and Fletcher 1984) for compressible viscous flow at low subsonic Mach numbers.

Compared with SIMPLE-type formulations on a staggered grid the use of quadratic velocity interpolation and linear pressure interpolation produces solutions of high accuracy for the velocity field as long as boundary layers that form at high Reynolds numbers are properly resolved (Castro et al. 1982). A failure to do this leads to an oscillatory solution of a cell-Reynolds nature (Sect. 9.3.1). It can be argued (Gresho and Lee 1981) that this is a virtue since it provides a warning that a physically important flow feature is not being properly resolved.

In practice the finite element method is often used with a relatively small number of elements (or equivalently nodal points) spanning the computational domain. For flows of moderate to high Reynolds numbers the resulting oscillatory solutions produced by the conventional finite element method has prompted considerable interest in Petrov-Galerkin (Fletcher 1984, Chap. 7) constructions that permit an effective "upwind" (Sect. 9.3) treatment of the convective terms. For example Brooks and Hughes (1982) base such a formulation on ensuring that the upwind treatment is directed along the local streamline, thereby avoiding numerical cross-stream diffusion (Sect. 9.5.3).

A very effective way of constructing accurate non-oscillatory finite element algorithms for high Reynolds number flows is based on identifying the oscillatory errors with higher-order terms in the Taylor series expansion in time of the corresponding pseudo-transient formulation. This is an extension of the ideas developed in Sects. 9.2–9.4. Such a construction is provided in a basic form by Baker (1983, Chap. 5) and developed as the Taylor weak statement (TWS) of the Galerkin finite element formulation (Baker et al. 1987).

It is possible to eliminate the explicit appearance of the pressure in the momentum equations by adopting a penalty function method (Temam 1968). In this method the continuity equation (7.1) is replaced by

$$\varepsilon p + \left(\frac{\partial u}{\partial x} + \frac{\partial v}{\partial y}\right) = 0 \ , \tag{17.87}$$

where ε is a small parameter, $10^{-9} \le \varepsilon \le 10^{-4}$ typically. This construction has some similarity with the artificial compressibility method (Sect. 17.2.1) but does not require a pseudotransient strategy. In practice (17.87), in either exact or discretised form, is used to eliminate the pressure from the momentum equations. Thus the solution is obtained in terms of the velocity field only. The pressure is recovered subsequently from (17.87). The penalty function formulation is widely used with the finite element method.

For the equivalent Stokes problem, i.e. the Navier–Stokes equations without the convective terms, it is demonstrated (Baker 1983, pp. 266–270) that the penalty function method can be derived by minimising a functional constructed to simultaneously satisfy the continuity and momentum equations. The extension to the Navier–Stokes equations follows by replacing the variational formulation with the Galerkin (weighted residuals) formulation (Sect. 5.1).

Two alternative penalty function formulations are used with the finite element method. First, (17.87) is substituted into the steady form of (17.2 and 3) and the Galerkin finite element formulation applied to the resulting equations. This construction has been used by Hughes et al. (1979) with the velocity field given a linear interpolation on quadrilateral elements.

However, in evaluating the integrals numerically in the equivalent of (17.84 and 85) Gauss quadrature (Zienkiewicz 1977) is used. In evaluating the penalty terms which replace the pressure terms it is necessary to use a lower or reduced order of Gauss quadrature so that the equivalent of (17.86) is non-singular. All other terms are evaluated with a sufficiently high-order Gauss quadrature scheme that the numerical integration is exact. Sani et al. (1981) demonstrate that the use of reduced integration is equivalent to the implicit use of lower-order interpolation for the pressure.

Alternatively, in the consistent penalty function method (Engelman et al. 1982), (17.87) is discretised using a conventional Galerkin formulation but with lower-order interpolation used for the pressure field and for the weight function. Subsequently the discrete pressure field arising from (17.84 and 85) is eliminated using the discrete form of (17.87). For quadrilateral elements with linear velocity interpolation and constant pressure interpolation the two formulations are identical at the discrete level.

The penalty formulation does provide inherent smoothing for the pressure field (Sani et al. 1981), although additional smoothing may also be required (Hughes et al. 1979). The penalty method is usually considerably more economical than the

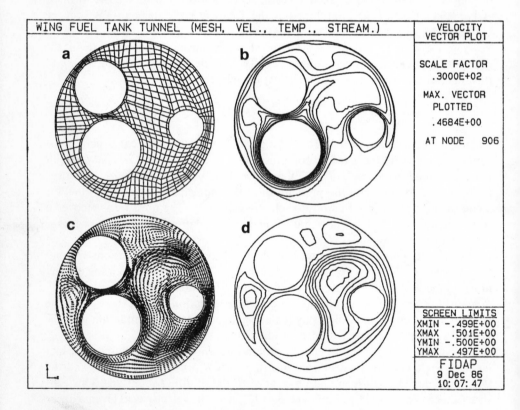

Fig. 17.11a–d. Natural convection in a wing fuel tank conduit. **a)** Finite element grid. **b)** Temperature contours. **c)** Velocity vectors. **d)** Streamlines (after Engelman, 1982; published with permission of Fluid Dynamics International)

mixed interpolation (u, v, p) formulation. For quadratically-interpolated velocity on elements with curved sides (to suit irregular geometries) the consistent penalty function formulation is more accurate (Engelman et al. 1982) than the use of reduced integration and theoretically better supported.

The finite element method lends itself to the construction of general-purpose codes for solving coupled fluid flow, heat transfer problems in complicated geometric domains. FIDAP (Fluid Dynamic Analysis Program) is such a general-purpose code and is described by Engelman (1982). A representative problem that can be successfully modelled by FIDAP is indicated in Fig. 17.11.

A conduit passing through a wing fuel tank contains three electrical wires at different temperatures. FIDAP determines the natural convection in the air gap surrounding the wires. Shown in Fig. 17.11 are the finite element grid, temperature contours, velocity vectors and streamlines for a Rayleigh number of 800 000. The solution indicates thermal plumes rising from the hot wire and dropping from the cold wire. The grid contains 2654 nodes and 624 nine-node quadrilateral elements.

17.3 Vorticity, Stream Function Variables

As an alternative to solving the governing equations in primitive variables it is possible to avoid the explicit appearance of the pressure by introducing the vorticity and stream function as dependent variables (Sect. 11.5.1), at least in two dimensions.

In two-dimensional flow the vorticity vector

$$\zeta = \text{curl } \mathbf{q} \tag{17.88}$$

has a single component, which is defined conventionally as

$$\zeta = \frac{\partial u}{\partial y} - \frac{\partial v}{\partial x} . \tag{17.89}$$

The transport equation for the vorticity (11.85) with the aid of the continuity equation (17.1) is

$$\frac{\partial \zeta}{\partial t} + \frac{\partial (u\zeta)}{\partial x} + \frac{\partial (v\zeta)}{\partial y} - \frac{1}{\text{Re}} \left(\frac{\partial^2 \zeta}{\partial x^2} + \frac{\partial^2 \zeta}{\partial y^2} \right) = 0 , \tag{17.90}$$

where the Reynolds number $\text{Re} = U_\infty L/v$. In two dimensions a stream function can be defined by

$$u = \frac{\partial \psi}{\partial y} \quad \text{and} \quad v = -\frac{\partial \psi}{\partial x} , \tag{17.91}$$

and substitution into (17.89) produces the following Poisson equation for the stream function:

$$\frac{\partial^2 \psi}{\partial x^2} + \frac{\partial^2 \psi}{\partial y^2} = \zeta \ . \tag{17.92}$$

Equations (17.90–92) constitute the governing equations for the vorticity stream function formulation of incompressible laminar flow. Strictly by substituting (17.91) into (17.90) it is possible to eliminate the explicit appearance of u and v. However, such a formulation may produce less accurate solutions although it does save the additional storage of u and v. Initial and boundary conditions to suit (17.90–92) are discussed in Sect. 11.5.1.

The system of equations (17.90–92) is applicable to both steady and unsteady laminar viscous flow. However, only the vorticity transport equation (17.90) depends explicitly on time. Consequently, for unsteady problems (17.92) implies that the stream function field must be determined to be compatible with the time-dependent vorticity distribution at every time-step.

For unsteady problems (17.90) is parabolic in time if u and v are known. Thus it can be marched efficiently in time using an ADI or approximation factorisation technique (Sect. 8.2). At each time step the discrete form of (17.92) is solved for ψ. Equation (17.92) is strongly elliptic if ζ is known and can be solved by iterative (Sect. 6.3) or direct methods (Sect. 6.2). Since (17.92) is a Poisson equation very efficient direct methods (Sect. 6.2.6) are available if the grid is uniform.

For steady flow problems, (17.91, 92) and the steady form of (17.90) are a system of elliptic partial differential equations. Since (17.90) is nonlinear it is necessary to employ an iterative algorithm. At each step of the iteration (17.90 and 92) are used to update the ζ and ψ solutions either sequentially or as a coupled system. Gupta and Manohar (1979) employ a sequential algorithm.

It is necessary to use under-relaxation in determining boundary values of the vorticity, to provide a Dirichlet boundary condition for the steady form of (17.90). The cause of this problem is that physical boundary conditions are available on ψ and $\partial \psi / \partial n$ but none on ζ. When numerical boundary conditions are constructed for ζ which satisfy the integral boundary condition (11.90), no under-relaxation is required (Quartapelle and Valz-Gris 1981), even though a sequential algorithm is used.

However, if the steady form of (17.90 and 92) are solved as a coupled system the two boundary conditions on ψ and $\partial \psi / \partial n$ are sufficient. Campion-Renson and Crochet (1978) use such a formulation with a finite element method to examine the flow in a driven cavity. No numerical boundary condition for ζ is required.

The pseudotransient strategy (Sect. 6.4) offers an alternative path to obtain the steady flow solution. To implement the pseudotransient approach (17.92) is replaced by

$$\frac{\partial \psi}{\partial \tau} - \left\{ \frac{\partial^2 \psi}{\partial x^2} + \frac{\partial^2 \psi}{\partial y^2} - \zeta \right\} = 0 \ . \tag{17.93}$$

When the steady state is reached (17.93) reverts to (17.92). The choice of the time-step $\Delta\tau$ that appears after discretisation of (17.93) provides an additional level of control over the pseudotransient iteration. The sequential versus coupled treatment of (17.90 and 93) is also relevant to the pseudotransient strategy. Typical examples are provided in the next section.

17.3.1 Finite Difference Formulations

In this section we consider a typical sequential and a typical coupled solution algorithm for the steady laminar flow in a driven cavity (Fig. 17.12). The lid of the cavity moves continuously to the right with a velocity $u=1$. No-slip boundary conditions on the velocity components u and v are equivalent, through (17.91), to the indicated boundary conditions on ψ and $\partial\psi/\partial n$.

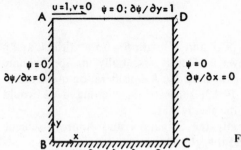

Fig. 17.12. Two-dimensional driven cavity

A sequential algorithm due to Mallinson and de Vahl Davis (1973) is described which is based on a pseudotransient solution of (17.90 and 93). In this formulation uniform-grid three-point centred difference formulae are introduced for first and second spatial derivatives. In the notation of Chap. 8,

$$\frac{\partial(u\zeta)}{\partial x}=L_x(u\zeta)_{j,k}+O(\Delta x^2) \ , \qquad \frac{\partial^2\zeta}{\partial y^2}=L_{yy}\zeta_{j,k}+O(\Delta y^2) \ , \quad \text{etc.},$$

where

$$L_x(u\zeta)_{j,k}=\frac{(u\zeta)_{j+1,k}-(u\zeta)_{j-1,k}}{2\Delta x} \ , \quad \text{and} \tag{17.94}$$

$$L_{yy}\zeta_{j,k}=\frac{\zeta_{j,k-1}-2\zeta_{j,k}+\zeta_{j,k+1}}{\Delta y^2} \ .$$

Mallinson and de Vahl Davis write the semi-discrete form of (17.90) as

$$\frac{1}{\varepsilon}\frac{\partial\zeta_{j,k}}{\partial t}=(A^x+A^y)\zeta_{j,k} \ , \quad \text{where} \tag{17.95}$$

$$A^x\zeta_{j,k}=(1/\text{Re})L_{xx}\zeta_{j,k}-L_x(u\zeta)_{j,k} \ ,$$
$$A^y\zeta_{j,k}=(1/\text{Re})L_{yy}\zeta_{j,k}-L_y(v\zeta)_{j,k} \ ,$$

and ε is a relaxation parameter that can be varied spatially. When all grid points are considered the following vector equation results:

$$\frac{\partial \zeta}{\partial t} = \varepsilon [\underline{A}^x + \underline{A}^y] \zeta \ . \tag{17.96}$$

The elements of the matrices $\underline{A}^x$ and $\underline{A}^y$ can be obtained from (17.94).

Equation (17.96) and an equivalent semi-discrete vector equation, based on (17.93), are advanced in time using an algorithm introduced by Samarskii and Andreev (1963),

$$[I - 0.5\varepsilon \, \Delta t \, \underline{A}^x] \Delta \zeta^* = \varepsilon \Delta t [\underline{A}^x + \underline{A}^y] \zeta^n \ ,$$

$$\tag{17.97}$$

$$[I - 0.5\varepsilon \, \Delta t \, \underline{A}^y] \Delta \zeta^{n+1} = \Delta \zeta^* \quad \text{and}$$

$$\zeta^{n+1} = \zeta^n + \Delta \zeta^{n+1} \ .$$

It is clear that (17.97) is equivalent to (8.23 and 24) with $\beta = 0.5$ and the u and v terms in $\underline{A}^x$, $\underline{A}^y$ evaluated at time-level n. This is essentially an approximate factorisation with Crank–Nicolson time differencing. A consideration of the modified Newton method (Sects. 6.4 and 10.4.3) suggests that setting $\beta = 1$ would produce a more rapid convergence to the steady state.

Mallinson and de Vahl Davis apply the Samarskii and Andreev scheme sequentially to (17.93 and 90). They find that the fastest convergence corresponds to $\Delta t \approx 0.8 \Delta x^2 = 0.8 \Delta y^2$ and $\Delta \tau \approx 50\varepsilon \Delta t$. De Vahl Davis and Mallinson (1976) use this algorithm to compare three-point central differencing and two-point upwind differencing for the convective terms in (17.90) for large Reynolds numbers. Clearly the higher-order upwind schemes (Sects. 9.3.2 and 17.1.5) could be incorporated into the present method with some modification of the implicit algorithm.

When solving (17.93) for the driven cavity problem the Dirichlet boundary condition for ψ is used. When solving (17.90) a Dirichlet boundary condition for ζ is constructed. How this is done is indicated in Sect. 17.3.2.

Rubin and Khosla (1981) solve (17.90 and 92) as a coupled system using a modified strongly implicit procedure (Sect. 6.3.3). To obtain a diagonally dominant system of coupled equations for large values of Re the following discretisation of $\partial(u\zeta)/\partial x$ is introduced:

$$\frac{\partial(u\zeta)}{\partial x} \approx \mu_x L_x^+ (u\zeta)_{j,k}^{n+1} + (1 - \mu_x) L_x^- (u\zeta)_{j,k}^{n+1} + 0.5\Delta x (1 - 2\mu_x) L_{xx}(u\zeta)_{j,k}^n \ , \tag{17.98}$$

where

$$L_x^+ (u\zeta)_{j,k} = \frac{[(u\zeta)_{j+1,k} - (u\zeta)_{j,k}]}{\Delta x} \ , \qquad L_x^- (u\zeta)_{j,k} = \frac{[(u\zeta)_{j,k} - (u\zeta)_{j-1,k}]}{\Delta x} \ ,$$

and $\mu_x = 0$ if $u_{j,k} \geq 0$ and $\mu_x = 1$ if $u_{j,k} < 0$. The above scheme due to Khosla and Rubin (1974) is an upwind scheme at the implicit level $(n+1)$. However, under steady-state conditions it reverts to a three-point centred finite difference scheme.

Using (17.98) and an equivalent form for $\partial(v\zeta)/\partial y$, but assuming u, $v > 0$, the discrete form of (17.90 and 92) can be written

$$\frac{\zeta_{j,k}^{n+1}}{\Delta t} + L_x^-(u\zeta)_{j,k}^{n+1} + L_y^-(v\zeta)_{j,k}^{n+1} - \frac{1}{\text{Re}}\{L_{xx} + L_{yy}\}\zeta_{j,k}^{n+1}$$

$$= \frac{\zeta_{j,k}^n}{\Delta t} - 0.5\Delta x\, L_{xx}(u\zeta)_{j,k}^n - 0.5\Delta y\, L_{yy}(v\zeta)_{j,k}^n \ , \tag{17.99}$$

$$\{L_{xx} + L_{yy}\}\psi_{j,k}^{n+1} - \zeta_{j,k}^{n+1} = 0 \ . \tag{17.100}$$

Equations (17.99 and 100) constitute a 2×2 system of equations which is diagonally dominant and couples together implicit $(n+1)$ values of ζ and ψ at grid points $(j-1, k)$, (j, k), $(j+1, k)$, $(j, k-1)$ and $(j, k+1)$. The velocity components in (17.99) are evaluated at the explicit (n) time level. If (17.99 and 100) at all interior nodes are considered collectively the resulting sparse 2×2 block matrix equation can be

RE=10000, UNIFORM GRID (257X257)

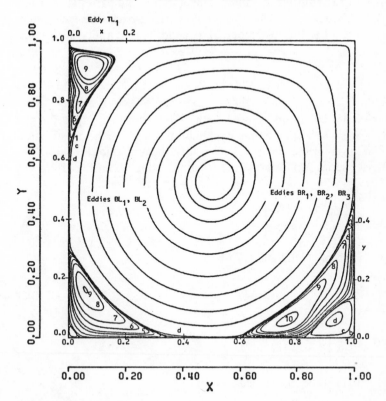

Fig. 17.13. Streamline pattern for flow in a driven cavity at $\text{Re} = 10\,000$ (after Ghia et al., 1982; reprinted with permission of Academic Press)

solved efficiently using the strongly implicit procedure (Sect. 6.3.3). The details are provided by Rubin and Khosla (1981). Because of the strong coupling between ζ and ψ at the implicit time level no under-relaxation is required for stability when implementing the vorticity boundary condition.

Ghia et al. (1982) combine the Rubin and Khosla formulation with multigrid (Sect. 6.3.5) to obtain the flow behaviour in a driven cavity (Fig. 17.12) for Reynolds numbers up to 10000 on a 257×257 uniform grid. A typical result is shown in Fig. 17.13. The flow is characterised by a primary eddy filling most of the cavity and a sequence of counterrotating corner eddies. Ghia et al. note that the use of multigrid produces an algorithm that is about four times more efficient than using the strongly implicit procedure conventionally on the finest grid.

17.3.2 Boundary Condition Implementation

The implementation of the boundary conditions for the ζ, ψ formulation will be discussed in this section. Most attention will be given to the construction of the vorticity boundary condition at the solid surface. However, the prescription of appropriate boundary conditions at inflow and outflow boundaries is also important and will be discussed in relation to the flow past a backward-facing step.

As indicated in Fig. 17.12 the no-slip boundary conditions at a solid surface are equivalent to

$$\psi = 0 \quad \text{and} \quad \frac{\partial \psi}{\partial n} = g \ . \tag{17.101}$$

The first boundary condition is used with the Poisson equation for the streamfunction (17.92). The second boundary condition is used in the construction of a boundary condition for the vorticity. This will be illustrated for the lid (AD in Fig. 17.12). A Taylor series expansion of the streamfunction about the grid point (j, k) on AD gives

$$\psi_{j,k-1} = \psi_{j,k} - \Delta y \left[\frac{\partial \psi}{\partial y} \right]_{j,k} + \frac{\Delta y^2}{2} \left[\frac{\partial^2 \psi}{\partial y^2} \right]_{j,k} + \ldots \ . \tag{17.102}$$

From the discrete form of (17.92) and (17.101a),

$$\zeta_{j,k} = \left[\frac{\partial^2 \psi}{\partial y^2} \right]_{j,k} , \tag{17.103}$$

$$\psi_{j,k} = 0 \quad \text{and} \quad \left[\frac{\partial \psi}{\partial y} \right]_{j,k} = g_j \ .$$

Consequently (17.102) can be rearranged to give

$$\zeta_{j,k} = \frac{2}{\Delta y^2} (\psi_{j,k-1} + \Delta y \, g_j) + O(\Delta y) \ . \tag{17.104}$$

This first-order formula was first used by Thom (1933) and has been used extensively since. Comparable formulae can be readily obtained for the other surfaces.

Since a second-order accurate discretisation is used in the interior it is desirable to use a second-order accurate implementation of the boundary conditions (Sect. 7.3). This can be achieved as follows.

A second-order implementation of (17.103) is

$$\zeta_{j,k} = \frac{\psi_{j,k-1} - 2\psi_{j,k} + \psi_{j,k+1}}{\Delta y^2} + O(\Delta y^2) \ . \tag{17.105}$$

In addition, a third-order accurate expressions for $[\partial\psi/\partial y]_{j,k}$ is

$$g_j = \left[\frac{\partial\psi}{\partial y}\right]_{j,k} = \frac{\psi_{j,k-2} - 6\psi_{j,k-1} + 3\psi_{j,k} + 2\psi_{j,k+1}}{6\Delta y} + O(\Delta y^3) \ . \tag{17.106}$$

The nodal value $\psi_{j,k+1}$ lies outside of the computational domain and is eliminated from (17.105 and 106) to give

$$\zeta_{j,k} = \frac{0.5}{\Delta y^2}(8\psi_{j,k-1} - \psi_{j,k-2}) + \frac{3g_j}{\Delta y} + O(\Delta y^2) \ . \tag{17.107}$$

This form is attributed to Jensen (1959) by Roache (1972) and is used by Pearson (1965) and Ghia et al. (1982).

Equation (17.107) produces more accurate solutions than the use of (17.104) in the comparative tests of Gupta and Manohar (1979). However, when used in a sequential algorithm more iterations are required using (17.107), and for large values of Re divergence may occur even when the boundary value of the vorticity is under-relaxed. When used in a coupled algorithm (17.107) causes no particular difficulty.

An alternative vorticity boundary condition for ζ is available in a pseudo-transient formulation,

$$\zeta_{j,k}^{n+1} = \zeta_{j,k}^n - \beta\{[\partial\psi/\partial n] - g\}_{j,k} \ . \tag{17.108}$$

This appears to provide a more direct implementation of the boundary condition (17.101b). The relaxation parameter β must be chosen appropriately (Israeli 1972) to ensure convergence. However, Peyret and Taylor (1983, p. 187) point out a rather direct link with a vorticity boundary value evaluation via (17.104), as follows.

At the $(n+1)$-th step of a pseudotransient formulation the boundary value for the vorticity is given by

$$\zeta_{j,k}^{n+1} = \gamma\zeta_{j,k}^* + (1-\gamma)\zeta_{j,k}^n \ , \tag{17.109}$$

where $\zeta^*_{j,k}$ is obtained from (17.104) and γ is a relaxation coefficient. Combining (17.104) and (17.109) to eliminate $\zeta^*_{j,k}$ gives

$$\zeta^{n+1}_{j,k} = \zeta^n_{j,k} + \frac{2\gamma}{\Delta y^2}(\psi^n_{j,k-1} + \Delta y\, g^n_j - 0.5\, \Delta y^2\, \zeta^n_{j,k})\ . \tag{17.110}$$

If $[\partial\psi/\partial n]_{j,k}$ in (17.108) is replaced by $(\psi_{j,k} - \psi_{j,k-1})/\Delta y$ the result is

$$\zeta^{n+1}_{j,k} = \zeta^n_{j,k} + \frac{\beta}{\Delta y}(\psi^n_{j,k-1} + \Delta y\, g^n_j)\ . \tag{17.111}$$

To $O(\Delta y)$, (17.110 and 111) are equivalent if $\beta = 2\gamma/\Delta y$.

To examine suitable computational boundary conditions on open boundaries it is convenient to consider the flow past a backward-facing step, Fig. 17.14. As noted in Sects. 11.5 and 11.6.4, open boundaries can be classified as inflow and outflow boundaries and the required number of physical boundary conditions are indicated in Table 11.5.

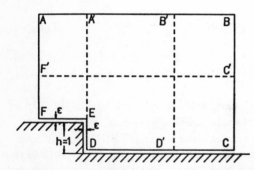

Fig. 17.14. Flow past a backward-facing step

In relation to the flow past a backward-facing step (Fig. 17.14), AF is an inflow boundary and BC is an outflow boundary. However, AB will be either an inflow or outflow boundary depending on the local sign of v_{AB}. The crucial feature of boundary AB is that it is remote from the backward-facing step and the local flow direction is almost parallel to AB. Such a boundary will be called a farfield boundary and appropriate boundary conditions will be indicated below that do not depend on specifically identifying it as an inflow or outflow boundary.

At an inflow boundary it is appropriate (Table 11.5) to specify all but one of the dependent variables for incompressible viscous flow. For flow past a backward-facing step it is appropriate to specify $u(y)$, $p(y)$ and to determine $v(y)$ from the interior solution. Thus in the stream function, vorticity formulation ψ is specified at inflow; specifying ζ is not recommended. Roache (1972) prefers to specify $\partial^2 v/\partial x^2 = 0$. On AF in Fig. 17.14, ζ is obtained from (17.89) as

$$\zeta_{1,k} = \frac{u_{1,k+1} - u_{1,k-1}}{2\Delta y} - \left[\frac{\partial v}{\partial x}\right]_{1,k} , \qquad (17.112)$$

if three-point centred difference formulae are used for discretisation. However,

$$\left[\frac{\partial^2 v}{\partial x^2}\right]_{1,k} = 0 \quad \text{implies}$$

$$\left[\frac{\partial v}{\partial x}\right]_{1,k} = \left[\frac{\partial v}{\partial x}\right]_{2,k} = -\frac{\psi_{1,k} - 2\psi_{2,k} + \psi_{3,k}}{\Delta x^2} .$$

Thus $\zeta_{1,k}$ is evaluated in (17.112) from the boundary values of u and the interior values of ψ. A similar construction is used by Fletcher and Srinivas (1983) except that $[\partial v/\partial x = -\partial^2\psi/\partial x^2]_{j,k}$ is obtained from the interior solution via a one-sided discretisation without invoking $\partial^2 v/\partial x^2 = 0$.

At outflow, BC in Fig. 17.14, Roache (1972) and Baker (1983) recommend

$$\frac{\partial \zeta}{\partial x} = 0 \quad \text{and} \quad \frac{\partial^2 \psi}{\partial x^2} = 0 . \qquad (17.113)$$

The second boundary condition is implemented as $\partial^2\psi/\partial y^2 = \zeta$, from (17.92). However, it is important that this boundary condition is compatible with the boundary conditions on DC and AB. Roache (1972) also recommends an alternative vorticity treatment of evaluating $\partial(u\zeta)/\partial x$ in (17.90) on BC using a two-point upwind formula and evaluating $[\partial^2\zeta/\partial x^2]_{JMAX,k} = [\partial^2\zeta/\partial x^2]_{JMAX-1,k}$. Then no boundary condition on ζ_{BC} is required. Fletcher and Srinivas (1983) obtain a very similar effect by deleting the term $\partial^2\zeta/\partial x^2$ from (17.90) on BC; this is justified on an order-of-magnitude basis. However, with this simplification (17.90) becomes spatially parabolic in the x direction and no boundary condition on ζ is required. An alternative interpretation, which is computationally equivalent, is to set $\partial^2\zeta/\partial x^2 = 0$ on BC.

The farfield boundary condition specification of $u = 1$ provides a Neumann boundary condition on ψ. For streamlined bodies it is possible to compute an approximate farfield solution by assuming inviscid flow everywhere, e.g. using a panel method (Sect. 14.1). This often permits a more precise boundary condition $u = U_\infty(x)$ and allows AB to be brought closer to the body. However, it is recommended that this procedure be used to provide Neumann boundary conditions, to avoid the possibility of an unphysical boundary layer adjacent to AB that a Dirichlet boundary condition specification may cause.

For the separating flow field caused by a backward-facing step it is better to keep AB sufficiently far away that $u = 1$ and $\zeta = 0$. An alternative frictionless "wind-tunnel" boundary condition is to specify $u = 1$, $v = 0$ on AB. This effectively imposes a Dirichlet boundary condition $\psi_{AB} = \psi_A$. The Dirichlet boundary condition for the vorticity is constructed from the interior ψ field in the same manner as (17.104). If these two boundary conditions are combined they also imply $\partial u/\partial y = \partial v/\partial x = 0$. As long as AB is sufficiently far from the step the global solution

will be relatively insensitive to the particular boundary condition specification on
AB. However, a poor choice may introduce a boundary layer or locally oscillatory
solution adjacent to AB.

17.3.3 Group Finite Element Formulation

In this subsection the group finite element formulation (Sect. 10.3) will be applied to
the incompressible laminar flow past a backward-facing step (Fig. 17.14). A
pseudotransient formulation applied to (17.90, 91 and 93) will be used to provide
the steady flow behaviour. Subsequently the related flow past a rearward-facing
cavity (Fig. 17.17) will be considered. When blowing and suction is introduced into
the cavity it is found that an unsteady solution can result.

The Galerkin finite element method, with bilinear interpolation on rectangular
elements (Sect. 5.3), is applied to (17.90). Approximate solutions, like (5.58), are
introduced for ζ and for the groups $u\zeta$ and $v\zeta$, as in (10.54). The result, in semi-
discrete form, is

$$M_x \otimes M_y \dot{\zeta} + M_y \otimes L_x u\zeta + M_x \otimes L_y v\zeta - \frac{1}{\mathrm{Re}}(M_y \otimes L_{xx} + M_x \otimes L_{yy})\zeta = 0 \ ,$$

$$(17.114)$$

where M_x, M_y are directional mass operators and L_x, L_{yy}, etc., are directional
difference operators (Vol. 1, Appendix A.2). On a rectangular but non-uniform grid
(Fig. 17.15) these operators have the form

$$M_x \equiv A\left\{\frac{1}{6}, \frac{1+r_x}{3}, \frac{r_x}{6}\right\} \ , \qquad M_y \equiv B\left\{\frac{r_y}{6}, \frac{1+r_y}{3}, \frac{1}{6}\right\}^T \ ,$$

$$L_x \equiv \frac{0.5A}{\Delta x}\{-1, 0, 1\} \ , \qquad L_y \equiv \frac{0.5B}{\Delta y}\{1, 0, -1\}^T \ , \qquad (17.115)$$

$$L_{xx} \equiv \frac{A}{\Delta x^2}\left\{1, -\frac{(1+r_x)}{r_x}, \frac{1}{r_x}\right\} \ , \qquad L_{yy} \equiv \frac{B}{\Delta y^2}\left\{\frac{1}{r_y}, -\frac{(1+r_y)}{r_y}, 1\right\}^T \ ,$$

where $A = 2/(1+r_x)$ and $B = 2/(1+r_y)$.

For a uniform grid ($r_x = r_y = 1$) the formulae given in (17.115) revert to those
given in Table 9.1.

Equation (17.114) represents a system of ordinary differential equations in time.
The following three-level algorithm is derived from (17.114) to advance the solution
in time:

$$M_x \otimes M_y \frac{(1+\gamma)\Delta\zeta^{n+1} - \gamma\Delta\zeta^n}{\Delta t} = \beta\,\mathrm{RHS}^{n+1} + (1-\beta)\mathrm{RHS}^n \ , \qquad (17.116)$$

where γ and β can be chosen to suit the application. For time-dependent problems
a suitable choice is $\gamma = 0$ and $\beta = 0.5$, which produces the Crank–Nicolson scheme.

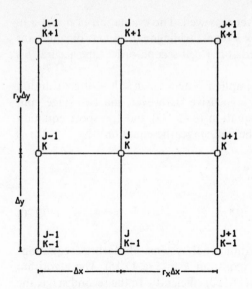

Fig. 17.15. **Fig. 17.15.** Nonuniform rectangular grid

When obtaining steady-state solutions using a pseudotransient approach a preferred choice is $\gamma = 0.5$, $\beta = 1.0$, since it gives faster convergence (Fletcher and Srinivas 1983). The above algorithm is also applied to the two-dimensional transport equation (9.87).

In (17.116) the following terms are defined:

$$\Delta \zeta^{n+1} = \zeta^{n+1} - \zeta^n, \qquad \Delta \zeta^n = \zeta^n - \zeta^{n-1} \quad \text{and} \tag{17.117}$$

$$\text{RHS} = (1/\text{Re})\{M_y \otimes L_{xx} + M_x \otimes L_{yy}\} - M_y \otimes L_x u\zeta - M_x \otimes L_y v\zeta .$$

To produce an economical algorithm while avoiding severe stability restrictions on the time-step it is necessary to obtain a linear system of equations from (17.116) for $\Delta \zeta^{n+1}$. This requires linearising RHS^{n+1}. This is done most efficiently by expanding as a Taylor series about time level n, i.e.

$$\text{RHS}^{n+1} = \text{RHS}^n + \left[\frac{\partial}{\partial \zeta}(\text{RHS})\frac{\partial \zeta}{\partial t} + \frac{\partial}{\partial u}(\text{RHS})\frac{\partial u}{\partial t} + \frac{\partial}{\partial v}(\text{RHS})\frac{\partial v}{\partial t}\right]\Delta t + \dots \tag{17.118}$$

and truncating after the terms shown. If $\partial \zeta / \partial t$ is replaced by $\Delta \zeta^{n+1}/\Delta t$, substitution of (17.118) into (17.116) gives

$$(1+\gamma)\left[M_x \otimes M_y - \Delta t \frac{\beta}{1+\gamma}\frac{\partial}{\partial \zeta}(\text{RHS})\right]\Delta \zeta^{n+1} = \Delta t \, \text{RHS}^{n,\beta} + M_x \otimes M_y \Delta \zeta^n . \tag{17.119}$$

In $\text{RHS}^{n,\beta}$ all terms are evaluated at time level n except u and v which are evaluated at $t^n + \beta \Delta t$. This is desirable for solving unsteady problems, but for steady problems, where the transient accuracy is not important, it is computationally more efficient to evaluate u and v at time-level n as well. The evaluation of u and v in RHS$^{n,\beta}$ produces scalar, rather than 3×3 block, tridiagonal systems of equations (17.120 and 121). If u and v are evaluated at $t^n + \beta \Delta t$ second-order time accuracy is preserved.

Equation system (17.119) is a linear implicit system for $\Delta \zeta^{n+1}$. A direct solution of (17.119) would be computationally expensive. However, the two-stage split schemes developed for the diffusion equation (8.45, 47) and transport equation (9.88, 89) are also applicable here. In the first stage, the equation

$$\left[M_x - \Delta t \frac{\beta}{1+\gamma} \left(\frac{1}{\text{Re}} L_{xx} - L_x u \right) \right] \Delta \zeta^*$$

$$= \frac{\Delta t}{1+\gamma} \text{RHS}^{n,\beta} + \frac{\gamma}{1+\gamma} M_x \otimes M_y \Delta \zeta^n \qquad (17.120)$$

provides a collection of independent tridiagonal systems of algebraic equations along each gridline in the x-direction (constant k in Fig. 17.15). The algorithm provided in Sects. 6.2.2 and 6.2.3 solves (17.120) efficiently. In the second stage, the equation

$$\left[M_y - \Delta t \frac{\beta}{1+\gamma} \left(\frac{1}{\text{Re}} L_{yy} - L_y v \right) \right] \Delta \zeta^{n+1} = \Delta \zeta^* \qquad (17.121)$$

is solved using the Thomas algorithm (Sect. 6.2.2) for each gridline in the y-direction (constant j in Fig. 17.15). In (17.120 and 121) u and v are functions of position and are operated on by the L_x and L_y operators respectively. This is not the case in (9.88 and 89).

The Galerkin finite element discretisation of (17.93) follows the same path as above. The following semi-discrete form is produced in place of (17.114):

$$M_x \otimes M_y \frac{\partial \psi}{\partial \tau} = \{ M_y \otimes L_{xx} + M_x \otimes L_{yy} \} \psi - M_x \otimes M_y \zeta . \qquad (17.122)$$

Applying the same splitting algorithm as for the vorticity transport equation produces the two-stage algorithm

$$\left(M_x - \Delta \tau \frac{\beta}{1+\gamma} L_{xx} \right) \Delta \psi^* = \frac{\Delta \tau}{1+\gamma} (M_y \otimes L_{xx} + M_x \otimes L_{yy}) \psi^m$$

$$- M_x \otimes M_y \left(\frac{\Delta \tau}{1+\gamma} \zeta + \frac{\gamma}{1+\gamma} \Delta \psi^m \right) \qquad (17.123)$$

and

$$\{ M_y - \Delta \tau [\beta/(1+\gamma)] L_{yy} \} \Delta \psi^{m+1} = \Delta \psi^* . \qquad (17.124)$$

In (17.123 and 124) $\Delta \tau$ is a pseudo-time-step that permits the iterative solution ψ^{m+1} to be obtained at each physical time-step Δt. For unsteady problems the iterative solution must continue until (17.92) is satisfied, but for steady problems it is acceptable to iterate (17.123 and 124) three or four times per physical time step Δt. Convergence to the solution of (17.92) occurs as the solution of (17.93) approaches the steady state.

At each time step Δt the Thomas algorithm, Sect. 6.2.2, can be applied directly to obtain the solution of (17.123 and 124) along gridlines in the j and k directions respectively.

Equation (17.91) relates the velocity field to the streamfunction solution. Applying a one-dimensional Galerkin formulation to (17.91) produces the following dependence of the velocity solution on the streamfunction solution:

$$M_y u = L_y \psi \quad \text{and} \quad M_x v = -L_x \psi \ . \tag{17.125}$$

These equations are tridiagonal along gridlines in the k and j directions respectively. Consequently they can be solved efficiently to provide the velocity field (u, v) once the streamfunction solution is known.

The implementation of the boundary conditions requires additional procedures. At a solid surface a pseudotransient form of (17.92) is used to provide the boundary value of ζ,

$$\frac{\partial \zeta}{\partial t} = \alpha \left(\frac{\partial^2 \psi}{\partial x^2} + \frac{\partial^2 \psi}{\partial y^2} - \zeta \right) \ . \tag{17.126}$$

Application of the Galerkin finite element method and an approximate factorisation produces a two-stage algorithm similar to (17.120 and 121). Thus for the first stage,

$$(\gamma + \alpha \beta \Delta t) M_x \Delta \zeta^* = \alpha \Delta t (M_y \otimes L_{xx} + M_x \otimes L_{yy}) \psi^n + \alpha \Delta t [M_y f(v) + M_x g(u)] \ , \tag{17.127}$$

and for the second stage,

$$M_y \Delta \zeta^{n+1} = \Delta \zeta^* \ . \tag{17.128}$$

The additional terms $f(v)$ and $g(u)$ arise from the application of Green's theorem to $\partial^2 \psi / \partial x^2$ and $\partial^2 \psi / \partial y^2$ when the Galerkin node is on the boundary (Sect. 8.4.2). For surface FE in Fig. 17.14, $f(v) = 0$ and $g(u) = u_{FE}/\Delta y$. If the computational boundary coincides with the surface of the step, $u_{FE} = 0$. However, Fletcher and Srinivas introduce a surface layer to avoid difficulties with the singular behaviour of the vorticity at E. As a result u_{FE} is non-zero.

In applying the Galerkin finite element method only two rectangular elements are used adjacent to a boundary, instead of four as in the interior. Consequently the mass and difference operators normal to the boundary have a different form to that in the interior (17.115). However, the tangential mass and difference operators have the same form as in the interior. Thus on surface FE (Fig. 17.14),

$$M_y \equiv \{\tfrac{1}{6}, \tfrac{1}{3}, 0\}^T \ , \quad L_{yy} = \frac{1}{\Delta y^2} \{1, -1, 0\}^T \ . \tag{17.129}$$

Equations (17.127 and 128) are tridiagonal and are included with the interior equations (17.120, 121) to provide a tridiagonal system of equations whose solution produces the updated vorticity field ζ^{n+1}. The relaxation parameter α must be limited to $\alpha \leqq 0.1$ to avoid instabilities with a sequential implementation. The other dependent variables, u, v and ψ, have Dirichlet boundary conditions at solid surfaces.

At the inflow boundary, AF in Fig. 17.14, u is specified as a boundary layer profile adjacent to F and takes the freestream value $u=1$ from the boundary layer outer edge to A. The streamfunction on AF is then obtained from (17.91a). Equations (17.127 and 128) are used to provide the vorticity on AF with $f(v) = v_{AF}/\Delta x$ and $g(u) = 0$. The operators M_y and L_{yy} appearing in (17.127, 128) are as indicated in (17.115). However, M_x and L_{xx} take the form

$$M_x \equiv \{0, \tfrac{1}{3}, \tfrac{1}{6}\} , \qquad L_{xx} = \frac{1}{\Delta x^2}\{0, -1, 1\} . \tag{17.130}$$

A Dirichlet boundary condition is required for v on AF to evaluate (17.125). This is provided by the interior ψ solution via (17.91b),

$$v_{AF} = \frac{1}{\Delta x}\left[\left(\frac{r_x+2}{r_x+1}\right)\psi_{1,k} - \left(\frac{r_x}{r_x+1}\right)\psi_{2,k} + \left(\frac{1}{r_x(1+r_x)}\right)\psi_{3,k} \right], \tag{17.131}$$

where $j=1$ coincides with boundary AF.

At the outflow boundary, BC in Fig. 17.14, the streamfunction is calculated using (17.123 and 124) with the addition of $-\Delta\tau\, M_y v_{BC}/\Delta x$. The operators M_y and L_{yy} in (17.123 and 124) are given by (17.115). The operators M_x and L_{xx} are given by

$$M_x \equiv \{\tfrac{1}{6}, \tfrac{1}{3}, 0\} \quad \text{and} \quad L_{xx} = \frac{1}{\Delta x^2}\{1, -1, 0\} . \tag{17.132}$$

Once the ψ solution on BC is available, v_{BC} is obtained from the interior ψ solution, equivalent to (17.131). The boundary condition $\partial^2\zeta/\partial x^2 = 0$ is imposed on BC. Consequently in evaluating (17.120) on BC, $L_{xx}\zeta = 0$. In addition the operators M_x and L_x are given by

$$M_x \equiv \{\tfrac{1}{6}, \tfrac{1}{3}, 0\} \quad \text{and} \quad L_x = \frac{0.5}{\Delta x}\{-1, 1, 0\} . \tag{17.133}$$

Corresponding operators in the y direction are given by (17.115).

At a farfield boundary, AB in Fig. 17.14, $u=1$ and $\zeta=0$. The stream function is obtained from (17.123 and 124) with the addition of the term $\Delta\tau\, M_x u_{AB}/\Delta y$ to the right-hand side of (17.123). The operators M_y and L_{yy} appearing in (17.123 and 124) take the form

$$M_y \equiv \{0, \tfrac{1}{3}, \tfrac{1}{6}\}^T \quad \text{and} \quad L_{yy} = \frac{1}{\Delta y^2}\{0, -1, 1\}^T . \tag{17.134}$$

As indicated in Fig. 17.14 the computational boundary is displaced from the solid surface by a thin layer. An order-of-magnitude analysis (Fletcher and Srinivas 1983) is used to evaluate u, v and ψ at the edge of the layer in terms of the vorticity ζ at the edge of the layer. The introduction of the surface layer permits the isolation of the corners, D and E in Fig. 17.14, from the computational domain. This is important because the vorticity has a singular behaviour at a sharp corner. A local analytic solution for the vorticity, after Lugt and Schwiderski (1965), adjacent to the sharp corner is used to determine local values for u, v and ψ. Details are provided by Fletcher and Srinivas (1983).

The overall procedure to generate the solution at each time-step is to solve (17.120 and 121) to obtain ζ^{n+1}, to solve (17.123 and 124) three or four times to obtain ψ^{n+1} and to solve (17.125) to obtain u^{n+1} and v^{n+1}. This procedure is repeated for consecutive time steps until the solution no longer changes. The result is the steady-state solution.

The present formulation has been used to compute the laminar flow past the two-dimensional step shown in Fig. 17.14. Typical results for the reattachment length of the flow behind the step are shown in Fig. 17.16. Good agreement with the experimental results of Sinha et al. (1981) and Goldstein et al. (1970) is apparent. The flow separates from the sharp edge (E in Fig. 17.14) and a separation bubble of slowly recirculating fluid forms behind ED. The boundary of the separation bubble is provided by the dividing streamline from E which reattaches to DC at positions depending primarily on the step-height Reynolds number (Fig. 17.16). The reattachment length behind the step, x_r, is a weak function of the upstream boundary layer thickness adjacent to F in Fig. 17.14. This is the reason for the separation of the two sets of experimental data.

The finest grid is used adjacent to E in Fig. 17.14. The grid grows in both x directions until $\Delta x = \Delta x_{max}$; thereafter $r_x = 1$. In the y direction, $\Delta y = \Delta y_{min}$ from

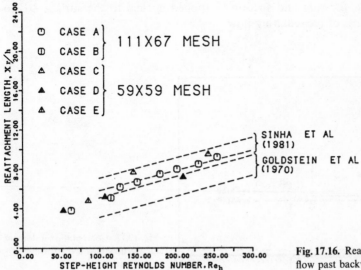

Fig. 17.16. Reattachment length for flow past backward-facing step

DC and $F'C'$ (edge of the inflow boundary layer). Between $F'C'$ and AB Δy grows with the geometric ratio r_y. For the finer grid, $r_x \leq 1.04$, $r_y \leq 1.07$. For the coarser grid, $r_x, r_y \leq 1.15$. Depending on the Reynolds number convergence to the steady state requires about 500–1000 time steps and corresponds to the steady-state residual, RHS in (17.117), being less than 1×10^{-5}.

The mass operators in equations like (17.114) are responsible for the greater accuracy of the finite element method, when compared with an equivalent finite difference method. The equivalent finite difference method can be obtained by lumping the mass operators. For example, M_x in (17.115) is replaced by

$$M_x = \{0, 1, 0\} \ .$$

Consequently by dividing through by $0.25(1 + r_x)(1 + r_y)$ the explicit appearance of the mass operators in the finite difference form of equations like (17.93) can be avoided.

The impact of mass lumping on the accuracy, stability and computational efficiency of the above algorithm is examined by Fletcher and Srinivas (1984). On a coarse grid (29×18) the mass operators must be retained adjacent to the computational boundaries to obtain a stable solution. Mass lumping in the interior does not affect the stability or seriously reduce the accuracy. In two dimensions interior mass lumping produces a small (18%) improvement in the economy. An operation count estimate suggests the improvement in the economy of interior mass lumping would be about 40% for a three-dimensional flow using a corresponding vorticity, velocity formulation (Sect. 17.4.2).

The present method can be extended to consider the flow past a rearward-facing cavity (Fig. 17.17). The cavity is formed by adding a lip, EG, to a backward-facing step. The introduction of the cavity displaces the primary bubble of recirculating fluid downstream and a secondary recirculation bubble can form within the cavity. Blowing and suction is applied normal to the surface DE to modify the patterns of recirculating flow.

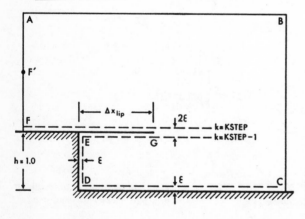

Fig. 17.17. Computational domain for the flow past a rearward-facing cavity

The governing equations and boundary conditions are the same as for the backward-facing step. The stream function ψ is set equal to zero on FGE. The stream function distribution ψ_{DE} is obtained by integrating the known blowing and suction distribution u_{DE}. On DC the stream function is constant, $\psi_{DC} = \psi_D$.

The vorticity at G (Fig. 17.17) is multivalued and a local analytic solution and a surface layer are used to match the local solution with the adjacent computational solution at the edge of the surface layer. Details are provided by Fletcher and Barbuto (1986a).

The introduction of the surface layer ($\varepsilon = 0.05\,\Delta y_{min}$) produces a very non-uniform grid downstream of G (Fig. 15.15) in the y direction. In applying the finite element discretisation for Galerkin nodes on $k = \text{KSTEP}$ (Fig. 17.18), the grid line $k = \text{KSTEP} - 1$ is ignored. Consequently the discretised equations have contributions from nodes on $k = \text{KSTEP} - 2$, KSTEP and KSTEP + 1. For Galerkin nodes on $k = \text{KSTEP} - 1$, the gridline $k = \text{KSTEP}$ is ignored and contributions to the discretised equations come from nodes on $k = \text{KSTEP} - 2$, KSTEP − 1 and KSTEP + 1. This procedure permits coupling of the local solutions and leads to locally smooth solutions.

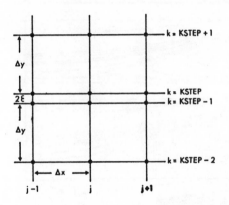

Fig. 17.18. Grid configuration downstream of lip

A typical flow pattern for a long lip, $EG/ED = 1.18$, is shown in Fig. 17.19 when clockwise blowing and suction is applied on surface DE. The velocity distribution u_{DE} is linear with a maximum value $|u_{DE}/U_\infty| = 0.6$, and blowing and suction is equal and opposite. The blowing and suction introduces a clockwise circulating cell which is isolated from the primary cell by a weak anticlockwise circulating cell. This flow pattern is quite steady.

If the lip is shortened to $EG/ED = 0.56$, a three-cell flow structure is produced but the flow is no longer steady (Fletcher and Barbuto 1986b). A typical sequence over a single period is shown in Fig. 17.20. Effectively the absence of the long lip prevents the secondary cell from stabilising. Since the flow pattern is unsteady some modification of the above algorithm is required. A typical procedure is provided by Peyret and Taylor (1983, p. 198). For the present algorithm it is necessary to introduce an iteration at each time step to ensure that the steady versions of (17.93 and 126) are satisfied.

Fig. 17.19. Flow pattern for rearward-facing cavity with a long lip, $Re_h = 217$, and clockwise blowing and suction

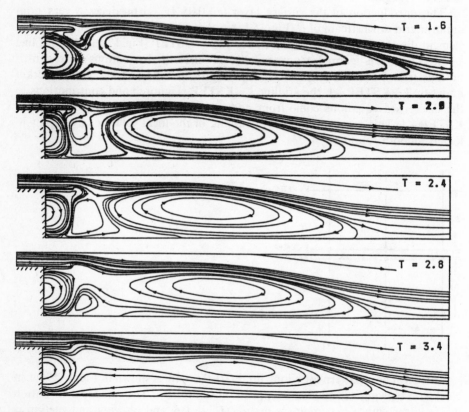

Fig. 17.20. Flow pattern for rearward-facing cavity with a short lip, $Re_h = 217$, and clockwise blowing and suction

17.3.4 Pressure Solution

In the stream function vorticity formulation the pressure does not appear explicitly. However, once the velocity solution is available the pressure solution can be obtained without difficulty. Techniques will be discussed here for steady flow; the extension for unsteady flow is straightforward.

The most direct means of computing the pressure is to treat the momentum equations (17.2 and 3) as ordinary differential equations in p. This technique is reasonably effective close to regions of known pressure, e.g. the freestream, and if the spatial pressure gradients are not large. However, the errors in the velocity field accumulate so that a long integration may imply a significant error. In addition if the pressure at a particular point is obtained by integrating along different paths some means of averaging or smoothing will need to be introduced to avoid a multivalued pressure solution. Such a technique is described by Raithby and Schneider (1979) as a modification to the SIMPLE algorithm (Sect. 17.2.3). For the flow over a backward-facing step Fletcher and Srinivas (1983) have used parallel integration of the momentum equations and normal extrapolation to obtain the pressure at the surface.

To obtain the pressure in the interior it is preferable to construct a Poisson equation from the momentum equations. In two dimensions, this can be written

$$\frac{\partial^2 p}{\partial x^2} + \frac{\partial^2 p}{\partial y^2} = 2\left(\frac{\partial u}{\partial x}\frac{\partial v}{\partial y} - \frac{\partial v}{\partial x}\frac{\partial u}{\partial y}\right) , \tag{17.135}$$

where the right-hand side of (17.135) is known from the stream function, vorticity solution. Equation (17.135) is applicable to both steady and unsteady flow.

Boundary conditions to suit (17.135) are usually Dirichlet boundary conditions in the freestream and Neumann boundary conditions at a solid surface. The Neumann boundary conditions are obtained from the normal momentum equation, which reduces to the following nondimensional form:

$$\frac{\partial p}{\partial n} = \frac{1}{Re}\frac{\partial \zeta}{\partial s} , \tag{17.136}$$

where s is measured along the boundary. For high Reynolds number flow parallel to a flat surface, (17.136) reduces to the boundary layer assumption $\partial p/\partial n = 0$. The solution of (17.135) must also satisfy the global integral constraint (17.16). This implies

$$\iint\left(\frac{\partial^2 p}{\partial x^2} + \frac{\partial^2 p}{\partial y^2}\right)dx\,dy = 0 = \int_c \frac{\partial p}{\partial n}\,ds . \tag{17.137}$$

For internal flow problems where a Neumann boundary condition is specified on all boundaries it is important to ensure that (17.137) is satisfied.

Since (17.135) is a Poisson equation any of the techniques suitable for linear strongly elliptic problems are available to solve the discrete form of (17.135). If the discretisation is undertaken on a uniform grid direct Poisson solvers (Sect. 6.2.6) are suitable. For both uniform and nonuniform grids the iterative techniques described in Sect. 6.3 are appropriate.

For external flows, like the flow over a backward-facing step, there is an advantage of working with the Bernoulli variable, H, instead of the pressure (11.49). In nondimensional form

$$H = c_p + u^2 + v^2 , \tag{17.138}$$

where the pressure coefficient $c_p = (p - p_\infty)/0.5\varrho U_\infty^2$. From the momentum equations, a Poisson equation for H replaces (17.135):

$$\frac{\partial^2 H}{\partial x^2} + \frac{\partial^2 H}{\partial y^2} = 2\left(\frac{\partial(u\zeta)}{\partial y} - \frac{\partial(v\zeta)}{\partial x}\right) . \tag{17.139}$$

Equation (17.139) is applicable to both steady and unsteady flow. Neumann and Dirichlet boundary conditions for H are obtained from the momentum equations. Where the flow is locally inviscid H is a constant. Consequently for flow about an isolated body it is possible to solve the discrete form of (17.139) with the farfield boundary much closer to the body than would be the case when solving (17.135). Equation (17.139) is solved to obtain the global pressure distribution for the flow past rearward-facing cavities (Fletcher and Barbuto 1986a, b).

For steady two-dimensional flow (17.135) or (17.139) need only be solved once after the velocity solution has been obtained. If the pressure is required for an unsteady flow it is necessary to solve (17.135) or (17.139) at every time step. In this case a primitive variable approach is often preferred to a stream function vorticity formulation.

17.4 Vorticity Formulations for Three-Dimensional Flows

In two dimensions the vorticity stream function formulation is often more efficient than a primitive variable formulation, primarily because the use of the stream function avoids explicit solution of the continuity equation (17.1). In three dimensions vorticity-related formulations lead to more dependent variables, typically six, than is the case for primitive variables, typically four. As a result three-dimensional vorticity-related formulations have not been used very often.

In this section two alternative formulations are examined. Both use the three-component vorticity transport equations and avoid the explicit appearance of the pressure. They differ in the choice of additional equations to obtain the velocity field.

17.4.1 Vorticity, Vector Potential Formulation

The extension of the vorticity stream function formulation (Sect. 17.3) to three-dimensional flow requires replacement of the stream function by a three-component vector potential and requires consideration of all three vorticity components.

The three-component vorticity transport equation, replacing (17.90), is

$$\frac{\partial \zeta}{\partial t} + \nabla \cdot (\mathbf{u}\,\zeta) - (\zeta \cdot \nabla)\mathbf{u} - \frac{1}{Re}\,\nabla^2 \zeta = 0 . \tag{17.140}$$

The structure of (17.140) is similar to that of (17.90) except that a new term $(\zeta \cdot \nabla)\mathbf{u}$ appears, which can be thought of as a vortex stretching term. In Cartesian coordinates the x-component of (17.140) is

$$\frac{\partial \zeta_x}{\partial t} + \frac{\partial}{\partial x}(u\zeta_x) + \frac{\partial}{\partial y}(v\zeta_x) + \frac{\partial}{\partial z}(w\zeta_x) - \zeta_x \frac{\partial u}{\partial x} - \zeta_y \frac{\partial u}{\partial y} - \zeta_z \frac{\partial u}{\partial z}$$

$$- \frac{1}{Re}\left(\frac{\partial^2 \zeta_x}{\partial x^2} + \frac{\partial^2 \zeta_x}{\partial y^2} + \frac{\partial^2 \zeta_x}{\partial z^2}\right) = 0 \ . \tag{17.141}$$

The three vorticity components are related to the velocity components by $\zeta = \mathrm{curl}\,\mathbf{u}$. However, to obtain the velocity field from the vorticity field it is necessary to introduce a vector potential $\boldsymbol{\psi}$, such that

$$\mathbf{u} = \mathrm{curl}\,\boldsymbol{\psi} \ , \quad \text{i.e.} \tag{17.142}$$

$$u = \frac{\partial \psi_z}{\partial y} - \frac{\partial \psi_y}{\partial z} \ , \quad v = \frac{\partial \psi_x}{\partial z} - \frac{\partial \psi_z}{\partial x} \ , \quad w = \frac{\partial \psi_y}{\partial x} - \frac{\partial \psi_x}{\partial y} \ .$$

Clearly the vector potential $\boldsymbol{\psi}$ is the three-dimensional extension of the scalar stream function in two dimensions ($\psi = \psi_z$, $\psi_x = \psi_y = 0$).

The three-dimensional equivalent of (17.92) is

$$\nabla \boldsymbol{\psi} = -\zeta \ . \tag{17.143}$$

Thus three-dimensional viscous incompressible flow is governed by (17.140, 142 and 143). Since each equation has three components the solution of three-dimensional flow is less economical using the vorticity, vector potential formulation than using primitive variables (Sects. 17.1 and 17.2). However, since (17.140) are transport equations and (17.143) are Poisson equations the same computational techniques are appropriate as in two dimensions.

For confined flows, such as the driven cavity problem, boundary conditions for the vector potential are given by Aziz and Hellums (1967) as

i) Surface $x = \mathrm{const}$: $\dfrac{\partial \psi_x}{\partial x} = \psi_y = \psi_z = 0$,

ii) Surface $y = \mathrm{const}$: $\dfrac{\partial \psi_y}{\partial y} = \psi_x = \psi_z = 0$, $\qquad\qquad$ (17.144)

iii) Surface $z = \mathrm{const}$: $\dfrac{\partial \psi_z}{\partial y} = \psi_x = \psi_y = 0$,

and for the vorticity:

i) Surface $x = \mathrm{const}$: $\zeta_x = 0$, $\qquad \zeta_y = -\dfrac{\partial w}{\partial x}$, $\qquad \zeta_z = \dfrac{\partial v}{\partial x}$,

ii) Surface $y = \mathrm{const}$: $\zeta_x = \dfrac{\partial w}{\partial y}$, $\qquad \zeta_y = 0$, $\qquad \zeta_z = -\dfrac{\partial u}{\partial y}$,

iii) Surface $z = \mathrm{const}$: $\zeta_x = -\dfrac{\partial v}{\partial z}$, $\zeta_y = \dfrac{\partial u}{\partial z}$, $\qquad \zeta_z = 0$. $\qquad$ (17.145)

The vorticity, vector potential formulation has been used by Aziz and Hellums (1967) and by Mallinson and de Vahl Davis (1977) to study three-dimensional natural convection in a box.

For problems with inflow and outflow the boundary conditions given by (17.144, 145) must be generalised. Although this is possible (Hirasaki and Hellums 1968) the result is cumbersome. A preferred procedure (Hirasaki and Hellums 1970) is to replace (17.142) with

$$\mathbf{u} = \text{curl } \boldsymbol{\psi} + \nabla\phi ,\tag{17.146}$$

where ϕ is an auxiliary potential (compare Sect. 17.2.2) introduced to provide a simpler prescription of the inflow, outflow boundary conditions. The satisfaction of continuity implies that

$$\nabla^2\phi = 0 .\tag{17.147}$$

The other governing equations remain as before. The boundary conditions for (17.147) are of Neumann type,

$$\frac{\partial\phi}{\partial n} = -n\cdot\mathbf{u} .\tag{17.148}$$

Thus a prescribed inflow/outflow velocity distribution enters through (17.148). At a solid surface (17.148) reduces to $\partial\phi/\partial n = 0$. In addition, boundary conditions (17.144 and 145) are applicable without further modification. Aregbesola and Burley (1977) have used the vorticity, vector potential, auxiliary potential formulation to study three-dimensional duct flows.

Wong and Reizes (1984) demonstrate that the introduction of the auxiliary potential no longer automatically satisfies continuity when the discrete form of the equations are considered. Consequently they prefer to replace (17.146) with

$$\mathbf{v} = \text{curl } \boldsymbol{\psi} + w_0 ,\tag{17.149}$$

where $w_0(x, y)$ is the specified inlet velocity distribution for a straight duct aligned parallel to the z axis. In this formulation (17.144) are applicable at solid surfaces and at an inflow boundary, $z = \text{const}$. At an outflow boundary, $z = \text{const}$. Wong and Reizes (1984) recommend the following boundary condition in place of (17.144):

$$\frac{\partial\psi_x}{\partial z} = \frac{\partial\psi_y}{\partial z} = 0 , \quad \frac{\partial\psi_z}{\partial z} = -\left(\frac{\partial\psi_x}{\partial x} + \frac{\partial\psi_y}{\partial y}\right) .\tag{17.150}$$

The boundary conditions on the vorticity are as indicated in (17.145). The numerical implementation of these boundary conditions is the same as in two dimensions (Sect. 17.3.2).

A vector potential formulation, assuming compressible, rotational inviscid flow is employed by Rao et al. (1989) to simulate the flow through a large low-speed wind tunnel. Vanes and screens are represented by actuator disk theory. Good agreement with experimental pressure data is indicated.

17.4.2 Vorticity, Velocity Formulation

In this formulation (Richardson and Cornish 1977; Dennis et al. 1979) the vorticity transport equations (17.140) are retained. However, from the definition of the vorticity, $\zeta = $ curl $\mathbf{u}$, and the continuity equation it is possible to derive the following Poisson equations for the velocity components:

$$\nabla^2 u = \frac{\partial \zeta_y}{\partial z} - \frac{\partial \zeta_z}{\partial y} \ , \qquad \nabla^2 v = \frac{\partial \zeta_z}{\partial x} - \frac{\partial \zeta_x}{\partial z} \ ,$$

$$\nabla^2 w = \frac{\partial \zeta_x}{\partial y} - \frac{\partial \zeta_y}{\partial x} \ . \tag{17.151}$$

In the present formulation (17.140 and 151) provide the governing equations. At solid surfaces boundary conditions are given by no slip, $u = v = w = 0$, and by (17.145) for the vorticity. For inflow boundaries it is appropriate to specify the velocity field; at outflow Neumann boundary conditions for the velocity components are specified (17.17). In addition simplifications to the vorticity transport equations (as in Chap. 16) may avoid the need to prescribe downstream boundary conditions.

Dennis et al. (1979) use modified exponential differencing (Dennis 1985) for the convective terms in the vorticity transport equations. Conventional three-point differencing is used for the second derivative terms in (17.140) and all the terms in (17.151). The discretised steady form of (17.140) and the discretised form of (17.151) form a global diagonally dominant system of equations. Dennis et al. solve these using successive over-relaxation. They compute the flow in a three-dimensional driven cavity for Reynolds numbers up to Re = 400 on a $25 \times 25 \times 25$ grid.

It is possible to consider a two-dimensional version of the vorticity, velocity formulation, namely the vorticity transport equation (17.90), the vorticity definition equation (17.89) and the continuity equation (17.1). Gatski et al. (1982) have used such a formulation to examine the driven cavity problem and more complex unsteady viscous flows (Gatski and Grosch 1985). Gatski et al. combine the discrete form of (17.1 and 89) into a global block matrix equation. This equation is not diagonally dominant and it is reported (Gatski et al. 1982) that the iterative algorithm to solve the block matrix equation is not very efficient. The extension of the algorithm to three-dimensional unsteady flow is provided by Gatski et al. (1989).

To solve the discrete form of (17.90), auxiliary variables are introduced for $\partial \zeta/\partial x$ and $\partial \zeta/\partial y$. This allows a form of differencing similar to the Keller box scheme (Sect. 15.1.3). The discrete vorticity equation is solved using a modified ADI procedure. The overall algorithm is sequential rather than coupled, i.e. the vorticity transport equation is solved separately from the vorticity definition, continuity equation combination.

Fasel (1976) has solved (17.90) in conjunction with (17.151) with $\zeta_x = \zeta_y = 0$ to examine transition phenomena in two-dimensional boundary layer flows. Orlandi (1987) uses a related scheme but includes the differentiated form of the continuity equation as well. Orlandi constructs a block ADI-like scheme in which all equations are coupled at each half time-step. For the first half time-step the coupled equations are (17.90, 151b) and

$$\frac{\partial^2 u}{\partial x^2} + \frac{\partial^2 v}{\partial x \partial y} = 0 \ . \tag{17.152}$$

For the second half time-step the coupled equations are (17.90, 151a) and

$$\frac{\partial^2 u}{\partial x \partial y} + \frac{\partial^2 v}{\partial y^2} = 0 \ . \tag{17.153}$$

Equations (17.152 and 153) are constructed by differentiating the continuity equation (17.1). It may also be noted that the differentiated form of the continuity equation is used in constructing (17.151).

It would appear that using a differentiated form of the incompressible continuity equation, $\partial D / \partial x = \partial D / \partial y = 0$, where D is the dilatation (11.13), does not guarantee that continuity is satisfied unless D is set to zero at least at one point. Orlandi does this by explicitly imposing (17.1) on the boundary. Orlandi notes that this guarantees exact mass conservation for the discretised equations.

Orlandi indicates that in practice this leads to a more efficient scheme since fewer iterations are required to produce a velocity field that satisfies continuity. Orlandi demonstrates the scheme for the driven cavity problem and the flow over a backward-facing step.

It may be concluded that vorticity-based formulations are not so efficient as primitive variables in three dimensions unless the vortex motion, in particular unsteady, is of special interest. In addition it may be noted that almost all practical turbulence modelling has been undertaken in terms of the primitive variables.

17.5 Closure

Historically vorticity stream function formulations have been a popular means of computing two-dimensional incompressible viscous flow. Although such formulations are economical the prescription of an effective solid-surface boundary condition for the vorticity is often a weak point. In addition the advantages of the economy of vorticity-related formulations do not carry over to three-dimensional flows.

Consequently the computation of incompressible viscous flow is more often undertaken using primitive variables. The major difficulty is in satisfying the continuity equation. This is handled implicitly in the MAC method (Sect. 17.1.2), the SIMPLE algorithm (Sect. 17.2.3) and the penalty finite element method (Sect. 17.2.4). Satisfaction of the continuity equation is handled more explicitly in the artificial compressibility (Sect. 17.2.1), auxiliary potential (Sect. 17.2.2) and traditional finite element method (Sect. 17.2.4).

The pressure is usually obtained by solving a Poisson equation. The Poisson equation may occur in the continuous form, as in the vorticity stream function formulation (Sect. 17.3) or in the discrete form, as in the MAC method (Sect. 17.1.2) and projection method (Sect. 17.1.4). It may also occur in disguise, as in the auxiliary potential function method (Sect. 17.2.2) and the SIMPLE algorithm (Sect. 17.2.3).

As long as three-point central differencing is used to discretise the convective and pressure gradient terms it is advantageous to introduce a staggered grid (Sect. 17.1.1), primarily to avoid an oscillatory pressure solution. However, the use of a staggered grid with generalised coordinates is rather cumbersome.

For flows with severe velocity gradients there is often an advantage in using higher-order differencing for the convective terms (Sect. 17.1.5) to obtain accurate solutions without excessive grid refinement. However, the overall scheme may be less robust, particularly if used with an explicit marching algorithm.

Different formulations for computing incompressible viscous flow can be applied with alternative means of discretisation. Thus the spectral method (Sect. 17.1.6) uses a projection algorithm. The group finite element method (Sect. 17.3.3) uses a pseudo-transient formulation very similar to that used in finite difference methods (Sect. 17.3.1). The finite volume method (Sect. 17.2.3) uses the SIMPLE algorithm, which is very similar to the use of an auxiliary potential function (Sect. 17.2.2).

No description is provided in this book of vortex methods which, in their simplest form, simulate incompressible viscous flows by the introduction of point vortices satisfying Laplace's equation (11.51). Such methods are reviewed by Leonard (1980, 1985) and are used to provide qualitative descriptions of complex unsteady separating flows (e.g. Oshima et al. 1986).

17.6 Problems

Primitive Variables: Unsteady Flow (Sect. 17.1)

17.1 Integrate (17.1) over the square domain $0 \le x \le 1, 0 \le y \le 1$, and demonstrate that (17.4) is satisfied. Subdivide the domain into four cells, $\Delta x = \Delta y = 0.5$, and demonstrate that the discrete equivalent also holds if u and v are averaged over each cell face (Fig. 17.1) and two-point differencing is used to evaluate derivatives, as in (17.5).

17.2 Show that the discrete Poisson equation for the pressure (17.13, 14) can be obtained from (17.12).

17.3 Confirm the analysis following (17.18 and 19), which demonstrates that homogeneous Neumann boundary conditions for the pressure are available with the MAC formulation.

17.4 By setting u equal to a constant, demonstrate that the discretisation scheme provided by (17.31–33) is equivalent to (9.71, 72).

17.5 Show that values of u^* and v^* are not required on the boundary in the CPSM algorithm by demonstrating that (17.41 and 42) follow from (17.39 and 40).

Primitive Variables: Steady Flow (Sect. 17.2)

17.6 Demonstrate that the Jacobians $\underline{A}$ and $\underline{B}$ are given by (17.48). Show, by direct substitution, that $\mathbf{F} = \underline{A}\mathbf{q} - u\underline{D}\mathbf{q}$ and $\mathbf{G} = \underline{B}\mathbf{q} - v\underline{D}\mathbf{q}$.

17.7 Compare the role of the auxiliary potential in Sect. 17.2.2 with the role of the pressure correction in the Hirt and Cook formulation (17.26–29) and the pressure correction in the SIMPLE algorithm (Sect. 17.2.3).

17.8 Determine specific expressions for $a_{j,k}^u$ and a_{nb}^u in (17.69) and for $a_{j,k}^p$ and a_{nb}^p in (17.77). Demonstrate that (17.77) is a discrete Poisson equation for δp.

17.9 Discuss how (17.69) would change if a three-point second-order upwind scheme, $q = 1.5$ in (9.53), were used to discretise the convective terms in (17.2).

Vorticity, Stream Function Variables (Sect. 17.3)

17.10 Show that the discretisation of $\partial(u\zeta)/\partial x$, given by (17.98), reverts to a three-point centred difference scheme if $(u\zeta)^{n+1} = (u\zeta)^n$, i.e. at steady state.

17.11 Derive the first-order and second-order vorticity boundary conditions at a solid surface (17.104 and 107).

17.12 Obtain the expressions for the mass and difference operators on a non-uniform grid (17.115) by applying the Galerkin finite element method in one dimension with linear interpolation (Sect. 5.4 and Vol. 1, Appendix A.2).

17.13 Introduce a three-level time discretisation into (17.114) to generate (17.116). Demonstrate that (17.116) can be linearised to give (17.119) and that (17.120 and 121) are consistent with (17.119) to $O(\Delta t^2)$.

17.14 Obtain (17.125) by applying a one-dimensional Galerkin formulation to (17.91). Carry out a Taylor series expansion to demonstrate that (17.125) is fourth-order accurate on a uniform grid. How accurate is it on a non-uniform grid?

17.15 Demonstrate, using the results of Sect. 8.4.2, that $g(u) = u_{FE}/\Delta y$ when (17.127) is derived at surface FE.

17.16 Derive the Poisson equation for the Bernoulli function (17.139).

Vorticity Formulations for 3D Flows (Sect. 17.4)

17.17 Develop a scalar three-stage approximate factorisation algorithm based on a three-point centred difference discretisation of (17.141) in terms of the correction to the vorticity component, $\Delta\zeta_x^{n+1}$.

17.18 Derive the Poisson equations for the velocity components (17.151).

18. Compressible Viscous Flow

In this chapter computational algorithms will be considered for solving flows governed by the full compressible Navier–Stokes equations, i.e. unsteady flow or flows with large areas of separation. Steady compressible viscous flows with a dominant flow direction and only small regions of separation can be handled with the techniques described in Chap. 16, particularly external flows around bodies in a supersonic freestream.

Areas of application include transonic flow around aircraft and through turbomachinery, and low speed duct flows involving significant heat transfer. For design purposes there is generally more interest in steady than unsteady compressible viscous flow. However, most solution algorithms are constructed around marching the unsteady equations in time. For steady flow problems this is just the pseudotransient formulation (Sect. 6.4).

Most compressible viscous flows are also turbulent. The inherent complexity of the governing equations motivates the use of relatively simple turbulence models. Most applications incorporate eddy viscosity turbulence models (Sect. 18.1.1), either algebraic or in conjunction with the k–ε formulation (Sect. 11.5.2). For flows that are subsonic or weakly transonic with large inviscid regions, accurate solutions can be obtained with an algebraic energy equation (Sect. 18.1.2) replacing the differential energy equation.

For unsteady problems it is computationally efficient to use explicit schemes as long as the time step is not unduly limited by stability considerations. The well-known MacCormack scheme is described in Sect. 18.2.1. Runge–Kutta schemes (Sect. 18.2.2), although explicit, produce stable and accurate solutions with larger time steps than the typical unit CFL number restriction (Sect. 9.1.2).

The need to use even larger time-steps, as is usually the case with the pseudo-transient construction, motivates the consideration of implicit schemes (Sect. 18.3). The implicit MacCormack scheme (Sect. 18.3.1) is a direct extension of the explicit MacCormack method. The other implicit schemes considered in Sect. 18.3 all introduce approximate factorisation in the sense of Sect. 8.2.

For problems of practical significance the computational domain will usually be of irregular shape which can be handled effectively using generalised coordinates (Sect. 18.4). The approximate factorisation treatment of multidimensional implicit algorithms is simplified if physically dissipative terms are retained only for directions normal to solid surfaces (Sects. 18.1.3 and 18.4.1).

If the compressibility of the flow is associated with motion (large Mach number) then a large Reynolds number and turbulent flow are also to be expected in many situations. Many of the computational algorithms available are close to being neutrally stable and it is necessary to deliberately include additional numerical dissipation (Sect. 18.5.1). This overcomes aliasing and nonlinear instability in those parts of the computational domain where the physical dissipation is very small.

If the flow is locally supersonic embedded shock waves are likely to be present. If these are weak accurate solutions can usually be obtained without modification to the basic computational algorithm, other than the inclusion of additional numerical dissipation. However, if the shocks are strong then the same techniques as described for the Euler equations (Sect. 14.2.6) are appropriate (Sect. 18.5.2).

18.1 Physical Simplifications

The governing equations for unsteady three-dimensional compressible viscous flow are given by (11.116) and (11.117). The required number and types of boundary conditions to suit these equations are discussed in Sect. 11.6.4.

For most compressible flows of practical importance that require consideration of the full Navier–Stokes equations, the flow is turbulent. Although direct simulation and large eddy simulation are conceptually possible, the capacity of present day computers has prompted greater interest in some form of turbulence modelling, typically at the eddy viscosity level (Sect. 18.1.1). Turbulence modelling provides a means of rendering a very complicated set of equations more tractable and permits flow behaviour in non-simple computational domains to be predicted without a prohibitive execution time.

If the flows under investigation are restricted to subsonic or transonic conditions with no external heating then the energy equation can be simplified by assuming that the total enthalpy remains constant. This assumption (Sect. 18.1.2) allows the differential energy equation to be replaced with an algebraic energy equation.

For compressible flows at large Reynolds numbers viscous and turbulence effects are only significant close to solid surfaces in the absence of massive separation. Consequently dissipative terms, associated with the direction normal to the surface only, need be retained. This is the essential idea behind the thin-layer approximation which is described in Sect. 18.1.3. The implementation of this approximation leads to simpler coding, particularly for implicit schemes (Sect. 18.3), and some gains in economy (Sect. 18.4.1).

A historically important simplification is the notion of subdividing the computational domain into zones such that simpler sets of equations, with faster equation solvers, may be used in certain zones. Traditionally the boundary layer equations (Chap. 15) are used adjacent to solid surfaces approximately parallel with the main flow and the Euler or potential equations are used for the region farther away. Modern developments of this idea are considered in Sects. 16.3.4–7.

For transonic flows around three-dimensional wings the complicated shock boundary layer interactions produce local regions of separated flow and it is appropriate to use the full Navier–Stokes equations close to the surface and the Euler equations in the farfield. An example of the type of flow that can be computed is indicated in Fig. 1.5. The matching of the solutions between the zones is discussed by Holst et al. (1986). This type of physical simplification will not be discussed in this chapter.

For turbulent compressible flow a direct application of the Reynolds averaging process as in Sect. 11.4.2 leads to the appearance of third-order moments, e.g. $\overline{\varrho' u' v'}$, containing density and temperature fluctuations as well as velocity fluctuations. However, the overall complexity of the Reynolds averaged Navier–Stokes equations can be reduced by introducing mass weighted velocities and thermal variables (Favre 1965),

$$\tilde{u} = \overline{\varrho u}/\bar{\varrho} \quad \text{and} \quad \tilde{T} = \overline{\varrho T}/\bar{\varrho} \ . \tag{18.1}$$

To carry out mass-weighted Reynolds averaging it is necessary to split the dependent variables into average and fluctuating parts,

$$u = \tilde{u} + u'' \ , \quad T = \tilde{T} + T'' \ .$$

The mass weighted splitting is applied to all variables except density and pressure, which are split conventionally

$$\varrho = \bar{\varrho} + \varrho' \ , \quad p = \bar{p} + p' \ .$$

The detailed procedures involved in mass averaging are described by Cebeci and Smith (1974, Chap. 2). Application of the averaging process described in Sect. 11.4.2 produces the mass-weighted Reynolds averaged Navier–Stokes equations which can be made identical with the laminar form at the stress and heat flux level (11.116), if some small fluctuating terms are neglected (Rubesin and Rose 1973). The stresses in (11.117) are replaced by, in Cartesian tensor notation,

$$\tau_{ij} = \mu \left(\frac{\partial \tilde{u}_i}{\partial x_j} + \frac{\partial \tilde{u}_j}{\partial x_i} - \frac{2}{3} \delta_{ij} \frac{\partial \tilde{u}_k}{\partial x_k} \right) - \overline{\varrho u_i'' u_j''} \ . \tag{18.2}$$

In the energy equation the heat fluxes become

$$\dot{Q}_i = -k \frac{\partial \tilde{T}}{\partial x_i} + c_p \overline{\varrho T'' u_i''} \ . \tag{18.3}$$

The averages over fluctuating quantities appearing in (18.2, 3) must be modelled as functions of the mean flow quantities in order to close the system of governing equations.

Turbulence modelling for compressible flow is reviewed by Marvin (1983) and Bradshaw (1977). For Mach numbers less than five, turbulence models developed for incompressible flow can be used almost without modification as long as the spatial and temporal variation of the mean density is allowed for.

To date, most modelling of turbulent compressible flow has been via the introduction of an eddy viscosity, i.e. the Reynolds stresses in (18.2 and 3) are related to mean flow quantities by

$$-\overline{\varrho u_i'' u_j''} = \mu_T \left(\frac{\partial \tilde{u}_i}{\partial x_j} + \frac{\partial \tilde{u}_j}{\partial x_i} - \frac{2}{3} \delta_{ij} \frac{\partial \tilde{u}_k}{\partial x_k} \right) - \frac{2}{3} \delta_{ij} \varrho k^{\text{te}} \tag{18.4}$$

and

$$-c_p \overline{\varrho T'' u_i''} = k_T \frac{\partial \tilde{T}}{\partial x_i} , \tag{18.5}$$

where k^{te} in (18.4) is the turbulent kinetic energy [k in (11.95)], and the term containing k^{te} is combined with the pressure in the momentum equations. The turbulent conductivity, k_T is related to the eddy viscosity by $k_T = c_p \mu_T / \mathrm{Pr}_T$, where Pr_T is the turbulent Prandtl number. For air $\mathrm{Pr}_T \approx 0.9$.

The introduction of the eddy viscosity and turbulent conductivity permits the same form of the governing equations to be used as for laminar flow if μ and k are replaced by $(\mu + \mu_T)$ and $(k + k_T)$. Within the eddy viscosity framework both algebraic models (Sect. 11.4.2) and two-equation models (Sect. 11.5.2) are available.

HaMinh et al. (1986) and Vandromme and HaMinh (1986) have combined a full Reynolds stress closure with the compressible Navier–Stokes equations. That is, full transport equations for the Reynolds stresses are included in the equation set. Full Reynolds stress closure is not discussed in this book.

For two-dimensional compressible turbulent flow, with the assumption that the Reynolds stresses and turbulent heat fluxes can be related to the mean flow via (18.4 and 5), the governing equations become (dropping the $^-$ and $^\sim$ to denote mean quantities)

$$\frac{\partial \mathbf{q}}{\partial t} + \frac{\partial \mathbf{F}}{\partial x} + \frac{\partial \mathbf{G}}{\partial y} = 0 , \quad \text{where} \tag{18.6}$$

$$\mathbf{q} = \begin{bmatrix} \varrho \\ \varrho u \\ \varrho v \\ E \end{bmatrix} , \quad \mathbf{F} = \begin{bmatrix} \varrho u \\ \varrho u^2 + p - \tau_{xx} \\ \varrho u v - \tau_{xy} \\ (E + p - \tau_{xx})u - \tau_{xy}v + \dot{Q}_x \end{bmatrix}$$

$$\mathbf{G} = \begin{bmatrix} \varrho v \\ \varrho u v - \tau_{xy} \\ \varrho v^2 + p - \tau_{yy} \\ (E + p - \tau_{yy})v - \tau_{xy}u + \dot{Q}_y \end{bmatrix} . \tag{18.7}$$

For an ideal gas,

$$p = (\gamma - 1)[E - 0.5\varrho(u^2 + v^2)] . \tag{18.8}$$

The stresses in the momentum equation are given by

$$\tau_{xx} = 2(\mu + \mu_T)\frac{\partial u}{\partial x} - \frac{2}{3}(\mu + \mu_T)\mathscr{D} - \frac{2}{3}\varrho k^{\text{te}} \ ,$$

$$\tau_{yy} = 2(\mu + \mu_T)\frac{\partial v}{\partial y} - \frac{2}{3}(\mu + \mu_T)\mathscr{D} - \frac{2}{3}\varrho k^{\text{te}} \ , \qquad (18.9)$$

$$\tau_{xy} = (\mu + \mu_T)\left(\frac{\partial u}{\partial y} + \frac{\partial v}{\partial x}\right) \ ,$$

and the heat fluxes by

$$\dot{Q}_x = -\left(k + \frac{c_p \mu_T}{\text{Pr}_T}\right)\frac{\partial T}{\partial x} \ , \qquad \dot{Q}_y = -\left(k + \frac{c_p \mu_T}{\text{Pr}_T}\right)\frac{\partial T}{\partial y} \ . \qquad (18.10)$$

In (18.9) $\mathscr{D}$ is the dilatation, i.e. $\mathscr{D} = \partial u/\partial x + \partial v/\partial y$.

For unsteady problems (18.6) requires the specification of all dependent variables **q** as initial data. Boundary conditions are required for both steady and unsteady flows. Boundary conditions for (18.6) are discussed in Sect. 11.6.4. At solid surfaces no-slip velocity boundary conditions and either specified temperature or heat transfer rate are required. Thus for a stationary surface,

$$u = v = 0 \ , \qquad T = T_s \quad \text{or} \quad k\frac{\partial T}{\partial n} = -\dot{Q}_n \ . \qquad (18.11)$$

For open boundaries, i.e. those through which flow takes place, it is useful to differentiate between inflow and outflow boundaries. In addition it is useful to distinguish between internal flows and external flows. For the external flow about a body in a uniform stream the viscous and turbulent terms are usually negligibly small at farfield (open) boundaries. Consequently, the governing equations reduce to the Euler equations and characteristic theory indicates the number and type of boundary conditions. References are provided in Sect. 14.2.8.

For internal flows or external flows where open boundaries occur close to solid surfaces, (turbulent) viscous and heat conduction effects cannot be ignored and the appropriate prescription of boundary conditions is not yet theoretically established. Some discussion is provided by Rudy and Strikwerda (1981), Bayliss and Turkel (1982) and Roe (1989). The number of required boundary conditions to suit the compressible Navier–Stokes equations is indicated in Table 11.6.

Thus at inflow all dependent variables should be specified. At outflow, boundary conditions for all but one dependent variable should be provided. Typically the outflow boundary conditions are specified as zero Neumann conditions ($\partial f/\partial x = 0$, where x is the exit flow direction). Also the governing equations at the exit can often be reduced to a simpler set with a corresponding reduction in the number of required boundary conditions (Chap. 16).

18.1.1 Eddy Viscosity Turbulence Modelling

Most computations involving the turbulent Navier–Stokes equations (Marvin 1983) use algebraic eddy viscosity formulae to represent μ_T in (18.9 and 10). The eddy viscosity formulae given by (11.77–79) for incompressible boundary layer flow can be extended to provide reasonably accurate predictions of mean flow quantities for separating compressible flow.

The eddy viscosity expression for boundary layer flow (11.77) is generalised to

$$\mu_T = \varrho l^2 \left[\left(\frac{\partial u}{\partial y} \right)^2 + \left(\frac{\partial v}{\partial x} \right)^2 \right]^{1/2} , \qquad (18.12)$$

where l is the mixing length given by (11.78). This mixing length prescription is used close to solid surfaces. In the outer part of boundary layers and in wakes the Clauser formulation is available:

$$\mu_T = 0.0168 \, \varrho u_e \delta^* I , \qquad (18.13)$$

where u_e is the velocity in the flow direction at the outer boundary of the viscous region. The displacement thickness is given by

$$\delta^* = \int_{y_{DS}}^{\delta} \left(1 - \frac{u}{u_e} \right) dy , \qquad (18.14)$$

where δ is the outer boundary of the viscous region and y_{DS} is the inner boundary. For a boundary layer y_{DS} is the solid surface coordinate; for a wake y_{DS} is the dividing streamline between the recirculating fluid (Fig. 17.19) and the outer flow. The intermittency factor I is given by

$$I = \left[1 + 5.5 \left(\frac{y - y_{DS}}{\delta} \right)^6 \right]^{-1} , \qquad (18.15)$$

and accounts for the effect of the intermittency of the turbulence on the average value of μ_T.

Although the above eddy viscosity expressions imply equilibrium between turbulence production and dissipation it is possible to allow for upstream effects empirically via the following relaxation procedure:

$$\mu_T = \alpha(\mu_T)_{up} + (1 - \alpha) \mu_T^i , \qquad (18.16)$$

where μ_T^i is evaluated from (18.12) or (18.13) and $(\mu_T)_{up}$ is the eddy viscosity at the intersection of the velocity vector projected upstream through grid point (j, k) and the nearest grid line (Fig. 18.1). The relaxation parameter α is typically given the value $\alpha = 0.3$.

An alternative algebraic eddy viscosity formulation is provided by Baldwin and Lomax (1978). Close to solid surfaces (18.12) is replaced by

$$\mu_T = \varrho l^2 |\zeta| , \qquad (18.17)$$

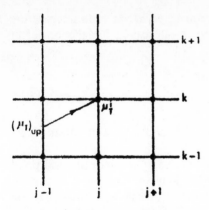

Fig. 18.1. Upstream influence on the eddy viscosity

where the vorticity $\zeta = \partial u/\partial y - \partial v/\partial x$. Away from solid surfaces the following formula is used in place of (18.13):

$$\mu_T = 0.0168\, \varrho V_0 L_0 \,, \tag{18.18}$$

where

$$V_0 = \min\left(F_{\max},\, 0.25\, q^2_{\text{dif}}/F_{\max}\right) \tag{18.19}$$

and

$$L_0 = 1.6\, y_{\max}\, I^k \,. \tag{18.20}$$

In (18.19) $F_{\max} = \max(|\zeta| l/\kappa)$ where the mixing length l and the von Karman constant κ are defined by (11.78). The parameter $y_{\max}$ is the value of y at which $F_{\max}$ occurs. The quantity q_{dif} is the difference between the maximum and minimum values of the absolute velocity. The Klebanoff intermittency factor I^k is given by

$$I^k = \left[1 + 5.5\left(\frac{0.3y}{y_{\max}}\right)^6\right]^{-1} \,. \tag{18.21}$$

The Baldwin–Lomax model is more robust than the Clauser formulation in regions of separated flow (Deiwert 1984). With either of the above turbulence models the turbulent Prandtl number is usually assumed constant, and for air $\mathrm{Pr}_T \approx 0.9$.

However an alternative model (Goldberg and Chakravarthy 1986), suitable for recirculating flow, obtains the eddy viscosity from an assumed Gaussian distribution of k normal to the wall and a turbulence length scale proportional to the thickness of the separated flow region. More recently (Goldberg and Chakravarthy 1990) this model has been combined with a one-equation (k) turbulence model to obtain the eddy viscosity in separated flow regions.

The k–ε turbulence model (Sect. 11.5.2) and other two-equation turbulence models have been combined with the compressible Navier–Stokes equations by Coakley (1983) and Horstman (1986), amongst others. Marvin (1983) reports that additional compressible terms in the k–ε model are usually insignificant up to a Mach number of five, and consequently can be dropped.

For turbulent flows very severe normal gradients of velocity and temperature occur close to solid surfaces. An appropriate local scaling for the turbulent boundary layer flow over surface FE in Fig. 17.14 is

$$y^+ = \frac{u_\tau y}{v} \quad \text{and} \quad u^+ = \frac{u}{u_\tau} , \tag{18.22}$$

where y is the normal direction, $u_\tau = (\tau_w/\varrho)^{1/2}$ and the wall shear stress $\tau_w = \mu \partial u / \partial y|_w$.

To properly resolve the velocity profile it is necessary to place the grid point nearest the wall at a location $y^+ < 5$, i.e. in the laminar sublayer. For a compressible flow accurate resolution of the temperature profile in the normal direction may require the nearest grid point to the wall to be located such that $y^+ < 2$.

To avoid needing such fine grids algebraic wall functions are often constructed by considering a thin layer adjacent to the wall and neglecting velocity and temperature gradients along the wall. If $\partial p / \partial x$, etc., are treated as externally determined parameters, the resulting "Couette" flow only involves normal derivatives and can be integrated to provide the solution at the outer edge of the thin layer. This can be used as the edge of the computational domain and the wall functions provide the boundary conditions. The thickness of the layer is chosen so that the computational boundary falls in the range $30 < y^+ < 200$.

Alternatively, a conventional grid can be used, i.e. terminating at the wall, but with the grid point adjacent to the wall, y_2, occurring in the range, $30 < y_2^+ < 200$. At this grid point the solution is provided by the wall function. For surface FE in Fig. 17.14 the wall function is the classical "law-of-the-wall", i.e.

$$u_2^+ = 5.0 + (1/\kappa)\ln y^+ , \tag{18.23}$$

where the von Karman constant $\kappa = 0.41$, typically. To convert u^+ into physical coordinates it is necessary to know τ_w and hence $\partial u / \partial y|_w$. This is evaluated from the interior solution using a one-sided discretisation.

Wall functions for non-boundary-layer flow, i.e. adjacent to DE in Fig. 17.14, can still be constructed from the Couette flow model. Patankar and Spalding (1970) provide a thorough description of the concept. Essentially the same approach is used to provide boundary conditions to be used with the k–ε turbulence model (Launder and Spalding 1974).

18.1.2 Constant Total Enthalpy Flow

For transonic viscous flow without external heat sources the temperature variation throughout the computational domain is small. Consequently, for steady flow the governing equations (18.6) can be simplified by replacement of the energy equation with an algebraic equation. This result can be obtained as follows.

The total enthalpy $H = (E + p)/\varrho$. Therefore, the steady energy equation (18.6, 7) can be written

$$\frac{\partial}{\partial x}(\varrho u H) + \frac{\partial}{\partial y}(\varrho v H) = \frac{\partial}{\partial x}(u\tau_{xx} + v\tau_{xy} - \dot{Q}_x)$$

$$+ \frac{\partial}{\partial y}(v\tau_{yy} + u\tau_{xy} - \dot{Q}_y) \ . \tag{18.24}$$

Making use of (18.9, 10) and the ideal gas enthalpy relationship

$$H = c_p T + \tfrac{1}{2}(u^2 + v^2) \tag{18.25}$$

allows (18.24) to be written as

$$\frac{\partial}{\partial x}\left[\varrho u H - \left(\frac{\mu}{\mathrm{Pr}} + \frac{\mu_T}{\mathrm{Pr}_T}\right)\frac{\partial H}{\partial x}\right] + \frac{\partial}{\partial y}\left[\varrho v H - \left(\frac{\mu}{\mathrm{Pr}} + \frac{\mu_T}{\mathrm{Pr}_T}\right)\frac{\partial H}{\partial y}\right] = \mathrm{RHS} \ , \tag{18.26}$$

where

$$\mathrm{RHS} = \frac{\partial}{\partial x}\left(\mu v\frac{\partial u}{\partial y} - \frac{2}{3}\mu u\frac{\partial v}{\partial y}\right) + \frac{\partial}{\partial y}\left(\mu u\frac{\partial v}{\partial x} - \frac{2}{3}\mu v\frac{\partial u}{\partial x}\right)$$

$$+ \frac{\partial}{\partial x}\left\{\left[\mu\left(\frac{4}{3} - \frac{1}{\mathrm{Pr}}\right) + \mu_T\left(\frac{4}{3} - \frac{1}{\mathrm{Pr}_T}\right)\right]u\frac{\partial u}{\partial x}\right\} + \frac{\partial}{\partial y}\left\{\left[\mu\left(\frac{4}{3} - \frac{1}{\mathrm{Pr}}\right)\right.\right.$$

$$\left.\left. + \mu_T\left(\frac{4}{3} - \frac{1}{\mathrm{Pr}_T}\right)\right]v\frac{\partial v}{\partial y}\right\} + \frac{\partial}{\partial x}\left\{\left[\mu\left(1 - \frac{1}{\mathrm{Pr}}\right) + \mu_T\left(1 - \frac{1}{\mathrm{Pr}_T}\right)\right]v\frac{\partial v}{\partial x}\right\}$$

$$+ \frac{\partial}{\partial y}\left\{\left[\mu\left(1 - \frac{1}{\mathrm{Pr}}\right) + \mu_T\left(1 - \frac{1}{\mathrm{Pr}_T}\right)\right]u\frac{\partial u}{\partial y}\right\} \ . \tag{18.27}$$

For high Reynolds number flows around bodies in a uniform stream, viscous and turbulence effects are confined to a thin layer close to the body and in the wake. An order-of-magnitude analysis for (18.27) indicates that only the term

$$\frac{\partial}{\partial y}\left\{\left[\mu\left(1 - \frac{1}{\mathrm{Pr}}\right) + \mu_T\left(1 - \frac{1}{\mathrm{Pr}_T}\right)\right]u\frac{\partial u}{\partial y}\right\} \tag{18.28}$$

is of comparable magnitude to terms in the momentum and continuity equations and on the left-hand side of (18.26). For purely laminar flow the assumption $\mathrm{Pr} = 1.0$ causes (18.28) to disappear. For turbulent flow $\mu_T \gg \mu$ and the assumption $\mathrm{Pr}_T = 1.0$ causes (18.28) to disappear. It may be noted that for air $\mathrm{Pr} = 0.7$ and $\mathrm{Pr}_T \approx 0.9$ so that a small contribution may still be expected. In a region of limited separation (18.28) is still expected to be the dominant term so that the above remarks apply. In the inviscid region, i.e. far from solid surfaces, all terms in (18.27) are negligible.

Therefore, for steady transonic flow it is a reasonable approximation to replace (18.26) with

$$\frac{\partial}{\partial x}\left[\varrho u H - \left(\frac{\mu}{\mathrm{Pr}} + \frac{\mu_T}{\mathrm{Pr}_T}\right)\frac{\partial H}{\partial x}\right] + \frac{\partial}{\partial y}\left[\varrho v H - \left(\frac{\mu}{\mathrm{Pr}} + \frac{\mu_T}{\mathrm{Pr}_T}\right)\frac{\partial H}{\partial y}\right] = 0 \ . \qquad (18.29)$$

Clearly (18.29) is satisfied by $H = \mathrm{const}$, or, from (18.25),

$$c_p T + 0.5(u^2 + v^2) = \mathrm{const} \ ,$$

which can be written as

$$\frac{\gamma - 1}{\gamma}\frac{p}{\varrho} + 0.5(u^2 + v^2) = \frac{\gamma - 1}{\gamma}\frac{p_\infty}{\varrho_\infty} + 0.5\, U_\infty^2 \ . \qquad (18.30)$$

Equation (18.30) provides an algebraic equation linking p, ϱ, u and v and this replaces the energy equation in (18.6 and 7). This type of formulation is discussed by Briley and McDonald (1977) and used by Fletcher and Srinivas (1985).

18.1.3 Thin Layer Approximation

As noted in Sect. 18.1.2, viscous and turbulence effects are only significant close to solid surfaces and in wake regions for flows at large Reynolds number. Unless massive separation is occurring in the approximate flow direction many of the dissipative terms, i.e. contributions to the τ and $\dot{Q}$ terms in (18.9 and 10), can be dropped from the governing equations on an order-of-magnitude basis. For three-dimensional flows it is usually not possible, due to computer memory limitations, to provide a fine enough grid in all directions to evaluate accurately all terms in the three-dimensional equivalent of (18.9 and 10).

These parallel features, one physical and one computational, are combined in the thin layer approximation (Baldwin and Lomax 1978). First a fine grid is used

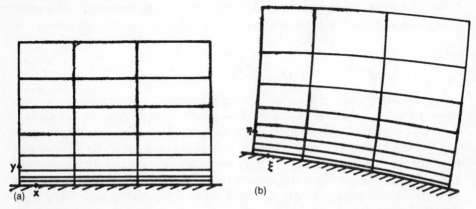

Fig. 18.2a, b. Grid refinement close to a solid surface. (a) Cartesian coordinates; (b) generalised coordinates

only in the direction normal to the surface (Fig. 18.2). The coarse grid parallel to the surface (x direction) is unable to represent accurately x derivatives associated with the τ and $\dot{Q}$ terms in (18.7). However, such terms can be deleted on an order-of-magnitude basis. The thin layer approximation of (18.6) to (18.10) is

$$\frac{\partial \mathbf{q}}{\partial t} + \frac{\partial \mathbf{F}^I}{\partial x} + \frac{\partial \mathbf{G}^I}{\partial y} + \frac{\partial \mathbf{G}^v}{\partial y} = 0 \ , \tag{18.31}$$

where

$$\begin{aligned}
\mathbf{F}^I &= \{ \varrho u, \varrho u^2 + p, \varrho uv, (E+p)u \}^T \\
\mathbf{G}^I &= \{ \varrho v, \varrho uv, \varrho v^2 + p, (E+p)v \}^T \quad \text{and} \\
\mathbf{G}^v &= -\{ 0, \tau_{xy}^m, \tau_{yy}^m, (\tau_{xy}^m u + \tau_{yy}^m v - \dot{Q}_y^m) \}^T \ .
\end{aligned} \tag{18.32}$$

In (18.32),

$$\tau_{xy}^m = (\mu + \mu_T)\frac{\partial u}{\partial y} \ , \qquad \tau_{yy}^m = \frac{4}{3}(\mu + \mu_T)\frac{\partial v}{\partial y} - \frac{2}{3}\varrho k^{\text{te}} \ , \quad \text{and}$$

$$\dot{Q}_y^m = -\left(k + c_p \frac{\mu_T}{\mathrm{Pr}_T} \right)\frac{\partial T}{\partial y} \ .$$

It may be noted that when the various contributions to $\mathbf{G}^v$ are substituted into (18.31) only y derivatives are present and cross derivatives are excluded. This simplifies the construction of implicit schemes (Sect. 18.4.1). The thin layer approximation is usually used in conjunction with generalised coordinates (Chap. 12 and Sect. 18.4.1) so that significant physical dissipation can be restricted to one co-ordinate direction for most computational geometries. However, for flow near the junction of two walls it is appropriate to obtain dissipative terms associated with the normal directions to both walls. The use of the thin layer approximation then neglects some cross-derivative terms that may be of comparable magnitude to terms retained. In practice the overall solution accuracy is not much affected (Hung and Kordulla 1984).

The thin layer approximation for steady flow can be interpreted as a reduced form of the Navier–Stokes equations (Chap. 16) and can be manipulated to generate a marching algorithm (Chang and Merkle 1989). Essentially, the same approximation is used in obtaining the spatial marching algorithm for supersonic viscous flow (Sect. 16.3.1). However, the further restriction to positive values of u is not usually used with the thin layer approximation. Instead the thin layer approximation is combined with the pseudotransient algorithm (Sect. 6.4) to obtain the steady-state solution. Consequently, flows with small regions of separation in the main flow direction can be accurately predicted. The thin layer approximation is widely applicable (Chausee 1984) and demonstrates good agreement with experimental data.

18.2 Explicit Schemes

In Sect. 18.2.1 the explicit MacCormack scheme is applied to the compressible Navier–Stokes equations. Although this scheme is very economical for a genuinely unsteady flow, the unit CFL time-step restriction makes it less suitable for obtaining steady solutions. Runge–Kutta schemes, permitting larger time steps, are described in Sect. 18.2.2.

18.2.1 Explicit MacCormack Scheme

The most widely used explicit scheme for the compressible Navier–Stokes equations is the MacCormack (1969) scheme, which is described for one-dimensional inviscid flow in Sect. 14.2.2 and for a time-like multidimensional inviscid flow problem in Sect. 14.2.4. The application of the MacCormack scheme to equations with second spatial derivatives will be illustrated here using the two-dimensional Burgers' equations (10.57, 58). The extension to the compressible Navier–Stokes equations, (18.6) to (18.10), introduces the complication of a mixed second derivative; this will be dealt with subsequently.

The two-dimensional Burgers' equations are written as in (10.52),

$$\frac{\partial \mathbf{q}}{\partial t} + \frac{\partial \mathbf{F}}{\partial x} + \frac{\partial \mathbf{G}}{\partial y} - \mathbf{S} = 0 \ , \quad \text{where} \tag{18.33}$$

$$\mathbf{q} = \begin{pmatrix} u \\ v \end{pmatrix} , \quad \mathbf{F} = \begin{pmatrix} u^2 - v\partial u/\partial x \\ uv - v\partial v/\partial x \end{pmatrix} ,$$

$$\mathbf{G} = \begin{pmatrix} uv - v\partial u/\partial y \\ u^2 - v\partial v/\partial y \end{pmatrix} \quad \text{and} \quad \mathbf{S} = \begin{pmatrix} 0.5u(u^2+v^2)/v \\ 0.5v(u^2+v^2)/v \end{pmatrix} . \tag{18.34}$$

MacCormack's scheme applied to (18.33) on a uniform grid has the form

Predictor stage:

$$\mathbf{q}_{j,k}^* = \mathbf{q}_{j,k}^n - \frac{\Delta t}{\Delta x}[\mathbf{F}_{j+1,k}^n - \mathbf{F}_{j,k}^n] - \frac{\Delta t}{\Delta y}[\mathbf{G}_{j,k+1}^n - \mathbf{G}_{j,k}^n] + \Delta t \mathbf{S}_{j,k}^n \tag{18.35}$$

Corrector stage:

$$\mathbf{q}_{j,k}^{n+1} = 0.5(\mathbf{q}_{j,k}^n + \mathbf{q}_{j,k}^*) - 0.5\frac{\Delta t}{\Delta x}[\mathbf{F}_{j,k}^* - \mathbf{F}_{j-1,k}^*]$$

$$- 0.5\frac{\Delta t}{\Delta y}[\mathbf{G}_{j,k}^* - \mathbf{G}_{j,k-1}^*] + 0.5\Delta t \mathbf{S}_{j,k}^* \ . \tag{18.36}$$

It may be noted that each spatial group, $\mathbf{F}$ or $\mathbf{G}$, is discretised with one-sided finite difference operators in the predictor and corrector stages. The overall scheme is second-order accurate in time and space as long as the derivatives appearing in the expressions for F and G, i.e. from (18.34), are differenced in the opposite direction to the differencing of F and G in (18.35 and 36). For example, in the predictor stage,

$$(F_1)_{j+1,k}^n = (u^2)_{j+1,k}^n - \frac{v(u_{j+1,k}^n - u_{j,k}^n)}{\Delta x} \quad \text{and} \tag{18.37}$$

$$(F_1)_{j,k}^n = (u^2)_{j,k}^n - \frac{v(u_{j,k}^n - u_{j-1,k}^n)}{\Delta x} \ .$$

For the Navier–Stokes equations (18.6) the predictor and corrector stages (18.35, 36) are directly applicable with $S = 0$. However the components of F and G (18.7–10) introduce cross derivatives. For example

$$G_2 = \varrho uv - (\mu + \mu_T)\left(\frac{\partial u}{\partial y} + \frac{\partial v}{\partial x}\right) \ . \tag{18.38}$$

The discretisation of $\partial u/\partial y$ is handled as in (18.37). The term $\partial v/\partial x$ is discretised using centred differences. Thus, in the corrector stage,

$$(G_2)_{j,k}^* = (\varrho uv)_{j,k}^* - (\mu + \mu_T)_{j,k}\left(\frac{u_{j,k+1}^* - u_{j,k}^*}{\Delta y} + \frac{0.5(v_{j+1,k}^* - v_{j-1,k}^*)}{\Delta x}\right)$$

and

$$(G_2)_{j,k-1}^* = (\varrho uv)_{j,k-1}^* - (\mu + \mu_T)_{j,k-1}\left(\frac{u_{j,k}^* - u_{j,k-1}^*}{\Delta y} + \frac{0.5(v_{j+1,k-1}^* - v_{j-1,k-1}^*)}{\Delta x}\right) \ . \tag{18.39}$$

The use of forward differencing in the predictor stage and backward differencing in the corrector stage can be reversed and can be different for different spatial directions. However, to retain second-order accuracy it is important to maintain symmetry of the differencing formulae between the predictor and corrector stages.

The stability of the MacCormack schemes applied to (18.33) and (18.6) is discussed by Peyret and Taylor (1983); precise results are not available. From the explicit nature of the scheme the inviscid part of the equations are expected to lead to a CFL-type restriction, similar to (9.11) and the viscous parts to a diffusion-type restriction (Sect. 7.1.1).

For the scalar equivalent of (18.33), Peyret and Taylor recommend the following necessary condition, with $\Delta x = \Delta y$, for stability:

$$\Delta t \leqq \frac{\Delta x^2}{4v + (|u| + |v|)\,\Delta x} \ . \tag{18.40}$$

For the laminar compressible Navier–Stokes equations (18.6) Peyret and Taylor (1983) recommend the corresponding condition for stability, with $\Delta x = \Delta y$:

$$\Delta t \leq \frac{\Delta x^2}{(2\mu/\mathrm{Re}\varrho)[2\gamma/\mathrm{Pr}+(2/3)^{0.5}]+[|u|+|v|+(2)^{0.5}a]\Delta x} , \tag{18.41}$$

where a is the speed of sound.

The form of the stability restrictions (18.40, 41) indicates that the time-step is more restricted in three dimensions than in two or one dimension.

MacCormack (1971) introduced a time-split version of the above scheme to avoid this difficulty. The time-split version introduces a sequence of one-dimensional spatial operators, similar to the procedure in Sect. 8.5. For (18.6) the one-dimensional operators can be written

$$\mathbf{q}_{j,k}^{**} = P_x(\Delta t_x)\mathbf{q}_{j,k}^* , \tag{18.42}$$

$$\mathbf{q}_{j,k}^{**} = P_y(\Delta t_y)\mathbf{q}_{j,k}^* , \tag{18.43}$$

where (18.42) is equivalent to

$$\mathbf{q}_{j,k}' = \mathbf{q}_{j,k}^* - \frac{\Delta t_x}{\Delta x}[\mathbf{F}_{j+1,k}^* - \mathbf{F}_{j,k}^*] \quad \text{and}$$

$$\mathbf{q}_{j,k}^{**} = 0.5(\mathbf{q}_{j,k}^* + \mathbf{q}_{j,k}') - 0.5\frac{\Delta t_x}{\Delta x}[\mathbf{F}_{j,k}' - \mathbf{F}_{j-1,k}'] .$$

An equivalent expression for (18.43) can be deduced from (18.35 and 36). The complete algorithm, replacing (18.35, 36), to advance the solution one time-step becomes

$$\mathbf{q}_{j,k}^{n+1} = P_y\left(\frac{\Delta t_y}{2}\right) P_x\left(\frac{\Delta t_x}{2}\right) P_x\left(\frac{\Delta t_x}{2}\right) P_y\left(\frac{\Delta t_y}{2}\right) q_{j,k}^n , \tag{18.44}$$

where the symmetric pattern of repeated spatial operators over $\Delta t/2$ is necessary to obtain an algorithm that is second-order accurate in time.

The stability of (18.44) is determined by the stability of the individual operators. Thus the equivalent restriction to (18.41) on P_x, for arbitrary Δx, Δy, is

$$\Delta t_x \leq \frac{\Delta x^2}{(\mu/\mathrm{Re}\varrho)[2\gamma/\mathrm{Pr}+(2/3)^{0.5}(\Delta x/\Delta y)]+[|u|+a]\Delta x} . \tag{18.45}$$

Clearly larger time steps are available. A further advantage is that different time steps can be associated with different coordinate directions. For flow past a slender body parallel to the x direction a fine grid in the y direction will be necessary to resolve the severe normal gradients of velocity and temperature associated with the surface boundary layer. The use of the time-split scheme (18.44) avoids the time step restriction on Δt_y affecting Δt_x.

The time-split scheme is effective, although it may require auxiliary procedures adjacent to boundaries (Peyret and Taylor 1983, p. 73). In addition, for larger values of the Reynolds number (18.45) the MacCormack scheme is not efficient as a means of obtaining steady-state solutions via a pseudo-transient formulation (Sect. 6.2.4) unless a multigrid strategy (Sect. 6.3.5) is also incorporated (Chima and Johnson 1985).

18.2.2 Runge–Kutta Schemes

The explicit MacCormack schemes described in Sect. 18.2.1 achieve second-order spatial accuracy very economically. However, particularly when obtaining steady solutions via a pseudotransient formulation for flows at large Reynolds number, the time-step restriction (18.45) becomes a severe impediment to overall efficiency.

When the split MacCormack scheme is used at high Reynolds number an examination of (18.45) indicates the time step is effectively limited to a Courant number, $C = (|u| + a) \Delta t_x / \Delta x$, of unity. Of course the large Reynolds number influence appears directly in the restriction on the size of the spatial grid to resolve the thin boundary layer.

By adopting the method of lines strategy (Sect. 7.4), of discretising spatially to reduce the governing equations to a system of time-dependent ordinary differential equations, it is advantageous to introduce Runge–Kutta time marching schemes since they allow larger Courant numbers than the MacCormack scheme. For example, for unsteady flow problems it would be appropriate to use the fourth-order Runge–Kutta scheme (7.53). This scheme is stable for a Courant number $C \leq 2\sqrt{2}$ if the dissipative terms are small enough.

If the residuals are approximately averaged, preferably implicitly (Jameson and Schmidt 1985), during the stages of the Runge–Kutta marching scheme it is possible to obtain stable solutions with even higher values of the Courant number. Various multistage Runge–Kutta schemes are possible. For steady-state multigrid calculations four and five-stage schemes are effective with structured grids (Jameson and Schmidt 1985) and with unstructured grids (Mavriplis et al. 1989). For unsteady inviscid transonic flow a three-stage scheme (Venkatakrishnan and Jameson 1988) is available which preserves the TVD property (14.81).

For steady solutions obtained via a pseudotransient formulation it is possible to use a rational Runge–Kutta (RRK) scheme (Wambecq 1978) of first or second order since large Courant numbers are possible. The rational Runge–Kutta time discretisation will be illustrated for the general equation

$$dq/dt = W(q) \ . \tag{18.46}$$

This equation can be interpreted as one component of (18.6) after a spatial discretisation. Satofuka et al. (1986), who use an RRK algorithm, recommend conventional three-point centred difference discretisation of the first and second spatial derivatives appearing in (18.6–10).

A two-stage RRK scheme, applied to (18.46), can be constructed as follows. Intermediate corrections are evaluated as

$$\Delta q^1 = \Delta t\, W(q^n)\ , \quad \Delta q^2 = \Delta t\, W(q^n + c\Delta q^1)\ , \quad \Delta q^3 = (1-b)\,\Delta q^1 + b\Delta q^2\ ,$$

and the solution is obtained from

$$q^{n+1} = q^n + \frac{2\Delta q^1(\Delta \mathbf{q}^1, \Delta \mathbf{q}^3) - \Delta q^3(\Delta \mathbf{q}^1, \Delta \mathbf{q}^1)}{(\Delta \mathbf{q}^3, \Delta \mathbf{q}^3)}\ . \tag{18.47}$$

In (18.47) $(\mathbf{e}, \mathbf{f})$ denotes a scalar product, i.e.

$$(\mathbf{e}, \mathbf{f}) = \sum_i e_i f_i\ ,$$

with i ranging over all grid points. The scalar products, which are evaluated once per time-step, provide weighting factors for the respective corrections, Δq^1 or Δq^3, at each grid point. Thus the method is explicit and economical. The RRK scheme (18.47) is first-order accurate in time unless $bc = -0.5$, for which it is second-order accurate. The scheme is $A(\alpha)$ stable (Fig. 7.10) if

$$bc \leqq \frac{-1}{2\cos\alpha(2-\cos\alpha)}\ . \tag{18.48}$$

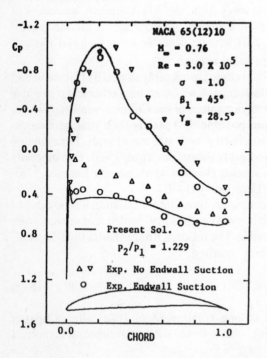

NACA 65(12)10
$M_\infty = 0.76$
$Re = 3.0 \times 10^5$
$\sigma = 1.0$
$\beta_1 = 45°$
$\gamma_s = 28.5°$

—— Present Sol.
$p_2/p_1 = 1.229$

△ ▽ Exp. No Endwall Suction
○ Exp. Endwall Suction

Fig. 18.3. Comparison with experimental pressure distribution 65(12)10 cascade (after Satofuka, 1986; reprinted with permission of AIAA)

When used with the Navier–Stokes equations, Satofuka et al. (1986) report using Courant numbers up to 4 in obtaining the steady flow solution about a compressor blade at a transonic Mach number, $M_\infty = 0.76$, and a Reynolds number of 3×10^5 with a Baldwin and Lomax turbulence model (Sect. 18.1.1). As indicated in Fig. 18.3 good agreement is achieved with the experimental pressure distribution based on the use of endwall suction. This solution was obtained with a 129×33 grid using generalised coordinates (Chap. 12).

Morinishi et al. (1991) combine the RRK scheme with multigrid acceleration and residual averaging (Jameson and Schmidt 1985) to predict transonic viscous flow about inclined two-dimensional airfoils. Convergence is achieved in about 500 steps.

18.3 Implicit Schemes

Notwithstanding the success of explicit schemes when used with Runge–Kutta time marching algorithms and multigrid, there is a preference for implicit schemes if only the steady-state solution is of interest. This is because implicit schemes can be constructed so that there is no formal time-step restriction in the linear stability sense (Sect. 4.3). In practice there is an effective time-step limitation but it is much less restrictive than for explicit schemes. The time-step limitation for implicit schemes may be due to nonlinear effects or due to accuracy requirements for unsteady problems or due to slow convergence of pseudotransient procedures for steady problems.

In this section four implicit algorithms will be described. First, MacCormack's bidiagonal scheme will be presented as a direct extension of the explicit scheme (Sect. 18.2.1). Second, the Beam and Warming (1978) implicit algorithm (Sect. 18.3.2) will be applied to the compressible Navier–Stokes equations. This algorithm is similar to the approximate factorisation algorithms described in Sects. 8.2, 8.3, 9.5, 10.4.2 and 14.2.8.

In Sect. 18.3.3 the approximate factorisation group finite element method (Sect. 17.3.3) is extended to compressible flow. A development of the approximate factorisation algorithm to introduce a further LU splitting is described in Sect. 18.3.4.

18.3.1 Implicit MacCormack Scheme

This method will be introduced as an extension to the explicit MacCormack scheme (Sect. 18.2.1) by applying it to the one-dimensional transport equation (9.56). To suit the present notation the one-dimensional transport equation is written as

$$\frac{\partial q}{\partial t} + a \frac{\partial q}{\partial x} - \mu \frac{\partial^2 q}{\partial x^2} = 0 \; . \tag{18.49}$$

The explicit MacCormack scheme (18.35, 36) applied to (18.49) can be written as follows.

Predictor stage:

$$\Delta q_j^{*,e} = -a\Delta t L_x^+ q_j^n + \mu \Delta t L_{xx} q_j^n ,$$

$$q_j^{*,e} = q_j^n + \Delta q_j^{*,e} ,$$

(18.50)

Corrector stage:

$$\Delta q_j^{n+1,e} = -a\Delta t L_x^- q_j^{*,e} + \mu \Delta t L_{xx} q_j^{*,e} ,$$

$$q_j^{n+1} = 0.5(q_j^n + q_j^{*,e} + \Delta q_j^{n+1,e}) ,$$

(18.51)

where L_x^- and L_x^+ are one-sided difference operators

$$L_x^- q_j = \frac{q_j - q_{j-1}}{\Delta x} , \qquad L_x^+ q_j = \frac{q_{j+1} - q_j}{\Delta x} ,$$

and L_{xx} is the second-order centred difference operator

$$L_{xx} q_j = \frac{q_{j-1} - 2q_j + q_{j+1}}{\Delta x^2} .$$

For stability the time step used in (18.50, 51) must be chosen so that

$$a + \frac{2\mu}{\Delta x} - \frac{\Delta x}{\Delta t} \leq 0 .$$

(18.52)

This condition is the one-dimensional analogue of (18.40).

MacCormack (1982) generates an implicit algorithm, equivalent to (18.50, 51) as follows.

Predictor stage:

$$\left(1 + \lambda \frac{\Delta t}{\Delta x}\right) \Delta q_j^{*,i} = \Delta q_j^{*,e} + \left(\lambda \frac{\Delta t}{\Delta x}\right) \Delta q_{j+1}^{*,i} ,$$

$$q_j^{*,i} = q_j^n + \Delta q_j^{*,i} ,$$

(18.53)

Corrector stage:

$$\left(1 + \lambda \frac{\Delta t}{\Delta x}\right) \Delta q_n^{n+1,i} = \Delta q_j^{n+1,e} + \left(\lambda \frac{\Delta t}{\Delta x}\right) \Delta q_{j-1}^{n+1,i} ,$$

$$q_j^{n+1} = 0.5(q_j^n + q_j^{*,i} + \Delta q_j^{n+1,i}) .$$

(18.54)

To evaluate (18.53 and 54) the corrections $\Delta q_j^{*,e}$ and $\Delta q_j^{n+1,e}$ are obtained from (18.50 and 51). In evaluating (18.51), $q_j^{*,e}$ on the right-hand side is replaced by $q_j^{*,i}$ determined from (18.53).

The parameter λ is chosen to ensure unconditional linear stability (Sect. 4.3). This requires

$$\lambda \geq \max\left[\left(a + \frac{2\mu}{\Delta x} - \frac{\Delta x}{\Delta t}\right), 0\right] . \tag{18.55}$$

Comparison with (18.52) indicates that if Δt is such that the explicit algorithm alone would be unstable, λ is a positive parameter such that the combined algorithm (18.50–54) is stable. If Δt is such that the explicit algorithm alone is stable, i.e. (18.52) is satisfied, λ is set to zero so that the implicit stages (18.53, 54) are not required.

Equation (18.55) is evaluated on a point-by-point basis so that for many problems the additional implicit steps are not required in parts of the computational domain. This contributes to the economy of the overall algorithm.

As long as $\mu \Delta t / \Delta x^2$ is bounded as Δt, Δx approach zero the implicit MacCormack algorithm is second-order accurate in time and space like the explicit MacCormack algorithm (18.50, 51). The retention of second-order time accuracy follows (MacCormack 1982) from the fact that the additional corrections in (18.53 and 54) are third-order accurate in time.

The bidiagonal nature of (18.53 and 54) means that the "implicit" corrections, $\Delta q^{*,i}$ and $\Delta q^{n+1,i}$, can be evaluated explicitly. Thus for the predictor step, (18.53) is evaluated as

$$\Delta q_j^{*,i} = \frac{\Delta q_j^{*,e} + (\lambda \Delta t / \Delta x) \Delta q_{j+1}^{*,i}}{1 + \lambda \Delta t / \Delta x} , \tag{18.56}$$

from the right-hand boundary, $j = NX$, where it is assumed a Dirichlet boundary condition is available for q_{NX}. Equation (18.56) is evaluated for decreasing values of j until the left-hand boundary is reached. The complete implicit algorithm applied to (18.49) consists of (18.50) followed by (18.53) for the predictor step and (18.51) followed by (18.54) for the corrector step.

The extension of the method to the compressible Navier–Stokes equations (18.6) is motivated by constructing an implicit algorithm of the form

$$\left[\underline{I} + \Delta t \left(\frac{\partial}{\partial x} \underline{A} + \frac{\partial}{\partial y} \underline{B}\right)\right] \Delta \mathbf{q}_{j,k}^{n+1} = \Delta \mathbf{q}_{j,k}^{n+1,e} , \tag{18.57}$$

where the Jacobians $\underline{A} = \partial \mathbf{F} / \partial \mathbf{q}$ and $\underline{B} = \partial \mathbf{G} / \partial \mathbf{q}$, and $\Delta \mathbf{q}_{j,k}^{n+1,e} = -\Delta t [\partial \mathbf{F} / \partial x + \partial \mathbf{G} / \partial y]_{j,k}$. The spatial derivatives on the left-hand side of (18.57) operate on the products $\underline{A} \Delta \mathbf{q}$ and $\underline{B} \Delta \mathbf{q}$. A comparison with (14.103) indicates that $\gamma = 0$ and $\beta = 1$ and in (18.57) the spatial discretisation is still to be introduced.

The total algorithm, equivalent to (18.50–54), can be written

Predictor stage:

$$\Delta \mathbf{q}_{j,k}^{*,e} = -\Delta t (L_x^+ \mathbf{F}_{j,k}^n + L_y^+ \mathbf{G}_{j,k}^n) \ ,$$

$$(\underline{I} - \Delta t L_x^+ \underline{A}^m)(\underline{I} - \Delta t L_y^+ \underline{B}^m)\Delta \mathbf{q}_{j,k}^{**,i} = \Delta \mathbf{q}_{j,k}^{*,e} \ , \tag{18.58}$$

$$\mathbf{q}_{j,k}^* = \mathbf{q}_{j,k}^n + \Delta \mathbf{q}_{j,k}^{**,i} \ ,$$

Corrector stage:

$$\Delta \mathbf{q}_{j,k}^{n+1,e} = -\Delta t (L_x^- \mathbf{F}_{j,k}^* + L_y^- \mathbf{G}_{j,k}^*) \ ,$$

$$(\underline{I} - \Delta t L_x^- \underline{A}^m)(\underline{I} - \Delta t L_y^- \underline{B}^m)\Delta \mathbf{q}_{j,k}^{n+1,i} = \Delta \mathbf{q}_{j,k}^{n+1,e} \ , \tag{18.59}$$

$$\mathbf{q}_{j,k}^{n+1} = 0.5(\mathbf{q}^n + \mathbf{q}^* + \Delta \mathbf{q}^{n+1,i})_{j,k} \ .$$

The one-sided differencing of $\mathbf{F}$ and $\mathbf{G}$ is the same as that described in Sect. 18.2. Where cross-derivatives occur, central differencing is used.

The modified Jacobians, $\underline{A}^m$ and $\underline{B}^m$, are related to the true Jacobians for inviscid flow, but are guaranteed to have positive eigenvalues. The Jacobians for inviscid flow can be factorised as in (14.107), i.e.

$$\underline{A} = \underline{T}_A^{-1} \underline{A}_A \underline{T}_A \quad \text{and} \quad \underline{B} = \underline{T}_B^{-1} \underline{A}_B \underline{T}_B \ , \tag{18.60}$$

where $\underline{A}_A$, $\underline{A}_B$ are diagonal matrices containing the eigenvalues of $\underline{A}$ and $\underline{B}$, i.e.

$$\mathrm{diag}\underline{A}_A = \{u, u+a, u, u-a\} \ , \quad \mathrm{diag}\underline{A}_B = \{v, v, v+a, v-a\} \ . \tag{18.61}$$

The matrices $\underline{T}_A$ and $\underline{T}_B$ are given by MacCormack (1982).

The modified Jacobians are formed as

$$\underline{A}^m = \underline{T}_A^{-1} \underline{D}^A \underline{T}_A \quad \text{and} \quad \underline{B} = \underline{T}_B^{-1} \underline{D}^B \underline{T}_B \ , \tag{18.62}$$

where $\underline{D}^A$ and $\underline{D}^B$ are diagonal matrices whose diagonal entries are of the form

$$D_{l,l}^A = \max\left[|\lambda_{Al,l}| + \frac{2v}{\varrho \Delta x} - \frac{0.5 \Delta x}{\Delta t}, \ 0. \right] \quad \text{and} \tag{18.63}$$

$$D_{l,l}^B = \max\left[|\lambda_{Bl,l}| + \frac{2v}{\varrho \Delta y} - \frac{0.5 \Delta y}{\Delta t}, \ 0. \right] \ , \tag{18.64}$$

where $v = \max(4\mu/3, \gamma\mu/\mathrm{Pr})$ and μ and Pr can be interpreted as their laminar or turbulent values as appropriate. Equations (18.63 and 64) can take only positive values and constitute a generalization of (18.55) for two-dimensional systems of equations. That is, if the relevant diagonal entries $D_{l,l}^A$ and $D_{l,l}^B$ are greater than zero the implicit steps in (18.58 and 59) are included. If $D_{l,l}^A$ and/or $D_{l,l}^B = 0$ the explicit contributions will be stable and the relevant implicit steps are not required.

The evaluation of the implicit steps proceeds in two parts. For the predictor stage an intermediate implicit correction $\Delta q_{j,k}^{*,i}$ is obtained by solving

$$\left[\underline{I}+\frac{\Delta t}{\Delta x}\,\underline{A}^m\right]\Delta\mathbf{q}_{j,k}^{*,i} = \Delta\mathbf{q}_{j,k}^{*,e}+\frac{\Delta t}{\Delta x}\,\underline{A}_{j+1,k}^m\,\Delta\mathbf{q}_{j+1,k}^{*,i}\ . \tag{18.65}$$

Starting with the $k = NY$ grid line and the right-hand boundary $j = NX$, (18.65) is applied for decreasing j values until the left-hand boundary is reached. Subsequently the process is repeated for decreasing values of k until the bottom boundary is reached.

Comparing (18.65) with (18.58) it is clear that

$$(\underline{I}-\Delta t\,L_y^+\,\underline{B}^m)\,\Delta q_{j,k}^{**,i} = \Delta q_{j,k}^{*,i}\ . \tag{18.66}$$

Consequently, $\Delta q_{j,k}^{**,i}$ is obtained from

$$\left(\underline{I}+\frac{\Delta t}{\Delta y}\,\underline{B}^m\right)\Delta\mathbf{q}_{j,k}^{**,i} = \Delta\mathbf{q}_{j,k}^{*,i}+\frac{\Delta t}{\Delta y}\,\underline{B}_{j,k+1}^m\,\Delta\mathbf{q}_{j,k+1}^{**,i}\quad(=\mathbf{W})\ . \tag{18.67}$$

Starting with the $j = NX$ grid line and the top boundary $k = NY$, (18.67) is applied for decreasing k values until the bottom boundary is reached. Subsequently the process is repeated for decreasing j values until the left-hand boundary is reached. Because of the block nature of $\underline{A}^m$ and $\underline{B}^m$ in (18.65, 66), each step would appear to require the solution of a 4×4 system of equations. However, this can be avoided by making use of (18.62). Thus (18.67) is written as

$$\left(\underline{I}+\frac{\Delta t}{\Delta y}\,(\underline{T}^B)^{-1}\,\underline{D}^B\,\underline{T}^B\right)\Delta\mathbf{q}_{j,k}^{**,i} = \mathbf{W}\ , \tag{18.68}$$

where $\mathbf{W}$ is the evaluation of the right-hand side of (18.67). Multiplying both sides of (18.68) by $\underline{T}^B$ and a little manipulation gives

$$\underline{T}^B\Delta\mathbf{q}_{j,k}^{**,i} = \left(\underline{I}+\frac{\Delta t}{\Delta y}\,\underline{D}^B\right)^{-1}\,\underline{T}^B\mathbf{W} = \mathbf{Y}\quad\text{and} \tag{18.69}$$

$$\Delta\mathbf{q}_{j,k}^{**,i} = (\underline{T}^B)^{-1}\mathbf{Y}\ . \tag{18.70}$$

It may be noted that $(\underline{T}^B)^{-1}$ is available analytically. MacCormack (1982) indicates that the various steps (18.68–70) which replace (18.67) can be coded very efficiently, requiring about 28 Fortran statements. An equivalent algorithm is available to solve (18.65).

The above algorithm applied to the compressible Navier–Stokes equations is second-order accurate in space, as for the model equation (18.49). MacCormack (1982) demonstrates the algorithm for a shock, boundary layer interaction sufficient to cause separation.

The algorithm is more suitable for obtaining steady solutions than unsteady solutions because of the approximate treatment of the viscous terms in constructing the modified Jacobians $\underline{A}^m$ and $\underline{B}^m$. For transonic flows about inclined aerofoils Kordulla and MacCormack (1982) find it necessary to include numerical dissipation in both the explicit and implicit stages of (18.58 and 59) if large time steps are taken.

From the execution times presented it appears that the fully implicit algorithm (18.58, 59) requires roughly the same execution time per grid point as the block tridiagonal algorithm to be described in Sect. 18.3.2. However, an advantage of the MacCormack scheme is that for many grid points the underlying explicit algorithm is stable. Consequently for many problems the overall efficiency of the implicit MacCormack algorithm is superior.

Hung and Kordulla (1984) apply the implicit MacCormack algorithm with the thin-layer approximation (18.31) in conjunction with finite-volume spatial discretisation and incorporate a one-dimensional splitting equivalent to that described in Sect. 18.2.1. They apply the method to the supersonic flow past a blunt fin on a flat plate at $M_\infty = 2.95$. The Reynolds number based on the freestream velocity and the fin diameter is 0.8×10^6. The Baldwin–Lomax algebraic eddy viscosity model (18.17–21) is used to account for turbulence effects. The equations are solved in generalised coordinates on a stretched C grid with 40, 32 and 32 points in the circumferential, radial and vertical directions, respectively.

The predicted pressures, Fig. 18.4, show excellent agreement with the measurements of Dolling and Bogdonoff (1982). The upstream pressure rise is associated with the occurrence of the primary horseshoe vortex, the axis of which is at $x/D \approx -0.75$. The downstream pressure rise is associated with passage through the bow shock and compression to stagnation conditions at $x/D = 0$.

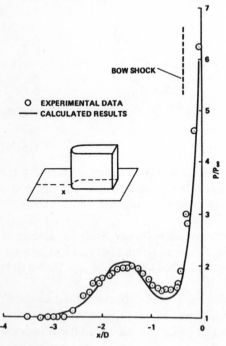

Fig. 18.4. Pressure on the flat plate along the line of symmetry (after Hung and Kordulla, 1984; reprinted with permission of AIAA)

The implicit MacCormack (1982) algorithm is effective with Dirichlet boundary conditions but the bidiagonal solver lacks flexibility in solving some other types of boundary conditions. MacCormack (1985) prefers to replace the bidiagonal solver with a line Gauss–Seidel iterative method (Sect. 6.3) or Newton's method (Sect. 6.1) after using a flux splitting construction (Steger and Warming 1981) to render (18.57) diagonally dominant.

18.3.2 Beam and Warming Scheme

This scheme is the forerunner of the approximate factorisation scheme (14.105, 106) applied to the Euler equations in Sect. 14.2.8. The Beam and Warming (1978) scheme and the Briley and McDonald (1977) scheme are both approximate factorisation schemes closely related to the schemes described in Sects. 8.2 and 10.4.2.

To implement the Beam and Warming scheme the compressible Navier–Stokes equations (18.6) are written as

$$\partial \mathbf{q}/\partial t = \text{RHS} \ , \tag{18.71}$$

where RHS contains all the spatial derivatives, i.e.

$$
\begin{aligned}
\text{RHS} &= -\frac{\partial \mathbf{F}}{\partial x} - \frac{\partial \mathbf{G}}{\partial y} \\[2mm]
&= -\frac{\partial}{\partial x}\left[\mathbf{F}^I - \mathbf{F}_1^v(\mathbf{q}, \mathbf{q}_x) - \mathbf{F}_2^v(\mathbf{q}, \mathbf{q}_y) \right] \\[2mm]
&\quad -\frac{\partial}{\partial y}\left[\mathbf{G}^I - \mathbf{G}_1^v(\mathbf{q}, \mathbf{q}_x) - \mathbf{G}_2^v(\mathbf{q}, \mathbf{q}_y) \right] \ .
\end{aligned}
\tag{18.72}
$$

In (18.72), the inviscid fluxes $\mathbf{F}^I$ and $\mathbf{G}^I$ coincide with $\mathbf{F}$ and $\mathbf{G}$ defined by (14.95). The other terms $\mathbf{F}_1^v$, $\mathbf{F}_2^v$, $\mathbf{G}_1^v$ and $\mathbf{G}_2^v$ account for the viscous terms in (18.7) after substituting (18.9 and 18.10), but grouped so that $\mathbf{F}_1^v, \mathbf{G}_1^v$ contain only x derivatives and $\mathbf{F}_2^v, \mathbf{G}_2^v$ contain only y derivatives.

Application of the generalised three-level scheme (Sect. 8.2.3) to (18.71) gives

$$(1+\alpha)\,\Delta\mathbf{q}^n - \alpha\Delta\mathbf{q}^n = \Delta t(\beta\,\text{RHS}^{n+1} + (1-\beta)\text{RHS}^n) \ , \quad \text{where} \tag{18.73}$$

$$\Delta\mathbf{q}^{n+1} = \mathbf{q}^{n+1} - \mathbf{q}^n \quad \text{and} \quad \Delta\mathbf{q}^n = \mathbf{q}^n - \mathbf{q}^{n-1} \ .$$

The parameters α and β are chosen to secure appropriate stability and accuracy properties. In (18.73) α is equivalent to γ in (8.26). In (18.73) RHS^{n+1} is a nonlinear function of $\mathbf{q}, \mathbf{q}_x, \mathbf{q}_y$. Linearisation is made about the nth time level as in (14.101 and 102), i.e.

$$\text{RHS}^{n+1} = \text{RHS}^n + \left(\frac{\partial\text{RHS}}{\partial\mathbf{q}}\right)\Delta\mathbf{q}^{n+1} + \left(\frac{\partial\text{RHS}}{\partial\mathbf{q}_x}\right)\Delta\mathbf{q}_x + \left(\frac{\partial\text{RHS}}{\partial\mathbf{q}_y}\right)\Delta\mathbf{q}_y + \cdots$$

$$\tag{18.74}$$

or

$$\text{RHS}^{n+1} = \text{RHS}^n - \frac{\partial}{\partial x}(\underline{A}\,\Delta\mathbf{q}^{n+1} - \underline{P}\Delta\mathbf{q}^{n+1} - \underline{R}\,\Delta\mathbf{q}_x^{n+1} - \Delta\mathbf{F}_2^{v,n+1})$$

$$- \frac{\partial}{\partial y}(\underline{B}\Delta\mathbf{q}^{n+1} - \underline{Q}\Delta\mathbf{q}^{n+1} - \Delta\mathbf{G}_1^{v,n+1} - \underline{S}\Delta\mathbf{q}_y^{n+1}) \ . \tag{18.75}$$

In (18.75) $\underline{A}$ and $\underline{B}$ are the inviscid Jacobians (14.99) and

$$\underline{P} = \frac{\partial\mathbf{F}_1^v}{\partial\mathbf{q}}\ , \quad \underline{R} = \frac{\partial\mathbf{F}_1^v}{\partial\mathbf{q}_x}\ , \quad \underline{Q} = \frac{\partial\mathbf{G}_2^v}{\partial\mathbf{q}} \quad \text{and} \quad \underline{S} = \frac{\partial\mathbf{G}_2^v}{\partial\mathbf{q}_y}\ .$$

Equation (18.75) can be simplified using

$$\underline{R}\,\Delta\mathbf{q}_x^{n+1} = (\underline{R}\Delta\mathbf{q}^{n+1})_x - \underline{R}_x\Delta\mathbf{q}^{n+1}\ ,$$

and similarly for $\underline{S}\Delta\mathbf{q}_y^{n+1}$. The terms $\Delta\mathbf{F}_2^{v,n+1}$ and $\Delta\mathbf{G}_1^{v,n+1}$ introduce cross-derivatives when (18.75) is substituted into (18.73). These terms are handled most easily by noting that

$$\Delta\mathbf{F}_2^{v,n+1} = \Delta\mathbf{F}_2^{v,n} + O(\Delta t^2)\ , \quad \Delta\mathbf{G}_1^{v,n+1} = \Delta\mathbf{G}_1^{v,n} + O(\Delta t^2)\ . \tag{18.76}$$

That is, to second order the implicit corrections to $\mathbf{F}_2^v$ and $\mathbf{G}_1^v$ can be replaced by the corresponding corrections for the previous time-step which are available explicitly. It is noted by Beam and Warming (1978) that (18.76) applied to the model equation (18.49) does not jeopardise the unconditional linear stability of the present algorithm, for two-dimensional flow. However it appears that the scheme may be unstable for three-dimensional compressible flows (Pan and Lomax 1988).

Substitution of (18.75 and 76) into (18.73) produces the following linear system of equations for $\Delta\mathbf{q}^{n+1}$:

$$\left[\underline{I} + \frac{\beta\Delta t}{1+\alpha}\left(\frac{\partial}{\partial x}(\underline{A} - \underline{P} + \underline{R}_x)^n - \frac{\partial^2}{\partial x^2}\underline{R}^n\right)\right.$$

$$\left. + \frac{\beta\Delta t}{1+\alpha}\left(\frac{\partial}{\partial y}(\underline{B} - \underline{Q} + \underline{S}_y)^n - \frac{\partial^2}{\partial y^2}\underline{S}^n\right)\right]\Delta\mathbf{q}^{n+1}$$

$$= \frac{\Delta t\,\text{RHS}^n}{1+\alpha} + \frac{\alpha}{1+\alpha}\Delta\mathbf{q}^n + \frac{\beta'\Delta t}{1+\alpha}\left(\frac{\partial}{\partial x}\Delta\mathbf{F}_2^{v,n} + \frac{\partial}{\partial y}\Delta\mathbf{G}_1^{v,n}\right)$$

$$= \Delta\mathbf{q}^m\ . \tag{18.77}$$

Analytic expressions for $-\underline{P} + \underline{R}_x$, $-\underline{Q} + \underline{S}_y$, $\underline{R}$ and $\underline{S}$ are provided by Beam and Warming. In (18.77) β' is set equal to β when solving unsteady problems since the overall scheme is then second order accurate in time if $\beta = \alpha + 0.5$. However, in solving steady problems via the pseudotransient formulation (Sect. 6.4) it is more efficient to use a scheme that is first-order accurate in time and to set $\beta' = 0$ to economise on the operation count. The spatial accuracy of (18.77) depends on the spatial discretisation; Beam and Warming recommend second-order central difference expressions.

Equation (18.77) is linear but globally coupled. To second-order in time the left-hand side of (18.77) can be approximately factorised in the same manner that replaces (14.103) with (14.104). The approximate factorisation leads directly to the following two-stage algorithm, equivalent to (14.105 and 106):

$$\left\{ \underline{I} + \frac{\beta \Delta t}{1+\alpha} \left[L_x (\underline{A} - \underline{P} + \underline{R}_x)^n - L_{xx} \underline{R}^n \right] \right\} \Delta \mathbf{q}^* = \Delta \mathbf{q}^m \tag{18.78}$$

and

$$\left\{ \underline{I} + \frac{\beta \Delta t}{1+\alpha} \left[L_y (\underline{B} - \underline{Q} + \underline{S}_y)^n - L_{yy} \underline{S}^n \right] \right\} \Delta \mathbf{q}^{n+1} = \Delta \mathbf{q}^* \ . \tag{18.79}$$

In (18.78, 79) $L_{xx} \underline{R}^n$]} $\Delta \mathbf{q}^*$ implies $L_{xx}(\underline{R} \Delta \mathbf{q}^*)$, etc.

In (18.78) $\Delta \mathbf{q}^m$ is the vector resulting from evaluating the right-hand side of (18.77). If L_x, L_{xx}, L_y and L_{yy} are three-point centred finite difference expressions, as in (9.85), (18.78) represents 4×4 block tridiagonal systems of equations along each gridline in the x direction and (18.79) represents 4×4 block tridiagonal systems of equations associated with each gridline in the y direction. The algorithm described in Sect. 6.2.5 is suitable for efficiently solving these systems of equations.

Beam and Warming (1978) note that if the dependence of μ and k on $\mathbf{q}$ is ignored then $-\underline{P} + \underline{R}_x = 0$ and $-\underline{Q} + \underline{S}_y = 0$ with a consequent simplification of (18.78 and 79). This is particularly appropriate with the pseudotransient formulation since the steady-state solution is independent of the treatment of the left-hand sides of (18.78, 79).

For large Reynolds number flows or if weak shocks are present it is recommended that higher-order, typically fourth-order, numerical dissipation terms be added (Sect. 18.5.1). The Beam and Warming algorithm is also applicable to purely hyperbolic equations (Beam and Warming 1976) like the Euler equations (Sect. 14.2.8). However, it appears that for the three-dimensional hyperbolic case the algorithm is only conditionally stable (Jameson and Turkel 1981).

A conceptually similar non-iterative approximate factorisation algorithm for the compressible Navier–Stokes equations is described by Briley and McDonald (1977). For large Reynolds number viscous flows the use of implicit algorithms is particularly appropriate with the pseudotransient formulation since CFL numbers much larger than unity are available. The relevant two-dimensional CFL numbers are $(|u| + a) \Delta t / \Delta x$ and $(|v| + a) \Delta t / \Delta y$. It is often found that the steady-state solution is reached in the minimum number of time steps if Δt is chosen on a local basis to be $O(10)$. Too large a value implies that the error introduced by the approximate factorisation will distort the transient path. Too small a value will allow the correct transient path to be followed but unnecessarily slowly.

18.3.3 Group Finite Element Method

For transonic conditions μ and k are approximately constant and the energy equation can be replaced by the algebraic equation (18.30). As a result it is convenient to replace (18.6–8) with a system of three equations

$$\frac{\partial \mathbf{q}}{\partial t} + \frac{\partial \mathbf{F}^I}{\partial x} + \frac{\partial \mathbf{G}^I}{\partial y} = \frac{\partial^2 \mathbf{R}}{\partial x^2} + \frac{\partial^2 \mathbf{S}}{\partial x \partial y} + \frac{\partial^2 \mathbf{T}}{\partial y^2} \; , \tag{18.80}$$

where $\mathbf{F}^I$ and $\mathbf{G}^I$ are the inviscid contributions to $\mathbf{F}$ and $\mathbf{G}$ in (18.7). With a suitable nondimensionalisation the various vectors in (18.80) are given by

$$\mathbf{q} = \{\varrho, \varrho u, \varrho v\}^T \; , \qquad \mathbf{F}^I = \{\varrho u, \varrho u^2 + p, \varrho uv\}^T \; ,$$

$$\mathbf{G}^I = \{\varrho v, \varrho uv, \varrho v^2 + p\}^T \; , \qquad \mathbf{R} = \left\{\frac{\theta \varrho}{\mathrm{Re}}, \frac{4\mu_e u}{3}, \mu_e v\right\}^T \; , \tag{18.81}$$

$$\mathbf{S} = \left\{0, \frac{\mu_e v}{3}, \frac{\mu_e u}{3}\right\}^T \; , \qquad \mathbf{T} = \left\{\frac{\theta \varrho}{\mathrm{Re}}, \mu_e u, \frac{4\mu_e v}{3}\right\}^T \; .$$

The effective viscosity is $\mu_e = 1/\mathrm{Re} + \mu_T$ in the above expressions. This form of the equations is suitable for laminar ($\mu_T = 0$) or turbulent flow. However, terms like $u\partial^2 \mu_T/\partial y^2 + (\partial u/\partial y)(\partial \mu_T/\partial y)$ in the x and y momentum equations have been dropped. These terms will only be significant immediately adjacent to a solid surface where the local velocity solution is usually obtained via the use of wall functions (18.23). The parameter θ in the $\mathbf{R}$ and $\mathbf{T}$ components is a dissipative term introduced to stabilise the discretised equations for flows with large values of Re. In equations (18.80, 81) there are four dependent variables u, v, ϱ and p. To close the system the nondimensional form of (18.30) is made use of, i.e.

$$1 + \gamma M_\infty^2 p = \varrho\{1 + 0.5(\gamma - 1)M_\infty^2[1 - (u^2 + v^2)]\} \; , \tag{18.82}$$

where M_∞ is the freestream Mach number and γ is the specific heat ratio.

The group finite element formulation (Sect. 10.3) is applied to (18.80) by introducing approximate or trial solutions for the groups in (18.81). For example, with bilinear interpolation in rectangular elements (Sect. 5.3),

$$\mathbf{F}^I = \sum_{m=1}^{4} \phi_m(x, y) \mathbf{F}_m^I \; , \tag{18.83}$$

where ϕ_m are bilinear interpolation functions (5.59) and $\mathbf{F}_m^I$ are nodal values of $\mathbf{F}^I$. Application of the Galerkin finite element method (Chap. 5) to (18.80) with (18.81) produces the following semi-discrete form:

$$M_x \otimes M_y \frac{\partial \mathbf{q}}{\partial t} + M_y \otimes L_x \mathbf{F}^I + M_x \otimes L_y \mathbf{G}^I = M_y \otimes L_{xx} \mathbf{R} + L_x \otimes L_y \mathbf{S}$$

$$+ M_x \otimes L_{yy} \mathbf{T} \; , \tag{18.84}$$

where the directional mass and difference operators are given by (17.115).

A three-level implicit algorithm can be constructed to march (18.84) in time, just as (17.120, 121) were constructed. The result is a two-stage algorithm of the form

$$\left[M_x - \frac{\beta}{1+\alpha} \Delta t \left(L_{xx} \frac{\partial \mathbf{R}}{\partial \mathbf{q}} - L_x(\underline{A}) \right) \right] \Delta q^*$$

$$= \frac{\Delta t}{1+\alpha} (\text{RHS})^a + \frac{\alpha}{1+\alpha} M_x \otimes M_y \Delta \mathbf{q}^n \qquad (18.85)$$

and

$$\left[M_y - \frac{\beta}{1+\alpha} \Delta t \left(L_{yy} \frac{\partial \mathbf{T}}{\partial \mathbf{q}} - L_y(\underline{B}) \right) \right] \Delta \mathbf{q}^{n+1} = \Delta \mathbf{q}^* \ . \qquad (18.86)$$

Since $\mathbf{q}$ and $\mathbf{F}$, etc. are three-component vectors, (18.85 and 86) constitute 3×3 block tridiagonal systems of equations associated with gridlines in the x and y directions respectively. These systems can be solved efficiently using the algorithm described in Sect. 6.2.5. The parameters α and β are the same as in (18.73) and the same as γ and β in (17.116). The Jacobians $\underline{A} = \partial \mathbf{F}^I / \partial \mathbf{q}$ and $\underline{B} = \partial \mathbf{G}^I / \partial \mathbf{q}$ are 3×3 matrices and equivalent to $\underline{A}$ and $\underline{B}$ defined by (14.99).

The augmented right-hand side, $(\text{RHS})^a$, in (18.85) is given by

$$(\text{RHS})^a = M_y \otimes L_{xx} \mathbf{R} + L_x \otimes L_y \mathbf{S} + M_x \otimes L_{yy} \mathbf{T} - M_y \otimes L_x \mathbf{F}^I$$

$$- M_x \otimes L_y \mathbf{G}^I + L_x \otimes L_y (\partial \mathbf{S} / \partial \mathbf{q}) \Delta \mathbf{q}^n \ . \qquad (18.87)$$

The final term in $(\text{RHS})^a$ arises from the need to evaluate that term explicitly. This is equivalent to the treatment of $\Delta \mathbf{F}_2^v$ in (18.76). For typical problems this does not significantly reduce the maximum time-step for which stable solutions can be obtained.

For the choice $\alpha = 0.5$, $\beta = 1.0$ solutions produced by (18.85 and 86) are second-order accurate in time and space on a uniform grid. On a uniform grid the presence of the mass operators has a smoothing effect and produces a fourth-order spatial discretisation of the inviscid terms $\partial \mathbf{F}^I / \partial x$ and $\partial \mathbf{G}^I / \partial y$.

The present algorithm has been used to obtain the steady flow behaviour past a backward-facing step (Fig. 17.14) for subsonic flow conditions ($M_\infty = 0.4$). Both laminar and turbulent flows have been considered. For turbulent flow the mixing length algebraic eddy viscosity model (18.12) is appropriate for regions adjacent to solid surfaces. In the outer boundary layer region and the wake region the modified Clauser model with upstream relaxation (18.13–16) is used. In the separated flow region behind the step (Fig. 17.14) the following expression for the eddy viscosity (Deiwert 1976) is preferable to (18.13):

$$\mu_T = 0.0168 \, \varrho u_e \delta^* \left(\frac{y}{y_{ds}} \right) \text{Dr} \ , \qquad (18.88)$$

where y is measured from the wall (CD in Fig. 17.14) and Dr is the van Driest damping factor $\text{Dr} = 1 - \exp(-y^+ / 26)$ and y^+ is given by (18.22).

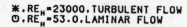

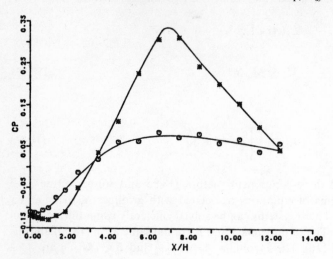

Fig. 18.5. Pressure distribution behind a step, $M_\infty = 0.40$

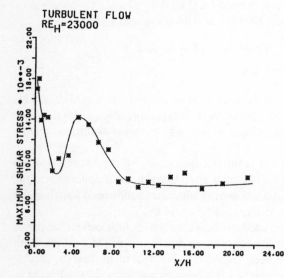

Fig. 18.6. Maximum shear stress distribution behind step, $M_\infty = 0.40$

For the flow over a backward-facing step the pressure distribution behind the step depends on whether the flow is laminar or turbulent. Typical computed pressure distributions are shown in Fig. 18.5. The pressure coefficient

$$C_p = \frac{p - p_\infty}{0.5 \varrho U_\infty^2} \quad \text{and} \quad RE_H = U_\infty H / v ,$$

where H is the step-height. The results shown in Fig. 18.5 were obtained with a 34×42 gridpoint distribution. The grid was uniform in the y direction but a variable grid was used in the x direction with a fine grid close to the step and a growing grid ($r_x = 1.2$) in both the upstream and downstream directions.

For turbulent flow the corresponding distribution of maximum shear stress is shown in Fig. 18.6. The shear stress distribution is in good qualitative agreement with experimental results (Eaton 1981). Further details and results are provided by Srinivas and Fletcher (1984).

18.3.4 Approximate LU Factorisation

For the compressible Navier–Stokes equations the construction of implicit schemes based on approximate factorisation (Sect. 8.2) leads to the solution of 4×4 block tridiagonal systems of equations associated with each gridline, for example (18.78) or (18.79). If the algebraic energy equation can be exploited the block size reduces to 3×3, for example (18.85 and 86). However, as indicated in Sect. 6.2.5 the operation count for the block Thomas algorithm is $O(5NM^3/3)$ where M is the order of the block. Clearly a means of avoiding solving *block* tridiagonal systems is desirable.

This can be achieved by constructing the implicit spatial operators from one-sided difference formulae so that an approximate LU factorisation becomes possible. Alternatively each approximate factorisation in equations like (18.78, 79) can be further factorised into approximate LU form. However, it is important that this additional approximate factorisation is achieved as accurately as possible. A failure to do so usually leads to a loss of temporal accuracy and for pseudo-transient steady-state solvers an increase in the number of iterations to reach the steady state.

An application of the approximate factorisation algorithm (Sect. 8.2.2), to the one-dimensional scalar convection-diffusion equation,

$$\frac{\partial q}{\partial t} + \frac{\partial F}{\partial x} = 0, \quad \text{where} \quad F = F^I_{(q)} - \mu \frac{\partial q}{\partial x} \,, \tag{18.89}$$

produces the following implicit algorithm:

$$[1 + \beta \Delta t (L_x A - \mu L_{xx})] \Delta q_j^{n+1} = \Delta t \, \text{RHS}^n = \Delta t (-L_x F_j^n + \mu L_{xx} q_j^n) \,, \tag{18.90}$$

where L_x, L_{xx} are three-point centred difference operators and $A = \partial F^I / \partial q$. On the left-hand side of (18.90) the second derivative operator L_{xx} can be replaced by

$$L_{xx} \Delta q_j^{n+1} = \frac{1}{\Delta x} (L_x^+ - L_x^-) \Delta q_j^{n+1} \,, \tag{18.91}$$

where L_x^+ and L_x^- are forward and backward two-point difference operators after (18.51).

To $O(\Delta t)$ the operator $L_x A \Delta q_n^{n+1}$ can be replaced with

$$L_x A \Delta q_j^{n+1} \rightarrow L_x^- (A_j^+ \Delta q_n^{n+1}) + L_x^+ (A_j^- \Delta q_j^{n+1}) \,, \quad \text{where} \tag{18.92}$$

$$A^+ = 0.5(A + |A|) \quad \text{and} \quad A^- = 0.5(A - |A|) \,. \tag{18.93}$$

Clearly depending on the sign of A either A^+ or A^- will be zero. Substituting into (18.90) gives the system of equations

$$\left\{1+\beta\Delta t\left[L_x^-\left(A_j^++\frac{\mu}{\Delta x}\right)-L_x^+\left(|A_j^-|+\frac{\mu}{\Delta x}\right)\right]\right\}\Delta q_j^{n+1}=\Delta t\,\mathrm{RHS}^n\ . \qquad (18.94)$$

To $O(\Delta t^2)$, (18.94) can be replaced with

$$\left[1+\beta\Delta tL_x^-\left(A_j^++\frac{\mu}{\Delta x}\right)\right]\left[1+\beta\Delta tL_x^+\left(|A_j^-|+\frac{\mu}{\Delta x}\right)\right]\Delta q_j^{n+1}=\Delta t\,\mathrm{RHS}^n\ . \tag{18.95}$$

Equation (18.95) is an LU decomposition since the first factor is lower triangular and the second factor is upper triangular. Equation (18.95) is solved in two stages.

$$\Delta q_j^*=\frac{\Delta t\,\mathrm{RHS}^n+\beta\dfrac{\Delta t}{\Delta x}\left(A_j^++\dfrac{\mu}{\Delta x}\right)\Delta q_{j-1}^*}{1+\beta\dfrac{\Delta t}{\Delta x}\left(A_j^++\dfrac{\mu}{\Delta x}\right)} \tag{18.96}$$

and

$$\Delta q_j^{n+1}=\frac{\Delta q_j^*+\beta\dfrac{\Delta t}{\Delta x}\left(|A_j^-|+\dfrac{\mu}{\Delta x}\right)\Delta q_{j+1}^{n+1}}{1+\beta\dfrac{\Delta t}{\Delta x}\left(|A^-|+\dfrac{\mu}{\Delta x}\right)}\ . \tag{18.97}$$

Equation (18.96) is solved successively from the left-hand boundary moving in the increasing j direction. Equation (18.97) is solved successively from the right-hand boundary moving in the decreasing j direction. It may be noted that the approximations introduced to evaluate the implicit terms have no effect on the steady-state solution, $\mathrm{RHS}^n=0$. Thus the approximate LU factorisation is a robust and economical algorithm for obtaining the steady-state solution.

An equivalent LU factorisation for the compressible Navier–Stokes equations can be obtained as follows. First (18.6) is written as

$$\frac{\partial\mathbf{q}}{\partial t}=\frac{\partial\mathbf{F}^v}{\partial x}+\frac{\partial\mathbf{G}^v}{\partial y}-\frac{\partial\mathbf{F}^I}{\partial x}-\frac{\partial\mathbf{G}^I}{\partial y}\ , \tag{18.98}$$

with $\mu_T=0$, $k_T=0$ for convenience of exposition. An approximate factorisation of the discrete form of (18.98) can be written

$$[\mathbf{I}+\beta\Delta tL_x\{\underline{A}-\underline{P}\}][\mathbf{I}+\beta\Delta tL_y\{\underline{B}-\underline{Q}\}]\Delta\mathbf{q}^{n+1}=\Delta t\,\mathrm{RHS}^n\ , \tag{18.99}$$

where

$$\text{RHS}^n = L_x(\mathbf{F}^v - \mathbf{F}^I) + L_y(\mathbf{G}^v - \mathbf{G}^I) \ ,$$

(18.100)

$$\underline{A} = \frac{\partial \mathbf{F}^I}{\partial \mathbf{q}} \ , \qquad \underline{B} = \frac{\partial \mathbf{G}^I}{\partial \mathbf{q}} \ , \qquad \underline{P} = \frac{\partial \mathbf{F}^v}{\partial \mathbf{q}} \quad \text{and} \quad \underline{Q} = \frac{\partial \mathbf{G}^v}{\partial \mathbf{q}} \ .$$

Equation (18.99) is similar to the Beam and Warming algorithm (18.78, 79) except that $\alpha = 0$ and the treatment of the viscous terms is different in (18.99). In forming $\underline{P}$ and $\underline{Q}$, terms that would lead to spatial cross-derivatives are ignored. This simplifies the algorithm and is an effective strategy when the steady-state solution is obtained via a pseudotransient constructions, since the simplification has no effect on the steady-state solution.

Equation (18.100) can also be considered to be equivalent to (18.90). To introduce an approximate LU factorisation into (18.99) it is necessary to extract the eigenvalues of $\underline{A}$ and $\underline{B}$, using (18.60). From (18.61) it is clear that both positive and negative values are to be expected. Consequently the eigenvalue matrices $\underline{\Lambda}_A$ and $\underline{\Lambda}_B$ are split into positive and negative components in the following way:

$$\underline{\Lambda}_A = \underline{\Lambda}_A^+ + \underline{\Lambda}_A^- \quad \text{and} \quad \underline{\Lambda}_B = \underline{\Lambda}_B^+ + \underline{\Lambda}_B^- \ ,$$

(18.101)

where $\underline{\Lambda}_A^+ = 0.5\{\underline{\Lambda}_A + |\underline{\Lambda}_A|\}$ and $\underline{\Lambda}_A^- = 0.5\{\underline{\Lambda}_A - |\underline{\Lambda}_A|\}$, as in (18.93) for the scalar case.

As a result of the splitting in (18.101), $L_x\underline{A}$ and $L_y\underline{B}$ in (18.100) can be replaced by contributions associated with positive and negative eigenvalues and one-sided difference formulae as in the scalar case. Thus

$$L_x\underline{A} \to L_x^- \, \underline{T}_A \underline{\Lambda}_A^+ \underline{T}_A^{-1} + L_x^+ \, \underline{T}_A \underline{\Lambda}_A^- \underline{T}_A^{-1} \ ,$$

and similarly for $L_y\underline{B}$. Since $\mathbf{F} = \underline{A}\mathbf{q}$ it is also possible to split at the flux vector level (Steger and Warming 1981) so that $\mathbf{F} = \mathbf{F}^+ + \mathbf{F}^-$. This is discussed briefly in Sect. 14.2.5 in relation to computing supersonic flows with shocks. In the present algorithm, due to Obayashi and Kuwahara (1986), the splitting is introduced solely to facilitate the LU factorisation on the implicit terms.

Due to the approximately implicit treatment of the viscous terms it is also possible to split $L_x\mathbf{P}$ and $L_y\mathbf{Q}$ as

$$L_x\underline{P} = L_{xx}\hat{\underline{P}} = (L_x^+ - L_x^-)\hat{\underline{P}}/\Delta x \ ,$$

(18.102)

and similarly for $L_y\underline{Q}$. Consequently, in (18.100)

$$L_x\{\underline{A} - \underline{P}\} \to L_x^- (\underline{\bar{T}}_A \underline{\Lambda}_A^+ \underline{\bar{T}}_A^{-1} + \hat{\underline{P}}/\Delta x) - L_x^+ (\underline{\bar{T}}_A |\underline{\Lambda}_A^-| \underline{\bar{T}}^{-1} + \hat{\underline{P}}/\Delta x) \ ,$$

(18.103)

and similarly for $L_y\{\underline{B} - \underline{Q}\}$. However in the Obayashi and Kuwahara (1986) formulation, $\hat{\underline{P}}/\Delta x$ is replaced by $\underline{\bar{T}}_A k \underline{I} \underline{T}_A^{-1}$, where k is chosen to ensure stability, i.e.

$$k = v/(\text{Re}\varrho\Delta x) \quad \text{and} \quad v = \max(2\mu, \gamma\mu/\text{Pr}) \ .$$

(18.104)

This approximation of $\hat{\underline{P}}$ mimics the scalar form (18.95). With this approximation and (18.103) the approximate factorisation (18.99) can be factorised further into LU form:

$$[\underline{I}+\beta\Delta tL_x^-\,\hat{\underline{A}}^+][\underline{I}-\beta\Delta tL_x^+\,\hat{\underline{A}}^-][\underline{I}+\beta\Delta tL_y^-\,\hat{\underline{B}}^+]$$
$$[\underline{I}-\beta\Delta tL_y^+\,\hat{\underline{B}}^-]\Delta\mathbf{q}^{n+1}=\Delta t\,\mathrm{RHS}^n\;,\tag{18.105}$$

where

$$\hat{\underline{A}}^\pm=\underline{T}_A^{-1}(|\underline{\Delta}_A^\pm|+k\underline{I})\,\underline{T}_A\;.\tag{18.106}$$

Each factor in (18.105) is bidiagonal and can be solved for "$\Delta\mathbf{q}$" with a single sweep in either the positive or negative x or y directions. Thus the complete algorithm to solve (18.105) becomes

$$[\underline{I}+\beta\Delta tL_x^-\,\hat{\underline{A}}^+]\Delta\mathbf{q}^*=\Delta t\,\mathrm{RHS}^n\;,$$
$$[\underline{I}-\beta\Delta tL_x^+\,\hat{\underline{A}}^-]\Delta\mathbf{q}^{**}=\Delta\mathbf{q}^*\;,\tag{18.107}$$
$$[\underline{I}+\beta\Delta tL_y^-\,\hat{\underline{B}}^+]\Delta\mathbf{q}^{***}=\Delta\mathbf{q}^{**}\;,\quad\text{and}$$
$$[\underline{I}-\beta\Delta tL_y^+\,\hat{\underline{B}}^-]\Delta\mathbf{q}^{n+1}=\Delta\mathbf{q}^{***}\;.$$

As an example, the evaluation of the second factor requires a sweep from right to left in the x direction. At each grid point the following 4×4 system of equations is solved for $\Delta\mathbf{q}_{j,k}^{**}$:

$$[\underline{I}+\beta\,\frac{\Delta t}{\Delta x}\,\hat{\underline{A}}_{j,k}^-]\Delta\mathbf{q}_{j,k}^{**}=\Delta\mathbf{q}_{j,k}^*+\beta\,\frac{\Delta t}{\Delta x}\,\hat{\underline{A}}_{j+1,k}^-\Delta\mathbf{q}_{j+1,k}^{**}\;.\tag{18.108}$$

Since $\hat{\underline{A}}$ can be factorised as in (18.106) the algorithm provided by (18.68–70) gives $\Delta\mathbf{q}_{j,k}^{**}$ very efficiently. The solution of (18.108) is repeated for all grid lines in the y direction. The other systems of equations in (18.107) are dealt with in a similar manner to that of (18.108).

Obayashi and Kuwahara (1986) apply the above algorithm to laminar shock boundary layer interaction and Fujii and Obayashi (1986) apply the algorithm to turbulent transonic flow about an aerofoil using a discretisation in generalised coordinates (Chap. 12 and Sect. 18.4).

The present scheme is combined with an effective rotated differencing procedure (Sect. 14.3.3) to generate a streamwise upwind scheme (Goorjian and Obayashi 1991) that is particularly effective in capturing oblique shocks. The method is able to accurately predict steady supersonic vortical viscous flow over a delta wing and unsteady transonic viscous flow over an oscillating rectangular wing.

18.4 Generalised Coordinates

To suit flows around bodies of smooth but otherwise arbitrary shape it is convenient to introduce generalised, body-conforming coordinates (Chap. 12). The

governing equations (18.6) for compressible viscous flow take a form not significantly more complicated than for Cartesian coordinates. For large Reynolds
number flows with only small regions of separation it is computationally expedient
to combine the use of generalised coordinates with the thin layer approximation
(Sect. 18.1.3). An advantage of the thin layer approximation is that it streamlines
considerably the implicit treatment of the viscous terms, particularly when constructing an approximate factorisation algorithm (Sect. 18.4.1).

The use of generalised coordinates leaves open the choice of discretisation in
the computational domain. In Sect. 18.4.2 the group finite element method is used
which introduces the explicit appearance of mass operators. In Sect. 18.4.2 it is
shown how an approximate factorisation algorithm may be constructed which
preserves the mass operator structure.

18.4.1 Steger Thin Layer Formulation

Using the techniques described in Chap. 12 the equations governing two-dimensional compressible viscous flow are expressed in generalised coordinates as

$$\frac{\partial \hat{\mathbf{q}}}{\partial t} + \frac{\partial \hat{\mathbf{F}}^I}{\partial \xi} + \frac{\partial \hat{\mathbf{G}}^I}{\partial \eta} = \frac{\partial \hat{\mathbf{R}}}{\partial \xi} + \frac{\partial \hat{\mathbf{S}}}{\partial \eta} \,, \tag{18.109}$$

where

$$\hat{\mathbf{q}} = J^{-1} \begin{bmatrix} \varrho \\ \varrho u \\ \varrho v \\ E \end{bmatrix}, \quad \hat{\mathbf{F}}^I = J^{-1} \begin{bmatrix} \varrho U^c \\ \varrho u U^c + \xi_x p \\ \varrho v V^c + \xi_y p \\ (E+p)U^c \end{bmatrix},$$

$$\hat{G}^I = J^{-1} \begin{bmatrix} \varrho V^c \\ \varrho u V^c + \eta_x p \\ \varrho v V^c + \eta_y p \\ (E+p)V^c \end{bmatrix}, \tag{18.110}$$

$$\hat{R} = \mathrm{Re}^{-1} J^{-1} \begin{bmatrix} 0 \\ \xi_x \tau_{xx} + \xi_y \tau_{xy} \\ \xi_x \tau_{xy} + \xi_y \tau_{yy} \\ \xi_x R_4 + \xi_y S_4 \end{bmatrix} \quad \text{and} \quad \hat{S} = \mathrm{Re}^{-1} J^{-1} \begin{bmatrix} 0 \\ \eta_x \tau_{xx} + \eta_y \tau_{xy} \\ \eta_x \tau_{xy} + \eta_y \tau_{yy} \\ \eta_x R_4 + \eta_y S_4 \end{bmatrix} .$$

In forming (18.109) it is assumed that $\xi = \xi(x, y)$, $\eta = \eta(x, y)$. The additional
contributions to (18.110) for the more general case $\xi = \xi(x, y, z, t)$, $\eta = \eta(x, y, z, t)$
are provided by Chaussee (1984) for three-dimensional flow. For simplicity of
presentation (18.6–10) are considered in their laminar form, μ_T, $\mathrm{Pr}_T = 0$, in this
section. In addition, the Cartesian velocity components u and v have been nondimensionalised with respect to a_∞, the freestream speed of sound, density with

respect to ϱ_∞ and the total energy with respect to $\varrho_\infty a_\infty^2$. Consequently the Reynolds number $\mathrm{Re} = \varrho_\infty a_\infty L/\mu_\infty$, where L is a characteristic length.

In (18.110) J is the Jacobian and is given by

$$J = \xi_x \eta_y - \xi_y \eta_x \ . \tag{18.111}$$

The various metric coefficients, ξ_x, etc., are evaluated numerically as in Sect. 12.2 once the grid is constructed. The contravariant velocity components U^c and V^c are related to the Cartesian velocity components u and v by

$$U^c = \xi_x u + \xi_y v \ , \qquad V^c = \eta_x u + \eta_y v \ . \tag{18.112}$$

The flux vectors $\hat{\mathbf{F}}^I$ and $\hat{\mathbf{G}}^I$ are linked to the corresponding Cartesian flux vectors $\mathbf{F}^I$ and $\mathbf{G}^I$ (14.95) by

$$\hat{\mathbf{F}}^I = \frac{\xi_x}{J} \mathbf{F}^I + \frac{\xi_y}{J} \mathbf{G}^I \quad \text{and}$$

$$\hat{\mathbf{G}}^I = \frac{\eta_x}{J} \mathbf{F}^I + \frac{\eta_y}{J} \mathbf{G}^I \ . \tag{18.113}$$

The terms in the energy equation, R_4 and S_4, are given by

$$R_4 = u\tau_{xx} + v\tau_{xy} + \frac{k}{(\gamma-1)\mathrm{Pr}} \frac{\partial a^2}{\partial x} \quad \text{and}$$

$$S_4 = u\tau_{xy} + v\tau_{yy} + \frac{k}{(\gamma-1)\mathrm{Pr}} \frac{\partial a^2}{\partial y} \ , \tag{18.114}$$

where the various stresses are obtained from (18.9) with $\mu_T = 0$ and $k^{\mathrm{te}} = 0$.

By handling the viscous terms at the shear stress level the need for direct evaluation of second derivative transformation parameters, e.g. ξ_{xx}, is avoided. It may be recalled (Sect. 12.2.4) that second derivatives are handled less accurately than first derivatives when introducing generalised coordinates. After discretisation the viscous stresses, τ_{xx}, etc., are defined at the grid points and a second transformation and discretisation is required to evaluate the viscous stresses from the velocity field.

For flows at large Reynolds number viscous effects are only significant close to solid surfaces and in the wake regions. Consequently as long as flows with massive separation in the main flow direction are not being considered it is advantageous to introduce the thin layer approximation (Sect. 18.1.3).

By an appropriate grid construction, e.g. a C grid around an isolated aerofoil, it is possible to ensure that the fine grid in one direction, say η, properly resolves the significant viscous terms both adjacent to the solid surface and in the wake. As indicated in Fig. 18.2 a coarse grid is used parallel to the surface (ξ direction). On such a coarse grid ξ derivatives associated with viscous terms cannot be accurately resolved. Consequently all ξ derivatives arising from the $\hat{R}$ and $\hat{S}$ terms in (18.109) are dropped. It is clear that the thin layer approximation is applied in the computational domain rather than in the physical domain.

The thin layer approximation replaces (18.109) with

$$\frac{\partial \hat{\mathbf{q}}}{\partial t} + \frac{\partial \hat{\mathbf{F}}^I}{\partial \xi} + \frac{\partial \hat{\mathbf{G}}^I}{\partial \eta} = \frac{\partial \hat{\mathbf{S}}}{\partial \eta} \ . \tag{18.115}$$

After substituting for τ_{xx}, etc., transforming to (ξ, η) coordinates and deleting ξ derivatives, $\hat{\mathbf{S}}$ takes the form

$$\hat{\mathbf{S}} = \mathrm{Re}^{-1} J^{-1} \begin{bmatrix} 0 \\ \mu(\eta_x^2 + \eta_y^2)u_\eta + (\mu/3)\eta_x(\eta_x u_\eta + \eta_y v_\eta) \\ \mu(\eta_x^2 + \eta_y^2)v_\eta + (\mu/3)\eta_y(\eta_x u_\eta + \eta_y v_\eta) \\ (\eta_x^2 + \eta_y^2)\{k/[(\gamma - 1)\mathrm{Pr}](a^2)_\eta + 0.5\mu(u^2 + v^2)_\eta\} \\ + (\mu/6)[\eta_x^2(u^2)_\eta + \eta_y^2(v^2)_\eta + 2\eta_x\eta_y(uv)_\eta] \end{bmatrix} . \tag{18.116}$$

An approximate factorisation algorithm, similar to the Beam and Warming scheme (Sect. 18.3.2), is constructed from (18.115). The result is

$$(\underline{I} + \beta \Delta t L_\xi \hat{\underline{A}} - \varepsilon_I J^{-1} L_{\xi\xi} J)(\underline{I} + \beta \Delta t L_\eta \hat{\underline{B}} - \beta \Delta t L_\eta J^{-1} \hat{\underline{M}} J - \varepsilon_I J^{-1} L_{\eta\eta}) \Delta \hat{\mathbf{q}}^{n+1}$$
$$= -\Delta t [L_\xi \hat{\mathbf{F}}^I + L_\eta \hat{\mathbf{G}}^I - L_\eta \hat{\mathbf{S}}]^n - \varepsilon_E J^{-1}[(\nabla_\xi \Delta_\xi)^2 + (\nabla_\eta \Delta_\eta)^2] J \hat{\mathbf{q}}^n \ . \tag{18.117}$$

Operators L_ξ and $L_{\xi\xi}$ are second-order central difference operators, as in (18.145) with $r_\xi = 1$. The Jacobians $\hat{\underline{A}} = \partial \hat{\mathbf{F}}^I / \partial \hat{\mathbf{q}}$ and $\hat{\underline{B}} = \partial \hat{\mathbf{G}}^I / \partial \hat{\mathbf{q}}$ are constructed from the Jacobians of the Cartesian flux vectors $\underline{A} = \partial \mathbf{F}^I / \partial \mathbf{q}$ and $\underline{B} = \partial \mathbf{G}^I / \partial \mathbf{q}$ by making use of (18.113). Thus

$$\hat{\underline{A}} = \frac{\xi_x}{J} \underline{A} + \frac{\xi_y}{J} \underline{B} \quad \text{and}$$

$$\hat{\underline{B}} = \frac{\eta_x}{J} \underline{A} + \frac{\eta_y}{J} \underline{B} \ . \tag{18.118}$$

The Cartesian Jacobians $\underline{A}$ and $\underline{B}$ are expressed in terms of the dependent variables by (14.99). The Jacobian $\hat{\underline{M}} = \partial \hat{\mathbf{S}} / \partial \hat{\mathbf{q}}$ arises in the linearisation of $\hat{\mathbf{S}}^{n+1}$ about $\hat{\mathbf{S}}^n$, i.e. as in (18.74). The Jacobian $\hat{\underline{M}}$ has the following elements:

$$\hat{\underline{M}} = \begin{bmatrix} 0 & 0 & 0 & 0 \\ m_{21} & \alpha_1 \partial \varrho^{-1}/\partial \eta & \alpha_2 \partial \varrho^{-1}/\partial \eta & 0 \\ m_{31} & \alpha_2 \partial \varrho^{-1}/\partial \eta & \alpha_3 \partial \varrho^{-1}/\partial \eta & 0 \\ m_{41} & m_{42} & m_{43} & m_{44} \end{bmatrix} , \tag{18.119}$$

where

$$m_{21} = -\alpha_1 \frac{\partial}{\partial \eta}(u/\varrho) - \alpha_2 \frac{\partial}{\partial \eta}(v/\varrho) \; ,$$

$$m_{31} = -\alpha_2 \frac{\partial}{\partial \eta}(u/\varrho) - \alpha_3 \frac{\partial}{\partial \eta}(v/\varrho) \; , \tag{18.120}$$

$$m_{41} = -\alpha_4 \frac{\partial}{\partial \eta}[-(E/\varrho^2) + (u^2 + v^2)/\varrho] - \alpha_1 \frac{\partial}{\partial \eta}(u^2/\varrho)$$

$$- 2\alpha_2 \frac{\partial}{\partial \eta}(uc/\varrho) - \alpha_3 \frac{\partial}{\partial \eta}(v^2/\varrho) \; ,$$

$$m_{42} = -\alpha_4 \frac{\partial}{\partial \eta}(u/\varrho) - m_{21} \; , \qquad m_{43} = -\alpha_4 \frac{\partial}{\partial \eta}(v/\varrho) - m_{31} \; ,$$

$$m_{44} = \alpha_4 \frac{\partial \varrho^{-1}}{\partial \eta} \; , \quad \text{and}$$

$$\alpha_1 = \mu \left(\frac{4}{3}\eta_x^2 + \eta_y^2\right) \; , \qquad \alpha_2 = \frac{\mu}{3}\eta_x\eta_y \; ,$$

$$\alpha_3 = \mu \left(\eta_x^2 + \frac{4}{3}\eta_y^2\right) \; , \qquad \alpha_4 = \frac{\gamma k}{\text{Pr}}(\eta_x^2 + \eta_y^2) \; . \tag{18.121}$$

Equation (18.121) is based on the assumption of laminar flow; the equivalent eddy viscosity form follows directly from (18.9, 10). In forming $\hat{M}$ the dependence of μ and k on the solution q is ignored. To include these terms would substantially increase the operation count of evaluating (18.117) for $\Delta\hat{q}^{n+1}$ without increasing the accuracy significantly for unsteady flows and not at all for steady flows.

Included in (18.117) are fourth-order explicit numerical dissipation terms $(\nabla_\xi \Delta_\xi)^2$ and second-order implicit terms $L_{\xi\xi}$, $L_{\eta\eta}$. The explicit dissipation terms are introduced to damp high-frequency waves arising from the nonlinear terms on a finite grid, i.e. aliasing, at large Reynolds number (Sect. 18.5.1). For small Reynolds number there is enough physical dissipation to control nonlinear instability. The algebraic form of $(\nabla_\xi \Delta_\xi)^2$ is given after (17.51).

If only the explicit dissipation term is included, the overall scheme is conditionally stable if $\varepsilon_E > 1/16$. This restriction can be removed completely by introducing fourth-order implicit dissipation terms on the left-hand side of (18.117) as well. However this would produce a pentadiagonal system of equations which is more costly to solve than the tridiagonal system arising from the use of three-point operators. However, since the addition of numerical dissipation terms should be as small as possible it is convenient to add second-order implicit dissipation terms to the left-hand side of (18.117) with $\varepsilon_I = 2\varepsilon_E$ and $\varepsilon_E = \Delta t$. At the body surface ε_I and ε_E are set to zero. Adjacent to boundaries the explicit fourth-order operator is replaced by the second-order Laplacian operator.

Equation (18.117) is solved in two stages, as with the Beam and Warming algorithm (18.78, 79),

$$(\underline{I}+\beta\Delta t L_\xi\underline{\hat{A}}-\varepsilon_I J^{-1}L_{\zeta\zeta}J)\Delta\hat{\mathbf{q}}^*_{j,k}=\Delta\hat{q}^e_{j,k} \tag{18.122}$$

and

$$(\underline{I}+\beta\Delta t L_\eta\underline{\hat{B}}-\beta\Delta t L_\eta J^{-1}\underline{\hat{M}}J-\varepsilon_I J^{-1}L_{\eta\eta})\Delta\hat{q}^{n+1}_{j,k}=\Delta\hat{q}^*_{j,k} \tag{18.123}$$

where

$$\Delta\hat{\mathbf{q}}^e_{j,k}=-\Delta t(L_\zeta\hat{\mathbf{F}}^I+L_\eta\hat{\mathbf{G}}^I-L_\eta\hat{\mathbf{S}})^n-\varepsilon_E J^{-1}[(\nabla_\xi\Delta_\xi)^2+(\nabla_\eta\Delta_\eta)^2]J\hat{\mathbf{q}}^n \ . \tag{18.124}$$

Equations (18.122) are 4×4 block tridiagonal systems of equations associated with grid lines in the ξ and η directions, respectively. The choices $\beta=0.5$ and 1.0 produce algorithms that are second and first-order accurate in time, respectively. When the pseudotransient construction (Sect. 6.4) is used to obtain the steady state solution the choice $\beta=1.0$ is more robust and more efficient.

The application of (18.122 and 123) with generalised coordinates is not significantly less efficient than the use of the Beam and Warming algorithm (18.78, 79) in Cartesian coordinates. This is primarily because the thin layer approximation and the assumption in forming $\hat{M}$ that μ and k do not depend on $\mathbf{q}$ lead to a significantly simpler treatment of the viscous terms.

In the early uses of this algorithm (Steger 1978, Pulliam and Steger 1980), the boundary conditions were implemented explicitly, i.e. $\hat{\mathbf{q}}^{n+1}$ is set equal to $\hat{\mathbf{q}}^n$ on the boundaries. This introduces a first-order error in time which is of no consequence in obtaining steady-state solutions. However, for unsteady problems or if the present algorithm is used in a space-marching context (Sect. 16.3.1) it is desirable to implement the boundary conditions implicitly so that second-order time accuracy is maintained.

At the body surface the no-slip condition provides two boundary conditions

$$u=v=0 \ , \tag{18.125}$$

and specified temperature or an adiabatic wall provides a third,

$$T=T_{\text{wall}} \quad \text{or} \quad \partial T/\partial n=0 \ , \tag{18.126}$$

where n is the direction normal to the body surface. For a grid which is locally orthogonal to the surface, as recommended in Chap. 13, the generalised coordinate direction η (Fig. 15.24) will coincide with the normal direction n.

For flows at high Reynolds number, the thin layer assumption implies

$$\frac{\partial p}{\partial n}=\frac{\partial p}{\partial \eta}=0 \ , \tag{18.127}$$

as long as the surface curvature is not too great. Combining (18.125, 127) and (18.8) implies that at the surface

$$\frac{\partial E}{\partial n} = \frac{\partial E}{\partial \eta} = 0 \ . \tag{18.128}$$

If this is implemented with a one-sided, first-order accurate finite difference expression, then

$$\frac{E_{j,2} - E_{j,1}}{\Delta \eta} = 0 \quad \text{or} \quad E_{j,1} = E_{j,2} \ , \tag{18.129}$$

where grid point $(j, 1)$ lies on the surface.

For an adiabatic wall, $\partial T / \partial \eta = 0$ can be combined with the ideal gas law $p = \varrho R T$ and (18.127) to give

$$\partial \varrho / \partial \eta = 0 \ , \tag{18.130}$$

which is implemented as

$$\varrho_{j,1} = \varrho_{j,2} \ . \tag{18.131}$$

If the wall temperature is specified, then the ideal gas law gives

$$\varrho_{j,1} = \frac{p_{j,1}}{R T_{\text{wall}}} \ . \tag{18.132}$$

The above boundary conditions need to be combined with the approximate factorisation algorithm (18.122–124). The first stage, (18.122), can be implemented without difficulty for $k = 1$ (i.e. along the body surface) as long as one-sided difference formulae are used to evaluate η derivatives in (18.124). For the second-stage the block equation formed at the wall can be written

$$\underline{B} \Delta \hat{\mathbf{q}}_{j,1}^{n+1} + \underline{C} \Delta \hat{\mathbf{q}}_{j,2}^{n+1} = \mathbf{D}^* \ , \quad \text{where} \tag{18.133}$$

$$\underline{B} = \begin{bmatrix} -J_1/J_2 & 0 & 0 & 0 \\ 0 & 1 & 0 & 0 \\ 0 & 0 & 1 & 0 \\ 0 & 0 & 0 & -J_1/J_2 \end{bmatrix} \ ,$$

$$\underline{C} = \begin{bmatrix} 1 & 0 & 0 & 0 \\ 0 & 0 & 0 & 0 \\ 0 & 0 & 0 & 0 \\ 0 & 0 & 0 & 1 \end{bmatrix} \quad \text{and} \quad \mathbf{D}^* = \{0,0,0,0\}^T \ . \tag{18.134}$$

The form of $\underline{B}$, $\underline{C}$ and D^* is appropriate to the adiabatic wall condition. For a specified wall temperature D^* is the same but $\underline{B}$ and $\underline{C}$ become

$$
\underline{B} = \begin{bmatrix} 1 & 0 & 0 & -1/RT_{\text{wall}} \\ 0 & 1 & 0 & 0 \\ 0 & 0 & 1 & 0 \\ 0 & 0 & 0 & -J_1/J_2 \end{bmatrix}, \quad \underline{C} = \begin{bmatrix} 0 & 0 & 0 & 0 \\ 0 & 0 & 0 & 0 \\ 0 & 0 & 0 & 0 \\ 0 & 0 & 0 & 1 \end{bmatrix}. \tag{18.135}
$$

Equation (18.133) is combined with the interior grid point equations to form tridiagonal systems on each η grid line (different j value).

In the farfield the flow is essentially inviscid and boundary conditions are constructed based on characteristic theory (Sect. 14.2.8). These farfield boundary conditions can also be embedded implicitly in the approximate factorisation algorithm (Rai and Chaussee 1984).

The three-dimensional version of the above algorithm has been used to obtain the flow behaviour about a hemispherical cylinder at 19° incidence for $M_\infty = 1.2$

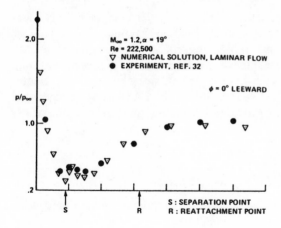

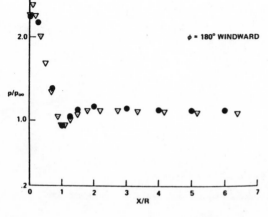

Fig. 18.7. Pressure distribution for hemispherical cylinder at M_∞ and $\alpha = 19°$ (after Pulliam and Steger, 1980; reprinted with permission of AIAA)

and Re = 222 500 based on cylinder diameter. The pressure variation along the length of the hemisphere cylinder is shown in Fig. 18.7.

These results were obtained on an exponentially-stretched grid with 48 points along the cylinder axis, 20 radial points and 12 circumferential points. Symmetry about a vertical plane through the hemispherical cylinder is assumed. The spherical grid has 30 axial and radial points and 12–18 circumferential points. This is a relatively coarse grid and to compensate for this the solutions shown in Fig. 18.7 were obtained with five-point fourth-order differencing of the convective terms on the right-hand side of (18.124). With $\beta = 1$ the algorithm is still unconditionally stable, with $\beta = 0.5$ it is unconditionally unstable.

The numerical solutions (Pulliam and Steger 1980) show good agreement with the experimental data of Hsieh (1976) except that the numerical results show a slightly more downstream separation (S in Fig. 18.7) over the hemispherical nose along the leeward line of symmetry. The point R denotes the predicted reattachment of the streamwise flow. The algorithm is also successful in predicting the complicated cross flow separation patterns (not shown). This example indicates the power of the present method and the effectiveness of the thin layer assumption even if streamwise separation occurs.

The above approximate factorisation of the thin layer compressible Navier–Stokes equations is also used by Thomas et al. (1990) to examine subsonic laminar flow past low aspect-ratio delta wings at high angle of attack. However Thomas et al. use upwind-biased spatial differencing based on the van Leer flux-vector splitting strategy (Sect. 14.2.5) after discretising with a finite volume technique that evaluates the metrics geometrically (Sect. 12.2.3). In addition the FAS multigrid algorithm (Sect. 6.3.5) with a four-level W cycle provides accelerated convergence to the steady state.

18.4.2 Approximate Factorisation Finite Element Method

For transonic viscous flow it is possible to accurately represent the flow behaviour by (18.80–82). In generalised coordinates (18.80) becomes

$$\frac{\partial \hat{\mathbf{q}}}{\partial t} + \frac{\partial \hat{\mathbf{F}}}{\partial \xi} + \frac{\partial \hat{\mathbf{G}}}{\partial \eta} - \frac{\partial^2 \hat{\mathbf{R}}}{\partial \xi^2} - \frac{\partial^2 \hat{\mathbf{S}}}{\partial \xi \partial \eta} - \frac{\partial^2 \hat{\mathbf{T}}}{\partial \eta^2} = 0 \ , \tag{18.136}$$

where

$$\hat{\mathbf{q}} = \left\{ \frac{\varrho}{J}, \frac{\varrho u}{J}, \frac{\varrho v}{J} \right\}^T \ , \tag{18.137}$$

$$\hat{\mathbf{F}} = \frac{1}{J} \begin{bmatrix} \varrho U_c \\ \xi_x p + \varrho u U_c + (4\xi_{xx}/3 + \xi_{yy}) u \mu_e + (\xi_{xy}/3) v \mu_e \\ \xi_y p + \varrho v U_c + (\xi_{xx} + 4\xi_{yy}/3) v \mu_e + (\xi_{xy}/3) u \mu_e \end{bmatrix} \ , \tag{18.138}$$

$$\hat{\mathbf{G}} = \frac{1}{J} \begin{bmatrix} \varrho V_c \\ \eta_x p + \varrho u V_c + (4\eta_{xx}/3 + \eta_{yy})u\mu_e + (\eta_{xy}/3)v\mu_e \\ \eta_y p + \varrho v V_c + (\eta_{xx} + 4\eta_{yy}/3)v\mu_e + (\eta_{xy}/3)u\mu_e \end{bmatrix} , \tag{18.139}$$

$$\hat{\mathbf{R}} = \frac{1}{J} \begin{bmatrix} \theta(\xi_x^2 + \xi_y^2)\varrho/\mathrm{Re} \\ (4\xi_x^2/3 + \xi_y^2)u\mu_e + \xi_x\xi_y(v/3)\mu_e \\ (\xi_x^2 + 4\xi_y^2/3)v\mu_e + \xi_x\xi_y(u/3)\mu_e \end{bmatrix} , \tag{18.140}$$

$$\hat{\mathbf{S}} = \frac{1}{J} \begin{bmatrix} 2\theta(\xi_x\eta_x + \xi_y\eta_y)\varrho/\mathrm{Re} \\ 2(4\xi_x\eta_x/3 + \xi_y\eta_y)u\mu_e + (\xi_x\eta_y + \xi_y\eta_x)(v/3)\mu_e \\ 2(\xi_x\eta_x + 4\xi_y\eta_y/3)v\mu_e + (\xi_y\eta_x + \eta_y\xi_x)(u/3)\mu_e \end{bmatrix} , \tag{18.141}$$

$$\hat{\mathbf{T}} = \frac{1}{J} \begin{bmatrix} \theta(\eta_x^2 + \eta_y^2)\varrho/\mathrm{Re} \\ (4\eta_x^2/3 + \eta_y^2)u\mu_e + \eta_x\eta_y(v/3)\mu_e \\ (\eta_x^2 + 4\eta_y^2/3)v\mu_e + \eta_x\eta_y(u/3)\mu_e \end{bmatrix} . \tag{18.142}$$

It may be noted that (18.80) corresponds to (12.62) and (18.136) corresponds to (12.69).

As was the case for Cartesian coordinates (18.80), it is possible to discretise (18.136) without regard for the detailed form of $\hat{\mathbf{F}}$, etc. Thus, using the group formulation, as in (18.83), with bilinear interpolation in rectangular elements,

$$\hat{\mathbf{F}} = \sum_{m=1}^{4} N_m(x, y)\hat{\mathbf{F}}_m . \tag{18.143}$$

Applying the Galerkin finite element method (Chap. 5) to (18.136), after substitution of expressions like (18.143), produces the result

$$M_\xi \otimes M_\eta \partial\hat{\mathbf{q}}/\partial t + M_\eta \otimes L_\xi\hat{\mathbf{F}} + M_\xi \otimes L_\eta\hat{\mathbf{F}}$$
$$- M_\eta \otimes L_{\xi\xi}\hat{\mathbf{R}} - L_\xi \otimes L_\eta\hat{\mathbf{S}} - M_\xi \otimes L_{\eta\eta}\hat{\mathbf{T}} = 0 , \tag{18.144}$$

which is structurally equivalent to (18.84). The directional mass and difference operators appearing in (18.144) are defined as

$$M_\xi \equiv A\left[\frac{1}{6}, \frac{1+r_\xi}{3}, \frac{r_\xi}{3}\right] , \qquad M_\eta \equiv B\left[\frac{r_\eta}{6}, \frac{1+r_\eta}{3}, \frac{1}{6}\right]^T$$

$$L_\xi \equiv \frac{0.5A}{\Delta\xi}[-1, 0, 1], \qquad L_\eta \equiv \frac{0.5B}{\Delta\eta}[1, 0, -1]^T \tag{18.145}$$

$$L_{\xi\xi} \equiv \frac{A}{\Delta\xi^2}\left[1, -\left(1+\frac{1}{r_\xi}\right), \frac{1}{r_\xi}\right], \qquad L_{\eta\eta} \equiv \frac{B}{\Delta\eta^2}\left[\frac{1}{r_\eta}, -\left(1+\frac{1}{r_\eta}\right), 1\right]^T$$

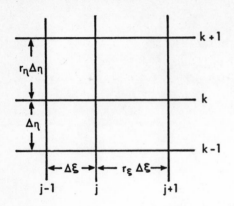

Fig. 18.8. Grid in (ξ, η) space

where

$$A = 2/(1 + r_\xi) \quad \text{and} \quad B = 2/(1 + r_\eta).$$

The operator definitions shown in (18.145) correspond to a nonuniform grid in (ξ, η) space (Fig. 18.8). This tends to produce less extreme values of the transformation parameters ξ_x, etc., than if a uniform rectangular grid $(r_\xi = r_\eta = 1)$ is used in (ξ, η) space. The equivalent two-stage block-tridiagonal algorithm to (18.85, 86) becomes

$$\left\{ M_\xi - \frac{\beta}{1+\gamma} \Delta t \left[L_{\xi\xi} \left(\frac{\partial \hat{\mathbf{R}}}{\partial \hat{\mathbf{q}}} \right) - L_\xi \left(\frac{\partial \hat{\mathbf{F}}}{\partial \hat{\mathbf{q}}} \right) \right] \right\} \Delta \hat{\mathbf{q}}^*$$

$$= \frac{\Delta t}{1+\gamma} (\text{RHS})^a + \frac{\gamma}{1+\gamma} M_\xi \otimes M_\eta \Delta \hat{\mathbf{q}}^n \; , \tag{18.146}$$

and

$$\left\{ M_\eta - \frac{\beta}{1+\gamma} \Delta t \left[L_{\eta\eta} \left(\frac{\partial \hat{\mathbf{T}}}{\partial \hat{\mathbf{q}}} \right) - L_\eta \left(\frac{\partial \hat{\mathbf{G}}}{\partial \hat{\mathbf{q}}} \right) \right] \right\} \Delta \hat{\mathbf{q}}^{n+1} = \Delta \hat{\mathbf{q}}^* \; , \tag{18.147}$$

where β and γ are parameters controlling the time-marching algorithm (8.26). The choice $\beta = 1.0, \gamma = 0.5$ gives the three-level fully implicit algorithm which is second-order accurate in time. The augmented right-hand side $(\text{RHS})^a$ includes the extra contribution associated with the explicit treatment of the cross-derivative term, i.e.

$$(\text{RHS})^a = (\text{RHS})^n + \beta L_\xi \otimes L_\eta \left(\frac{\partial \hat{\mathbf{S}}}{\partial \hat{\mathbf{q}}} \right) \Delta \hat{\mathbf{q}}^n \; , \tag{18.148}$$

and

$$(\text{RHS})^n = M_\eta \otimes L_{\xi\xi} \hat{\mathbf{R}} - L_\xi \otimes L_\eta \hat{\mathbf{S}} - M_\xi \otimes L_{\eta\eta} \hat{\mathbf{T}} - M_\eta \otimes L_\xi \hat{\mathbf{F}} - M_\xi \otimes L_\eta \hat{\mathbf{G}} \; . \tag{18.149}$$

Equations (18.146, 147) produce 3×3 block tridiagonal systems of equations associated with each ξ and η gridline respectively. These are solved efficiently using the algorithm described in Sect. 6.2.5.

The above algorithm is described in more detail by Srinivas and Fletcher (1985) and in a broader context by Fletcher and Srinivas (1985).

For the flow past an asymmetric trailing edge (Fig. 18.9) the use of generalised coordinates produces a computational domain in which the finite thickness trailing edge becomes a zero-thickness flat plate, i.e. D coincides with B in the computational domain.

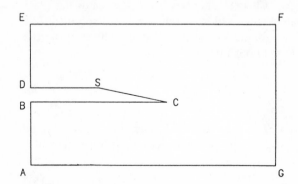

Fig. 18.9. Asymmetric trailing edge geometry; flow left to right

Solutions have been obtained (Srinivas and Fletcher 1986) using the above algorithm for the flow conditions $M_\infty = 0.4$ and $Re = 26 \times 10^6$. Turbulence effects are represented by an algebraic eddy viscosity model (18.12–16). The above algorithm has been used in a pseudotransient sense to obtain steady-state solutions. For the velocity distributions adjacent to and downstream of the trailing

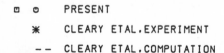

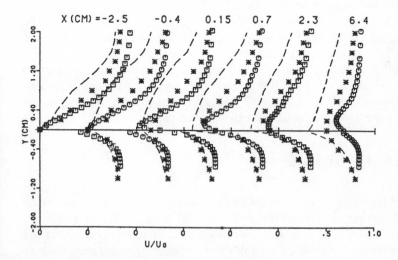

Fig. 18.10. Velocity distribution behind the trailing edge

edge (Fig. 18.10) convergence is assumed when the r.m.s. value of $(RHS)^4$ in (18.148) is less than 10^{-3}. For a $41(x, \xi)$ by $82(y, \eta)$ grid this requires approximately 1000 time-steps.

At the solid surfaces $DSCB$ both velocity components are set to zero. Adjacent to the solid surfaces, wall functions, Sect. 18.1.1, are used to specify the normal variation of the tangential velocity component. At the inlet, AB and ED in Fig. 18.9, $u = 1.0$ and $v = 0$ far from the body. Close to the body a $1/7$ power law flat plate boundary layer profile appropriate to the local Reynolds number is specified. The pressure on the inflow boundary is obtained from the interior solution using the discrete form of the characteristic relationship

$$\frac{\partial p}{\partial x} - \varrho a \frac{\partial u}{\partial x} = 0 \ . \tag{18.150}$$

On freestream boundaries, AG and EF, $u = 1.0$ and $v = 0$. The pressure is obtained from the interior solution via the discrete form of the characteristic relationship

$$\frac{\partial p}{\partial y} - \varrho a \frac{\partial v}{\partial y} = 0 \ . \tag{18.151}$$

At the downstream boundary, FG in Fig. 18.9,

$$\frac{\partial u}{\partial x} = 0 \ , \quad \frac{\partial v}{\partial x} = 0 \quad \text{and}$$

$$\frac{\partial p}{\partial t} - \varrho a \frac{\partial u}{\partial t} + 0.3(p - p_\infty) = 0 \ . \tag{18.152}$$

Equation (18.152) is a non-reflecting boundary condition that helps to accelerate convergence to the steady-state (Sect. 14.2.8).

In Fig. 18.10, x is measured from the trailing edge (C in Fig. 18.9). The computational results obtained on a $41(x)$ by $82(y)$ grid, are seen to be in good agreement with the experimental results of Cleary et al. (1980). The computational results of Cleary et al. are based on a finite difference solution using a two-equation turbulence model on a $60(x)$ by $100(y)$ grid. However a later finite difference solution (Horstman 1983) on a 79 by 82 grid with a modified turbulence model does achieve better agreement with the experimental results.

18.5 Numerical Dissipation

In Sect. 9.2 the truncation error, introduced in replacing the continuous governing equation with a discrete equivalent, is interpreted as a source of numerical dissipation and dispersion. It is demonstrated that large amounts of either are undesirable, particularly numerical dissipation if it is of comparable magnitude or larger than the physical dissipation.

However, it is also noted in Sect. 9.2 that dissipation, numerical or physical, does attenuate small wavelengths very effectively with a much smaller impact on large wavelengths. This last feature can be exploited to put numerical dissipation to good use in computing high Reynolds number flows or flows with shock waves.

In analysing flow behaviour it is often useful to carry out a Fourier analysis of the physical variables, e.g. velocity components or pressure, to determine the amplitude associated with different wave-numbers. A Fourier analysis can be made over time or space as appropriate. The wave number can be interpreted as an inverse time or length scale.

The instantaneous velocity distribution through a shock wave, e.g. obtained by solving the one-dimensional Euler equations, is depicted in Fig. 18.11a. A Fourier analysis of $u(x)$ can be written

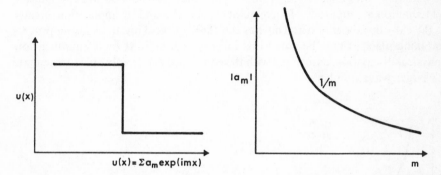

Fig. 18.11a, b. Spectral analysis of a discontinuous function. (a) Velocity profile; (b) spectral analysis

$$u(x) = \sum_{m=-N}^{N} a_m \exp(imx) \; , \tag{18.153}$$

where m is the wave number (Sect. 9.2) and a_m is the amplitude of the mth component. The corresponding spectral analysis, i.e. the variation of amplitude with wave number, is shown in Fig. 18.11b. Clearly, large amplitudes are associated with small wave numbers, i.e. large length scales. Small but finite amplitudes are associated with large wave numbers, i.e. small length scales. It can be seen that in representing a discontinuous velocity distribution as produced by a shock wave, all scales are present.

Solutions obtained on a finite grid can only support a finite number of wave numbers in the discrete Fourier series (18.153). A grid of NX equally spaced points will support wave numbers up to $N = (NX - 1)/2$ if the solution is represented by (18.153). Clearly, a shock wave "contains" wave numbers greater than N with finite amplitude.

It may be recalled (Sect. 10.1.1) that it is the nonlinear convective terms that lead to steepening of the profile and in the absence of dissipation to the presence of discontinuous profiles or "shocks". However, when the u solution is interpreted as a Fourier series, the convective term $u\partial u/\partial x$ is seen to introduce a product

$$u\frac{\partial u}{\partial x} \to i\sum_m a_m e^{imx} \sum_l la_l e^{ilx} \ ,$$

which implies the appearance of wave numbers $m-l$ and $m+l$. Thus a discrete unsteady solution, containing wave numbers up to $m=N$, constantly generates larger wave numbers at each time step. The amplitude of wave numbers greater than N add, spuriously, to wave numbers lower than N. Reconstruction of the solution indicates that an aliasing error is being introduced. If unchecked this aliasing error will cause nonlinear instability.

This process is shown schematically in Fig. 18.12. Wave numbers greater than N (coinciding with the cut-off wave number $m=\pi/\Delta x$) are referred to as subgrid wave numbers. Numerical dissipation is deliberately introduced so as to cause a rapid reduction in amplitude of wave numbers close to, and by implication, greater than the cut-off wave number. This has the effect of blocking the aliasing process from taking place. It may be noted that for a typical high Reynolds number flow the physical dissipation, associated with the eddy viscosity, is only able to attenuate much larger wave numbers.

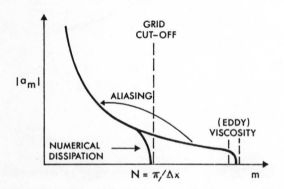

Fig. 18.12. Numerical dissipation of subgrid amplitudes

For flows with shocks the numerical dissipation is applied locally, i.e. at the shock. This can be introduced explicitly as in Sect. 14.2.7. However, the flux-limiting schemes described in Sects. 14.2.6 and 18.5.2 can also be interpreted as a conventional (central) differencing plus numerical dissipation. For high Reynolds number flow, away from shocks, the level of numerical dissipation required to control aliasing is much smaller than for shocks, primarily because the amplitudes in the equivalent of Fig. 18.11 are much smaller. In regions where physical dissipation is significant, i.e. in shear layers, it is necessary to ensure that the numerical dissipation is much less than the physical dissipation. In essentially inviscid regions the numerical dissipation should be sufficiently small so that the balance between other terms in the governing equation is not changed, i.e. that the solution is unaffected.

18.5.1 High Reynolds Number Flows

For flows at high Reynolds numbers past a stationary body, viscous and turbulence effects only influence the solution close to the body. Turbulence, particularly through eddy viscosity turbulence models (Sect. 18.1.1), provides a direct physical dissipation mechanism in the momentum equations. Laminar viscous effects are negligible by comparison. To resolve the severe normal velocity gradients a fine grid is required and the magnitude of the physical dissipation is sufficient to control the aliasing indicated in Fig. 18.12.

However, away from the body the eddy viscosity, and hence the physical dissipation, is negligible. In this region the solution is determined by the balance between the convection and pressure gradient terms. To control aliasing and nonlinear instability it is necessary to add numerical dissipation.

It may be recalled from Sect. 17.1.1 that, with symmetric differencing of the pressure gradient term and use of the same grid-point locations for all dependent variables, there is a tendency for the pressure solution to be oscillatory. Since there is no term in the compressible Navier–Stokes equations which provides direct pressure dissipation, oscillatory pressure solutions are also to be expected away from the body if the physical dissipation, which acts through the velocity field, is insufficient.

MacCormack and Baldwin (1975) endeavour to control potentially oscillatory pressure and velocity fields by adding a dissipative term of the following form to the right-hand side of (18.6):

$$\varepsilon_E(\Delta x)^4 \frac{\partial}{\partial x}\left(\frac{(|u|+a)}{4p}\left|\frac{\partial^2 p}{\partial x^2}\right|\frac{\partial \mathbf{q}}{\partial x}\right) . \tag{18.154}$$

This term introduces a fourth-order error and can be interpreted as a smoothing term on $\mathbf{q}$ whose coefficient is proportional to $|\partial^2 p/\partial x^2|$. If oscillations in p develop, the smoothing on $\mathbf{q}$ is increased. Since p is coupled with the $\mathbf{q}$ solution through (18.8) the p field will also be smoothed. For stability it is necessary that $0 \leq \varepsilon_E \leq 0.5$ when used with the MacCormack explicit scheme, Sect. 18.2.1. MacCormack and Baldwin add (18.154) to the P_x operators in the split scheme (18.44) used to study shock boundary layer interactions.

A more widely used form of numerical dissipation is to add

$$-\varepsilon_E\left((\Delta x)^4\frac{\partial^4 \mathbf{q}}{\partial x^4}+(\Delta y)^4\frac{\partial^4 \mathbf{q}}{\partial y^4}\right) \tag{18.155}$$

to the right-hand side of (18.6). On a uniform grid $(\Delta x^4)\partial^4\mathbf{q}/\partial x^4$ is discretised as

$$(\Delta x^4)\frac{\partial^4 \mathbf{q}}{\partial x^4} \approx \mathbf{q}_{j-2,k}-4\mathbf{q}_{j-1,k}+6\mathbf{q}_{j,k}-4\mathbf{q}_{j+1,k}+q_{j+2,k} , \tag{18.156}$$

and similarly for $(\Delta y^4)\partial^4\mathbf{q}/\partial y^4$. In generalised coordinates this type of numerical dissipation is included in the Steger approximate factorisation implicit scheme

(18.122–124) as

$$-\varepsilon_E J^{-1}[(\nabla_\zeta \Delta_\zeta)^2 + (\nabla_\eta \Delta_\eta)^2] J\hat{\mathbf{q}} \;, \tag{18.157}$$

on the right-hand side of (18.117). It may be recalled from Sect. 9.1.2 that a fourth derivative term on the right-hand side of (18.6) introduces positive dissipation if its coefficient is negative.

The fourth-order dissipation (18.155) is also used in the artificial compressibility scheme (17.51) for incompressible flow, in the spatial marching scheme (16.145) for supersonic viscous flow and in obtaining steady solutions of the Euler equations, Sect. 14.2.8.

18.5.2 Shock Waves

If shock waves are present in the computational domain the "background" numerical dissipation described in Sect. 18.5.1 is insufficient to prevent dispersive oscillations close to the shock wave, in the essentially inviscid parts of the domain. Close to a solid surface the physical dissipative mechanisms reduce the severe gradients associated with the shock and additional numerical dissipation is not required, unless the shock is very strong.

The explicit addition of numerical dissipation (artificial viscosity) to smooth shock profiles is described in Sect. 14.2.3. This can cause a considerable spreading out of the shock profile (Fig. 14.18). FCT algorithms and TVD schemes (Sect. 14.2.6) can be interpreted as adding numerical dissipation until shock profiles are monotonic and then selectively removing it to sharpen the shock profile. In this section a typical TVD scheme will be examined from the perspective of a three-point central differencing (second-order) with the addition of numerical dissipation. This type of interpretation lends itself to the construction of more efficient implicit algorithms for computing steady compressible turbulent flows when shocks are present.

A nonlinear scalar conservation law is written as

$$\frac{\partial p}{\partial t} + \frac{\partial f(\varrho)}{\partial x} = 0 \;, \tag{18.158}$$

and $u(\varrho) = \partial f / \partial \varrho$.

The TVD scheme (14.88, 89) can be written

$$\varrho^{n+1} = \varrho^n - \frac{\Delta t}{\Delta x}(\tilde{f}^n_{j+1/2} - \tilde{f}^n_{j-1/2}) \;, \quad \text{where} \tag{18.159}$$

$$\tilde{f}_{j+1/2} = 0.5(f_j + f_{j+1}) - 0.5[\phi(r)C]_{j+1/2}\Delta f_{j+1/2}$$
$$- 0.5\sigma[1 - \phi(r)]_{j+1/2}\Delta f_{j+1/2} \;, \tag{18.160}$$

$C_{j+1/2} = u_{j+1/2}\Delta t/\Delta x$, $\sigma = \mathrm{sgn}\,C_{j+1/2}$ and $\phi(r)$ is the limiter. Various choices for $\phi(r)$ are possible; one particular choice is provided by (14.86).

If (18.160) and an equivalent expression for $\tilde{f}_{j-1/2}$ are substituted into (18.159) it is apparent that the term, $0.5(f_j + f_{j+1})$, contributes to a three-point centred difference discretisation of $\partial f / \partial x$. The other terms in (18.160) contribute to numerical dissipation when substituted into (18.159).

Equation (18.159) is an explicit algorithm that is suitable for unsteady problems. For problems governed by the steady Euler or compressible Navier–Stokes equations it is desirable to use an implicit algorithm and to use a form of numerical flux that is independent of the time-step so that the steady-state solution will be independent of the time-step.

Yee (1987) achieves these goals by replacing (18.159) with

$$\varrho_j^{n+1} + \beta \frac{\Delta t}{\Delta x}(\tilde{f}_{j+1/2}^{n+1} - \tilde{f}_{j-1/2}^{n+1}) = \varrho_j^n - (1-\beta)\frac{\Delta t}{\Delta x}(\tilde{f}_{j+1/2}^n - \tilde{f}_{j-1/2}^n) , \qquad (18.161)$$

where β has the same role as in (18.73). Equation (18.160) is replaced with

$$\tilde{f}_{j+1/2} = 0.5(f_j + f_{j+1}) - 0.5|u_{j+1/2}|[1 - \phi(r)]_{j+1/2}\Delta\varrho_{j+1/2} . \qquad (18.162)$$

It is found that for (18.161) to be TVD, (14.81), it is necessary to introduce a CFL-like restriction

$$|u_{j+1/2}|\frac{\Delta t}{\Delta x} < \frac{2}{3(1-\beta)} . \qquad (18.163)$$

To compute steady solutions it is recommended that $\beta = 1$ is used in (18.161) with the result that (18.163) is satisfied for any choice of Δt.

To make use of a tridiagonal solver (Sect. 6.2.3) it is necessary to linearise (18.161) about time-level n to generate a linear system of equations for $\Delta\varrho^{n+1}$. This produces the tridiagonal system

$$B_j^1 \Delta\varrho_{j-1}^{n+1} + B_j^2 \Delta\varrho_j^{n+1} + B_j^3 \Delta\varrho_{j+1}^{n+1} = -\frac{\Delta t}{\Delta x}(\tilde{f}_{j+1/2}^n - \tilde{f}_{j-1/2}^n) , \qquad (18.164)$$

where

$$B_j^1 = 0.5\beta\frac{\Delta t}{\Delta x}(-u_{j-1} - E_{j-1/2}) , \qquad B_j^2 = 1 + 0.5\beta\frac{\Delta t}{\Delta x}(E_{j-1/2} + E_{j+1/2}) ,$$

$$\qquad (18.165)$$

$$B_j^3 = 0.5\beta\frac{\Delta t}{\Delta x}(u_{j+1} - E_{j+1/2}) \quad \text{and} \quad E_{j+1/2} = |u_{j+1/2}|[1 - \phi(r)]_{j+1/2} .$$

Due to the nature of the coefficients B_j^1, B_j^2, B_j^3, (18.164) is a five-point scheme although tridiagonal in $\Delta\varrho$. For purely steady-state calculations, obtained via the pseudotransient construction (Sect. 6.4), it is more economical (Yee 1987) to drop the limiter in E in (18.165) so that $E_{j+1/2} = |u_{j+1/2}|$. However, the limiters are retained on the right-hand side of (18.164) so that the scheme is spatially second-order accurate.

The TVD scheme extends to systems of equations like the Euler equations (Sect. 14.2.6) by splitting each component equation into a system of characteristic fields as in (14.91–92). The above algorithm is applied to each characteristic field. When the contributions from the characteristic fields are added together, (18.164) is replaced by a block tridiagonal scheme.

This can be illustrated for the one-dimensional Euler equations (14.43)

$$\frac{\partial \mathbf{q}}{\partial t} + \frac{\partial \mathbf{F}}{\partial x} = 0 \ . \tag{18.166}$$

The characteristic fields are obtained by replacing (18.166) with

$$\frac{\partial \mathbf{q}}{\partial t} + \underline{A} \frac{\partial \mathbf{q}}{\partial x} = 0 \ , \tag{18.167}$$

where the Jacobian $\underline{A} = \partial \mathbf{F}/\partial \mathbf{q}$. The elements of $\underline{A}$ for the equivalent two-dimensional problem are given by (14.99). The Jacobian $\underline{A}$ is factorised as in (18.60):

$$\underline{A} = \underline{T}_A^{-1} \underline{A}_A \underline{T}_A \ , \tag{18.168}$$

where the diagonal matrix $\underline{A}_A$ contains the eigenvalues of $\underline{A}$, i.e. $\mathrm{diag}\underline{A}_A = \{u, u+a, u-a\}$.

The equivalent of the implicit algorithm (18.164, 165) to solve (18.166) is

$$\underline{B}_j^1 \Delta \mathbf{q}_{j-1}^{n+1} + \underline{B}_j^2 \Delta \mathbf{q}_j^{n+1} + \underline{B}_j^3 \Delta \mathbf{q}_{j+1}^{n+1} = -\frac{\Delta t}{\Delta x} (\tilde{\mathbf{F}}_{j+1/2} - \tilde{\mathbf{F}}_{j-1/2}) \ , \tag{18.169}$$

where

$$\underline{B}_j^1 = 0.5\beta \frac{\Delta t}{\Delta x} (-\underline{A}_{j-1} - \underline{E}_{j-1/2})^n \ , \qquad \underline{B}_j^2 = \underline{I} + 0.5\beta \frac{\Delta t}{\Delta x} (\underline{E}_{j-1/2} + \underline{E}_{j+1/2})^n \ , \tag{18.170}$$

$$\underline{B}_j^3 = 0.5\beta \frac{\Delta t}{\Delta x} (\underline{A}_{j+1} - \underline{E}_{j+1/2})^n \ , \quad \text{and}$$

$$\tilde{\mathbf{F}}_{j+1/2} = 0.5(\mathbf{F}_j + \mathbf{F}_{j+1} - \underline{T}_{A,j+1/2} \Phi_{j+1/2}) \ . \tag{18.171}$$

The term $\Phi_{j+1/2}$ is equivalent to the second term on the right-hand side of (18.162) and contains the contributions $\phi_{j+1/2}^l$ from each (lth) characteristic field

$$\phi_{j+1/2}^l = |\lambda_{A,j+1/2}^l|[1 - \phi^l(r)]_{j+1/2} \underline{T}_{A,j+1/2}^{-1}(\mathbf{q}_{j+1} - \mathbf{q}_j) \ . \tag{18.172}$$

The terms $\underline{E}$ in (18.170) are given by

$$\underline{E}_{j\pm1/2} = [\underline{T}_A \underline{\Omega} \underline{T}_A^{-1}]_{j\pm1/2} \ , \quad \text{where} \tag{18.173}$$

$$\mathrm{diag}\,\underline{\Omega}_{j\pm1/2} = \{|\lambda_A^l|(1 - \phi^l(r))\}_{j\pm1/2} \ . \tag{18.174}$$

Equation (18.169) is block tridiagonal and can be solved efficiently using the algorithm described in Sect. 6.2.5. If (18.171) is substituted into (18.169) it is clear that the spatial discretisation can be interpreted as central differencing of the flux **F** plus additional numerical dissipation constructed from the individual characteristic fields.

Essentially the same algorithm can be extended to multi-dimensions by introducing an approximate factorisation algorithm, as in Sect. 18.3.2. Yee (1987) discusses means of diagonalising the left-hand side of (18.169) so that the steady state may be reached more economically.

For flows with irregular computational domains the present formulation can be expressed in generalised coordinates (Yee and Harten 1985) leading to an algorithm closely related to that in Sect. 18.4.1, except for the TVD numerical dissipation treatment of the convective terms. For inviscid steady flow about a NACA 0012 aerofoil inclined at an angle of $7°$ and with $M_\infty = 1.2$ Yee obtains solutions using the above algorithm on a 163×49 grid. A comparison is made with the ARC2D code of Pulliam and Steger (1985), which is a slightly refined version of the algorithm described in Sect. 18.4.1. As might be expected, the present algorithm provides a better solution in the region of the bow shock and the trailing edge shock that forms on the top surface of the aerofoil. The disparity between the two methods becomes greater when the freestream Mach number is increased to $M_\infty = 1.8$. The disparity occurs in the immediate vicinity of the shocks and is due to the increase in local shock strength.

Away from the shock regions the solutions produced by the two methods are very similar; in particular, the surface pressure distributions are identical. It should be noted that the evaluation of the TVD numerical dissipation, even if this is only done on the equivalent of the right-hand side of (18.169), adds significantly to the overall execution time.

For unsteady flows it is more accurate to use an explicit scheme, $\beta = 0$ in (18.169–174). Yee (1986) has considered the passage of a planar shock past a NACA 0018 aerofoil at $30°$ incidence. The shock speed is $M_s = 1.5$ and the solutions are obtained on a 299×79 C-grid. The density solutions are compared with the experimentally determined density interferograms of Mandella and Bershader (1987) in Fig. 18.13. The capability of accurately predicting the complex shock interactions is apparent.

It may be noted that, by interpreting TVD schemes as providing *additional* numerical dissipation to conventionally centred differenced inviscid flux terms it is straightforward to modify existing computer codes to incorporate additional numerical dissipation (Yee 1987). To make this an economical modification it is necessary to test the local gradients and to introduce the TVD dissipation only when the gradients are sufficiently severe, i.e. in regions where shocks occur. Munz (1988) provides a systematic comparison of the numerical dissipation of many TVD schemes for a model two-dimensional convection equation (rotating cone problem).

TVD schemes to discretise the inviscid terms are readily combined with centred differencing of the viscous and turbulent terms in the compressible Navier–Stokes

Interferograms   Curved Shock   Planar Shock

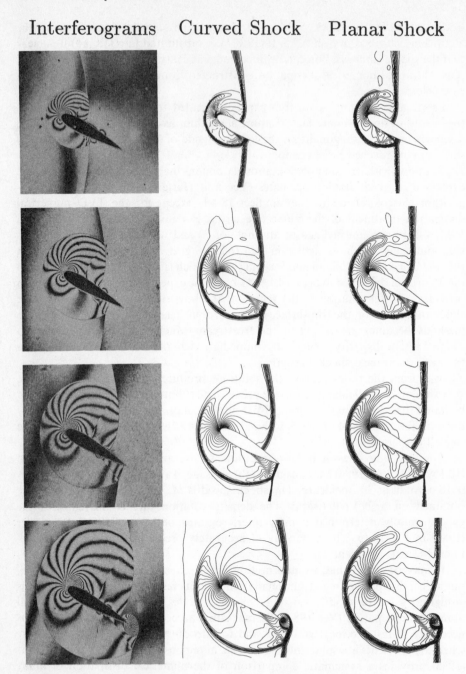

2nd-order Symmetric TVD Scheme

Fig. 18.13. Density contours for the shock passage past a NACA 0018 aerofoil at 30° incidence (after Yee, 1986)

equations leading to block tridiagonal algorithms comparable to (18.169). The pseudotransient solution of the multidimensional compressible steady Navier–Stokes equations with TVD discretisation may be via approximate factorisation (Yee 1987), block bidiagonal (Lombard et al. 1986) or relaxation schemes (Chakravarthy 1987).

A very readable review of contemporary TVD schemes for inviscid and viscous, steady and unsteady hypersonic flow is provided by Yee et al. (1990).

Additionally, the TVD algorithm may be based on flux vector splitting similar to that discussed in Sect. 14.2.5. Walters et al. (1986) describes an algorithm to compute solutions to the compressible Navier–Stokes equations that combines the flux-vector splitting of van Leer (1982) with either approximate factorisation or relaxation schemes to advance the pseudotransient solution. The flux-vector splitting algorithm can also be interpreted (MacCormack 1984) as a numerically dissipative modification to a centred difference treatment of the inviscid fluxes.

18.6 Closure

Computational solutions of the full compressible Navier–Stokes equations, for other than very simple computational domains, are a relatively recent phenomenon. The various algorithms available up to 1975 are lucidly summarised by Peyret and Viviand (1975). Most of these algorithms were explicit. Within four years, (MacCormack and Lomax 1979), the situation had changed significantly, with many very efficient implicit algorithms (Sects. 18.3.2 and 18.4.1) becoming available. This, coupled with the development of generalized-coordinate algorithms (Sect. 18.4.1), facilitated the solution for complicated domains exemplified by the flow around a two-dimensional aerofoil (Steger 1978).

Ten years after Peyret and Viviand, Shang (1985) indicates that, primarily due to the advance in computer hardware (Chap. 1), flows around most of the components that make up an aircraft can be predicted reasonably efficiently. It is expected that computations of the compressible turbulent flowfield about complete aircraft (Jameson 1989) will soon be sufficiently economical to form the backbone of the design process.

It seems likely that such large-scale computations will be based broadly on the algorithms described in this chapter. TVD schemes, and possibly adaptive grids (Kim and Thompson 1990), are expected to provide the primary means of obtaining sharp shock simulation.

For steady flow predictions the main area for future improvement is in the acceleration of the convergence process to the steady state. Although multigrid techniques are helpful it is not clear if they are as effective as for the Euler equations alone due to the wider range of grid scales in the computational domain, particularly if a zonal strategy is exploited (Holst et al. 1986; Kaynak and Flores 1989). Alternatives to the approximate factorisation construction, such as Newton's method (Sect. 6.1) and preconditioned conjugate gradient methods (Sect. 6.3.4), may provide more efficient iterative or smoothing schemes to reach the steady state

(Venkatakrishnan et al. 1991). However, the effectiveness of such techniques will depend on using discretisation constructions that reinforce diagonal dominance.

The other major area of future development is the improvement of turbulence models. Presently most of the impetus in this area comes from computing incompressible flows. From a design perspective algebraic eddy viscosity models (Sect. 18.1.1) do give accurate predictions of mean flow quantities, such as the surface pressure distribution, as long as the flow is attached. But where substantial areas of separated and/or unsteady flow are to be expected more elaborate turbulence models, involving the direct solution for the Reynolds stresses (HaMinh et al. 1986), will be required to augment the compressible Navier–Stokes equations. However, the structure of the additional equations modelling the turbulence is similar to that of the Navier–Stokes equations so that no significant changes in the form of the computational algorithm are expected.

What is more likely to alter the optimal choice for the computational algorithm are developments in computer architecture. As computers based on parallel processors, perhaps one per grid point, become more widespread (Ortega and Voigt 1985) simpler explicit (Sect. 18.2) algorithms will probably be more cost effective. Oran et al. (1990) have applied a contemporary version of the explicit FCT algorithm (Sect. 14.2.7) to two and three-dimensional time-dependent compressible flows on a Connection Machine with 16 K processors. Computational speeds are achieved that are five to six times faster than on a single-processor Cray YMP.

Also the embedding of CFD algorithms into global optimisation procedures that automate the design process will probably place more emphasis on Newton-like iterative techniques and less emphasis on rapid convergence from arbitrary starting solutions.

In this chapter the discretisation process has been described mainly in terms of finite difference and finite element techniques. However finite volume techniques are also very effective and widely used (Deiwert 1984). There have not been as many applications of spectral methods to the compressible Navier–Stokes equations (Zang et al. 1989), primarily because of the difficulty with embedded shocks.

18.7 Problems

Physical Simplifications (Sect. 18.1)

18.1 Show that the continuity equation in two-dimensional turbulent flow can be written as,

(a) with conventional Reynolds averaging,

$$\frac{\partial \bar{\varrho}}{\partial t} + \frac{\partial (\overline{\varrho u})}{\partial x} + \frac{\partial (\overline{\varrho' u'})}{\partial x} + \frac{\partial (\overline{\varrho v})}{\partial y} + \frac{\partial (\overline{\varrho' v'})}{\partial y} = 0 \ , \tag{18.175}$$

(b) with mass-weighted Reynolds averaging,

$$\frac{\partial \bar{\varrho}}{\partial t} + \frac{\partial (\bar{\varrho} \tilde{u})}{\partial x} + \frac{\partial (\bar{\varrho} \tilde{v})}{\partial y} = 0 \ . \tag{18.176}$$

18.2 Apply a "Couette" model to the two-dimensional incompressible turbulent x-momentum equation and show that close to a solid surface ($y = 0$)

$$\tau - \tau_w = \frac{\partial p}{\partial x} y \ , \tag{18.177}$$

where the local shear stress $\tau = (v + v_T)\partial u/\partial y$ and $\partial p/\partial x$ is assumed constant across the layer. Introduce a mixing length representation $v_T = (\kappa y)^2 |\partial u/\partial y|$, and assuming $v_T \gg v$, show that

$$\kappa y \frac{\partial u}{\partial y} = \left(\tau_w + \frac{\partial p}{\partial x} y\right)^{1/2} \Big/ \varrho^{1/2} \ . \tag{18.178}$$

Show that (18.23) can be obtained from (18.178) if $\partial p/\partial x = 0$.
Develop a discretised form of (18.178) which will permit the u velocity component to be obtained at the grid point adjacent to the wall.

18.3 Obtain (18.26) and (18.27) from the steady energy equation (18.24). Carry out an order-of-magnitude analysis, as in Sect. 16.1.1, and show that (18.26) can be simplified to give (18.29).

18.4 Delete all x-derivatives associated with viscous terms in (18.6–10) and show that the thin layer approximation (18.31, 32) is obtained.

Explicit Schemes (Sect. 18.2)

18.5 For the explicit MacCormack scheme (18.35, 36) carry out a Taylor series expansion, at the $\mathbf{F}$, $\mathbf{G}$ level, and show that the leading term in the truncation error is $O(\Delta t^2, \Delta x^2)$.

18.6 For the explicit MacCormack scheme applied to (18.49), show that a von Neumann stability analysis (Sect. 4.3) indicates the following restriction on Δt for a stable solution:

$$\Delta t \le \frac{\Delta x^2}{2\mu + a\Delta x} \ .$$

18.7 For the time-split MacCormack method write (18.43) applied to (18.6) in an appropriate predictor, corrector format.

18.8 Consider the application of the Wambecq scheme (18.47) to the diffusion equation

$$\frac{\partial u}{\partial t} - \mu \frac{\partial^2 u}{\partial x^2} = 0 \ . \tag{18.179}$$

Show that, if $b = -0.5$ and $c = 1$,

$$\Delta u^1 = \Delta t\, \mu L_{xx} u^n \ , \qquad \Delta u^2 = \Delta t\, \mu L_{xx} u^* \quad \text{and}$$

$$\Delta u^3 = \Delta t\, \mu L_{xx}(1.5 u^n - 0.5 u^*) \ ,$$

where $u^* = u^n + \Delta u^1$ and L_{xx} is the three-point centred-difference representation of $\partial^2 u/\partial x^2$. Assume that the scalar products are such that $(\mathbf{e}, \mathbf{f}) = n e_i f_i$, i.e. all elements of $\mathbf{e}$ and $\mathbf{f}$ are equal. Then show that the Wambecq scheme becomes

$$\Delta u^{n+1} = \Delta t \frac{\mu (L_{xx} u^n)^2}{L_{xx}(1.5u^n - 0.5u^*)} \ . \tag{18.180}$$

Implement the true Wambecq scheme and (18.180) applied to (18.179) and determine empirically if there are any limits on Δt for stable solutions.

Implicit Schemes (Sect. 18.3)

18.9 Show that the additional (implicit) terms in (18.53 and 54) can be interpreted as third-order (in time) perturbations to (18.50 and 51).

18.10 If λ is given by (18.55) demonstrate that (18.53 and 54) are unconditionally stable.

18.11 Starting with (18.58), derive the algorithm given by (18.65, 67, 69 and 70).

18.12 Apply the Beam and Warming algorithm to the two-dimensional transport equation (9.81), and show that the equivalent of (18.78 and 79) is given by (9.88 and 89) with $M_x = M_y^T = \{0, 1, 0\}$.

18.13 Starting with (18.84) derive the approximate factorisation algorithm given by (18.85–87).

18.14 Expand the left-hand and right-hand sides of (18.92) as Taylor series and identify the additional dissipative terms introduced on the right-hand side. Consequently comment on the suitability of the overall algorithm (18.96, 97) for unsteady problems.

Generalised Coordinates (Sect. 18.4)

18.15 Show that the elements of $\hat{\mathbf{S}}$ in (18.116) are obtained by dropping ξ derivatives (as part of the thin layer approximation) in the expression for $\hat{\mathbf{S}}$ given by (18.110).

18.16 Derive the coefficient m_{22} in (18.119).

18.17 By using the results of Sect. 12.3, demonstrate that (18.138) and (18.140) can be obtained from (18.80–82).

Numerical Dissipation (Sect. 18.5)

18.18 Add a fourth-order dissipative term $\varepsilon_E \Delta x^4 \partial^4 \bar{T}/\partial x^4$ to the left-hand side of the one-dimensional transport equation (9.56). Apply a general two level implicit scheme

$$\frac{\Delta T^{n+1}}{\Delta t} = \beta \mathrm{RHS}^{n+1} + (1 - \beta)\mathrm{RHS}^n \ ,$$

where RHS contains all of the spatial terms discretised with centred finite difference expressions, three-point for $\partial \bar{T}/\partial x$ and $\partial^2 \bar{T}/\partial x^2$ and five-point (18.156) for $\partial^4 \bar{T}/\partial x^4$.

Apply a von Neumann analysis to determine the limits on Δt as a function of ε_E and β, for a stable solution, particularly when α/u is very small.

18.19 Substitute (18.160) into (18.159) and show that the result can be written as a conventional centred difference construction plus additional dissipative terms. Demonstrate that this is also the case when (18.162) is substituted into (18.161).

References

Chapter 11

Aris, R. (1962): *Vectors, Tensors and the Basic Equations of Fluid Dynamics* (Prentice-Hall, Englewood Cliffs, N.J.)

Batchelor, G.K. (1967): *An Introduction to Fluid Dynamics* (Cambridge University Press, Cambridge)

Bird, G. (1976): *Molecular Gas Dynamics* (Oxford University Press, Oxford)

Cebeci, T., Bradshaw, P. (1977): *Momentum Transfer in Boundary Layers* (Hemisphere–McGraw-Hill, Washington, D.C.)

Cebeci, T., Bradshaw, P. (1984): *Physical and Computational Aspects of Convective Heat Transfer* (Springer, New York, Berlin, Heidelberg)

Chu, C.K. (1978): Adv. Appl. Mech. **18**, 285–331

Crochet, M.J., Davis, A.R., Walters, K. (1984): *Numerical Simulation of Non-Newtonian Flow* (Elsevier, Amsterdam)

Eckert, E.R.G., Drake, R.M. (1972): *Analysis of Heat and Mass Transfer* (McGraw-Hill, New York)

Gresho, P.M. (1991): "Incompressible Fluid Dynamics: Some Fundamental Formulation Issues", Annu. Rev. Fluid Mech. **23** (to appear)

Gustafson, K.E. (1980): *Partial Differential Equations and Hilbert Space Methods* (Wiley, New York)

Gustafsson, B., Sundstrom, A. (1978): SIAM J. Appl. Math. **35**, 343–357

Hughes, W.F., Gaylord, E.W. (1964): *Basic Equations of Engineering Science* (McGraw-Hill, New York)

Hussaini, M.Y., Zang, T.A. (1987): Annu. Rev. Fluid Mech. **19**, 339–367

Launder, B.E., Spalding, D.B. (1974): Comput. Methods Appl. Mech. Eng. **3**, 269–289

Liepmann, H.W., Roshko, A. (1957): *Elements of Gasdynamics* (Wiley, New York)

Lighthill, M.J. (1963): In *Laminar Boundary Layers*, ed. by L. Rosenhead (Oxford University Press, Oxford) pp. 1–45

Marvin, J.G. (1983): AIAA J. **21**, 941–955

Milne-Thomson, L.M. (1968): *Theoretical Hydrodynamics*, 5th ed. (Macmillan, London)

Nordstrom, J. (1989): J. Comput. Phys. **85**, 210–244

Oliger, J., Sundstrom, A. (1978): SIAM J. Appl. Math. **3**, 419–446

Panton, R.L. (1984): *Incompressible Flow* (Wiley, New York)

Patel, V.C., Rodi, W., Scheuerer, G. (1985): AIAA J. **23**, 1308–1319

Peyret, R., Taylor, T.D. (1983): *Computational Methods for Fluid Flow*, Springer Ser. Comput. Phys. (Springer, Berlin, Heidelberg)

Quartapelle, L., Valz-Gris, F. (1981): Int. J. Numer. Methods Fluids **1**, 129–144

Richardson, S.M., Cornish, A.R.H. (1977): J. Fluid Mech. **82**, 309–340

Rodi, W. (1980): *Turbulence Models and Their Application in Hydraulics* (I.A.H.R., Delft)

Rodi, W. (1982): AIAA J. **20**, 872–879

Rogallo, R.S., Moin, P. (1984): Annu. Rev. Fluid Mech. **16**, 99–137

Rosenhead, L. (1963): *Laminar Boundary Layers* (Oxford University Press, Oxford)

Schlichting, H. (1968): *Boundary Layer Theory*, 6th ed. (McGraw-Hill, New York)

Sherman, F.S. (1990): *Viscous Flow* (McGraw-Hill, New York)

Simpson, R.L. (1981): J. Fluids Eng. **103**, 520–533

Streeter, V.L., Wylie, E.B. (1979): *Fluid Mechancis*, 7th ed. (McGraw-Hill, New York)

Thompson, K.W. (1990): J. Comput. Phys. **89**, 439–461

Tobak, M., Peake, D. (1982): Annu. Rev. Fluid Mech. **14**, 61–85
van Wylen, G.J., Sonntag, R. (1976): *Fundamentals of Classical Thermodynamics* (Wiley, New York)
von Schwind, J.J. (1980): *Geophysical Fluid Dynamics for Oceanographers* (Prentice-Hall, Englewood Cliffs, N.J.)
Wong, A., Reizes, J. (1984): J. Comput. Phys. **55**, 98–114.

Chapter 12

Aris, R. (1962): *Vectors, Tensors and the Basic Equations of Fluid Dynamics* (Prentice-Hall, Englewood Cliffs, N.J.)
Burns, A.D., Wilkes, N.S. (1987): " A Finite Difference Method for the Computation of Fluid Flows in Complex Three Dimensional Geometries", Harwell Laboratory Report AERE-R 12342 (Harwell, U.K.)
Eiseman, P.R., Stone, A.P. (1980): SIAM Rev. **22**, 12–27
Kerlick, D.G., Klopfer, G.H. (1982): "Assessing the quality of curvilinear coordinate meshes by decomposing the Jacobian matrix", in *Numerical Grid Generation*, ed. by J.F. Thompson (North-Holland, Amsterdam) pp. 787–807
Steger, J.L. (1978): AIAA J. **16**, 679–686
Thompson, J.F. (1984): AIAA J. **22**, 1505–1523
Thompson, J.F., Warsi, Z.U.A., Mastin, C.W. (1985): *Numerical Grid Generation, Foundations and Applications* (North-Holland, Amsterdam)
Vinokur, M. (1989): J. Comput. Phys. **81**, 1–52

Chapter 13

Ahlberg, J.H., Nilson, E.N., Walsh, J.L. (1967): *Theory of Splines and Their Applications* (Academic, New York)
Anderson, O.L., Davis, R.T., Hankins, G.B., Edwards, D.E. (1982): In *Numerical Grid Generation*, ed. by J.F. Thompson (North-Holland, Amsterdam) pp. 507–524
Cooley, J.W., Tuckey, J.W. (1965): Math. Comput. **19**, 297–301
Davis, R.T. (1979): "Numerical methods for coordinate generation based on Schwarz–Christoffel transformations", AIAA Paper No. 79–1463
Eiseman, P.R. (1979): J. Comput. Phys. **33**, 118–150
Eiseman, P.R. (1982a): In *Numerical Grid Generation*, ed. by J.F. Thompson (North-Holland, Amsterdam) pp. 193–234
Eiseman, P.R. (1982b): J. Comput. Phys. **47**, 331–351
Eiseman, P.R. (1982c): J. Comput. Phys. **47**, 352–374
Eiseman, P.R. (1985): Annu. Rev. Fluid Mech. **17**, 487–522
Eiseman, P.R. (1987): Comput. Methods Appl. Mech. Eng. **64**, 321–349
Eriksson, L.E. (1982): AIAA J. **20**, 1318–1319
Forsythe, G.E., Malcolm, M.A., Moler, C. (1977): *Computer Methods for Mathematical Computations* (Prentice-Hall, Englewood Cliffs, N.J.)
Gordon, W.J., Hall, C.A. (1973): Int. J. Numer. Methods Eng. **7**, 461–477
Gordon, W.J., Thiel, L.C. (1982): In *Numerical Grid Generation*, ed. by J.F. Thompson (North-Holland, Amsterdam) pp. 171–192
Ives, D.C. (1976): AIAA J. **14**, 1006–1011
Ives, D.C. (1982): In *Numerical Grid Generation*, ed. by J.F. Thompson (North-Holland, Amsterdam) pp. 107–136
Kennon, S.R., Dulikravich, G.S. (1986): AIAA J. **24**, 1069–1073
Kim, H.J., Thompson, J.F. (1990): AIAA J. **28**, 470–477
Mavriplis, D.J. (1990): AIAA J. **28**, 213–221

McNally, W.D. (1972): "Fortran Program for Generating a Two-Dimensional Orthogonal Mesh Between Two Arbitrary Boundaries", NASA TN-D6766

Milne-Thomson, L. (1968): *Theoretical Hydrodynamics* (Macmillan, London)

Moretti, G. (1980): "Grid Generation Using Classical Techniques", NASA CP 2166, 1–35

Nakahashi, K., Egami, K. (1991): "An Automatic Euler Solver Using an Unstructured Upwind Method", Comput. Fluids (to appear)

Rizzi, A., Eriksson, L.E. (1981): "Transfinite Mesh Generation and Damped Euler Equation Algorithm for Transonic Flow around Wing-Body Configurations", AIAA Paper 81-0999

Roberts, G.O. (1971): Lecture Notes in Physics, Vol. 8 (Springer, Berlin, Heidelberg) pp. 171–176

Rubbert, P.E., Lee, K.D. (1982): In *Numerical Grid Generation*, ed. by J.F. Thompson (North-Holland, Amsterdam) pp. 235–252

Ryskin, G., Leal, L.G. (1983): J. Comput. Phys. **50**, 71–100

Smith, R.E. (1982): In *Numerical Grid Generation*, ed. by J.F. Thompson (North-Holland, Amsterdam) pp. 137–170

Steger, J.L., Sorenson, R.L. (1980): "Use of Hyperbolic Partial Differential Equations to Generate Body-Fitted Coordinates", NASA CP 2166, pp. 463–478

Temperton, C. (1979): J. Comput. Phys. **31**, 1–20

Theodorsen, T., Garrick, I.E. (1933): "General Potential Theory of Arbitrary Wing Sections", NACA TR 452

Thomas, P.D. (1982): In *Numerical Grid Generation*, ed. by J.F. Thompson (North-Holland, Amsterdam) pp. 667–686

Thompson, J.F. (1982): In *Numerical Grid Generation*, ed. by J.F. Thompson (North-Holland, Amsterdam) pp. 1–31

Thompson, J.F. (1984): AIAA J. **22**, 1505–1523

Thompson, J.F. (1985): Appl. Num. Math. **1**, 3–28

Thompson, J.F. (1988): AIAA J. **26**, 271–272

Thompson, J.F., Thames, F.C., Mastin, C.W. (1977a): J. Comput. Phys. **24**, 274–302

Thompson, J.F., Thames, F.C., Mastin, C.W. (1977b): "Boundary-Fitted Curvilinear Coordinate Systems for Solution of Partial Differential Equations on Fields Containing any Number of Arbitrary Two-Dimensional Bodies", NASA CR-2729

Thompson, J.F., Warsi, Z.U.A., Mastin, C.W. (1982): J. Comput. Phys. **47**, 1–108

Thompson, J.F., Warsi, Z.U.A., Mastin, C.W. (1985): *Numerical Grid Generation, Foundations and Applications* (North-Holland, Amsterdam)

Trefethen, L.N. (1980): SIAM J. Sci. Stat. Comput. **1**, 82–102

Vinokur, M. (1983): J. Comput. Phys. **50**, 215–234

Woods, L.C. (1961): *The Theory of Subsonic Plane Flow* (Cambridge University Press, Cambridge)

Yu, N.J., Chen, H.C., Su, T.Y., Kao, T.J. (1990): "Development of a General Multiblock Flow Solver for Complex Configurations", in Notes Num. Fluid Mech. **29** (Vieweg, Wiesbaden) pp. 603–612

Chapter 14

Abbett, M.J. (1973): Proc. 1st AIAA Computational Fluid Dynamics Conf., AIAA, New York, pp. 153–172

Anderson, W.K., Thomas, J.L., Whitfield, D.L. (1988): AIAA J. **26**, 649–654

Ballhaus, W.F., Jameson, A., Albert, J. (1978): AIAA J. **16**, 573–579

Bayliss, A., Turkel, E. (1982): J. Comput. Phys. **48**, 182–199

Book, D.L., Boris, J.P., Hain, K. (1975): J. Comput. Phys. **18**, 248–283

Book, D.L. (ed.) (1981): *Finite-Difference Techniques for Vectorized Fluid Dynamics Calculations*, Springer Ser. Comput. Phys. (Springer, Berlin, Heidelberg)

Boris, J.P., Book, D.L. (1973): J. Comput. Phys. **11**, 38–69

Boris, J.P., Book, D.L. (1976): Methods Comput. Phys. **16**, 85–129

Brebbia, C.A. (1978): *The Boundary Element Method for Engineers* (Pentech Press, London)

Carmichael, R.L., Erikson, L.L. (1981): "PAN AIR – A Higher Order Panel Method for Predicting Subsonic or Supersonic Linear Potential Flows about Arbitrary Configurations", AIAA Paper 81–1255

Catherall, D. (1982): AIAA J. **20**, 1057–1063

Caughey, D.A. (1982): Annu. Rev. Fluid Mech. **14**, 261–283

Chakravarthy, S.R. (1983): AIAA J. **21**, 699–706

Chakravarthy, S.R. (1986): "Algorithmic Trends in Computational Fluid Dynamics" in Proc. Int. Symp. Comp. Fluid Dynamics, ed. K. Oshima (Japan Soc. of Comp. Fluid Dynamics, Tokyo) Vol. 1, pp. 163–173

Chakravarthy, S.R., Anderson, D.A., Salas, M.D. (1980): "The Split-Coefficient Matrix Method for Hyperbolic Systems of Gas Dynamic Equations", AIAA Paper 80–0268

Chakravarthy, S.R., Osher, S. (1983): "High Resolution Applications of the Osher Upwind Scheme for the Euler Equations", AIAA Paper 83–1943

Chakravarthy, S.R., Osher, S. (1985): Lect. Appl. Math. **22**, 57–86

Chima, R.V., Johnson, G.M. (1985): AIAA J. **23**, 23–32

Cole, J.D. (1975): SIAM J. Appl. Math. **29**, 763–787

Colella, P., Woodward, P.R. (1984): J. Comput. Phys. **54**, 174–201

Dadone, A., Magi, V. (1986): AIAA J. **24**, 1277–1284

Dadone, A., Moretii, G. (1988): AIAA J. **26**, 409–424

Dadone, A., Napolitano, M. (1983): AIAA J. **21**, 1391–1399

Dadone, A., Napolitano, M. (1985): Comput. Fluids **13**, 383–395

Davis, R.L., Ni, R.H., Bowley, W.W. (1984): AIAA J. **22**, 1573–1581

Deconinck, H., Hirsch, C.H. (1985): In *Advances in Computational Transonics*, ed. by W.G. Habashi (Pineridge Press, Swansea) pp. 733–775

Ecer, A., Akay, H.U. (1985): In *Advances in Computational Transonics*, ed. by W.G. Habashi (Pineridge Press, Swansea) pp. 777–810

Fletcher, C.A.J. (1975): AIAA J. **13**, 1073–1078

Fletcher, C.A.J. (1984): *Computational Galerkin Methods*, Springer Ser. Comput. Phys. (Springer, Berlin, Heidelberg)

Fletcher, C.A.J., Morton, K.W. (1986): "Oblique Shock Reflection by the Characteristic Galerkin Method", in Proc. Ninth Australasian Fluid Mechanics Conference, ed. by P.S. Jackson (University of Auckland, Auckland), pp. 411–415

Flores, J., Holst, T.L., Kwak, D., Batiste, D.M. (1983): "A New Consistent Spatial Differencing Scheme for the Transonic Full Potential Equation", AIAA Paper 83–0373

Flores, J., Barton, J., Holst, T., Pulliam, T. (1985): 9th Int. Conf. Numer. Methods Fluid Dynamics, ed. by Soubbaramayer, J.P. Boujot, Lecture Notes in Physics, Vol. 218 (Springer, Berlin, Heidelberg) pp. 213–218

Goorjian, P. (1985): In *Advances in Computational Transonics*, ed. by W.G. Habashi (Pineridge Press, Swansea) pp. 215–255

Habashi, W.G. (ed.) (1985): *Advances in Computational Transonics* (Pineridge Press, Swansea) pp. 23–58

Hafez, M.M. (1985): In *Advances in Computational Transonics*, ed. by W.G. Habashi (Pineridge Press, Swansea) pp. 23–58

Hall, M.G. (1984): R.A.E. Tech. Rep. 84013

Harten, A. (1983): J. Comput. Phys. **49**, 357–393

Harten, A., Lax, P.D., van Leer, B. (1983): SIAM Rev. **25**, 35–61

Hemker, P.W. (1986): In 10th Int. Conf. Numer. Methods in Fluid Dynamics, ed. by F.G. Zhuang, Y.L. Zhu, Lecture Notes in Physics, Vol. 264 (Springer, Berlin, Heidelberg) pp. 308–313

Hess, J.L. (1975): Comput. Methods Appl. Mech. Eng. **5**, 145–196

Hess, J.L. (1990): Annu. Rev. Fluid Mech. **22**, 255–274

Hess, J.L., Smith, A.M.O. (1967): Prog. Aeronaut. Sci. **8**, 1–138

Holst, T. (1979): AIAA J. **17**, 1038–1045

Holst, T. (1985): In *Advances in Computational Transonics*, ed. by W.G. Habashi (Pineridge Press, Swansea) pp. 59–82

Holst, T., Ballhaus, W.F. (1979): AIAA J. **17**, 145–152

Holt, M. (1984): *Numerical Methods in Fluid Dynamics*, 2nd ed., Springer Ser. Comput. Phys. (Springer, Berlin, Heidelberg)

Hughes, T.J.R., Mallet, M. (1985): Finite Elements in Fluids **6**, 339–353

Hussaini, M.Y., Zang, T.A. (1987): Annu. Rev. Fluid Mech. **19**, 339–367

Isaacson, E., Keller, H.B. (1966): *Analysis of Numerical Methods* (Wiley, New York)

Jameson, A. (1978): "Transonic Flow Calculations", in *Computational Fluid Dynamics*, ed. by H.J. Wirz, J.J. Solderen (Hemisphere, Washington, D.C.) pp. 1–87

Jameson, A. (1979): "Acceleration of Transonic Potential Flow Calculations on Arbitrary Meshes by the Multiple Grid Method", AIAA Paper 79–1458

Jameson, A. (1983): Appl. Math. Comput. **13**, 327–356

Jameson, A., Baker, T. (1986): In 10th Int. Conf. Numer. Methods in Fluid Dynamics. ed. by F.G. Zhuang, Y.L. Zhu, Lecture Notes in Physics, Vol. 264 (Springer, Berlin, Heidelberg) pp. 334–344

Jameson, A., Schmidt, W., Turkel, E. (1981): "Numerical Solution of the Euler Equations by Finite Volume Methods using Runge–Kutta Time Stepping Schemes", AIAA Paper 81–1259

Jaswon, M.A., Symm, G.T. (1977): *Integral Equation Methods in Potential Theory and Elastostatics* (Academic, London)

Johnson, G.M. (1983): Appl. Math. Comput. **13**, 357–380.

Klopfer, G.H., Nixon, D. (1984): AIAA J. **22**, 770–776

Kraus, W. (1978): "Panel Methods in Aerodynamics", in *Numerical Methods in Fluid Dynamics*, ed. by H.J. Wirz, J.J. Smolderen (Hemisphere, Washington, D.C.) pp. 237–297

Kuethe, A.M., Chow, C.Y. (1976): *Foundations of Aerodynamics* (Wiley, New York)

Kutler, P., Lomax, H. (1971): J. Spacecr. & Rockets, **8**, 1175–1182

Kutler, P., Warming, R.F., Lomax, H. (1973): AIAA J. **11**, 196–204

Lapidus, A. (1967): J. Comput. Phys. **2**, 154–177

Lax, P., Wendroff, B. (1960): Commun. Pure Appl. Math. **13**, 217–237

Lerat, A., Peyret, R. (1975): Rech. Aerosp. **1975-2**, 61–79

Lerat, A., Sides, J. (1982): *Proc. Conf. Numerical Methods in Aeronautical Fluid Dynamics*, ed. by P.L. Roe (Academic, London) pp. 245–288

Liepmann, H., Roshko, A. (1957): *Elements of Gas Dynamics* (Wiley, New York)

MacCormack, R.W. (1969): "The Effect of Viscosity in Hypervelocity Impact Cratering", AIAA Paper 69–354

Mavriplis, D.J. (1990): AIAA J. **28**, 213–221

McDonald, B.E. (1989): J. Comput. Phys. **82**, 413–428

Moretti, G. (1979): Comput. Fluids **7**, 191–205

Morton, K.W., Sweby, P.K. (1987): J. Comput. Phys., **73**, 203–230

Morton, K.W., Paisley, M.F. (1989): J. Comput. Phys. **80**, 168–203

Mulder, W.A. (1985): J. Comput. Phys. **60**, 235–252

Murman, E.M. (1973): Proc. 1st AIAA Comp. Fluid Dyn. Conf. (AIAA, New York) pp. 27–40

Napolitano, M. (1986): In 10th Int. Conf. Numer. Methods in Fluid Dynamics, ed. by F.G. Zhuang, Y.L. Zhu, Lecture Notes in Physics, Vol. 264 (Springer, Berlin, Heidelberg), pp. 47–56

Ni, R.H. (1982): AIAA J. **20**, 1565–1571

Osher, S., Solomon, F. (1982): Math. Comp. **38**, 339–374

Paul, J.C., LaFond, J.G. (1983): "Analysis and Design of Automobile Forebodies using Potential Flow Theory and a Boundary Layer Separation Criterion", SAE Paper 830999

Peyret, R., Taylor, T.D. (1983): *Computational Methods for Fluid Flow*, Springer Ser. Comput. Phys. (Springer, Berlin, Heidelberg)

Pulliam, T.H., Chaussee, D. (1981): J. Comp. Phys., **39**, 347–363

Pulliam, T.H. (1985): "Implicit Finite Difference Methods for the Euler Equations", in *Recent Advances in Numerical Methods for Fluids*, Vol. 4 (Pineridge Press, Swansea)

Pulliam, T.H. (1985): In Advances in Computational Transonics, ed. by W.G. Habashi (Pineridge Press, Swansea) pp. 503–542

Rackich, J.V., Kutler, P. (1972): "Comparison of Characteristics and Shock Capturing Methods with Application to the Space Shuttle Vehicle", AIAA Paper 72–191

Rai, M.M., Chaussee, D.S. (1984): AIAA J. **22**, 1094–1100

Richtmyer, R.D., Morton, K.W. (1967): *Difference Methods for Initial-Value Problems* (Interscience, New York)

Rizzi, A. (1981): In Notes on Numerical Fluid Mechanics, Vol. 3, ed. by Rizzi, A., Viviand, H. (Vieweg, Braunschweig)

Rizzi, A., Eriksson, L.E. (1982): AIAA J. **20**, 1321–1328

Rizzi, A., Viviand, H. (eds.) (1981): *Numerical Methods for the Computation of Inviscid Transonic Flows with Shock Waves*, Notes on Numerical Fluid Mechanics, Vol. 3 (Vieweg, Braunschweig)

Roe, P.L. (1981): J. Comput. Phys. **43**, 357–372

Roe, P.L. (1986): Annu. Rev. Fluid Mech. **18**, 337–365

Roe, P.L., Baines, M.J. (1982): "Algorithms for Advection and Shock Problems", in Proc. Fourth GAMM Conf. Numer. Meth. Fluid Mechanics, ed. by H. Viviand (Vieweg, Braunschweig)

Rubbert, P.E., Sarris, G.R. (1972): "Review and Evaluation of a Three-Dimensional Lifting Potential Flow Analysis Method for Arbitrary Configurations", AIAA Paper 72–188

Rudy, D.H., Strikwerda, J.C. (1981): Comput. Fluids **9**, 327–338

Salas, M., Jameson, A., Melnik, R. (1983): "A Comparative Study of the Nonuniqueness Problem of the Potential Equation", AIAA Paper 83–1888

Sod, G.A. (1978): J. Comput. Phys. **27**, 1–31

Steger, J.L., Pulliam, T.H., Chima, R.V. (1980): "An Implicit Finite Difference Code for Inviscid and Viscous Cascade Flow", AIAA Paper 80–1427

Steger, J.L., Warming, R.F. (1981): J. Comput. Phys. **40**, 263–293

Sweby, P.K. (1984): SIAM J. Numer. Anal. **21**, 995–1011

Thompson, K.W. (1987): J. Comput. Phys. **68**, 1–24

Thompson, K.W. (1990): J. Comput. Phys. **89**, 439–461

Tinoco, E.N., Chen, A.W. (1986): Prog. Astronaut. Aeronaut. **102**, 219–255

Turkel, E. (1985): In Ninth International Conference on Numerical Methods in Fluid Dynamics, ed. by Soubbaramayer, J.P. Boujot, Lecture Notes in Physics, Vol. 218 (Springer, Berlin, Heidelberg), pp. 571–575

Van Leer, B. (1974): J. Comput. Phys. **14**, 361–370

Van Leer, B. (1979): J. Comput. Phys. **32**, 101–136

Van Leer, B. (1982): In 8th Int. Conf. Numer. Methods Fluid Dynamics, ed. by E. Krause, Lecture Notes in Physics, Vol. 264 (Springer, Berlin, Heidelberg) pp. 677–683

Vinokur, M. (1989): J. Comput. Phys. **81**, 1–52

Viviand, H. (1981): In 7th Int. Conf. Numer. Methods in Fluid Dynamics, ed. by W.C. Reynolds, R.W. MacCormack, Lecture Notes in Physics, Vol. 141 (Springer, Berlin, Heidelberg) pp. 44–54

Warming, R.F., Beam, R.M. (1976): AIAA J. **14**, 1241–1249

Woodward, P., Colella, P. (1984): J. Comput. Phys. **54**, 115–173

Yang, J.Y., Lombard, C.K., Bardina, J. (1986): "Implicit Upwind TVD Schemes for the Euler Equations with Bidiagonal Approximate Factorisation", in Proc. Int. Symp. Comp. Fluid Dynamics, ed. by K. Oshima (Japan Soc. of Comp. Fluid Dynamics, Tokyo) Vol. 1, pp. 174–183

Yee, H. (1981): "Numerical Approximation of Boundary Conditions with Application to Inviscid Equations of Gas Dynamics", NASA TN 81265

Yee, H. (1986): In 10th Int. Conf. Numer. Methods in Fluid Dynamics, ed. by F.G. Zhuang, Y.L. Zhu, Lecture Notes in Physics, Vol. 264 (Springer, Berlin, Heidelberg) pp. 677–683

Yee, H., Warming, R.F., Harten, A. (1985): J. Comput. Phys. **57**, 327–360

Yee, H., Klopfer, G.H., Montagne, J.L. (1990): J. Comput. Phys. **88**, 31–61

Young, D.P., Melvin, R.G., Bieterman, M.B., Johnson, F.T., Samant, S.S., Bussoletti, J.E. (1990): J. Comput. Phys. (to appear)

Yu, N.J., Chen, H.C., Su, T.Y., Kao, T.J. (1990): "Development of a General Multiblock Flow Solver for Complex Configurations", in Notes Num. Fluid Mech. **29** (Vieweg, Wiesbaden) pp. 603–612

Zalesak, S.T. (1979): J. Comput. Phys. **31**, 335–362

Zalesak, S.T. (1987): *Advances in Computer Methods for Partial Differential Equations, VI*, eds. R. Vichnevetsky and R.S. Stepleman (IMACS, Rutgers University)

Chapter 15

Baker, A.J. (1983): *Finite Element Computational Fluid Mechanics* (McGraw-Hill, New York)

Blottner, F.G. (1975a): "Computational Techniques for Boundary Layers", AGARD LS-73, Nato, Brussels, pp. 3.1–3.51

Blottner, F.G. (1975b): Comput. Methods Appl. Mech. Eng. **6**, 1–30

Bradshaw, P. (1967): National Physical Laboratory, Aero Report 1219

Carter, J.E. (1981): "Viscous-Inviscid Interaction Analysis of Transonic Turbulent Separated Flow", AIAA Paper 81–1241

Cebeci, T. (1975): AIAA J. **13**, 1056–1064

Cebeci, T., Bradshaw, P. (1977): *Momentum Transfer in Boundary Layers* (McGraw-Hill, New York)

Coles, P., Hirst, E. (eds.) (1968): *Computation of Turbulent Boundary Layers* Vol. 2 (Stanford University, Stanford)

Dwyer, H. (1981): Annu. Rev. Fluid Mech. **13**, 217–229

Fleet, R.W., Fletcher, C.A.J. (1983): "Application of the Dorodnitsyn Boundary Layer Formulation to Turbulent Compressible Flow", in Eighth Australasian Fluid Mechanics Conference, ed. by R.A. Antonia (Newcastle University, Newcastle), pp. 1C1–1C4

Fletcher, C.A.J. (1983): Comput. Methods Appl. Mech. Eng. **37**, 225–243

Fletcher, C.A.J. (1984): *Computational Galerkin Methods*, Springer Ser. Comput. Phys. (Springer, Berlin, Heidelberg)

Fletcher, C.A.J. (1985): Int. J. Numer. Methods Fluids **5**, 443–462

Fletcher, C.A.J., Fleet, R.W. (1984a): Int. J. Numer. Methods Fluids **4**, 399–419

Fletcher, C.A.J., Fleet, R.W. (1984b): Comput. Fluids **12**, 31–45

Fletcher, C.A.J., Fleet, R.W. (1987): J. Appl. Mech., **54**, 197–202

Fletcher, C.A.J., Holt, .M. (1975): J. Comput. Phys. **18**, 154–164

Fletcher, C.A.J., Holt, M. (1976): J. Fluid Mech. **74**, 561–591

Gear, C.W. (1971): Commun. ACM **14**, 185–190

Isaacson, E., Keller, H.B. (1966): *An Analysis of Numerical Methods* (Wiley, New York)

Keller, H. (1978): Annu. Rev. Fluid Mech. **10**, 417–433

Krause, E. (1967): AIAA J. **7**, 1231–1237

Krause, E. (1973): "Numerical Treatment of Boundary Layer Problems", AGARD LS-64, Brussels, Nato pp. 4.1–4.21

McLean, J.D., Randall, J.L. (1979): "Computer Program to Calculate Three-Dimensional Boundary Layer Flows over Wings with Wall Mass Transfer", NASA CR-3123

Noye, J. (1983): "Finite Difference Techniques for Partial Differential Equations" in *Computational Techniques for Differential Equations*, ed. by J. Noye (North-Holland, Amsterdam)

Patankar, S.V., Spalding, D.B. (1970): *Heat and Mass Transfer in Boundary Layers*, 2nd ed. (Intertext, London)

Peyret, R., Taylor, T.D. (1983): *Computational Methods for Fluid Flow*, Springer Ser. Comput. Phys. (Springer, Berlin, Heidelberg)

Reynolds, W.C. (1976): Annu. Rev. Fluid Mech. **8**, 183–208

Rosenhead, L. (1964): *Laminar Boundary Layers* (Oxford University Press, Oxford)

Schiff, L., Steger, J.L. (1980): AIAA J. **18**, 1421–1430

Schlichting, H. (1968): *Boundary Layer Theory*, 6th ed. (McGraw-Hill, New York)

Yeung, W.S., Yang, R.J. (1981): J. Appl. Mech. **48**, 701–706

Zienkiewicz, O.C. (1977): *The Finite Element Method*, 3rd ed. (McGraw-Hill, London)

Chapter 16

Anderson, O.L. (1980): Comput. Fluids **8**, 391–441

Armfield, S.W., Fletcher, C.A.J. (1986): Int. J. Numer. Methods Fluids **6**, 541–556

Armfield, S.W., Cho, N.H., Fletcher, C.A.J. (1990): AIAA J. **28**, 453–460

Baker, A.J. (1983): *Finite Element Computational Fluid Mechanics* (McGraw-Hill, New York)

Berger, S.A., Talbot, L., Yao, L.S. (1983): Annu. Rev. Fluid Mech. **15**, 461–512

Briley, W.R. (1974): J. Comput. Phys. **14**, 8–28

Briley, W.R., McDonald, H. ((1979): "Analysis and Computation of Viscous Subsonic Primary and Secondary Flows", AIAA Paper 79–1453

Briley, W.R., McDonald, H. (1984): J. Fluid Mech. **144**, 47–77

Brown, G.M. (1960): AIChE J. **6**, 179–183

Carter, J.E. (1981): "Viscous Inviscid Interaction Analysis of Transonic Turbulent Separated Flow", AIAA Paper 81–1241

Carter, J.E., Wornom, S. (1975): "Solutions for Incompressible Separated Boundary Layers Including Viscous Inviscid Interaction", NASA SP-347

Catherall, D., Mangler, K.W. (1966): J. Fluid Mech. **26**, 163–182

Cebeci, T., Bradshaw, P. (1977): *Momentum Transfer in Boundary Layers* (McGraw-Hill, New York)

Cebeci, T., Stewartson, K., Whitelaw, J.H. (1984): In *Numerical and Physical Aspects of Aerodynamic Flows II*, ed. by T. Cebeci (Springer, New York, Berlin, Heidelberg)

Chang, C.-L., Merkle, C.L. (1989): J. Comput. Phys. **80**, 344–361

Chaussee, D.S. (1984): In *Computational Methods in Viscous Flow*, ed. by W. Habashi (Pineridge Press, Swansea) pp. 301–347.

Chen, Z.B., Bradshaw, P. (1984): AIAA J. **22**, 201–205

Cooke, C.H., Dwoyer, D.M. (1983): Int. J. Numer. Methods Fluids **3**, 493–506

Dash, S.M., Sinha, N. (1985): AIAA J. **23**, 153–155

Davis, R.T., Rubin, S.G. (1980): Comput. Fluids **8**, 101–132

Degani, D., Schiff, L.B. (1986): J. Comput. Phys. **66**, 173–196

Drazin, P.G., Reid, W.H. (1981): *Hydrodynamic Stability* (Cambridge University Press, Cambridge)

Fletcher, C.A.J. (1989): In *Advances in Fluid Mechanics*, ed. by W.F. Ballhaus and M.Y. Hussaini (Springer, Berlin, Heidelberg) pp. 57–68

Ghia, U., Ghia, K.N., Studerus, C.J. (1977): Comput. Fluids **5**, 205–218

Ghia, K.N., Sokhey, J.S. (1977): J. Fluids Eng. **99**, 640–648

Ghia, U., Ghia, K.N., Goyal, R.K. (1979): "Three-Dimensional Viscous Incompressible Flow in Curved Polar Ducts", AIAA Paper 79–1536

Ghia, U., Ghia, K.N., Rubin, S.G., Khosla, P.K. (1981): Comput. Fluids **9**, 123–142

Goldstein, S. (1948): Q. J. Mech. Appl. Math. **1**, 48–69

Gustafson, K.E. (1980): *Partial Differential Equations and Hilbert Space Methods* (Wiley, New York)

Helliwell, W.S., Dickson, R.P., Lubard, S.C. (1981): AIAA J. **19**, 191–196

Himansu, A., Rubin, S. (1988): AIAA J. **26**, 1044–1051

Horstman, C.C. (1984): In *Numerical and Physical Aspects of Aerodynamics Flows II*, ed. by T. Cebeci (Springer, New York, Berlin, Heidelberg) pp. 113–124

Hsieh, T. (1976): "An Investigation of Separated Flow About a Hemisphere–Cylinder at 0 to 19 deg Incidence in the Mach Number Range, 0.6 to 1.5", AEDC-TR-76-112

Humphrey, J.A.C., Taylor, A.M.K., Whitelaw, J.H. (1977): J. Fluid Mech. **83**, 509–527

Humphrey, J.A.C., Whitelaw, J.H., Yee, G. (1981): J. Fluid Mech. **103**, 443–463

Humphrey, J.A.C., Iacovides, H., Launder, B.E. (1985): J. Fluid Mech. **154**, 357–375

Johnston, W., Sockol, P. (1979): AIAA J. **17**, 661–663

Kaul, U.K., Chaussee, D.S. (1985): Comput. Fluids **13**, 421–441

Klineberg, J.M., Steger, J.L. (1974): "On Laminar Boundary Layer Separation", AIAA Paper 74–94

Kreskovsky, J.P., Shamroth, S.J. (1978): Comput. Methods Appl. Mech. Eng. **13**, 307–334

Kreskovsky, J.P., Briley, W.R., McDonald, H. (1984): AIAA J. **22**, 374–382

Kuchemann, D. (1978): *The Aerodynamic Design of Aircraft* (Pergamon, Oxford)

Lawrence, S.L., Chaussee, D.S., Tannehill, J.C. (1989): AIAA J. **27**, 1175–1183

Le Balleur, J.C. (1978): Rech. Aerosp. **183**, 65–76

Le Balleur, J.C. (1981): Rech. Aerosp. **1981-3**, 21–45

Le Balleur, J.C. (1984): In *Computational Methods in Viscous Flows*, ed. by W. Habashi (Pineridge Press, Swansea) pp. 419–450.

Le Balleur, J.C., Peyret, R., Viviand, H. (1980): Comput. Fluids **8**, 1–30

Leonard, B.P. (1979): Comput. Methods Appl. Mech. Eng. **19**, 59–98

Lighthill, M.J. (1958): J. Fluid Mech. **4**, 383–392

Lin, T.C., Rubin, S.G. (1973): Comput. Fluids **1**, 37–57

Lin, A., Rubin, S.G. (1981): AIAA J. **20**, 1500–1507

Lock, R.C. (1981): "A Review of Methods for Predicting Viscous Effects on Aerofoils of Transonic Speeds", AGARD-CP-291, Lecture 2

Lomax, H., Mehta, U.B. (1984): In *Computational Methods in Viscous Flows*, ed. by W. Habashi (Pineridge Press, Swansea) pp. 1–50

Lubard, S.C., Helliwell, W.S. (1974): AIAA J. **12**, 965–974

Moore, J., Moore, J.G. (1979): J. Fluids Eng. **101**, 415–421

Nakayama, A. (1984): *Numerical and Physical Aspects of Aerodynamic Flows II*, ed. by T. Cebeci (Springer, New York, Berlin, Heidelberg) pp. 233–253

Neumann, R.D., Patterson, J.L., Sliski, N.J. (1978): "Aerodynamic Heating to the Hypersonic Research Aircraft, X-24C", AIAA Paper 78-37

Patankar, S.V., Spalding, D.B. (1972): Int. J. Heat Mass Transfer **15**, 1787–1806

Platfoot, R.A., Fletcher, C.A.J. (1989): In Proc. Int. Sym. Comp. Fluid Dynamics ed. by M. Yasuhara and K. Oshima (Univ. of Tokyo Press, Tokyo) pp. 40–45

Povinelli, L.A., Anderson, B.H. (1984): AIAA J. **22**, 518–525

Pratap, V.S., Spalding, D.B. (1976): Int. J. Heat Mass Transfer **19**, 1183–1187

Pulliam, T.H., Steger, J.L. (1980): AIAA J. **18**, 159–167

Rackich, J.V., Davis, R.T., Barnett, M. (1982): In 8th Int. Conf. Numer. Methods in Fluid Dynamics, ed. by E. Krause, Lecture Notes in Physics, Vol. 170 (Springer, Berlin, Heidelberg) pp. 420–426

Rhie, C.M. (1985): Comput. Fluids **13**, 443–460

Rubin, S.G. (1981): *Numerical and Physical Aspects of Aerodynamic Flows I*, ed. by. T. Cebeci (Springer, New York, Berlin, Heidelberg) pp. 171–186

Rubin, S.G. (1984): In *Computational Methods in Viscous Flows*, ed. by W. Habashi (Pineridge Press, Swansea) pp. 53–100

Rubin, S.G. (1985): In 9th Int. Conf. Numer. Methods in Fluid Dynamics, ed. by Soubbaramayer, J.P. Boujot, Lecture Notes in Physics, Vol. 218 (Springer, Berlin, Heidelberg) pp. 62–71

Rubin, S.G. (1988): Comput. Fluid. **16**, 485–490

Rubin, S.G., Khosla, P.K., Saari, S. (1977): Comput. Fluids **5**, 151–173

Rubin, S.G., Lin, A. (1980): Isr. J. Technol. **18**, 21–31

Rubin, S.G., Reddy, D.R. (1983): Comput. Fluids **11**, 281–306

Rudman, S., Rubin, S.G. (1968): AIAA J. **6**, 1883–1890

Schiff, L.B., Steger, J.L. (1980): AIAA J. **18**, 1421–1430

Schlicting, H. (1968): *Boundary Layer Theory*, 6th ed. (McGraw-Hill, New York)

Senoo, Y., Kawaguchi, N., Nagata, T. (1978): Bull. JSME **21**, 112–119

So, K.L. (1964): "Vortex Decay in Conical Diffuser", Report No. 75, M.I.T. Gas Turbines Lab.

Sparrow, E.M., Hixon, C.W., Shavit, G. (1967): J. Basic Eng. **89**, 116–121

Steger, J.L. (1978): AIAA J. **16**, 679–686

Stewartson, K. (1974): Adv. Appl. Mech. 14, 145–239

Stuart, J.J. (1963): In *Laminar Boundary Layers*, ed. by L. Rosenhead (Oxford University Press, Oxford)

Tannehill, J.C., Venkatapathy, E., Rakich, J.V. (1982): AIAA J. **20**, 203–210

Thwaites, B. (ed.) (1960): *Incompressible Aerodynamics* (Oxford University Press, Oxford)

Veldman, A.E.P. (1981): AIAA J. **19**, 79–85

Veldman, A.E.P. (1984): In *Computational Methods in Viscous Flows*, ed. by W. Habashi (Pineridge Press, Swansea) pp. 343–364

Vigneron, Y.C., Rackich, J.V., Tannehill, J.C. (1978): "Calculation of Supersonic Viscous Flow over Delta Wings with Sharp Subsonic Loading Edges", NASA Tech. Memo 78500

Whitfield, D.L., Thomas, J.L. (1985): In *Computational Methods in Viscous Fl$\hat{B}l$ kPds*, ed. by W. Habashi (Pineridge Press, Swansea) pp. 451–474

Wigton, L.B., Holt, M. (1981): "Viscous–Inviscid Interaction in Transonic Flow", AIAA Paper 81-1003

Chapter 17

Amsden, A.A., Harlow, F.H. (1970): J. Comput. Phys. **6**, 322–325
Aregbesola, Y.A.S., Burley, D.M. (1977): J. Comput. Phys. **24**, 398
Aziz, K., Hellums, J.D. (1967): Phys. Fluids **10**, 314–324
Baker, A.J. (1983): *Finite Element Computational Fluid Mechanics* (McGraw-Hill, New York)
Baker, A.J., Kim, J.W., Freels, J.D., Orzechowski, J.A. (1987): Int. J. Numer. Methods Fluids **7**, 1235–1260
Brachet, M.E., Meiron, D.I., Orszag, S.A., Nickel, B.G., Morf, R.H., Frisch, U. (1983): J. Fluid Mech. **130**, 411–452
Briley, W.R. (1974): J. Comput. Phys. **14**, 8–28
Briley, W.R., McDonald, H. (1984): J. Fluid Mech. **144**, 47–77
Brooks, A.N., Hughes, T.J.R. (1982): Comput. Methods Appl. Mech. Eng. **30**, 199–259
Campion-Renson, A., Crochet, M.J. (1978): Int. J. Numer. Methods Eng., **12**, 1809–1818
Castro, I.P., Cliffe, K.A., Norgett, M.J. (1982): Int. J. Numer. Methods Fluids **2**, 61–88
Cazalbou, J.B., Braza, M., HaMinh, H. (1983): In Proc. 3rd Int. Conf. Numer. Methods Laminar and Turbulent Flow, Seattle (Pineridge Press, Swansea) pp. 786–797
Cho, N.-H., Fletcher, C.A.J., Srinivas, K. (1991): "Efficient Computation of Wing Body Flows", Twelfth Int. Conf. Num. Meth. in Fluid Dynamics, ed. by K. Morton, Lecture Notes in Physics (to appear) (Springer, Berlin, Heidelberg)
Chorin, A.J. (1967): J. Comput. Phys. **2**, 12–26
Chorin, A.J. (1968): Math. Comput. **22**, 745–762
Connell, S.D., Stow, P. (1986): Comput. Fluids **14**, 1–10
Davis, R.W., Moore, E.F. (1982): J. Fluid Mech. **116**, 475–506
Davis, R.W., Moore, E.F., Purtell, L.P. (1984): Phys. Fluids **27**, 46–57
Dennis, S.C.R. (1985): In 9th Int. Conf. Numer. Methods in Fluid Dynamics, ed. by Soubbaramayer, J.B. Boujot, Lecture Notes in Physics, Vol. 218 (Springer, Berlin, Heidelberg) pp. 23–36
Dennis, S.C.R., Ingham, D.B., Cook, R.N. (1979): J. Comput. Phys. **33**, 325–339
de Vahl Davis, G., Mallinson, G.D. (1976): Comput. Fluids **4**, 29–43
de Vahl Davis, G., Jones, I. (1983): Int. J. Numer. Methods Fluids **3**, 227–248
Deville, M.O. (1974): J. Comput. Phys. **15**, 362–374
Dodge, P.R. (1977): AIAA J. **15**, 961–965
Easton, C.R. (1972): J. Comput. Phys. **9**, 375–379
Engelman, M.S. (1982): Adv. Eng. Software **4**, 163–175; also FIDAP Theoretical and User's Manuals, Fluid Dynamics International, Evanston, Illinois
Engelman, M.S., Sani, R.L., Gresho, P.M., Bercovier, M. (1982): Int. J. Numer. Methods Fluids **2**, 25–42
Fasel, H. (1976): J. Fluid Mech. **78**, 353–383
Fletcher, C.A.J. (1982): Comput. Methods Appl. Mech. Eng. **30**, 307–322
Fletcher, C.A.J. (1984): *Computational Galerkin Methods*, Springer Ser. Comput. Phys. (Springer, Berlin, Heidelberg)
Fletcher, C.A.J., Srinivas, K. (1983): Comput. Methods Appl. Mech. Eng. **41**, 299–322
Fletcher, C.A.J., Srinivas, K. (1984): Comput. Methods Appl. Mech. Eng. **46**, 313–327
Fletcher, C.A.J., Barbuto, J. (1986a): Appl. Math. Model. **10**, 176–184
Fletcher, C.A.J., Barbuto, J. (1986b): "Backward-Facing Cavity Flow with Suction and Blowing", in Proc. Int. Symp. Comp. Fluid Dynamics, ed. by K. Oshima (Japan Soc. of Comp. Fluid Dynamics, Tokyo) Vol. 1, pp. 245–251
Fletcher, C.A.J., Bain, J.G. (1991): "An Approximate Factorisation Explicit Method for CFD", Comput. Fluids **19**, 61–74
Gatski, T.B., Grosch, C.E., Rose, M.E. (1982): J. Comput. Phys. **48**, 1–22
Gatski, J.B., Grosch, C.E. (1985): In 9th Int. Conf. Numer. Methods in Fluid Dynamics, ed. by Soubbaramayer, J.P. Boujot, Lecture Notes in Physics, Vol. 218 (Springer, Berlin, Heidelberg) pp. 235–239

Gatski, T.B., Grosch, C.E., Rose, M.E. (1989): J. Comput. Phys. **82**, 298–329
Ghia, K.N., Sokhey, J.S. (1977): J. Fluids Eng. **99**, 640–648
Ghia, K.N., Hankey, W.L., Hodge, J.K. (1979): AIAA J. **17**, 298–301
Ghia, U., Ghia, K.N., Shin, C.T. (1982): J. Comput. Phys. **48**, 387–411
Goda, K. (1979): J. Comput. Phys. **30**, 76–95
Goldstein, R.J., Ericksen, V.L., Olson, R.M., Eckert, E.R.G. (1970): J. Basic Eng. **92**, 732–741
Gottlieb, D., Orszag, S.A. (1977): *Numerical Analysis of Spectral Methods* (SIAM, Philadelphia)
Gottlieb, D., Hussaini, M.V., Orszag, S.A. (1984): In *Spectral Methods for Partial Differential Equations*, ed. by R.G. Voigt, D. Gottlieb, M.Y. Hussaini (SIAM, Philadelphia) pp. 1–54
Gresho, P.M., Lee, R. (1981): Comput. Fluids **9**, 223–253
Gresho, P.M., Sani, R.L. (1987): Int. J. Numer. Methods Fluids **7**, 1111–1145
Gupta, M.M., Manohar, R.P. (1979): J. Comput. Phys. **31**, 265–288
Haltiner, G.J., Williams, R.T. (1980): *Numerical Prediction and Dynamic Meteorology*, 2nd ed. (Wiley, New York)
Harlow, F.H., Welch, J.E. (1965): Phys. Fluids **8**, 2182–2189
Harlow, F.H., Shannon, J.P. (1967): Science **157**, 547–550
Hirasaki, G.J., Hellums, J.D. (1968): Q. Appl. Math. **26**, 331
Hirasaki, G.J., Hellums, J.D. (1970): Q. Appl. Math. **28**, 293
Hirt, C.W., Cook, J.L. (1972): J. Comput. Phys. **10**, 324–340
Hirt, C.W., Nichols, B.D., Romero, N.C. (1975): "SOLA-A Numerical Solution Algorithm for Transient Fluid Flows", Los Alamos Scientific Lab. Rep. LA-5852
Hood, P. (1976): Int. J. Numer. Methods Eng. **10**, 379–399
Hood, P., Taylor, C. (1974): "Navier–Stokes Equations Using Mixed Interpolation", in *Finite Element Methods in Flow Problems* (University of Alabama Press, Huntsville) pp. 121–132
Hortmann, M., Peric, M., Scheuerer, G. (1990): Int. J. Num. Method Fluids **11**, 189–207
Hughes, T.J.R., Liu, W.K., Brooks, A. (1979): J. Comput. Phys. **30**, 1–75
Hussaini, M.Y., Zang, T.A. (1987): Annu. Rev. Fluid Mech. **19**, 339–367
Israeli, M. (1972): Stud. Appl. Math. **51**, 67–71
Jensen, V.G. (1959): Proc. R. Soc. London, Ser. A **249**, 346–366
Kato, S., Murakami, S. (1986): "Three-Dimensional Numerical Simulation of Turbulent Air Flow in a Ventilated Room, by means of Two-Equation Model", in Proc. Int. Symp. Comp. Fluids Dynamics, ed. by K. Oshima (Japan Soc. of Comp. Fluid Dynamics, Tokyo) Vol. 2, pp. 217–228
Khosla, P.K., Rubin, S.G. (1974): Comput. Fluids **2**, 207–209
Khosla, P.K., Rubin, S.G. (1983): AIAA J. **21**, 1546–1551
Kim, J., Moin, P. (1985): J. Comput. Phys. **59**, 308–323
Korczak, K.Z., Patera, A.T. (1986): J. Comput. Phys. **62**, 361–382
Ku, H.C., Hatziavramidis, D. (1985): Comput. Fluids **13**, 99–113
Ku, H.C., Taylor, T.D., Hirsh, R.S. (1987a): Comput. Fluids, **15**, 195–214
Ku, H.C., Hirsh, R.S., Taylor, T.D. (1987b): J. Comput. Phys., **70**, 439–462
Ku, H.C., Hirsh, R.S., Taylor, T.D., Rosenberg, A.P. (1989): J. Comp. Phys. **83**, 260–291
Kwak, D., Chang, J.L.C., Shanks, S.P., Chakravarthy, S.R. (1986a): AIAA J. **24**, 390–396
Kwag, D., Rogers, S.E., Kaul, U.K., Chang, J.C.L. (1986b): "A Numerical Study of Incompressible Juncture Flows", NASA Tech. Memo 88319
Leonard, A. (1980): J. Comput. Phys. **37**, 289–335
Leonard, A. (1985): Annu. Rev. Fluid Mech. **17**, 17–24
Leonard, B.P. (1979): Comput. Methods Appl. Mech. Eng. **19**, 59–98
Leschziner, M.A., Dimitriadis, K.P. (1989): Comput. Fluids **17**, 371–396
Lugt, H.J., Schwiderski, E.W. (1965): Proc. R. Soc. London, Ser. A **285**, 382–399
Macaraeg, M., Streett, C.L. (1986): Appl. Numer. Methods, **2**, 95–108
Mallinson, G.D., de Vahl Davis, G. (1973): J. Comput. Phys. **12**, 435–461
Mallinson, G.D., de Vahl Davis, G. (1977): J. Fluid Mech. **83**, 1–31
Moin, P., Kim, J. (1980): J. Comput. Phys. **35**, 381–392
Orlandi, P. (1987): Comput. Fluids, **15**, 137–149
Orszag, S.A. (1980): J. Comput. Phys. **37**, 70–92

Orszag, S.A., Kells, L.C. (1980): J. Fluid Mech. **96**, 159–205

Orszag, S.A., Patera, A.T. (1981): Phys. Rev. Lett. **47**, 832–835

Orszag, S.A., Patera, A.T. (1983): J. Fluid Mech. **128**, 347–385

Ortega, J.M., Voigt, R.G. (1985): SIAM Rev. **27**, 149–240

Oshima, K., Oshima, Y., Izutsu, N., Ishii, Y., Noguchi, T. (1986): In 10th Int. Conf. Numer. Methods in Fluid Dynamics, ed. by F.G. Zhuang, Y.L. Zhu, Lecture Notes in Physics, Vol. 264 (Springer, Berlin, Heidelberg) pp. 511–515

Patankar, S.V. (1980): *Numerical Heat Transfer and Fluid Flow* (Hemisphere, Washington, D.C.)

Patankar, S.V., Spalding, D.B. (1972): Int J. Heat Mass Transfer **15**, 1787–1806

Patel, N.R., Briggs, D.G. (1983): Numer. Heat Transfer **6**, 383–394

Patel, M.K., Markatos, N.C. (1986): Int. J. Numer. Methods Fluids **6**, 129–154

Patera, A.T. (1984): J. Comput. Phys. **54**, 468–488

Pearson, C.E. (1965): J. Fluid Mech. **21**, 611–622

Perng, C-Y., Street, R.L. (1989): Int. J. Numer. Methods Fluids **9**, 341–362

Peyret, R., Taylor, T.D. (1983): *Computational Methods for Fluid Flow*, Springer Ser. Comput. Phys. (Springer, Berlin, Heidelberg)

Phillips, R.E., Schmidt, F.W. (1985): Numer. Heat Transfer **8**, 573–594

Pollard, A., Siu, A.L. (1982): Comput. Methods Appl. Mech. Eng. **35**, 293–313

Quartapelle, L., Valz-Gris, F. (1981): Int. J. Numer. Meth. Fluids, **1**, 129–144

Raithby, G.D., Schneider, G.E. (1979): Numer. Heat Transfer **2**, 417–440

Raithby, G.D., Galpin, P.F., Van Doormal, J.P. (1986): Numer. Heat Transfer **9**, 125–142

Rao, K.V., Steger, J.L., Pletcher, R.H. (1989): AIAA J. **27**, 876–884

Richardson, S.M., Cornish, A.R.H. (1977): J. Fluid Mech. **82**, 309–340

Roache, P.J. (1972): *Computational Fluid Dynamics* (Hermosa, Albuquerque, N.M.)

Rogers, S.E., Kwak, D. (1990): AIAA Journal **28**, 253–262

Rubin, S.G., Khosla, P.K. (1981): Comput. Fluids **9**, 163–180

Sakamoto, Y., Matuo, Y. (1980): Appl. Math. Model. **4**, 67–72

Samarskii, A.A., Andreev, V.B. (1963): USSR Comput. Math. Math. Phys. (Engl. Transl.) **3**, 1373–1378

Sani, R.L., Gresho, P.M., Lee, R.L., Griffiths, D.F. (1981): Int. J. Numer. Methods Fluids **1**, 17–43, 171–204

Schneider, G.E., Raithby, G.D., Yovanovich, M.M. (1978): Numer. Heat Transfer **1**, 433–451

Shyy, W., Tong, S.S., Correa, S.M. (1985): Numer. Heat Transfer **8**, 99–113

Sinha, S.N., Gupta, A.K., Oberai, M.M. (1981): AIAA J. **19**, 1527–1530

Srinivas, K., Fletcher, C.A.J. (1984): Int. J. Numer. Methods Fluids **4**, 421–439

Steger, J.L., Kutler, P. (1977): AIAA J. **15**, 581–590

Swartztrauber, P.N. (1974): SIAM J. Numer. Anal. **11**, 1136–1150

Takemoto, Y., Yamabe, H., Abe, Y., Minami, I. (1985): Trans. Jpn. Soc. Irrig. Drain. Reclam. Eng. **118**, 23–31

Takemoto, Y., Nakamura, Y., Yamabe, H., Abe, Y., Minami, I. (1986): Trans. Jpn. Soc. Irrig. Drain. Reclam. Eng. **121**, 57–65

Takemoto, Y., Nakamura, Y. (1986): "A Three-Dimensional Incompressible Flow Solver", in 10th Int. Conf. Numer. Methods in Fluid Dynamics, ed. by F.G. Zhuang, Y.L. Zhu, Lecture Notes in Physics, Vol. 264 (Springer, Berlin, Heidelberg) pp. 594–599

Temam, R. (1968): Bull. Soc. Math. Fr. **96**, 115–152

Temam, R. (1969): Arch. Ration. Mech. Anal. **32**, 377–385

Thom, A. (1933): Proc. R. Soc. London, Ser. A **141**, 651–666

Van Doormaal, J.R., Raithby, G.D. (1984): Numer. Heat Transfer **7**, 147–163

Viecelli, J.A. (1971): J. Comput. Phys. **8**, 119–143

Wong, A.K., Reizes, J.A. (1984): J. Comput. Phys. **55**, 98–114

Yashchin, D., Israeli, M., Wolfshtein, M. (1984): In Proc. of Computational Techniques and Applications Conference, CTAC-83, ed. by B.J. Noye, C.A.J. Fletcher (North-Holland, Amsterdam) pp. 533–552

Zang, T.A., Streett, C.L., Hussaini, M.Y. (1989): "Spectral Methods for CFD", VKI Lecture Series 89-04, Von Karman Institute

Zienkiewicz, O.C. (1977): *The Finite Element Method in Engineering Science*, 2nd ed. (McGraw-Hill, New York)

Chapter 18

Baldwin, B.S., Lomax, H. (1978): "Thin Layer Approximation and Algebraic Model for Separated Turbulent Flows", AIAA Paper 78–257

Bayliss, A., Turkel, E. (1982): J. Comput. Phys. **48**, 182–199

Beam, R.M., Warming, R.F. (1976): J. Comput. Phys. **22**, 87–110

Beam, R.M., Warming, R.F. (1978): AIAA J. **16**, 393–402

Bradshaw, P. (1977): Annu. Rev. Fluid Mech. **9**, 33–54

Briley, W.R., McDonald, H. (1977): J. Comput. Phys. **24**, 372–397

Cebeci, T., Smith, A.M.O. (1974): *Analysis of Turbulent Boundary Layers* (Academic, New York)

Chakravarthy, S.R. (1987): "Some Recent Advances in CFD Algorithms", unpublished presentation at Int. Symp. Computational Fluid Dynamics, ed. by G. de Vahl Davis, C.A.J. Fletcher (North-Holland, Amsterdam)

Chang, C.-L., Merkle, C.L. (1989): J. Comp. Phys. **80**, 344–361

Chaussee, D.S. (1984): In *Computational Methods in Viscous Flows*, ed. by W.G. Habashi (Pineridge Press, Swansea) pp. 255–279

Chima, R.V., Johnson, G.M. (1985): AIAA J. **23**, 23–32

Cleary, J.W., Viswanath, P.R., Horstman, C.C., Seegmiller, H.L. (1980): "Asymmetric Trailing-Edge Flows at High Reynolds Number", AIAA Paper 80–1396

Coakley, T.J. (1983): "Turbulence Modelling Methods for the Compressible Navier–Stokes Equations", AIAA Paper 83–1693

Deiwert, G.S. (1976): AIAA J. **14**, 735–740

Deiwert, G.S. (1984): In *Computational Methods in Viscous Flows*, ed. by W.G. Habashi (Pineridge Press, Swansea) pp. 281–308

Dolling, D.S., Bogdonoff, S.M. (1982): AIAA J. **20**, 1674–1680

Eaton, J.K. (1981): "Summary of Computations for Case 0421: Backward-facing Step Flow", in AFOSR-HTTM Stanford Conference on Complex Turbulent Flows, ed. by S.J. Kline, B. Cantwell, G.M. Lilley (Stanford University, Stanford)

Favre, A. (1965): J. Mec. **4**, 361–390

Fletcher, C.A.J., Srinivas, K. (1985): *Finite Elements in Fluids*, Vol. 4 (Wiley, New York) pp. 115–133

Fujii, K., Obayashi, S. (1986): "The Development of Efficient Navier–Stokes Codes for Transonic Flow Field Simulations", preprint for Int. Symp. Comp. Fluid Dynamics, ed. by K. Oshima (Japan Soc. of Comp. Fluid Dynamics, Tokyo) pp. 398–409

Goldberg, U.C., Chakravarthy, S.R. (1986): AIAA J. **24**, 1711–1713

Goldberg, U.C., Chakravarthy, S.R. (1990): AIAA J. **28**, 1005–1009

Goorjian, P.M., Obayashi, S. (1991): "Streamwise Upwind Algorithm Development for the Navier–Stokes Equations", 12th Int. Conf. Numer. Meth. in Fluid Dynamics, ed. by K.W. Morton, Lecture Notes in Physics (to appear) (Springer, Berlin, Heidelberg)

HaMinh, H., Rubesin, M.W., Vandromme, D., Viegas, J.R. (1986): "On the Use of Second-order Closure Modelling for the Prediction of Turbulent Boundary Layer/Shock Wave Interactions", in Proc. Int. Symp. Comp. Fluid Dynamics, ed. by K. Oshima (Japan Soc. of Comp. Fluid Dynamics, Tokyo) Vol. 1, pp. 192–204

Holst, T.L., Thomas, S.D., Kaynak, U., Grundy, K.L., Flores, J., Chaderjian, N.M. (1986): "Computational Aspects of Zonal Algorithms for Solving the Compressible Navier–Stokes Equations in Three Dimensions", in Proc. Int. Symp. Comp. Fluid Dynamics, ed. by K. Oshima (Japan Soc. of Comp. Fluid Dynamics, Tokyo) Vol. 1, pp. 113–122

Horstman, C.C. (1983): "Numerical Simulation of Turbulent Trailing-Edge Flows", in *Numerical and Physical Aspects of Aerodynamic Flows II*, ed. by T. Cebeci (Springer, New York, Berlin, Heidelberg)

Horstman, C.C. (1986): AIAA J. **24**, 1433–1440

Hsieh, T. (1976): "An Investigation of Separated Flow About a Hemisphere-Cylinder at 0–90 deg. Incidence in the Mach Number Range from 0.6 to 1.5″, AEDC-TR-76-112

Hung, C.M., Kordulla, W. (1984): AIAA J. **22**, 1564–1572

Jameson, A. (1989): Science **245**, 361–371

Jameson, A., Schmidt, W. (1985): Comp. Meth. Appl. Mech. Eng. **51**, 467–493

Jameson, A., Turkel, E. (1981): Math. Comput. **37**, 385–397

Kaynak, U., Flores, J. (1989): Comput. Fluids **17**, 313–332

Kim, H.J., Thompson, J.F. (1990): AIAA J. **28**, 470–477

Kordulla, W., MacCormack, R.W. (1982): In 8th Int. Conf. Numer. Methods in Fluid Dynamics, ed. by E. Krause, Lecture Notes in Physics, Vol. 170 (Springer, Berlin, Heidelberg) pp. 286–295

Launder, B.E., Spalding, D.B. (1974): Comput. Methods Appl. Mech. Eng. **3**, 269–289

Lombard, C.K., Bardina, J., Venkatapathy, E., Yang, J.Y., Luh, R.C.C., Nagaraj, N., Raiszadeh, F. (1986): In 10th Int. Conf. Numer. Methods in Fluid Dynamics, ed. by F.G. Zhuang, Y.L. Zhu, Lecture Notes in Physics, Vol. 264 (Springer, Berlin, Heidelberg) pp. 435–441

MacCormack, R.W. (1969): "Effect of Viscosity in Hypervelocity Impact Cratering", AIAA Paper 69-354

MacCormack, R.W. (1971): In Lecture Notes in Physics, Vol. 8 (Springer, Berlin, Heidelberg) pp. 151–163

MacCormack, R.W. (1982): AIAA J. **20**, 1275–1281

MacCormack, R.W. (1984): In *Computational Methods in Viscous Flows*, ed. by W.G. Habashi (Pineridge Press, Swansea) pp. 225–254

MacCormack, R.W. (1985): "Current Status of Numerical Solutions of the Navier–Stokes Equations", AIAA Paper 85-0032

MacCormack, R.W., Baldwin, B.S. (1975): A Numerical Method for Solving the Navier–Stokes Equations with Application to Shock-Boundary Layer Interactions, AIAA Paper 75-1

MacCormack, R.W., Lomax, H. (1979): Annu. Rev. Fluid Mech. **11**, 289–316

Mandella, M., Bershader, D. (1987): "Quantitative Study of the Compressible Vortices: Generation, Structure and Interaction with Airfoils", AIAA Paper 87-0328

Marvin, J.G. (1983): AIAA J. **21**, 941–955

Mavriplis, D.J., Jameson, A., Martinelli, L. (1989): "Multigrid Solution of the Navier–Stokes Equations on Triangular Meshes", AIAA 89-0120

Morinishi, K., Satofuka, N. (1991): "Convergence Acceleration of a Rational Runge–Kutta Scheme for Euler and Navier–Stokes Equations", Comput. Fluids (to appear)

Munz, C.-D. (1988): J. Comput. Phys. **77**, 18–39

Obayashi, S., Kuwahara, K. (1986): J. Comput. Phys. **63**, 157–167

Oran, E.S., Boris, J.P., Brown, E.F. (1990): "Fluid Dynamic Computations on a Connection Machine–Preliminary Timings and Complex Boundary Conditions", AIAA Paper 90-0335

Ortega, J.M., Voigt, R.G. (1985): SIAM Rev. **27**, 147–240

Pan, D., Lomax, H. (1988): AIAA J. **26**, 163–171

Patankar, S.V., Spalding, D.B. (1970): *Heat and Mass Transfer in Boundary Layers*, 2nd ed. (Intertext Books, London)

Peyret, R., Taylor, T.D. (1983): *Computational Methods in Fluid Flow*, Springer Ser. Comput. Phys. (Springer, Berlin, Heidelberg)

Peyret, R., Viviand, H. (1975): "Computation of Viscous Compressible Flow Based on the Navier–Stokes Equations", AGARDograph 212

Pulliam, T.H., Steger, J.L. (1980): AIAA J. **18**, 159–167

Pulliam, T.H., Steger, J.L. (1985): "Recent Improvements in Efficiency, Accuracy and Convergence for Implicit Approximate Factorization Algorithms", AIAA Paper 85-0360

Rai, M.M., Chaussee, D.S. (1984): AIAA J. **22**, 1094–1100

Roe, P.L. (1989): Comput. Fluids **17**, 221–231

Rubesin, M.W., Rose, W.C. (1973): "The Turbulent Mean-Flow, Reynolds-Stress and Heat Flux Equations in Mass-Averaged Dependent Variables", NASA TM-X-62248

Rudy, D.H., Strikwerda, J.C. (1981): Comput. Fluids **9**, 327–338

Satofuka, N., Morinishi, K., Tamaki, T., Shimizu, A. (1986): "Computation of Two-dimensional Transonic Cascade Flow Using a New Navier–Stokes Solver", AIAA Paper 86–1381

Shang, J.S. (1985): J. Aircr. **22**, 353–370

Srinivas, K., Fletcher, C.A.J. (1984): Int. J. Numer. Methods Fluids **4**, 421–439

Srinivas, K., Fletcher, C.A.J. (1985): Int. J. Numer. Methods Fluids **5**, 463–481

Srinivas, K., Fletcher, C.A.J. (1986): Z. Angew. Math. Phys. **37**, 53–63

Steger, J.L. (1978): AIAA J. **16**, 679–686

Steger, J.L., Warming, R.F. (1981): J. Comput. Phys. **40**, 263–293

Thomas, J.L., Krist, S.T., Anderson, W.K. (1990): AIAA J. **28**, 205–212

Vandromme, D., HaMinh, H. (1986): J. Comput. Phys. **65**, 386–409

van Leer, B. (1982): In 8th Int. Conf. Numer. Methods Fluid Dynamics, ed. by E. Krause, Lecture Notes in Physics, Vol. 170 (Springer, Berlin, Heidelberg) 507–512

Venkatakrishnan, V., Jameson, A. (1988): AIAA J. **26**, 974–981

Venkatakrishnan, V., Saltz, J.H., Mavriplis, D.J. (1991): "Parallel Preconditioned Iterative Methods for the Compressible Navier–Stokes Equations", 12th Int. Conf. Numer. Meth. in Fluid Dynamics, ed. by K.W. Morton, Lecture Notes in Physics (to appear) (Springer, Berlin, Heidelberg)

Walters, R.W., Thomas, J.L., van Leer, B. (1986): In 10th Int. Conf. Numer. Methods in Fluid Dynamics, ed. by F.G. Zhuang, Y.L. Zhu, Lecture Notes in Physics, Vol. 264 (Springer, Berlin, Heidelberg) pp. 628–635

Wambecq, A. (1978): Computing **20**, 333–342

Yee, H.C. (1986): In 10th Int. Conf. Numer. Methods in Fluid Dynamics, ed. by F.G. Zhuang, Y.L. Zhu, Lecture Notes in Physics, Vol. 264 (Springer, Berlin, Heidelberg) pp. 677–683

Yee, H.C. (1987): J. Comput. Phys., **68**, 151–179

Yee, H.C., Harten, A. (1985): "Implicit TVD Schemes for Hyperbolic Conservation Laws in Curvilinear Coordinates", AIAA Paper 85–1513

Yee, H.G., Klopfer, G.H., Montagné, J.-L. (1990): J. Comput. Phys. **88**, 31–61

Zang, T.A., Streett, C.L., Hussaini, M.Y. (1989): "Spectral Methods for CFD", VKI Lecture Series 89–04, Von Karman Institute

Subject Index

The numbers I or II preceding the page numbers refer to the respective volume of *Computational Techniques for Fluid Dynamics*

Contents of

Computational Techniques for Fluid Dynamics 1
Fundamental and General Techniques

Printing: COLOR-DRUCK DORFI GmbH, Berlin
Binding: Buchbinderei Lüderitz & Bauer, Berlin